THE UNITY
OF SCIENCE IN
UNIFICATION THOUGHT

To my
True Parents

With Thanks.

The Unity of Science in Unification Thought

ISBN: 978-1-304-72048-1

Printed and Distributed by:
LULU.COM

Published by:

Unification Thought Institute
Seiyaku Bldg. 4F
5-13-2 Shinjuku
Shinjuku-ku Tokyo 160-0022, Japan

THE UNITY
OF SCIENCE IN
UNIFICATION THOUGHT

by
Dr. Richard L. Lewis

UNIFICATION THOUGHT INSTITUTE
KOREA • JAPAN • USA

Contents

VOLUME 1.
QUANTUM
FOUNDATIONS
OF BIOLOGY

UNITY
OF THE
SCIENCES

Science is, almost indisputably, one of the most positive of the remarkable developments that have emerged in the last 500-or-so years since the Renaissance. The fruits, for good or ill, of scientific insights into how the world actually works have earned its practitioners a magisterial authority reserved in earlier ages for the revealers of mystical truth.

A bedrock belief of all the sciences—it can be considered the basic philosophical prerequisite for a discipline to be counted as a science—is that there is an objective reality "out there" to be studied. Moreover, it is the same objective reality for all of us. The holy grail of science is to come up with an accurate description of this objective reality.

While words can do a lot, the most accurate descriptions in science are couched in terms of mathematical shorthand. For example, two key insights by Newton and Einstein are succinctly described as: $E = mc^2$ and $F = ma$. Unless we have to, however, we will try to stick to words to get the point across.

Hierarchy of Science

To those unacquainted with its inner workings, scientists can seem to be a part of a vast, monolithic entity—an almost-priesthood with magic powers (and possibly-suspect motives, as attested by the plethora of evil-scientists with British accents in the movies).

To the many workers focused on the endless developments within their own subspecialty of a science, however, science seems less a unified entity than a multitude of relatively independent disciplines:

"The statement 'chemistry and biology are branches of physics' is not true. It is true that in chemistry and biology one does not encounter any new physical principles. Nevertheless, the systems on which the old principles act differ in such a drastic and qualitative way in the different fields that it is simply not useful to regard one as a branch of another. Indeed the systems are so different that 'principles' of new kinds must be developed...."[1]

For all this sense of independence, however, the autonomy of each discipline to develop its own conceptual framework is constrained by the pecking order in science. The rule is simple: a scientist is free to construct any theory so long as it does not contradict what has been established as an accurate description at a lower level in the hierarchy. The chemist is not free to contradict the concepts of physics, the biochemist must respect the rules of the chemist, and a biological theory cannot contradict biochemistry. For example, while neurologists have great latitude to develop concepts to explain the phenomena they encounter in the brain, they are not free to contradict the principles of cell interaction established in biology. Similarly, an evolution theorist cannot contradict the principles of biology—evolution depends on biological processes.

A scientist who wishes to excel at a discipline needs, at the minimum, to have a good grounding in the discipline just below: the evolutionist must know his biology; the chemist his physics. This is a one-way street, however, for you do not need to know anything about the levels above to do well in a discipline. A physicist can excel without knowing any biology whatsoever, for instance, which might explain the dearth of quantum concepts in mainstream genetics and the development of the body let alone evolution and the workings of the nervous system.

The physicists have no one beneath them in the hierarchy to acknowledge; their only constraint is that their theoretical constructs should be mathematically sound or, better yet, "elegant." To paraphrase a well-known eminence's stinging rejection of an aspirant's theory: It is so mathematically ugly that it is not even wrong!

Just why mathematics—a construct of human minds over many centuries—should have this uncanny ability to describe the natural world so accurately is not at all clear. "Opinions range from those who maintain that human beings have simply invented mathematics to fit the facts of experiment, to those who are convinced that there is a deep and meaningful significance behind nature's mathematical face."[2]

Mathematics, of course, is much broader than just its descriptive role in science and can describe constructs that have no—so far—use in describing objective reality, the constructs in nature. Mathematics is also self-contained; it has nothing more basic beneath it (except a faith in logic).

Whatever the rationale; all scientists aspire to put their disciplines on a firm mathematical foundation—to be a "hard" science—rather than being vague and suggestive—to be second-classed as a "soft" science. To have to resort to vague and shifting English etc. words and, a sure sign of fluffiness, endless hand-waving.

A simple analogy to the hierarchical nature of science is the Empire State Building just blocks from where I have worked for a score of years. The foundation, the basement is fundamental physics. Up go the floors, blending into chemistry then biochemistry then genetics then development to the floors in the 100s dealing with evolution, brain function etc.

Two science foundations

Newton is rightly considered the Father of Science as we know it. The themes he developed in classical physics have appeared throughout the scientific structure. Therefore, while biology might not be a branch of physics, the basic Newtonian concepts of classical science permeate biology.

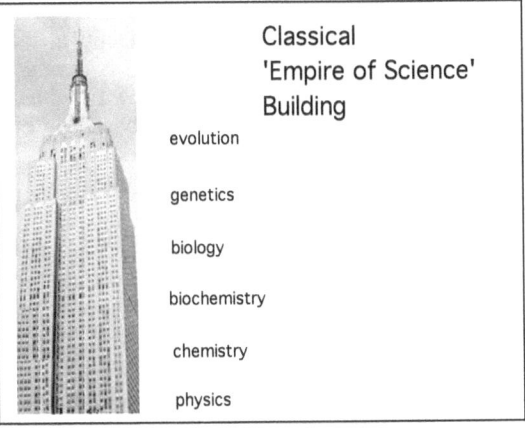

Classical
'Empire of Science'
Building

evolution

genetics

biology

biochemistry

chemistry

physics

Of course, one is philosophically free to drop the hierarchical constraints in constructing a theory of how the world works; but the construct will be something other than science as it is practiced today. The classic historical example of this is the attempt to explain living systems by the introduction of a "vital force" in one guise or another. While there are many philosophical constructs that embrace this as an acceptable explanation, none of them are part of biology because particles, atoms and molecules can be understood without a vital force and, if electrons and quarks don't have it, neither do the atoms and molecules they comprise, nor do cells or higher organisms.

On the other hand, we can expect the converse to be true. If particles, atoms and molecules have some aspect essential to their structure and function, then we might expect some biological systems to involve this aspect as well.

While all scientists accept this pecking order, there are currently two quite different physics to be found at the foundations of the scientific edifice.

The conceptual framework in which physics started out, and the one that is still used in the biological sciences, is described by many adjectives: Newtonian, classical, nineteenth-century, old-fashioned, high school, etc.

Classical physics, however, has been completely replaced by the quantum revolution. For the classical worldview was found to be almost totally inadequate. The more sophisticated replacement framework, the one physics currently embraces, is also multi-monikered: post-Newtonian, New Physics, quantum mechanics, twentieth-century, modern, post-grad, etc.

The new physics is based on the theories and explanations of quantum physics that successful explained a wide variety of phenomena that the old physics was incapable of dealing with. We will list these shortly.

The quantum perspective now pervades all of physics and it has been remarkably successful in dealing with things as different as the first microsecond of the Big Bang and the workings of lasers and superconductors.

The remarkable success of the new physics makes it unlikely that its concepts will be completely replaced by future theoretical developments. It is, of course, possible that they will suffer the same fate as the Newtonian concepts—and they were equally successful in their own day—and turn out that they are phenomena of a much deeper and sophisticated reality.

The new physics is, indeed, so radically weird to the classical mind that it is very difficult to accept the basic concepts at face value As one wit put it: not only is reality stranger than you think, it is stranger than you can think. And we are stuck with the weird quantum view which has gone from one success to the next throwing off a plethora of goodies based on electronics such as my Mac with its laser-run CD burner and DVD reader.

"Perhaps, someday, an experiment will be performed that contradicts quantum mechanics, launching physics into a new era, but it is highly unlikely that such an event would restore our classical version of reality. Remember that nobody, not even Einstein, could come up with a version of reality less strange than quantum mechanics, yet one, which still explained all the existing data. If quantum mechanics is ever superseded, then it seems likely we would discover the world to be even stranger."[3]

Therefore, science, at the commencement of the third millennium, is not just multi-disciplinary; it is a discipline with something of a split personality. In the hierarchy of physics, chemistry, biochemistry, biology and evolution, the switch-over from one science system to the other is to be found somewhere between physical and biological chemistry.

So, while the biology of our era is proud of its firm foundations in the "hard" sciences (those amenable to mathematical rigor), the physics in which it is rooted is the classical physics of Darwin's day. "It is most ironic that today's perceived conjunction between physics and biology, so fervidly embraced by biology in the name of unification, so deeply entrenched in a philosophy of naive reductionism, should have come long past the time when the physical hypotheses on which it rests have been abandoned by the physicists."[4]

There is still, of course, the sense that science should be a unified structure: "How does nature encompass and mold a billion galaxies, a billion, billion stars— and also the earth, teeming with exuberant life? New insights into how nature operates come from parallel advances in particle physics and in molecular biology; advances that make it possible to examine fundamental physical and biological processes side by side. The resulting stereoscopic view deep into the past reveals a previously hidden, unifying logic in nature: its paradigm for construction."[5]

Quantum 'Empire of Science' Building

This is the task of this work, to establish the basic quantum principles, based on the new physics, which are applicable to all levels of the scientific edifice.

On a personal note: I have been fascinated by all the sciences since early schooldays and chose the interdisciplinary biochemistry for my graduate education. (Like my inspiration, Isaac Asimov.) When it came to choose the topic for my Ph.D. thesis I came up with "The Impact of the Quantum Revolution on Evolutionary Thought." To both my and my advisor's surprise, I could not find any impact. The change in the basement had yet to be communicated to the top floors.

I wrote this somewhat 'negative' thesis and my late advisor encouraged me to expand it into a book. The book you are reading more than twenty years on.

To say that pre-twentieth century scientists were content with Newtonian physics, chemistry, etc. is an understatement. One eminence, commenting on the state of classical science at the end of the nineteenth century—at its apogee just before the 'unexplainable weird' became apparent—declared that all that now remained was mopping up, getting ever-increasing accuracy and more and more decimal places. He was oh-so wrong.

Scientists were, almost literally, dragged kicking-and-screaming into accepting the quantum worldview because the only deity in science insisted upon it. That deity is experiment. For no matter how elegant, mathematically-sound, politically-correct, etc. a theory might be, if it contradicts experiment it is crumpled up and thrown into the wastebasket.

Basement Problems

Genetics has been called the Plastics![6] of our age. For the science of genetics is still in a state as alchemy was to current chemical prowess. The DNA-protein connection was established just a half-century ago, and the possibilities that are opening up, even with our primitive understanding, seem endless.

Just a few possibilities are:

In the near future, repairing genetic defects, therapeutic cloning, ordering up a 20-year younger twin, etc. In decades: designing one's children, artificial wombs—and if we are not wise, all the monstrosities that strife can give birth to.

Who can tell where we will go with genetic engineering as the technologists move in behind the conceptual advances in understanding. One thing is certain, however; there are many Nobel Prizes and mega-dollar IPOs waiting for plucking. And lawsuits; and laws being fiddled with.

Somewhat spoiling this triumphal, exponential advance, however, is a grubby little secret: The conceptual edifice being constructed by the geneticists is lacking a solid foundation. This is nothing to do with the glamorous DNA, which gets all the press, but down in the very the basement of genetics, the realm of the proteins. Proteins lack the glamour of DNA, yet they do almost all the actual work.

If nucleic acids are the white-collar hierarchy on the upper floors, then proteins are the blue-collar handymen from the basement.

The management of even the most complex of organisms is founded on this sequence of cause-and-effect in the bottom-most basement of our Empire State of the life sciences:

Higher control levels release patterns from DNA onto RNA which is translated into a linear chain of aminoacids which folds and compacts to a protein with an active site that fits a molecule.

Protein Folding

All but one of these steps are well understood, leaving just the "protein folding" step as a major mystery 50 years into the genetic revolution. Protein folding is the technical term for the last step in making an active enzyme, for example. For all proteins are first spun out as a long, sticky thread that has to fold up into the precise 3-D shape. The precise shape that is the active protein.

In more complicated situations, it seems that the chains have to interact with other proteins to fold correctly.

Even the simple, unaided situation is, however, a puzzle.   Scientists have already taken into account all the known interactions such as hydrogen bonding, hydrophobic & hydrophilic interactions, 'metal–ion chelation', and 'steric hindrance' and calculated the predicted forms. But here, so far, they have hit a snag. The problem is that: "calculations designed to predict the three-dimensional structure of proteins … invariably give far too many solutions. In the literature on protein folding, this is known as the 'multiple-minimum' problem."[7] There are so many solutions it would not be possible for a protein to test all of these until it finds the right one, it would take too long. A small chain of 150 amino-acids testing 10^{12} different configurations each second would take about 10^{26} years—a billion, billion times the age of the universe—to find the 'correct' configuration. Yet the refolding of a denatured enzyme takes place in less than a minute.

Naturally, this problem has attracted the attention of many workers. A recent review of advance in this field noted, "It is not yet possible to predict a three-dimensional structure from just the amino-acid sequence, except by homology with a protein of known structure. Nevertheless, understanding the basic rules of protein architecture is now well advanced, and it is becoming possible to design folded structures de novo."[8]

While our understanding of the internal systems involved in protein folding is currently minimal. one thing is very clear: they all involve linear chains of amino acids. Occasionally a chain will be linked in a circle, even more rarely a peptide side chain will hang off the main chain. But in large, all the peptides and proteins of life are linear chains. Admittedly these chains are often linked, but the bonds linking them are not peptide bonds (well, perhaps rarely) but the thioesters bond involving the rather unusual sulfur - sulfur bond. The problem with such linearity is that amino acids are just as likely to form branches with their side chains—quite a few have amino or carboxyl groups on their side chains and these can participate in peptide bond formation.

This is exactly what happens in natural metabolism. In an environment that favors the peptide bond, a mix of amino acids will form all sorts of branching chains as the side groups participate in the peptide bond forming. Such tangles of non-linear chains are called proteinoid.

a-a-a-a-a-a-a-a-
  protein

              -a-a-a
a-a-a-a-a-a-a-
  proteinoid   -a-a-a
        -a-a-a

Even random linkage of amino-acids can produce molecules with interesting properties (such as the microspheres of Sidney Fox and his collaborators) such as "catalytic activity, membrane-like properties, electrical activity, sensitivity to light…"[9]

Proteinoid is a not-unlikely product of natural metabolism and some workers have proposed it as being central to proto-metabolism. If so, however, then the proteinoid has left about as much fossil evidence as has clay, namely very little.

A familiar example would be folding a plane sheet of paper into an intricate origami bird. There are a lot of steps that have to be done right to make it happen. In the same way, a nascent protein chain has to fold naturally (and usually without assistance) and properly in the same way to get to the desired end, the active form.

Unfortunately for classical science, there is no well-accepted explanation of just how a linear chain of ami-noacids folds precisely and quickly into its active form.

Many protein enzymes can be reversibly unfolded, or denatured, by elevated temperatures (a boiled egg is irre-versible denaturation). Warm the enzyme solution and the aminoacid chain un-folds—it returns to the unfolded form of its ribosomal nativity. The enzymatic activity totally disappears. Cool the solution, and the enzyme is reborn; the chain refolds into the exact same form as it had before and the enzymatic activity fully returns.

Now, while this does not sound too mysterious, in the conceptual framework of classical science it verges on the miraculous.

Aminoacid Desire

In order to give a broad overview of protein folding, I am going to resort to anthropomorphism—it makes things so much simpler to explain without having to use technical jargon. (If this gets irritating, just mentally translate "desire" into high energy, low probability state; and "mutual satisfaction" into bound, low energy, high probability state.)

Each of the 20 varieties of aminoacids has a set of 'desires' it seeks to 'satisfy' chemically. In the natal, extended state, each one of the aminoacids in the chain clamors and insists on satisfying its needs with a complementary partner or a ménage-a-many: Positive seeks negative charge; Water-hater seeks same for deep dehydration; Water-lover seeks ice princess; Active hydrogen-bonder desires passive partner; Sulfur looking for same to cohabit; Any aromatics out there? Etc. etc.

Some aminoacids have many needs; some have just one; some are complex and massive, others simple and small. Some are strident in their demands while other are moderate in their requirements. Odd proline has a kink, two cysteines like to cross-link, while glycine, the simplest, makes no demands at all.

There are a set of different properties that the amino acids have in different amounts:

Some participate and encourage the natural tendency of the backbone to wind up in an alpha helix, while others hate to do this and prefer straightness in their neighborhood. There is also another way the chain can fold, the beta sheet in which chains lie parallel to each other, some One love this, some do not. Proline has a kink in it **making the turns in beta sheets and ending any alpha-helix**. The 'blanks' are easy either way; they can be straight or happily dance a helix or pleat a sheet.

Water: Some aminoacids are very good at providing forms for water to participate in while others provide only 'hostile' forms that repel water. Some are acid,
some are basic. Some are strong acids and or bases whose charges have to be satisfied. Some are H-bond donors, others are acceptors. Cysteine has a -SH group at
the end of a hydrocarbon string. This likes to bond with a -SH on another chain
forming a disulphide bond, $-SS-$ and so "cross links" chains. (Insulin, a familiar
protein, has four chains all linked by such 'disulphide' linkages. Aromatic aminoacids have bulky rings that are most comfortable when they can stack together
like pancakes.

Each aminoacid can find satisfaction with many partners, i.e., they are promiscuous, or perhaps better put, generalists. Some swing both ways, especially where
water is concerned. Aminoacids will accept anyone with the right charms.

A similar list of 'desires' for the nucleotides is much simpler. They are the opposite of the generalist aminoacids. Only one partner, and one only, will satisfy a
nucleotide's monogamous desire.

In the following discussion, DNA is going to lose some of its star power. In
fact, we will hardly mention it at all. Rather, our focus will be on RNA in all its
many guises.

Only two chemical differences distinguish RNA and DNA, and both serve to
make DNA more inert and long-term stable than RNA (suitable for shipping down
the generations). The two differences are

The ribose backbone in DNA lacks a hydroxyl and its h-bonding ability. One of
the four nucleotides has one extra oily-spot, or CH_3- radical added to it.

DNA has less tendency to interact with water and a greater tendency to self-
interact than DNA and is a lot more stable and inert. As we shall see, DNA plays a
role similar to the shiny CD that I received from Microsoft with Office 2004 on it,
and the Word that I am creating this book with.

In and of itself, it is rather boring. It is inert, which is an excellent trait for
something being sent from Redmond down the somewhat hostile environment that
is the US Postal Service.

Insert it into the Mac, however, and it springs to 'life.' The stored program becomes an active program, a sophisticated linear construct running on the operating
system and blossoming into the faultless writing tool that is this Word. (Yes, I am
looking for financial backing from a generous sponsor and for such am willing to
overlook my long, bumpy and expensive history with MS Word since v1.0, and the
tortuous separation from my beloved, elegant, slim 5.1a, now just a memory.)

While DNA is as the compact disk, RNA is the program, the operating system,
and the PowerPC chip. The rest of the computer is protein.

Therefore, the focus will be on RNA. In fact, while all of the RNAs we will
encounter will be copied off a DNA, I will probably forget to mention it and take
that fact for granted.

Where DNA has a T, RNA has a U. I will generically use U, even when discussing DNA, to simplify things.

Each of the four nucleotides, N, has a complement $\underline{N}$ that it will avidly embrace, while it is actively repelled by the other three that are not its one-and-only.

Except for protein-mimics that can pry them temporarily apart as in duplication and transcription—no other partner will do for a nucleotide. Unlike the aminoacids, the nucleotides are picky specialists. Actually, proteins acting as nucleotide mimics, do pry apart their relationships apart temporarily such.

N	N
A	U
U	A
C	G
G	C

Back to proteins. Constraining the possible hook-ups is the chain that binds them. The desires of the needy-neighbors intrude and have to be taken into account—compromises have to be made. Moreover, the chain itself has needs: it's happy to self-interact into coils and sheets if given a little encouragement.

There are also a multitude of water molecules enveloping and interacting with the chain; and water molecules have a driving need to be as ice-like as possible. While individually small, their overwhelming numbers make them major players in the final configuration.

The final, unique configuration that the chain folds into maximizes the overall satisfaction of almost everyone: the aminoacids, the chain as well as the clinging coat of water. Each of the billions of identical chains folds and compacts to exactly the same configuration, the 'active' form of the protein.

As in politics, however, not every constituent can be satisfied—the best possible compromise can still leave a few aminoacids frustrated. These unsatisfied few, the excluded-from-the-party aminoacids, end up having to seek chemical satisfaction with the multifarious molecules in the milieu about them.

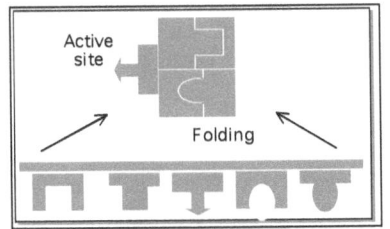

This frustrated minority is the source of the catalytic, manipulative abilities that characterize proteins. In the classical worldview, we have the 'lock-and-key' metaphor to guide us: we can think of the aminoacids as having "bumps and hollows" that fit together like, yes, lock and key.

In a few proteins, such as the albumin in egg white and blood, almost every aminoacid is happy, and the protein is inert. This is as close as living systems get to storing aminoacids.

Calcium flip

Less spectacular, but of tremendous importance for the later discussion, is that a folded protein is not a static thing. It is dynamic and can change abruptly.

A common example involves the calcium ion, a tiny but intense source of positive charge. Normally the concentration of calcium is kept extremely low inside a cell, while it is high outside the cell wall. This means that every single aminoacid chain, fresh off the ribosome, has folded in the total absence of calcium ions.

Almost all cells are sensitive to being prodded in a way that is of particular importance to the cell. When the sentinels in the cell wall receive this important message, they open the gates and allow calcium to flood into the cell interior.

For most proteins, this is of no consequence. But for dozens of proteins it matters a great deal. When calcium appears, their current form suddenly becomes improbable and a configuration including calcium becomes very probable. The chain jumps to this new form including calcium. (In our anthropomorphism: a gorgeous woman arrives and relationships shift and a new balance is reached.)

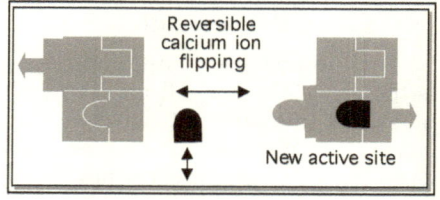

Unlike the original configuration, this calcium-plus form has a very active site. It immediately gets to work as a pebble starts an avalanche, and the whole cell is quickly informed that the summons has come. Dozens of different processes are hit on by the rapidly-activated horde and the cell 'responds' to the important signal received by the cell wall sentinel.

This is basically, what the muscle cells are doing in my typing fingers. A muscle cell is jolted awake by a neuron; calcium floods in, the muscle proteins flip to the short form; the cell contracts; my finger moves, the calcium is rapidly pumped out; the proteins flip back to the long form; the cell relaxes and awaits the next jolt from my brain.

Dozens of times a second, back and forth the aminoacid chain flips from one distinct form to the other. Clearly, whatever the process is by which the chain finds its final form is very fast acting indeed.

The same type of argument can be applied to the 'folding' of nucleic acid strands except that the 'needs' of nucleotide bases are singular and fussy: they only find satisfaction with their complementary base: no other. Aminoacids are generalists; nucleotide bases are specialists.

In addition, while protein folding (usually) involves the chain collapsing in upon itself, the "folding" of nucleic acids (usually) involves folding by lining up with another chain. Just as quickly as does a cooling protein fold, so do multi-thousand strands of nucleotides align with their complements in a cooling solution and coil up neatly in a double helix.

The DNA helix is actually quite dynamic and can be profoundly altered by inviting such things as proteins or testosterone into its configuration.

The Classical Commute

In classical science there is a concept that is taken for granted; it is so commonsensical that you undoubtedly agree with it. This 'belief' is that in order to go from point A to point B you have to cover all the points in-between.

This classical concept implies that the chain smoothly and continuously writhes and twists around before it settles into the 'correct' configuration. A newly-minted chain moves and twists about, testing the possible configurations for overall satisfaction, before settling down into the configuration that makes everyone happy.

In the conceptual framework of classical science (the one taught as "science" in high school) there is no way around it: each aminoacid is going to have to physically move—dragging the chain along with it—to each of the other aminoacids in turn to check out the possibilities of a liaison. And the aminoacids do not politely take turns—they are all actively hunting for satisfaction at the same time tugging at the neighbors to follow.

Now there are hundreds of aminoacids in a typical protein. Clearly there are a lot of different configurations that are possible, each with its associated level of overall satisfaction. So how does the chain find the best route from unfolded to fully-folded?

Taking all these aspects into account, classical physics allows us to estimate how long a simple enzyme should take to fold from the extended configuration into its unique, active form.

The result of this calculation is an eon upon eon of years measured in numbers with hundreds of digits (a million has just six, the age of the universe has just ten.)

The problem is that: "calculations designed to predict the three-dimensional structure of proteins … invariably give far too many solutions. In the literature on protein folding, this is known as the 'multiple-minimum' problem."[10]

If the aminoacid chain has to find the quick route through a vast "configuration space" it should, on average, take almost forever to do it. Yet, the actual time taken by proteins to correctly fold into their active, compact form is measured in fractions of seconds. In the reversible denaturation we discussed earlier, all the trillions upon trillions of identical chains, on cooling, fold into the active form very quickly. Yet, theory predicts a google of years for just one to make it. And, as my fingers are typing, the muscle proteins are happily flipping from short to long to short again in milliseconds.

Quite a failure of theory!

Time for another metaphor: that of the jigsaw puzzle. Take an assembled 100-piece, chunky, wooden puzzle and attach all the pieces to a length of string. Now break it up and agitate vigorously. Time how long it takes for the puzzle to reassemble.

Common sense tells us not to wait up; the chance of spontaneous reassembly, while possible, is so utterly improbable as to make winning MegaMillions look like a sure thing. The technical name for the mathematical treatment of these combinatorial possibilities is called the Traveling Salesman Problem, which sounds like a joke but is considered a serious field of study.

Yet this, in essence, is the best suggestion that classical science can come up with to explain protein folding.

There are problems with this scientific enigma:

It is demeaning-to-the-trade for the scientific community not to understand such a key step at the very foundations of genetics.

It stymies research for cures of such diseases as Mad Cow and Alzheimer's. The culprits here appear to be prions, proteins that have folded into a malignant form instead of a healthy form (whose function in the brain is obscure). The manipulation of molecules this rogue prion performs is simple: it coerces a normally-folded protein to turn to the rogue side. (A mode of replication used by vampires.)

The two prions then go off to corrupt fresh meat; and then there were four prions—exponential growth, unless a silver bullet can be found.

A possible solution is provided by the revolution that occurred in physics starting 100 years ago. In the next few sections, we will take a look at quantum physics and some of its implications for the rest of the sciences. Finally, we will return to protein folding, quantum concepts in hand, and suggest a mechanism that can be experimentally tested.

The Reluctant Revolution

The conceptual framework with which physics started out, and the one that is still in use in the biological sciences, is described by many adjectives: Newtonian, classical, nineteenth-century, old-fashioned, orthodox, conventional, etc. It can be epitomized for T-shirts as: "All is matter in motion responding to forces."

This conceptual framework was abandoned—except as a useful, if gross approximation—with great reluctance in last century because it was found to be <u>utterly,</u> and <u>totally,</u> inadequate. The more sophisticated replacement, the one physics currently embraces, is also multi-monikered: post-Newtonian, New Physics, quantum mechanics, twentieth-century, modern, way-out, totally weird, etc.

The search for a more comprehensive explanation that could deal with the experimental challenges to the classical view took physicists deeper into the nature of objective reality.

"In a sense, the difference between classical and quantum mechanics can be seen to be due to the fact that classical mechanics took too superficial a view of the world: it dealt with appearances. However, quantum mechanics accepts that appearances are the manifestation of a deeper structure ... and that all calculations must be carried out on this substructure."[11]

The new physics reached its apotheosis in the "adding endless little arrows" over-history' methodology perfected by Richard Feynman. This perspective is also called quantum electro-dynamics (QED), the official name for the theory that describes the behavior of electrons and photons in terms of internal probability.

QED is extraordinarily successful and accurate. Feynman has modestly stated that: "The theory of quantum electrodynamics has now lasted more than fifty years and has been tested more and more accurately over a wider and wider range of conditions. At the present time, I can proudly say that there is no significant difference between experiment and theory! ... To give you a feeling for the accuracy [of the quantum description of the electron]: if you were to measure the distance from Los Angeles to New York to this accuracy, it would be exact to the thickness of a human hair. That's how delicately quantum electrodynamics has, in the last fifty years, been checked—both theoretically and experimentally."[12]

The concepts and theories of quantum physics are so exquisitely successful in dealing with such a wide range of phenomena—including the furnace of the Big Bang, the graceful aging of our sun, the nature of the elements, and the workings of DVDs—that they have no serious contender.

The success of the new physics makes it unlikely that its concepts will be completely replaced by future theoretical developments. It is, of course, possible that they will

1	2	3	4	5	6
−( )	i²	0i−1	exp(iπ/2 + iπ/2)	e^{iπ/2} × e^{iπ/2}	??

suffer the same fate as the Newtonian concepts—and they were equally successful in their own day—and turn out that they are artifacts of a much deeper and sophisticated reality.

Quantum physics also graciously explains why treating atoms as solid little balls, and things made of atoms as solids, was a very workable and useful approximation. Classical physics does very well in its domain and the classical approximation is still useful. Houston put a man on the moon using simple Newtonian equations; the extra accuracy Einstein's tensor equations of General Relativity would have provided was as unnecessary as telling a carpenter you want your bookshelves 10.50269288 inches apart.

Laying down the tracery on a silicon chip, however, does require such accuracy. Such happened to the early pioneers when they began resolving phenomena at the atomic level. The classical concepts turned out to be blurred-out, external approximations of a deeper, internal aspect to objective reality.

The classical concepts, so useful for billiard balls, were totally impotent to describe the atomic phenomena being explored by the pioneers at the turn of the last century.

Something had to change, and change it did, slowly. Each concept along the way had to win acceptance against powerful opposition because each concept was so counter-intuitive and bizarre.

The new physics is, indeed, so radically weird to the classical mind that it is very difficult to accept the basic concepts. As one wit put it: not only is reality stranger than you think, it is stranger than you can think. And we are stuck with the weird quantum view, which has gone from one success to the next throwing off a plethora of goodies based on electronics such as my Mac with its CD burner and DVD reader.

"Perhaps, someday, an experiment will be performed that contradicts quantum mechanics, launching physics into a new era, but it is highly unlikely that such an event would restore our classical version of reality. Remember that nobody, not even Einstein, could come up with a version of reality less strange than quantum mechanics, yet one that still explained all the existing data. If quantum mechanics is ever superseded, then it seems likely we would discover the world to be even stranger."[13]

Hard Science

Therefore, science, as it enters the new millennium, is not just multi-disciplinary; it is a discipline with something of a split personality. In the hierarchy of physics, chemistry, biochemistry, biology and evolution, the switch-over from one science system to the other is to be found somewhere between physical and biological chemistry.

These are 'hard sciences' in the sense that their concepts are precisely expressed in the universal language of mathematics.

Like all sophisticated languages, mathematics can express the same concept in many ways, from a simple outline to an elaborate filigree. This is not to say that math-speak does not have pitfalls for the unwary.

For instance, it will take a graduate to whom math-speak is a second language in which they are fluent and can think in as easily as they can in English (etc.) to instantly see exactly what the next operator is in this sequence. See if you get it instantly as well. Did you get the answer?

It's −1. In fact, every single one of those math-speak-words is just a different way, depending on circumstance, of saying minus-one. The expression, $e^{i2n\pi}$, over all the integers n = 0 → ∞ is just another way of saying +1, and others can be even more intimidating! Don't worry: we will actually only call upon the simplest of these.

Incidentally, the $e^{i2n\pi}$ math-speak word for +1 is called the *exponential* or *transcendental* operator or function. This is why I will occasionally refer to quantum math as transcendental without implying anything fuzzily New Age.

The only reason I mention all this is that we are shortly going to encounter an equation that looks so fearsome you might, if unable to speak math, give up in despair of ever comprehending such a monster and throw down the book. Now, when you see it, please think "it's probably just saying '2+2=4' in rococo flourishes" and do not give up.

Newtonian Science

Newton is rightly considered the Father of Science as we know it. The themes he developed in classical physics have appeared throughout the scientific structure—biology might not be a branch of physics, but physics is certainly at the foundations of biology.

Of course, one is philosophically free to drop the hierarchical constraints in constructing a theory of how the world works; but the construct will be something other than science as it is practiced today. The classic historical example of this is the attempt to explain living systems by the introduction of a "vital force" in one guise or another. While there are many philosophical constructs that embrace this as an acceptable explanation, none of them are part of biology because particles, atoms and molecules can be understood without a vital force and, if electrons and quarks don't have it, neither do the atoms and molecules they comprise, nor neither do cells nor higher organisms.

So, while the biology of our era is proud of its firm foundations in the "hard" sciences (those amenable to mathematical rigor), the physics in which it is rooted is the classical physics of Darwin's day. "It is most ironic that today's perceived conjunction between physics and biology, so fervidly embraced by biology in the name of unification, so deeply entrenched in a philosophy of naive reductionism, should have come long past the time when the physical hypotheses on which it rests have been abandoned by the physicists."[14]

There is still, of course, the sense that science should be a unified structure: "How does nature encompass and mold a billion galaxies, a billion, billion stars—and also the earth, teeming with exuberant life? New insights into how nature operates come from parallel advances in particle physics and in molecular biology; advances that make it possible to examine fundamental physical and biological processes side by side. The resulting stereoscopic view deep into the past reveals a previously hidden, unifying logic in nature: its paradigm for construction."[15]

To say that pre-twentieth scientists were content with Newtonian physics, chemistry etc. is an understatement. Scientists were, almost literally, dragged kicking-and-screaming into accepting the quantum worldview because the only deity in science insisted upon it. That deity is experiment. For no matter how elegant, mathematically-sound, politically-correct etc. a theory might be, if it contradicts experiment it is crumpled and into the wastebasket.

Hopefully, by the end of the next section, you will be convinced that the new physics is truly and radically NOT the science you thought it was.

Shock and Confusion

In the Appendix: Slit Experiment, I deal with the actual experiments that so utterly confounded the physicists of a century ago. Here, just to put things in perspective, I would like to give a feel for the shock-horror these scientists felt when they saw extraordinary experimental results that insisted that all their preciously-won-since-Newton classical theories about reality had to be thrown into the wastebasket.

To do this I will tell a short story:

In the Big House, four executions are scheduled to take place by firing squad. The squad, all armed with machine-guns, is in one room, and a post to restrain the prisoner is in another. Between the two rooms are two very large windows in the wall that can be covered with heavy steel shutters.

On a whim, the warden decides to use the shutters to test his classical expectations. He was pretty certain as to what would happen but was prepared to test his theories against experiment:

The first experiment had both shutters closed. This 'control' lived up to expectations. The shutters over the holes stopped the bullets from reaching the prisoner and his life was spared.

The second and third setups had just one hole shuttered—first with the left open, then with the right. This experiment also "lived" up to theoretical expectations: The prisoners were each shredded by the hail of high-velocity bullets streaming through the void of the open window.

It was the fourth setup that violated all expectations. With both windows open, no bullets reached the prisoner. Not a one of the mighty hail of bullets reached the deafened and terrified prisoner. Just to be sure, the warden kept the gun firing extra time.

He found it hard believe his own eyes. Two voids stopped the bullet; while just one open window did not. Two empty openings were as effective a bullet shield as two steel shutters!!

The warden just had to know what was going on so he repeated the both-window open execution, but this time knocked holes in the walls so he could see in to watch the magic of 'nothing' stop bullets like solid steel.

Ratcheting up the warden's total stupefaction and torment, however, this time, as he was watching, the bullets behaved as expected. They poured through the holes and the prisoner was shredded very, very quickly.

The astonishment of the warden at this unexpected result and the mental gymnastics he went through trying to digest this result gives you a sense of the state of physics at the start of the twentieth century. To be true, the experiments that they had to explain did not involve bullets and criminals but to the scientists shooting electrons and atoms at detectors through slits, they might just as well have been.

This is, in essence, is what was observed in the slit experiments performed by the pioneers. Can you feel how horribly perplexed they were trying to digest such a phenomenon. The experiment violated all expectations on the most fundamental of levels.

Slit Experiment

We will now take a look at the experiment that played a pivotal role in the quantum revolution, the slit experiment, the equivalent of our weird execution. This experiment actually incorporates almost every aspect of quantum weirdness.

Much of the early history of experiments that lead to quantum mechanics involved trying to figure out if the basic stuff of matter was made of particles or waves. In classical science there is a clear distinction between a particle and a wave. A particle stays together as it moves through space while a wave spreads out.

In classical physics there was a simple "slit" experiment that could tell if something was a particle or a wave. The essentials are simple: fire the thing at a barrier with holes in it and watch what happens.

We can predict what happens when we fire particles at a barrier by considering a cannon firing balls at a wall in which there is a slit. We expect to see balls imbedded in the walls and a pile of balls on the other side that made it through the hole. The first key point is that each ball will arrive at a certain location on the far side.

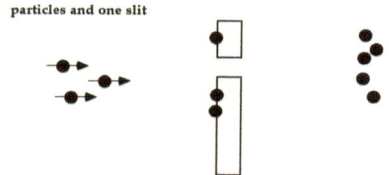

We repeat the experiment after opening another slit in the wall.

We now expect to find that more particles have traveled through the two holes combined.

We do not expect that opening another slit will prevent balls making it through the first slit to the other side.

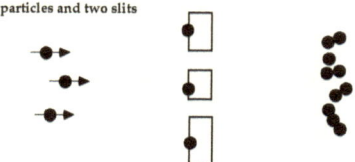

We expect something quite different to happen when we aim a wave at the slits—perhaps an ocean lapping against a wall with inlets to waters beyond. With one slit open the waves pass through and we expect to see the wave energy deposited in a diffuse zone on the further shore. This is quite different from the particles which arrive at specific locations.

With two slits open we expect something called "interference" to happen. On the far side of the barrier there are now two waves, and the crests and troughs of one wave will overlap those of the other.

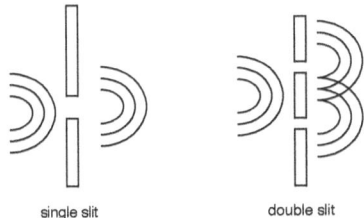

single slit   double slit

When the crests and troughs are "in phase" they will constructively interfere and combine their effects into extra big crests and troughs. When they are "out of phase," they interfere destructively and can even cancel each other out—there is no wave there at all.

CONSTRUCTIVE INTERFERENCE

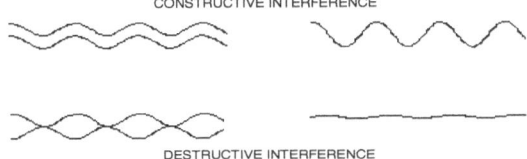

DESTRUCTIVE INTERFERENCE

The end result of this interference is that we expect the wave detectors on the far side of the barrier to register places where there is constructive interference—and much energy is deposited—and places where destructive interference occurs and little energy arrives there. When waves are involved, we will not be surprised if opening two slits register zero at a detector that fires when either slit is open.

Opening both slits creates a destructive interference at the detector and it registers no energy arriving in the wave.

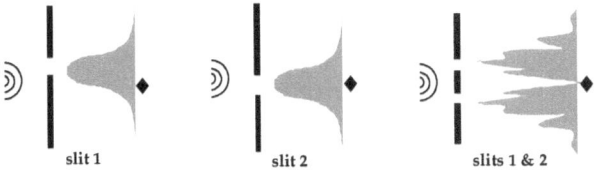

slit 1   slit 2   slits 1 & 2

With waves, we can expect "nothing" to block them from reaching certain places. This is quite different what we expect to happen when particles are being projected at single and double slits.

Particle or wave

The slit experiments were thought to be a simple way to distinguish between particles and waves because the predictions are so clear-cut.

	Worldline	Interference
wave	diffuse	constructive and destructive
particle	distinct	none

The distinction is very clear and scientists used such slit experiments to answer such questions as: Is light a wave or a particle? Is an electron a particle or a wave? Is an atom a wave or a particle?

It is with such slit experiments that physicists attempted to answer the question: Is light a particle phenomenon or a wave phenomenon?

Since the 1800s, it was known that when light passes through narrow slits it exhibits wave-like properties—it spreads out and exhibits interference patterns.

In experiments with light there were detectors that fired when either was open but not when they were both open.

The results eemed clear-cut—light was a wave.

For many years, such experiments were taken as convincing evidence that light was a wave phenomenon. Early physics had this wave occurring in an aether that pervaded all space; later thinking replaced this with undulation in an almost-equally-enigmatic electromagnetic field.

Newton had suggested that light was composed of particulate "corpuscles", but "regular" particles would certainly not be expected to behave in such a fashion. Opening both slits should make it easier for more particles to get from one side to the other. If light were classical particles, the expectation would be that more of them would get through when both slits were open than would get through when just one or the other slit was open.

Such behavior, however, can easily be explained by interference, and the consensus for many years was that light is a wave.

Compelling evidence gradually accumulated, however, that light could not be just a continuous wave phenomenon because it behaved, in many situations, as if it was composed of discrete particles, now called photons. For instance, when photons travel singly through the apparatus they behave as particles and fire a single detector—they do not arrive diffusely as we expect a wave to do.

Another example is that a high-energy photon can bounce off an electron, just like two pool balls colliding (the Compton Effect).

The upshot of this and many other experiments is that the photon has to be considered just as much a particle as is an electron.

Light, it seemed, was both a particle and a wave and, for a period, this dichotomy engendered various explanations of light such as a wave-particle wavicle—a dual-natured thing that was simultaneously a particle and a wave—or, even more contrived, that the instruments used to examine light determined whether you saw a particle or a wave.[16]

This fuzzy thinking soon became unnecessary. Explaining the slit-experiment pattern with just waves became untenable when sophisticated single-photon detectors were developed (just a little better than the three-photon sensitivity of the human eye). Exactly the same pattern could be produced over a long period of time when just one photon at a time passed through the apparatus. In the simple two-slit experiment, a single photon passing through the apparatus can fire the detector when either slit is open but never when both are open.

Here light is behaving as a particle at the beginning and end of the experiment—the emitter and the detector deal in single photons—and as an interfering wave while going through the apparatus—a single photon interferes with itself.

The basic systems of matter seem to have both particle and wave aspects. We will shortly see that the wave aspect is the internal wavefunction—hence the name—while the particle aspect is the external structure and interactions of the system.

The slit experiment has one more surprise to offer the theoretician. One obvious way to figure out if a photon is passing through the slit experiment as a particle or a wave is to put little detectors in the slits. Detectors that tell you if a photon passed through their slit.

The good news is that the experiment has been done and, yes, photons do pass through one slit or the other—i.e. as particles—and not through both as would waves. The bad news is that somehow, no matter how subtle the detectors, the characteristic pattern of the interference is no longer there. The photons behave as regular particles and arrive just as little cannon balls would be expected to.

Somehow, putting in the slit detectors makes this a quite different experiment, one that involves only particles behaving in a classical way.

Particle and wave

Coming from the other end of the "wavicle spectrum," similar interference patterns can be created in experiments with the decidedly-particulate electron—in either the all-together or the one-at-a-time situations. The electron can also behave as a particle or as a wave.

In the spring of 1991, four different laboratories independently demonstrated the interference of atoms which are indisputably bits of matter. "The first to report was Professor Jürgen Mlynek.... The sketch of [his] apparatus might have come from Young's own papers: the experiment itself was a repetition of the original 1803 version, with the crucial difference that the slits were irradiated not by sunlight but by a stream of material particles.... The most mysterious feature of the experiment... is the fact that each atom traversed the apparatus alone, uninfluenced by the jostle of other particles."[17]

This is, as noted, equivalent to the teleportation of stuff that is decidedly matter—scaling this up a zillion times, that is.

Revolution Step by Step

Try yourself, using the physics you picked up in high school, to come up with a reasonable explanation for nothing acting like steel. Don't spend a long time at it, however, genius has tried and endlessly failed.

One thing was clear. There was, and is, no way to explain such a thing with the "commonsense" notions at the heart of Newtonian physics. The path of science history from these first puzzling slit experiments to some sort of confident understanding spanned almost a century. To say that scientists were "forced" into the quantum description is not hyperbole.

The transition from the old to the new stretched over many decades and, even in these enlightened times, there is still debate about 'what it all means.' The quantum revolution was indeed a most reluctant revolution.

The 20th century was a time of transition "when the classical model of the mechanical universe became untenable and began to be modified by a patchwork of rules involving the energy quanta introduced by Planck in 1900."[18]

It was with great reluctance that scientists faced up to the implications of these changes: that their description of objective reality was horribly inadequate.

The establishment of the current worldview of physics was not based on theoretical speculation; the current view vanquished the old not for theoretical reasons but because that ultimate arbiter of science, experiment, insisted on it.

"The quantum era had arrived but it did not bring an end to controversy. The interpretation of the new quantum kinematics was, and still is, a source of both conceptual discussion and experimental exploration of its consequences in places where it contradicts deep-rooted intuitions of physicists and others, especially for questions of physical reality and causality. So far, all the experimental tests have decided in favor of the quantum kinematics. More than that cannot be said."[19]

Scientists are compelled to accept the quantum view—sometimes with profound discomfort—because it always, without fail, agrees with experiment while the classical view, just as consistently, does not.

Complex numbers

Not only is the sequence of cause-and-effect more sophisticated in quantum physics; so is the math. In grade school we are taught the law of signs which might as well end up with: Minus times minus is a plus, for reasons we will not discuss. It really is tough to make sense of this rule using regular numbers.

Fortunately, mathematicians, starting in the Renaissance, came to accept that the math of the regular numbers (the one we learn in school) is incomplete. Such real numbers, as they are called, cannot deal, for instance, with the square root of negative numbers, and these square roots pop up all over in pure and applied mathematics.

The completion of the number realm used in math was the expansion of the domain of numbers into the imaginary and the complex. In essence, while regular numbers allow for a measure of the size of things, complex numbers measure a size and direction at the same time. You might usefully recall a concept mentioned earlier at this point—that the cause-of-probability has size and direction.

The real numbers, the familiar ones, actually do have a direction to them. But there are only two of them—a direction of 0°, which is what the positive numbers have, and a direction of 180° which is the direction of the negative numbers going in the opposite direction.

Complex numbers are basically the same as these real numbers; the only difference is that they can have any angle of rotation from 0° to 360°.

When you multiply numbers with a direction you add their angles together. With this, the explanation of 'minus times minus is a plus' is as simple as two half-rotations bring us back to where we started:

$$180° + 180° = 360° = 0°$$

It is instructive to remember how difficult the concept of negative numbers seemed until quite recently. Take five oranges away from four oranges. How many oranges are left? Weird question. Eventually mathematicians realized that allowing for negative numbers introduced no contradictions, and in fact empower their calculation skills.

In general, numbers-with-direction have a size or magnitude, p, and an angle or amplitude. These are the set of *complex* numbers that have both a size and a direction united into, $z = p@a$. Equivalently, a complex number can be expressed as its *real* and *imaginary* rectangular components, $z = x + yi$.

(Note that zero can be considered either to have no direction at all or all directions at the same time—it ends up the same thing). From +1 and −1 you can easily construct all the real numbers such as 2, 3, 3.5, 4 etc., and -2, -3, -3.5, -4, etc.

Rules

The two basic rules for manipulating complex numbers with size and direction are very simple:

1) To add, put the numbers head-to-tail thus adding the rectangular components together. 2) To multiply, add the angles and multiply the sizes. Subtraction and division are just the inverse of these.

With this definition of multiplication, and being aware that a full rotation of 360° about the origin brings us back to 0°, we see the emergence of the rules for multiplying positive and negative numbers.

Positive numbers have 0° direction. When multiplying them, we add the two zeros together to get zero, so multiplying two positives—adding two zero angles—results in a size with zero direction, a positive number.

$$+1 \ \times \ +1 = (1@0°) \times (1@0°) = (1 \times 1)@(0° + 0°)$$
$$= 1@0° = +1$$

Negative numbers have a 180° direction. Adding 180° to 180° gets us to 360°, all the way around to zero. So multiplying two negative numbers results in a size with 0° direction, again a positive number.

$$-1 \ \times \ -1 = (1@180°) \times (1@180°) = (1 \times 1)@(180° + 180°)$$
$$= 1@0° = +1$$

Multiplying a positive and a negative—adding 0° to 180°—results in a size with 180° direction, a negative number.

$$+1 \times -1 = (1@0°) \times (1@180°) = (1 \times 1)@(0° + 180°)$$
$$= 1@180° = -1$$

This is how modern math deals with plus and minus numbers. It is perhaps not obvious, at this point, just what advantages this number-with-direction viewpoint has over more simple ways of dealing with positive and negative numbers. Yet this simple-to-grasp perspective will help immensely in comprehending the complex numbers with all sorts of directions.

The positive and negative "line" of numbers are all lumped together as the real numbers—the ones that lie on the axis through the zero point.

These real numbers are quite sufficient, and eminently useful, for measuring the external aspects of the world which is why we learn about them in elementary school. To be sure, the term "real" is in the Platonic sense as numbers, as entities, are actually quite abstract. To a mathematician at least, the number "two" has an abstract existence that is independent of "two things". Recognition of abstract entities is more difficult than concrete ones, and, for all the "obvious" utility of the real numbers in describing many of the quantitative aspects of our world, seeing that "two sticks" and "two daughters" had something in common took time and was an historical advance. The number "zero," the last integer to be recognized, wasn't fully acknowledged until the twelfth century.[20]

The fruit of all this effort, however, was most constructive as the real numbers bear their title well; they are eminently suited to describing the quantitative way in which many real things behave.

Imaginary numbers

While it had appeared briefly in earlier mathematics, it was only after the Renaissance that mathematicians finally confronted the fact that the mathematics of the real numbers was incomplete. Simply put, the real numbers were incapable of dealing with the square-roots of negative numbers (let alone their cube-roots, etc.).

Now finding the square-root of a number is considered an elementary operation in math. They are as common, and as important, as are addition and multiplication. For example, the final step in solving the fundamental Pythagoras equation involves finding the square-root.

As often as not, however, in solving their equations, mathematicians ended up with negative numbers under the square-root sign. If they were lucky, these "unsolvable, meaningless, imaginary" numbers would cancel out, and a solution could be obtained.

As often as not, however, such cancellation did not occur, and the equation was deemed unsolvable or as having an "imaginary" solutions (a put-down that was later adopted.)

One thing is clear, the square-root of a minus number cannot be any of the real numbers, for both positive and negative ones give positive numbers when squared.

The solution is simple in hindsight. Once we allow for rotation of numbers by units of 180° we can start thinking about numbers with a direction that is not 0° or 180° but something else.

We are looking for a number-with-direction that, when multiplied by itself, gives a negative number, a number-with-direction with an angle of 180°.

When we multiply two numbers-with-direction we add the two angles and multiply the two sizes (always a real, positive number). We want to solve this equation for the two unknowns:

$p@a \times p@a \quad = \quad p^2@2a = 1@180°$

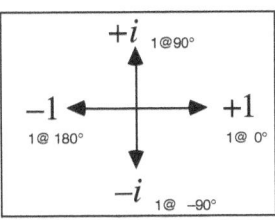

There are just two solutions:

$1@90° \times 1@90° = 1@180°$

$1@{-}90° \times 1@{-}90° = 1@180°$

These two numbers are so useful that they have their own symbols, i and –i. The imaginary numbers as they are called in that they hover directly above and below zero on the real line.

Just as +1 and –1 are the basis for all the infinity of the real numbers, so +i and –i are the basis for an infinity of imaginary numbers such as 2i, 1/3i, 3.5i, 4i etc. and -2i, -1/3i, -3.5i, -4i etc. It was these imaginary numbers, with directions +90° and –90°, that allowed for solutions to the mathematicians dilemma.

Little arrows

From here, it is no big leap for us to consider numbers that are not limited to having a directions just multiples of a right angle from 0°. What about numbers with any size and any direction? These are called the complex numbers and are central to the way the new physics describes the world.

We, of course, have the benefit of hindsight—the leap to these "complex numbers" took many a genius years of struggle to get the concept clear. A complex number can have any size, any direction. We will use the symbol p for the size of a complex number and a for its direction—and diagram it as an arrow with size p and angle a. It is these "little arrows" that feature so prominently in the math of quantum physics.

Complex numbers were discovered by mathematicians long before their remarkable usefulness in physics was understood. They are now as useful as the real numbers—inasmuch asmost of the fundamental equations of 20th century science use complex numbers. Both scientists and technologists would be lost without them—try understanding quantum mechanics or AC circuits without them—it's totally impossible. Like trying to do sophisticated arithmetic without a zero.

Just for completeness, we shall briefly look at the various forms that complex numbers take, each with its particular usefulness.

We have already mentioned the polar forms, describing the arrow in terms of size and direction. The rectangular form is the two components of the arrow, its projections on the real and imaginary axis—a combination of a real and an imaginary number.[21]

$z = p\, e^{i\pi a} = p@a$ polar form

$\quad = x + yi$ rectangular form

A complex number is, in the terminology discussed earlier, a 1D-extension in a 1D Hilbert space, the simplest of all.

The most common form in quantum physics is the exponential form. This is even more so for electronic AC theory where, somewhat confusingly, they use little-j to signify the square-root of minus-one. This tradition started because little-i was already firmly-established as signifying the negative intensity of the electric current. So j is used: $z = p\, e^{j\pi a}$

where 'e' is the 'natural' exponential base, a real, transcendental number that starts 2.17.... and 'a' is measured in radians, not degrees. Technically, as an angle greater than 360° restarts the numbers at 0, this 'a' is actually, (a mod 2π) radians.

This is a natural extension of the concept of the exponential function: when you raise a number to a real power you alter its size; when you raise a number to an imaginary power you alter its rotation.

The two forms are related by Pythagoras, $p^2 = x^2 + y^2$, and also trigonometry, *tan* a $= y/x$. Mathematicians, scientists and technicians always use radians. This 'natural' measure of angle is based on there being 2π radians in a full circle of rotation. The conversion table is: a positive real has $a = 0 = 360° = 2\pi$, a negative real has $a = 180° = \pi$. An *imaginary* number has $a = \pm90° = \pm\pi/2$. So all the following expressions are equivalent and interchangeable:

$$-1 = 1@180° = -1 + 0i = e^{\pi i \pi}$$

$$\sqrt{-i} = 1@90° = 0 + 1i = e^{\pi i \pi/2}$$

We shall only explore the very fringe of complex number math—just sufficient to appreciate how they so-perfectly describe the cause-of-probability in the new physics. The first thing is that complex numbers come in 'families' of four. They all have the same magnitude, it is the angles that relate them. This "family" of four related little arrows often appear together in quantum descriptions.

    (1) the number p@ a

    (2) the negative p@ a+180°

We mentioned earlier that, unlike +1 and −1, the twins +i and −i are virtually indistinguishable. This is why they often appear together in equations. The conjugate of a complex number simply replaces i with −i. Equivalently, put a minus sign in front of the angle. So the conjugates of the twins above are:

    (3) the conjugate p@ −a

    (4) the negative conjugate p@ 180°− a

While this sounds complicated, the diagram shows how simple their relations are. Think of the horizontal real axis as a mirror. Then the conjugate is just the "reflection" of the number in this mirror. Similarly, the negative conjugate is its reflection in the imaginary, vertical mirror. And the negative is the reflection of both conjugates in both mirrors.

While the properties of the complex numbers are fabulous and enthralling to the mathematician, the only complex math we will use as examples will involve this simple family of complex numbers combining with each other in different ways.

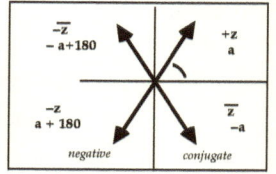

For example, demonstrating yet again what Nobel laureate Eugene Wigner called "the unreasonable effectiveness of mathematics in the natural sciences," a combination of complex number and conjugate describes exactly the connection between the internal and external aspects of the new physics while the combination of a complex number with its negative describes how "nothing" can shield a target from a "something."

Key Properties

Complex numbers are able to completely describe all the properties of the probability amplitudes and quantum cause-of-probabilities in general. Complex numbers are "unreasonably effective" in describing the internal extension of matter. The properties of complex numbers fundamental to quantum physics are multiplication, addition and conjugation. Other behaviors, such as subtraction and division of complex numbers, are defined by mathematicians, but are not needed in describing the behavior of the quantum cause-of-probability.

Addition: Adding complex numbers is most simply accomplished in the rectangular formation—just add the real and imaginary components separately.

$$x + yi \quad + \quad u + vi \quad = \quad (x + u) \quad + \quad (y + v)i$$

This is equivalent to putting the arrows head to tail in a diagram—the result is the arrow that joins the start and finish. The size of the final arrow depends on the angles—adding size 2 to size 2 can give a variety of sizes, including size 4.

Adding a complex number to its negative is adding plus and minus equal quantities so the result is zero:

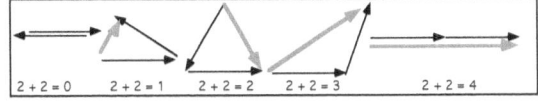

$$(x + yi) \quad + \quad (-x \quad - yi) \quad = (x - x) \quad + \quad (yi - yi) \quad = \quad 0$$

This exact canceling of adding a number and its negative will prove important when we describe combinations that have exactly zero resultant size.

Multiplication: Multiplication of complex numbers is simplest in the arrow, or polar, formulation—an arrow with a size and a direction. We have already encountered this rule: the magnitudes (sizes) multiply each other while the angles (amplitudes) sum together.

$$p@a \quad x \quad q@b \quad = \quad pq \quad @ \quad a + b$$

It is impossible to have a probability greater that 100% even in the quantum world. For this reason, in quantum physics the size of the arrows is never greater than unity. So multiplication either leaves the size alone—if the multiplicand has size exactly one—or it is less-than-one and so shrinks the final size. Thus the occasional reference to multiplication in QED as shrink-and-turn of little arrows.

Multiplication of complex numbers is used in modern physics to describe how sequences of cause-of-probabilities combine with each other.

Collapse to real: The last key property is that of multiplying a complex number with its twin, its conjugate.

$$p@ a \quad x \quad p@ -a = \quad p \; x \; p @ \; a - a = \quad p^2$$

Almost by definition, the angles always cancel out so this operation always gives a real, positive number with an direction of exactly $0°$. This combination of complex numbers is called the absolute square. We will refer to this as this the "squaring" of a complex number into a regular number with just size but no direction—a scalar, as the mathematicians would have it.

We will usually symbolize the absolute square of a complex number with P (think probability) or by any of the other, all equivalent, representations. Two things are of note here.

Cancellation: If a QPF and its negative are added together, the size of the result is exactly zero. If p is zero, P is zero. The probability is zero; it is forbidden. This is what underlies the bullet-proof nothing we encountered earlier.

If p = 0 then P = 0

Internal Amplification: If we double a QPF, we double its size while the angle stays the same. Doubling the size of the probability amplitude quadruples the probability:

z = p@a 5p@a 10 p@a

z^2 = P 25 P 100 P

We will refer to this as internal amplification of probability. It this that underlies the high probability of the 'contented' state of paired electrons that drives chemical interaction. Clearly, if we are dealing with a gazillion amplitudes adding, the amplification of probability that results is going to be a gazillion-squared. This is exactly what happens in a laser where the probability of all the photons jumping into exactly the same state is so overwhelmingly-large that they all jump there at the same time and a laser beam of coherent light zaps out. This is why it is called Light Amplification by Stimulated Emission of Radiation—a phenomenon Einstein predicted long before it was actually observed.Chess Pixels

Even through the quantum revolution is now a century old and is perfectly described by the mathematics, translating the math into a natural language is tricky and usually controversial. The basic firmly-established equation, for instance, that accurately describes an atom is:

This is the fancy, baroque way of putting it. Later, we will generalize this into something much simpler. Take heart from the following equation that is almost as bad: This is actually just a fancy, and occasionally useful, way of stating that 2/2 = 1.

$$ - \frac{d^2\psi}{dx^2} = \frac{2m}{\hbar^2}(E - V(x))\psi $$

$$ i^{200} - i^2 / (e^{\pi 2i} + e^{\pi\,200i}) = e^0 $$

There are only a limited number of solutions to this equation: The "1s orbital" of hydrogen is the simplest of these "eigenfunctions," while the largest, uranium, has electrons in the "7s orbital."

This equation of Schrödinger's is as firmly established in science as anything is. Yet, ask a quantum physicist, "What does it mean?" and you will get a variety of translations, some emphasizing uncertainty, some wave-particle duality, others non-local causality, etc.

For an excellent introduction to this equation I recommend the book, translated from the Russian, from where the following quote was lifted:

"This all implies that electrons exist in the atom not as particles but as waves, whose nature was not quite clear at first, even to Schrödinger himself. What was clear to him, however, was that whatever the nature of these electron waves, their motion must obey a wave equation. Schrödinger derived such an equation. It looks like this:

"The equation says absolutely nothing to those who see it for the first time. It induces curiosity or even a

$$ - \frac{d^2\psi}{dx^2} = \frac{2m}{\hbar^2}(E - V(x))\psi $$

nebulous feeling of instinctive objection (without serious grounds for the later.

"... Physicists were quick to appreciate the advantages of wave mechanics—its universality, elegance, and simplicity. Ever since they have almost abandoned [more chunky methods]."[22]

I am going to translate the math into familiar, anthropomorphic terms. And make no apologies for it; even attempt a justification for such apparent laziness towards the end. We will need just two aspects of the revolution encoded in this intimidating hieroglyphic. They are fundamental and decidedly non-classical: quantum pixels and quantum probability.

Quantum Pixels

First, the little 'h-bar' on the right side of the equation. This involves a key concept in both classical and quantum physics: that of the action, the fundamental measure of existence.

The concept of "action" and the affiliated "principle of least action," were developed by Lagrange and others in the eighteenth century as an alternative formulation of Newton's equations of motion, the basis of classical physics. The action equations are more cumbersome than Newton's in simple situations and, consequently, never caught on in classical physics—the action equation that describes the motion of a pendulum, for instance, is much more mathematically challenging than its simple equation of motion.

In complicated classical situations as well as quantum physics, however, the superiority of the action formulation is overwhelmingly apparent.

Action is a such fundamental measure of the state of systems that, in a sense, the task of science is to discover all the factors that contribute to the action of a system and the "action equation" that describes how they combine:

"Physics can be formulated with the action principle. A given body of physics is mastered if we can find a formula that empowers us to determine the action for any history... The action principle turns out to be universally applicable in physics. All physical theories established since the time of Newton may be formulated in terms of action...

"Our search for physical understanding boils down to determining one formula. When physicists dream of writing down the entire theory of the physical universe on a cocktail napkin, they mean to write down the action of the universe. [The accompanying illustration is a contemporary action equation; 's' is the total action.] It would take a lot more room to write down all the equations of motion...

"The action, in short, embodies the structure of physical reality."[23]

It is an action equation that describes the combined influence of all the many interactions on the changes in the overall history of the system. This is as true for quantum physics as it is for classical physics.

$$S = \int d^4x\, \sqrt{g}\, \left[\tfrac{1}{\kappa} R + \tfrac{1}{g^2} F^2 + \bar{\psi}\, \not{D}\, \psi + (D\varphi)^2 + V(\varphi) + \bar{\psi}\, \varphi\, \psi \right]$$

The Path of Least Resistance

Next we will deal with just what is the cause of the probability amplitude, the cause-of-the-cause-of-probability, if you will.   The cause of the probability amplitude involves something that we have already discussed. A path of history, a series of interactions, generates action, the scientific measure of resistance.

The rule is that systems tend to follow the path of least resistance. At first encounter it does not seem to be the sort of thing we expect scientists to state about the world. Certainly the statement might seem more at home in many a Californian subculture.

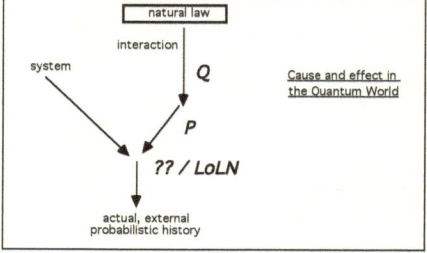

We will now proceed, however, to explain how the statement, Systems tend to follow the path of least resistance. has a precise—i.e. mathematical—scientific description of how the world works.

In classical science, systems always follow the path of least resistance, in the new they tend to follow that path. The new science introduces with this qualification an element of choice that is quite lacking in the classical view. For a tendency to do the "right" thing implies an occasional lapse into the "wrong" thing. We will later in the discussion liken this to black-and-white external states connected with waves-of-gray internal states.

Action and interaction

In physics, then, the action is a consequence of interaction. The equation just mentioned is an example of the way a scientist would describes the overall consequence of interaction contributing action along each path of history. The simple-looking symbols are actually highly sophisticated mathematical entities—tensors, matrices, path integrals and other such esoteric shorthand—that must be invoked in order to measure the action for a particular interaction. It was this complexity that precluded the action formulation from wide acceptance in simple mechanics.

You will notice, however, that the final step in solving the equation is the grade-one step of adding the six numbers together. This final sum gives the overall action. This refreshingly-simple step occurs because each of the terms in the action equation calculates the action generated by a particular interaction—one for the gravitational interaction, one the electromagnetic interaction, etc.—and the overall action is the simple sum of the actions generated by each interaction. Hence the simple final step in solving the equation.

The action equation for any system will be similar: the final action will be the sum of the action generated by each interaction the system is capable of—which, as noted, will depend on its coupling substructure, the subsystems it is capable of coupling with.

For each level in the scientific hierarchy—corresponding to the unique interactions found at each level in the material hierarchy—there will be a corresponding set of action equations that give the action that the system's interactions generate.

Luckily, in many instances, the situation is simpler than might be expected knowing the full hierarchical structure of systems. For instance, while the structure of atoms contains subsystems that can couple with the gluons and pions of the strong force, the equations that describe the action—and thus the history—of atoms and molecules does not include this. For none of these subsystems are involved in the interactions of atoms, they are sequestered in the nucleus. Neither the strong nor the weak force appear in the formulation that allows the path of least resistance to be calculated.

Universal and specific

The application of the quantum perspective to figuring out what any system will do is conceptually simple:

a) measure the interactions along each possible path

b) measure the action generated by each interaction

c) calculate the total action along each possible path

d) Compare them all: the path of least action will be the one the system will follow.

It is the necessity of examining all possible paths that makes the action way of looking at things complicated in practice.

Natural law in the quantum world thus has three aspects:

First is the universal impulse to follow the path of least action

Second, are the specific laws that determine just how much action each particular interaction generates when it is indulged in.

Third are the laws governing the development of the internal wavefunction, and how it is connected to external space-time.

The Principle of Least Action

First, the universal aspect of natural law, the PLA. The "Principle of Least Action" is one of the "givens" in our universe—it cannot be derived from any simpler principle (as if anything could be simpler!). It is sometimes referred to humorously as the Law of Cosmic Laziness; less insultingly as Cosmic Parsimony; and with dignity as the Zeroth Commandment that "Thou Shalt Not Generate Unnecessary Action." (We note here for comparison that the quantum perspective we will soon examine is not that different in that it asserts that a system will tend to follow the history of least action rather than stating that it will follow the path of least resistance..)

Without knowing anything about gravity, for instance, it is possible to "explain" why a ball falls to the ground: the gravitational interaction in staying put or moving upwards generates more action than moving straight downwards. If you know the correct equation, you can prove this. (This rule is so general that it is also the best answer to: "Why did the chicken cross the road?" as both classical and quantum science agree that it must have been because crossing the road generated the least action. Such "explanations," however, have about as much explanatory power as "God Did It" without knowing the equation—known for the interactions of a ball but not for a chicken.)

Path Integrals

Second, the specific aspect of natural law.

Quantum physics considers action of fundamental importance. "The fundamental law of quantum physics states that the probability amplitude of a given path being followed is determined by the action corresponding to that path."[24]

The probability amplitude, measured by complex little arrows, is the tendency to follow a particular history. It is consequence of the action along that particular path. This tendency has a size and a direction "pointing" in an internal dimension as already noted.

The connection between the size and direction of this internal tendency and the action along a path of history involves a somewhat sophisticated mathematical construct called a path integral. We will not go into this in any detail, just pick up on the main details.

The action itself can also be thought to have a cause—the amount of action generated by a particular interaction is a given in this universe—it is what we call a natural law. In both classical and quantum physics, the amount of action involved is determined and can be described by equations. Both classical and quantum science agree that natural law determines the action. Each interaction has a natural law that determines just what the action will be.

The basic rule in both classical and quantum science is that each interaction the system is capable of contributes to the overall action along a path of history:

The innocent-looking symbols for each interaction in the equation we looked at earlier represents "path integrals" that sum the action over each path of history: from the state the system is in to the state it could end up in.

A simple illustration of an action integral is to let the height of a curve be the action at that point in the history, so that the area under the curve represents the action of the complete path: the path integral as it is called.

The connection between the path integral over a path and the probability amplitude for that path is: The magnitude is inversely proportional to the area under the curve—the bigger the area the smaller the size of the arrow. Its angle or amplitude is derived from the perimeter of the curve. This is sophisticated math so let's leave it at that. This is where the probability amplitude, a quantum cause-of-probability measured by little internal arrows, come from.

In mathematics, the connection between a curve-with-area and a magnitude-with-direction is dealt with as the relationship between vectors and their cross products.[25] We need not, fortunately, go any further into this aspect of the math.

The general principle here is that the greater the resistance (action) along a path of history, the weaker the tendency to go that way. In quantum physics, the rule is that systems <u>tend</u> to follow the path of least resistance. It is this tendency that is the internal extension of a system.

Thus, the internal "graph" of the action, from which comes the probability amplitudes, is fully determined. If we know the action equations, we can fully know the internal tendency to change state, the probability amplitudes involved.

Non-locality

The connection between action and probability amplitude raises a very interesting question: How does the electron "know" what the action along the path is going to be before it actually travels it? How does the electron at A "know" that it will generate 47 units of action going to B and only 26 going to C without actually traversing the paths first?

While this ability to "probe the future" was also implicit in the classical action equations, it could not be dealt with satisfactorily, so was ignored (though the philosophers had a field day).

Just how a system can "know" the nature of all the possible paths open to it and unerringly pick the one with the least action is currently receiving a lot of attention under the rubric of "non-local" causality.

Certain experiments on the polarization of twined photons, for instance, can only be explained if they are able to communicate with each other about their state through some subtle agency which can convey information at speeds vastly in excess of the speed of light: "Does this non-locality actually operate at the quantum level so that two photons…, although far apart from the perspective of the scientist in his laboratory, are at another level connected? Such nonlocal connections could, in fact, stretch throughout the entire universe."[26]

We are not going to delve into this non-locality as it only really has significance in when systems are not interacting and most natural systems engage in incessant, continuous interactions at all times. It will help dispel doubt, however, when we get to discuss cells and fields of interaction that, at least, envelop a cell even though the external components exist on a much smaller, molecular level.

We can speculate, however, that this phenomena is explainable with probability amplitudes just because the little arrow is not pointing in external space-time, it is pointing in an internal dimension that is not space-time constrained. The internal extension is not constrained to the spatial extension of the system. Then internal development and change—described by the addition, multiplication and collapse of complex numbers—are communicated throughout what we will call the wavefunction independent of time.

This is consonant with experiment and theory. A single electron has a wavefunction throughout the universe—the only reason we can even think of a single electron as having a location—and single electrons have been trapped in quantum wells—is that its wavefunction is effectively zero throughout most of the universe except in the quantum orbital—it cannot "teleport" itself out of such a deep well.

Wavefunctions are not spatially limited by the speed of light in the two-open slit experiment: as the electron is leaving the source, its wavefunction is already interfering with itself at the detector.

Again though, this is not all that important as interaction interrupts constantly. With each collapse in the wavefunction—a probable event actually happens—a new wavefunction is established by the system in this new state and history progresses.

So we can expect that the influence of the internal aspect of systems will not be directly limited by time and space—constraints like the speed of light that so limit external things, etc.—but indirectly by sequential collapse in interaction and the passage of time.

Development of the wavefunction

The way the internal state of a system changes over time is usually called the "development of the wavefunction."

The Natural Laws in the new physics are the principles which govern the development of the wavefunction. For the elementary particles, the natural law is described by the Schrödinger Wave Equation. This equation describes exactly how the wavefunction changes and develops over time. "The most important lesson to be learned from Schrödinger's equation is that the time evolution of a quantum system is continuous and deterministic."[27]

Solving this equation for the electron, for example, enables one to calculate exactly what the wavefunction will be in the future. Unfortunately, this equation is fiendishly difficult to solve with current mathematical techniques so that only relatively simple situations, such as an electron and proton interacting to form a hydrogen atom, are fully solvable.

If this rule-of-internal-law holds for fundamental particles, we can expect it to hold for all systems composed of them. Each level in the material hierarchy will have its own natural law running the internal aspect of things by determining the amount of resistance generated by interaction—by systems externally sharing their subsystems.

Our current scientific understanding of the probability amplitude does not stretch much beyond simple molecules. With appropriate simplifications, perhaps as far as macromolecules. It most certainly, however, does not reach up into the realms of biology, genetics and evolutionary theorizing. The rumblings in the basement have yet to shake the battlements. But, even though the details of the action equation at these sophisticated levels are not known, we can expect that they are there.

It should be noted that, while all physics and some basic chemistry is formulated in terms of least action, the "laws" that biochemists, biologists and evolutionists (not to mention the soft sciences) are rarely, if ever, formulated as a Principle of Least Resistance. All physicists can comment is that they must then either be wrong or mere approximations of a more subtle level of natural law which can be so formulated.

We now have a sequence of cause and effect: probability from probability amplitude, probability amplitude from integrated action to the action being determined by natural laws.

We will not get into the detailed math description at this point. For our purposes, it is sufficient to think of the principle of least action operating on each path's action to associate it with a probability amplitude for that path.

PLA (action along paths) = p@a for path 1

   p@a for path 2, etc.

We can add this to our diagram of quantum concepts and their basic mathematical description.

Yes, I know it's looking complicated. But once we have all the details of each conceptual step well-established we will do some drastic simplification and condensation of the picture. But for now, the detail.

We are now ready to tackle the other branch in the sequence of quantum cause and effect, that sitting off to the left in the diagram, the "solid matter" that gets to do the choosing.

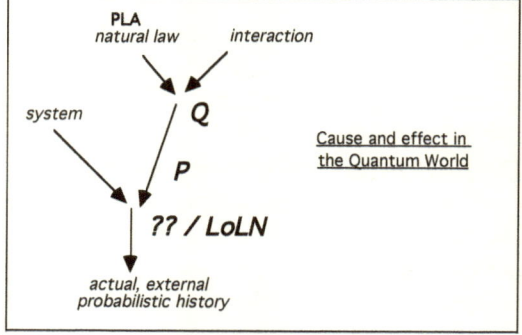

Bits of Action

In classical physics, energy-in-time and the amount of action was, and still is by non-scientists, considered to be continuous. This turned out to be incorrect.

This involved a question first tackled extensively by the philosophers of Classical Greece: How finely can you divide the things in objective reality?

There are basically only two ways to go: you can go on dividing forever; or you cannot go on dividing forever. In math-speak: reality is continuous—the forever case—or reality is discrete, its 'atomic'—you have to stop somewhere down there.

Continuous or Discrete

Take water, for example. It would seem that, no matter how small a drop of water one could imagine, it would still be water. If you cut an apple in half it is still apple, as is a tiny sliver. There is a seeming continuity here.

Take my son, however, and cut him in two and things are quite different. Half a teenager is no longer a person. Things here are discrete.

A very simple example is the difference between climbing a slope and climbing a stair.

The slope is a continuous set of states: It makes sense to say I am 1.0 foot up the hill, now I am 1.5 feet up the hill, now, getting slower, I am exactly 1.8765 feet up the hill, etc.

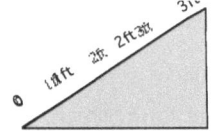 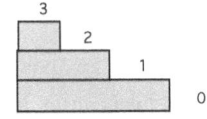

The steps, however, are a discrete set of states. It makes sense to say I am on step 1, now I am going from step 1 to step 2, now I am on step 2. It does not make sense to say that I am on step 1.5, let alone step 1.8765.

We can expand this into a 2-D illustration with multiple possible paths.

Continuous graph: A simple 2-D illustration of continuity is that of graph paper. I put a chess piece—the mighty Queen—down on the paper. The readout of a laser ultra-GPS ruler tells me that its center is, to an accuracy of four decimal places, 1.000 inches north, 1.00 inches east of the origin. With an ultra robot arm, it is no problem to shift the Queen by a circuitous and continuous route to exactly 1.876 inches north and 2.123 inches east, and then to 2.8888 N and 3.7171 or anywhere else for that matter. Moreover, it covers an innumerable number of points on its journey, as there is an infinite 2-D continuum of places to go.

Discrete chess: My favorite illustration of 2-D discrete steps is that of chess. A common opening move is to boldly claim the center by moving the pawn on 'state' e2 a 2-square jump to e4. Much less common would be the more timid 1-step move to e3 (the ghost). In standard notation this is:

1. e2 – e4 ... (or 1. e2 – e3...)

Either move is a single step in 2-D space—it is discrete. If I were to tell my computer's chess program to open with

1. e2 – e3.125

it will beep, sulk and most probably crash. For there is no such in-between state; only the before-state of e2 and the final-state e4 are relevant.

This is the essence of the discrete jump. The time it took to move is irrelevant, as is the path taken by the piece (it might have been dropped on the floor), as is just where in the final square it ends up resting—"more to the top and over just a little" is not meaningful in chess.

This is actually a good example of a quantum jump. 'Quantum' comes for the Greek for "a little bit." An electron, for instance, can be in one quantum state or another but not anywhere in between them, and the jump takes no heed of time-and-space constraints (AKA teleportation). And, like the knight move in chess, nothing can block it in.

Even the notation for the quantum electron states in atoms has similarities to chess notation. This, for example, is how scientists describe the quantum jump responsible for the bilious, if illuminating, yellow of sodium streetlights—as electrons by the gazillions monotonously making the quantum jumps:

$$3s - 3p \; 3p - 3s$$

ad infinitum, back and forth. Much, much less common, is the quantum jump $4s - 3s$ that is a lovely violet color. Just like in chess, however, asking a quantum physicist the color of the quantum jump:

$$3s - 3.125 \; p$$

will just prompt disparaging mutters such as "modern education!" and "hopeless, hopeless!"

Classical science soon found many of the things that they had assumed were continuous were actually discrete.

Cutting an apple has limits: when you get down to a single cell it is still apple, but dividing that cell in half is just like cutting my son in two—the two halves are no longer apple, they are cell debris. A drop of water can be divided much further until reaching a single molecule—cutting that no longer gives you water but atoms of hydrogen and oxygen. Matter was discrete.

But other things, including time, space, energy, light, spin, etc., all seemed to be continuous in nature, and were decidedly so in classical physics.

This turned out to be wrong. All of them turned out to be quite discrete and chess-like. The reason for the apparent continuity was resolution: a vast chessboard seen at a distance will not be checkered at all, it will just look an even gray and apparently continuous.

Physics describes pixels in the real world that are decidedly small by our standards, which is why, from our perspective, the jerkiness, is not apparent.

The 'pixels' of reality are so tiny, however, that the classical approximations of continuity—much the easier to describe in mathematics—still serve us very well. It is only at 'natural' resolutions that the quantized 'squares' of reality—of existence itself as action—become apparent.

So, from a distance, my PowerBook's screen seems continuous; the curves to the black letters seem crisp and sharp. Up real close, however, I can see square pixels and a bad case of the jaggies.

In a similar fashion, when scientists developed sophisticated devices to zoom in on reality, they found that everything came in pixels, nothing was continuous. Space, time, energy, charge, gravity, spin—you name it, they all came in pixels of a certain size.

Luckily for our sanity, reality is ultra-ultra-high resolution; the pixels are so, so tiny that we do not consider existence to be jerky.

Back to the Schrödinger equation of the atom and the little 'h' that appears on the right-hand side as its inverse squared. This conversion factor, h, for the action into pixels is known as Planck's Constant and, to a high accuracy, it's:

0.000000000000000000000000000000000000000000

0000000000000000000000033511346   lbs secs

Tiny indeed, which is why reality seems so smooth. We just can't see the jaggies no matter how hard we concentrate. The famous uncertainty relation is related to this 'pixel' aspect of quantum unexpectedness.

This pixelation of existence—the real world about us—has odd consequences.

The tick of time is very, very small. Even quite big bits of matter when multiplied by such a small number can still remain under the pixel of existence limit. So there is no reason why the bits of matter should not appear for a few q-ticks of time.

If a speck of matter were to appear out of nothing for just a q-tick or two before dematerializing then it would not "officially" exist. There is then no reason then, in a pixelated reality why specks of matter should not appear out of nothing and then disappear back to nothing. No "real pixel" of existence/action is created so that classical, approximate laws—such as the conservation of mass/energy—are not "really" violated.

We can define a virtual crime analogy in the penal code: stealing a wallet and then returning it unaltered to its rightful owner before the theft is noticed is, technically, a violation of the laws protecting private property. But it's not a 'real' crime, a pixel of law-enforcement response, so to speak, is not generated, it's a virtual crime. The bigger the item you steal, the less time you have before returning it and avoiding detection.

Now one of the simple rules of quantum physics is called the totalitarian principle: That which is not forbidden is compulsory—if the probability is not exactly zero, then it has a not-zero probability, no matter how small.

If there is nothing forbidding a speck of matter from popping into the universe for a q-tick or so then it is compelled to do so. And it does so. Naturally you cannot directly detect such fluctuations of the vacuum into matter and back—they would have to officially exist for that—but such "virtual" particles have been experimentally confirmed by their indirect influence on real things such as electrons in atoms. Note that the smaller the lump the longer it can hang around before generating a bit of existence and thus violating the conservation of mass and energy. For instance, the empty vacuum is actually a froth of virtual electrons and their antimatter counterparts (and all sorts of other things that are not forbidden).

Like virtual particles, indirect effects of virtual pick-pocketing might be observed. In our analogy, crossing Grand Central Station at rush hour with intense virtual pick-pocketing along the way might cause an otherwise-unexplainable fraying of the wallet pocket. In our society, Luckily, virtual pick-pocketing is not compulsory, or at least, not that I've noticed.

Limit to knowledge

In classical science there was the possibility of ever increasing accuracy in the measurement and knowledge of paired aspects of things such as position and momentum or duration and energy

So, for instance, the ratio π in mathematics has been calculated to an accuracy of billions of decimal places. In classical theory there was no reason to think that, given sufficient technical advance, position and velocity or duration and energy of things could also be accurate to an ever-increasing number of decimal places.

This, again, turned out to be incorrect, there is a limit to what can be known. This is the Uncertainty Principle that gets a lot of attention in most books about quantum science.

The Uncertainty Principle limits how accurately such paired attributes can be known: the better you know the one the less you can be certain about the other. Measure with accuracy the momentum of the pea-sized electron in a Yankee Stadium-scaled atom—which can easily be done—and its position could be anywhere within the stadium.

The combined precision of known momentum and position is measured in units of Planck's Constant. So this limit-to-knowledge is actually just another way of looking at the pixels of existence—you can only pin down things down to a single pixel; fractions of this are not 'real' and are thus unknowable.

Einstein's Nobel Insight

In all areas of the new physics, it became clear that the world obeyed the quantum, chess-like rules of the discrete, not graph-like continuous ones. We shall give a brief example that involved Einstein's prize-winning contribution to the Quantum Revolution. (Not for relativity.)

It involved the nature of energy. Is it continuous or discrete? It certainly seems to be so: classical physics was quite capable of measuring changes in energy to an accuracy of a dozen or so decimal places of a watt.

In classical physics the assumption, rarely noticed, was that energy was continuous, you could spread it out as thinly as you wished. We are all familiar with the fact that light intensity falls off with distance: the glaring headlights that cause temporary blindness just 10 yards away are hardly noticeable 10 miles away on a Great Plains interstate. What about at the distance of the moon? The next galaxy? The ends of the universe? If energy is continuous, with sufficiently-sensitive ultra-instruments, we should be able to follow the intensity as it gets closer and closer to zero without it ever reaching zero exactly.

Einstein won his Nobel for showing this to be incorrect: energy actually comes in distinct, and decidedly discrete packets.

He won by being the first to successfully explain a phenomenon that is often used these days in automatic door openers—the photoelectric effect.

The electrons in some metals are very loosely held and float freely, hardly held at all by any atom (a fact that underlies electricity and the shiny look of all metals). Just a little light suffices to kick huge numbers of electrons from some metals.

Einstein came up with the prize-winning explanation of the following, quite unexpected, result of experimenting with this photoelectric effect:

A metal is exposed to red or blue light and the number of electrons kicked out is measured. One light source is more intense than the other:

Red light: a 1,000,000,000-watt searchlight.
Blue light: a 1-watt Christmas tree decoration.

The classical expectation, if light energy is continuous, is that the billion-watt red light will kick out a lot more electrons than the feeble one-watt blue light.

Unfortunately, classical expectations were exactly wrong; the result that Einstein successfully explained was the ineffectiveness of intense red light compared to the effectiveness of the blue:

Red Blue

0 1,000,000,000,000

Einstein received his Nobel for coming up with the following combination of ideas already 'in the air' as his genius flourished:

1. Energy is discrete: it comes in distinct, particle-like, packets called photons—bits-of-light so to speak. It is utter nonsense to speak of half-a-photon of blue light, let alone 0.0123 of a red one. This is the nature of the quantum, chess-like jumps that characterize objective reality.

2. A single packet of blue light has more energy than a photon of red light.

3. There is a quantum jump between the bound and the free state with no in-between states. Chess-like e2-e3 behavior again.

This brilliantly explains everything as we can see with this simple diagram.

. The red light is not energetic enough to lift the bound-state electron into the free-state (the chance of two reds getting absorbed at exactly the same time before the red energy is reflected or turned into heat is infinitesimal.) Just one blue photon is quite sufficient, however,

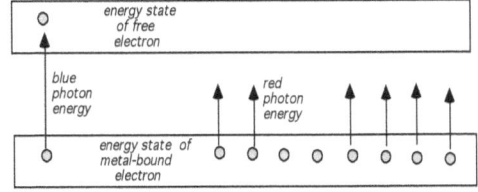

to jump an electron into the excited, freed state. And even a 1-watt light is gazillions of photons (so they can trigger the opening of your garage door, perhaps).

Simple; once some genius has figured it out first.

Pixels of spacetime

Even the continuous map has its limitations, it turns out, as on a really, really tiny scale even space-time itself is discrete with pixels we can call q-ticks and q-spans.

The quantum q-tick of time (officially the Planck Time) is about:

1/1,000…(40 zeros)…000 of a second.

It is as much nonsense to state that something takes only 1/2 a q-tick or 3.12345 q-ticks as a pawn in chess moving to e3.5.

The quantum q-span (officially the Planck Length) of space is about:

1/1,000…(36 zeros)…000 of a meter.

Our universe is about 15 billion years old and stretches about 15 billion light-years in each direction. I can figure out how many pixels there are on my Mac monitor by multiplying its dimensions X and Y:

number across x   number down = 1280 x 854

= 1,093,120 pixels

In the same way, we could calculate the total number of spacetime pixels in our historical universe by multiplying T, X, Y and Z:

number q-ticks in 10 billion years   x

(number of q-spans in 10 billion light years)^3

You can go figure it if you want but, be warned, it's a really, really big number.

Classical science has no concepts that can deal with discrete pixels of energy, space and time.

As is apparent, it is this chess-like stepwise 'quantum' nature that has given its name to the entire revolution.

The quantum world generates the appearance of a reality described by classical science. Just as well, a world with big pixels would be like looking at a really old movie. And people would just disappear in one place and appear in another. Travel would be sickening as the scenery flicks from one scene to another and we traverse the stepped landscape like those little moles popping up in the fairground game of hit the mole.

The Law of Large Numbers

Mathematicians chanced upon the Law of Large Numbers (LoLN) first when trying to describe games of chance, gambling, gaming, etc. Basically, the LoLN is the common-sense-notion that, in the long run, events reflect their probability.

In classical physics, this law is useful in statistics but is not considered to be all that fundamental. Some classical examples of the LoLN at work:.

If the chance of tossing a head with a fair coin is 50% then the chance of throwing three heads in a row is 12.5%, ten-in-a row is about 1/2,000, twenty-five-in a row is about 100 million to one—odds regularly encountered in state lotteries. The chance of a hundred-heads in a row is about one in a trillion-trillion-trillion—essentially, but not exactly, zero.

Its like the rationale I use to justify spending a $1 on a MegaMillions ticket for a $100,000,000.00 jackpot. The difference between having exactly zero chance of having $100 million dollars—my non-ticket-owning prospects—and the chance of having $100 million dollars not being exactly zero—the ticketed state—is worth every penny. Buying more than one ticket, let alone a fistful, however, increases this minuscule probability by so very little that I rarely buy more than one.

The law of large numbers guarantees that, given that sufficient people play, someone, somewhere, is going to throw "twenty-five heads in a row" and make that quantum jump to mega-wealth. Not all the lotteries in all of history would have a winner for the hundred-in-a-row, however. The number of attempts is just way too small. Atoms such as uranium, in their superabundance, exhibit detectable radioactivity even though the probability for each atom is so, so tiny.

On the other hand, the law of large numbers guarantees that, if you toss a coin a sufficiently-ridiculous number of times you will, eventually, get a hundred heads in row. If you buy a large enough number of lottery tickets the LoLN guarantees you will win the jackpot. (Beware the fine print to the innocuous-enough phrase, a large enough number. This number is large enough to make even the Pentagon's annual budget seem a pittance.

This aspect of the LoLN is just that which is not forbidden is compulsory in another guise: Either something has a probability of exactly zero to an infinite number of decimal places or its probability is not exactly zero even if a gazillion decimal places are involved.

There is one more aspect of the law of large numbers, one that we will often see in action.

Compare the probability of throwing a head with the actual number of heads we get when we throw real dice. If we just throw a few coins we might see a significant deviation from the probability: a not-unlikely four heads in a row is 100% heads, nothing like the probability of 50%. If we throw 1,000 tosses we will probably end up with something close to 50% e.g. 561 heads: 439 which is a ratio of 56% which is much more like it. Reality is reflecting probability with more accuracy the larger the numbers get. How about a trillion tosses: the heads will be 50% to many places of decimals.

It is this aspect of the law of large numbers, and this alone, that prevents the gazillion atoms of air in this room from all ever being in the other half of the room at the same time, leaving me gasping in my half of the room in a total vacuum. For, just like a fair coin, air molecules move so fast and freely that every microsecond they make a choice: this side of the room or that side of the room with essentially 50-50 probability. The numbers involved are so huge that its 50% in each half of the room to dozens of decimal places.

The basic rule is: the more you do it, the more reality will reflect the probabilities. The law of large numbers is also the reason why gambling, sorry, gaming casinos do not quail when a 'whale' wins a few million dollars. It has been taken into account. For, the in-built house advantage—and it varies from about 1% for blackjack to a usurious 30% in keno—is guaranteed by the LoLN to be the return on the gross. The only caveat is that there be a large-enough number of gamblers with disposable wealth flowing in the doors, which does not seem to be a problem. Over the long run, the house is guaranteed to make its 1% on blackjack and 30% on keno. The probability gets 'fleshed out,' so to speak, given large enough numbers for the LoLN to kick in.

In classical physics, the law of LN is useful but not fundamental. In quantum physics, however, it plays an essential role.

For instance, it is the LoLN in the new physics that is responsible for the classical—and now relegated to a 'useful approximation'—concept of 'solid matter.' It is how, as we shall see, a light, pea-sized electron hovering about a grapefruit-sized, massive proton can seem to be, and behaves as if it were, a Yankee Stadium-sized 'solid' atom as the electron teleports about within the bounds of the atom

Hierarchy

Of course, classical and quantum science do not disagree about everything; they are in complete agreement about the external aspects of matter.

One basic agreement is that all things are composed of simpler things. A basic tabulation of the hierarchy of non-living systems and their constituent subsystems is remarkably compact.

As far as the non-living systems are concerned, this taxonomy of systems is complete except for one not-so-minor detail—it seems to only embrace about 10% of the universe: the visible part. It has been established, through astro-

SYSTEMS	SUBSYSTEMS
Molecules	Atoms
Atoms	Nuclei and electrons (cool)
Atomic nuclei	Nucleons
Nucleons	Quarks
Quarks & electrons	?

nomical observation and cosmological theory, that the other 90% of the universe is made up of "dark matter" which is not visible (which accounts for why no one noticed it until recently).

This must surely be the coup de grace to the historical trend that—not content with moving the earth and its inhabitants from being at the very center of the universe to being a minor planet about a star among 100 billion others in our galaxy which is just one of 100 billion others visible to the Hubble—has relegated the grandeur of our multi-galactic visible universe to being a minor component of a much larger reality which, as yet, we are only vaguely aware of.

While there is no consensus as to what this dark matter actually is, most theories limit its possibilities to some sort of exotic elementary particle or, less convincingly, to various kinds of non-luminous "regular" matter.[28]

The previous was written in the 1990s. Things are now even stranger. I was informed at the proofing stage of this manuscript that, "I believe the current tally of the universe is 70% dark energy [Einstein's cosmological constant in reverse with a vengeance], 25% dark matter and 5% normal matter."[29]

Plus one hundred billion photons and neutrinos for each atom, I might add.

In either case, the dark matter is either made of quarks and electrons or can be lumped in with them as "fundamental systems whose structure is currently unknown."

Fundamental particles

While it is true that the structure of the fundamental particles is currently not fully known, modern physics does have some idea of what subsystems are to be found inside an electron or quark system. As this is required for the discussion of interaction in the next chapter, we will take a moment to explore this frontier of late-twentieth century physics.

These systems on the bottom rung of the hierarchy—the indivisible "atoms" of our age—do not seem to have any inner structure when probed with high-energy collider "microscopes" and, to some, are at rock bottom in the material hierarchy and exceptions to the principle of "systems of interacting subsystems." Their appearance as featureless points, however, is more plausibly explained as a limitation on current experimental methods, which can only "see" structures on the scale of 10^{-16} meter. There is speculation that there is inner structure on a much smaller scale:

Just as "the proton... [is] formed from three quarks... the electron ... [could be] formed from three very heavy new subquarks, all tightly bound .... Might not a subquark then be composed of three even heavier sub-subquarks or sub2-quarks? Extrapolation almost forces one to postulate a progression of new subX-quarks, smaller and smaller,... held together by new, stronger and stronger forces...."[30]

However: "Following this frenzy [of resolving bare quarks] we seem to have hit the experimental basement. Even if you turn up the magnification by an additional factor of 1,000—as you can at Fermilab or CERN—there appears to be no more layers of matter, no further strata. Bedrock down until '?' "[31]

Superstrings

Another line of speculation on the structure of the elementary particles is Superstring Theory, which theorizes that what we call an electron is a distortion in a space-time continuum with 26 or so exotic dimensions (or is it just 10, I forget) in addition to the four familiar dimensions of space and time.[32] A theory that abandons the concept of classical physics that space-and-time are a featureless stage upon which matter particles move. In this scenario, the subsystems of the "elementary" particles are vibrational modes and topological constructs in this poly-dimensional stuff.

We are not aware of these extra-dimensional extensions, the theory explains, because they exist on a scale on the order of 10^{-34} meter, extraordinarily small even on the subatomic scale. Properties of particles, such as electric charge, are the result of particular kinds of twists and deformations. The obvious "next question" as to what dimensions are made of—be they the space, time or the exotic variety—has yet to be answered convincingly.

One of the compelling reasons why Superstring Theory is being taken seriously—even though it seems to violate Occam's Razor of no unnecessary hypothesis—is that it is consistent with Einstein's work, which established that the familiar dimensions of space-time have a topological structure, a curvature that is perceived as the phenomenon of gravity. Our universe is basically flat—with local curvatures, such as that about the sun, being pretty mild—except for exceptions in the vicinity of neutron stars, black holes, etc. This is why we can, on a dark night with a good telescope, see for quintillions upon quintillions of miles in all directions.

The exotic dimensions, in this plausible theory, also have a curvature to them, though in their case it is anything but mild. A super-gravity puckers them up to "universes" on a scale so small that, to them, an electron is as the earth is to a grain of sand. In the Big Bang origin of the universe, all the dimensions started off with this tremendous curvature and just why the exotic dimensions remained crumpled up while the familiar three vastly expanded, will presumably emerge as some theory gets a better grip on what gravity is.

The Vacuum

What is known about the inner structure of the "fundamental" particles is based on the totally non-common-sense perspective that quantum physics has found to be the best description of what "nothing" is—the nature of the vacuum. The common-sense view is that a vacuum—a volume of space empty of all particles—is just nothing, the empty stage upon which particles move about.

The classical view, similar to common sense, is that "something" and "nothing" could not be more different. The quantum view, however, is that they are actually so similar that "nothing" easily turns into "something," and just as easily turns back to "nothing" again.

Classical physics viewed reality as particles of material existing in the nothingness of the vacuum. For example, an electron (clearly a "something") was considered as moving in the absolute emptiness of space (the epitome of "nothing"). In the classical framework, the vacuum is essentially different from matter; it's what you are left with when all matter is removed.

The quantum-mechanical description of a particle such as an electron, however, is so similar to the quantum-mechanical description of the "empty" underlying space/time in which the particle moves that the vacuum has a distinct tendency (technically, a "probability amplitude") to change into particles. Such materialization is always in the form of a particle-pair, a particle and its anti-particle such as an electron and positron (anti-electron). As a positron is just as much a material particle as is the electron, we are quite correct to view this as the creation of matter out of nothing.

As might be expected at this point, when a particle and an antiparticle recombine, we end up with no particle; we are back to the vacuum.

The reason why such materialization always involves particle-pairs is geometrically reasonable in superstring theory: the undeformed vacuum when "twisted" will produce complementary curves, such as "left and right" handed, which we label "particle and antiparticle." When a right- and left-handed twist gets together, you end up with no twist at all. In non-string theories, the rationale for pair formation involves conservation laws such as charge.

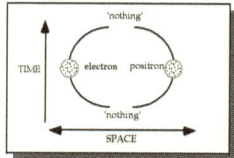

This is as simple as the sequence 0=1−1=0.

As twisting space-time takes energy, one constraint on this materialization is that it cannot create a quantum or unit of existence (as will be described in the section on Planck's Constant and the "action"). In the absence of an energy source, this restriction limits the time such a particle-pair can exist to extremely small fragments of a second before reverting back to nothing again.

This ephemeral existence—much too brief for our current techniques to directly observe—earns such particle-pairs the designation, "virtual particles." As far as quantum physics is concerned, however, these virtual particles are the same as the less-ephemeral "normal" particles; just not very long lived.

In the quantum view, "nothing" is so similar to "something" that they easily interconvert. This strange creation of "matter" out of "no matter" is clearly at odds with the 19th-century view in which matter most certainly would never appear out of nowhere. Yet, the quantum view is thoroughly supported by experimental evidence and is the only explanation of phenomena such as the Casimir Effect[33] in which the virtual particles exert a measurable pressure. As we shall see, this "something-out-of-nothing" phenomenon is central to the current explanations for the interaction of subatomic particles. As might be expected, the creation of matter out of nothing is thought to have played a central role in the origin of matter in the Big Bang, a discussion of which can be found in Appendix: Origin of Matter.

Contradicting common sense, the vacuum is a bad example of nothing; it has a structure, often described as a "quantum foam," made of these ephemeral, virtual particle-pairs. This vacuum structure is integral to the structure of regular particles and is included in the subsystems making up fundamental particles.

QUANTUM PROBABILITY

The other thing we note in the equation is the Greek letter psi, that odd looking 'Y' that appears on both sides.

$$ -\frac{d^2\psi}{dx^2} = \frac{2m}{\mathbf{h}^2}\left(E - V(x) \right)\psi $$

This mathematically represents an aspect of reality that goes by many names in natural language: the wavefunction, the 'final probability amplitude,' the internal cause of quantum probability, etc.

The wavefunction is not something that can be measured with regular numbers. Only complex numbers will do. Belying their name, these numbers are actually quite simple to understand and I have included a primer in the appendix for those unfamiliar with their delightful properties.

The crowning triumph for the developers of the new physics was the establishment of a highly accurate mathematical description of the electron and photon using such complex numbers to measure the objective extension of things in an internal space.

The techniques of QED are technically known as "Feynman diagrams," but are often called "adding little arrows"[34] because complex numbers are usually diagrammed that way.

In the new physics, such "little arrows" describe the probability amplitude. Their combining by addition and multiplication describes the wavefunction while their square describes the transition from the internal world of the probability amplitude to the external world of probability (which is always real and positive.)

I cannot resist a "what does it all mean; how to translate the math into English" comment here: If the probability amplitude is an arrow measuring an important aspect of objective reality, just exactly where are they pointing? The physicists call wherever it is they are pointing an extension in an "internal" space to distinguish it from the external space-time extensions so well described by classical physics.

Simple Rules

QED was the first theory to use probability amplitudes to deal successfully with the experimental challenges to the classical worldview. While QED can get mathematically forbidding after just a few pages, at heart, it is remarkably simple—it is basically the iteration of just three rules. Using these simple rules, QED can calculate the probability of all the possible histories—or sequences of events—that could happen to a system.

The simple rules[35] are:

Rule 1. Add complex numbers: If the history can occur by many different paths, <u>add</u> the probability amplitude for each path to get the final probability amplitude for the history.

Rule 2. Multiply complex numbers: If the event occurs by a series of sequential steps, <u>multiply</u> the probability amplitude for each step to get the final probability amplitude for the event.

Do this for every possible thing that could happen. This final combination of little arrows is the wavefunction, the psi in the Schrödinger equation.

The final step and the simplest:

Rule 3. Square the final set of complex numbers, the wavefunction: Transforms the internal probability amplitude into an external probability. In quantum science, the probability of an event is the absolute square of the probability amplitude for the event.

All three involve complex numbers and their combinatorial properties—a real number only pops up at the very end with step three. This is basically how science describes the quantum cause-of-probability.

Too Many Cooks

This is what must be the best illustration of the difference between a probability and probability amplitude:

"An amplitude is less definite than a probability: it is a sort of tendency, but as we saw it is a tendency that can help or hinder, it may be positive or negative. If you had several cooks in the kitchen each with a certain probability of making soup then the more cooks there were the more soup you would expect to get … [This situation could be measured by real numbers such that if the probability for cook 1 to make soup is the real number P, and that for cook 2 is also P, then probability of soup for dinner P + P = 2P.] However, as everybody knows cooks are not like this. Cooks should rather be said to have a [probability] amplitude to make soup. Even two cooks may interfere."[36]

Real numbers are incapable of describing this, but complex numbers can.

If the tendencies of the cooks are in opposite directions, one is the negative of the other, their sum is zero and the probability of soup for dinner is zero, not twice as likely as might be expected if one didn't know the contradictory attitude of some cooks and how their temperaments combine in complex ways.

$(+1p^2) = 1P = (-1p)^2$

$(+1p + -1p)^2 = 0$

This is so much nicer than my 'weird execution" scenario, but it does not convey the genuine mystification of classical science when confronted with the phenomena earlier.

It is just this behavior of complex numbers compared to real ones that encapsulates the difference between quantum and classical descriptions.

Mandelbrot Set

Underlying this description of wavefunction form are complex numbers and the way they combine with each other. The "shape" to the slit experiment is the wave-like way complex numbers interfere with each other. A short diversion is in order here to note that complex numbers seem to have an innate tendency to create interesting and sophisticated forms.

We have already noted that everything about the internal extension can be described by the mathematics of adding and multiplying complex numbers. A well-popularized development in mathematics during the last few decades specifically deals with the form-creating properties of complex numbers. Perhaps the most famous of these developments involves what has been called "the most complex object in mathematics"—the Mandelbrot Set—which has such a striking and complex form that is has been featured on the cover of Scientific American and graced countless books, computer screens and dorm walls.

The Mandelbrot Set is created by massively iterating the squaring and adding of complex numbers. This is an example of a simple operator—you drop in a complex number and out pops a new number. Drop that one in to the operator and out comes another number. Repeat ad infinitum. It is not immediately obvious that this simple process could have anything to do with form but it was discovered that, when a complex number is run through this iteration, it can be classified by the two basic things that can happen:

1. The number gets larger and larger and moves off, with increasing rapidity, towards infinity. Numbers which behave like this are not in the Mandelbrot Set. The speed with which they race off to infinity is often used in coloring the set.

2. The number does not get larger and larger. These numbers belong to the Mandelbrot Set (ignoring the complication of connected sets and the disconnected sets, or dusts, at the boundary.) The number does change each time it is processed by the operator but it remains within certain limits—called Julia sets—a bounded set that gets filled in as the numbers jump around. This is reminiscent of subsystems filling in a wavefunction. The boundaries to these Julia sets have all sorts of delightful forms to them.

The fascination with the Mandelbrot Set, which can be considered the catalog of all Julia sets, is that the boundary between the numbers in the set and the numbers not in the set has an abundance of complex forms to it. In the following magnifications of the Mandelbrot Set, the coordinates of numbers in the set are colored black, those not in the set are white. Each square is a successive enlargement of the previous one.[37]

The horizontal (real) axis of the initial view is from −1.5 to +0.5, the vertical (imaginary) axis from − i to +i. The final view is centered on the complex number 0.08378791+0.65584142i at a magnification of 34 million relative to the first view—the original is now solar system-sized and there is no end in sight!

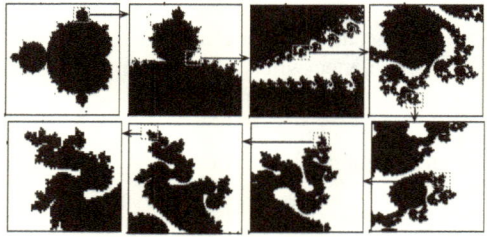

It has been proved that the depth of form is infinite, no matter how much you magnify the set—to billions or trillions of decimal places—the forms keep on emerging. Most astonishing is the emergence of miniature Mandelbrot sets at great magnification—and the whole process repeats itself all over again if not exactly. A poet might say that very nature of complex numbers seems to be pregnant with an abundance of form. End of digression and back to the real world and the way that wavefunctions—described by the addition and multiplication of complex numbers—determine the varied forms of natural systems.

This is what the 4f orbitals look like: the different shade lobes are the wavefunction going this way and that way internally[38]. Intricate shapes and fine detail, a consequence of the internal blending of waves, is something to be expected, according to quantum science.

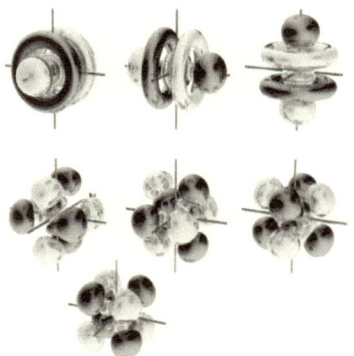

Internal Amplification

This transition from internal to external is referred to as "the collapse of the wavefunction" and is still a topic of lively, and sometimes bitter, debate as to "what it means" a century after it was discovered.

Yet, the math itself is so simple. Every quantum calculation always ends with the final step of squaring—transforming an internal probability amplitude, unobservable and measured with complex numbers, into an external probable history measured by observers with regular, real numbers.

This transformation is like squaring the familiar inch. An inch is just an ultra-thin line. Square it, however, and it turns into a postage stamp square inch, something quite different from two thin lines by themselves.

Everything interesting happens on the level of the internal wavefunction—describable only with complex numbers: only at the very end does the composite collapse into a real number by squaring. This squaring has an amplifying effect on how probabilities combine in the new physics.

This is an illustration of how increasing the size of a probability amplitude by a modest amount increases the probability exponentially:

$p \longrightarrow p^2 \longrightarrow 1\ P$

$2p \longrightarrow (2p)^2 \longrightarrow 4\ P$

$100p \longrightarrow (100p)^2 \longrightarrow 10{,}000\ P$

$1000p \longrightarrow (1000p)^2 \longrightarrow 1{,}000{,}000\ P$

What if this internal amplification applies to our linear chain of aminoacids and thousands of water molecules? The quantum probability would be a million-fold, while classical concepts only suggest a thousand. As a wage earner, I am all-too-aware of the difference between a thousand dollars and a cool million.

Many of the "not-common-sense" phenomena in the quantum world are a consequence of this internal amplification: we just don't expect probability to behave in this exponential way. (If Lady Luck behaved this way, buying lots and lots of MegaMillions tickets would actually make a great deal of sense. But she doesn't.)

Principle of Least Action

Naturally, a question comes to mind; where does this intangible probability amplitude come from, what is its cause. We mentioned that Planck's constant involved "the action"—and this is where the wavefunction comes from.

Quantum physics considers action of fundamental importance. "The fundamental law of quantum physics states that the probability amplitude of a given path being followed is determined by the action corresponding to that path."[39]

The probability amplitude, measured by complex numbers, is like a tendency to follow a particular history. This tendency is consequence of the action along that particular path, a tendency with a size and a direction "pointing" in an internal dimension, not an external one of time or space.

The connection between the size and direction of this internal tendency and the action along a path of history involves a somewhat sophisticated mathematical construct called a path integral. We need not go into this in any detail, thankfully.

Briefly put then, we can say that the cause of quantum probability form—itself the cause of the probability of what will happen—is the Principle of Least Action. This is the basic law of the universe and is the ultimate cause of what happens in the universe eventually.

To summarize:

In classical science, the action determines what happens

In the new science, the action determines the *probability* of what happens.

It's a subtle, but highly significant difference.

It allows, for instance, the concept of autonomy to appear in science at the level of subatomic particles.

Autonomy

Neither the internal nor the external probabilities are directly observable. We do not observe probabilities, after all, we witness events and interactions.

This is the final step in the new physics. The step from the probability of an event happening to the event actually happening. This is the collapse of the wave-function into an observable state.

This is a basic question asked by scientists: "What determines what actually happens?"

This has a simple answer in classical science: Natural laws determine what happens.

In the quantum world, however, natural laws determine probability, and only probability; nothing more, nothing less.

Natural law, in the new science, does not determine what happens. Surely, our classically-raised minds complain, there must be something determining what actually happens, some process that can be described by mathematics, no matter how sophisticated that might be.

Quantum physics takes a quite unexpected turn here: it asserts that the connection between probability and actuality cannot be pinned-down in an equation that predicts what happens.

There is a mathematical description of this step in the new physics, of course, but it is another non-classical concept, that of the random choice operator.

This is basically total randomness, which, almost by definition, cannot be described by an equation or a program.

Controversy about the "meaning" of this failure to take responsibility in the quantum view abounds, on a par with the "meaning" of the little arrows pointing internally. They, at least, can be described by definite equations.

We have already seen that the quantum view is different in that it introduces quantum probability forms, QPF, into the mix.

This is another big difference between the two views: Quantum physics has dropped the concept of determinism—there is no well-defined aspect of reality that determines what happens, given the probabilities.

Quantum science avers that there is nothing, other than the probability, that determines what happens externally. In dealing with this we will encounter concepts that, to the classically-trained, are almost as wrong-headed as admitting that matter has an internal extension.

To hearten the reader through this classically-disturbing section, I will mention two points surrounding the controversy about the "collapse of the wavefunction" as this transformation from probability-of-happening to actually-happening is often called:

• Scientists only feel panic about "what it all means" when thinking in regular language. The mathematical formulation is flawless—the problem is translating the perfect-description math into imperfect everyday language. As we shall see, however, the math required to describe the collapse of the wavefunction is less challenging than complex numbers—and I hope these numbers feel as natural as − 1 by now. So, while by the end of this section we will be forced, by language, to use such provocative terms as "autonomy" and "free choice," hopefully the mathematical concept of a "random choice operator" will be associated with the words and not some vague philosophical or cultural concepts with all their attendant baggage.

• The failure of predictive ability is really only a problem when dealing with a small number of events. When huge numbers are involved—as they often are in most natural systems we will be looking at—the Law of Large Numbers takes over, so to speak, and ends up turning probability into actual history. This is a well-defined formulation, there are lots of nice, well-defined equations involved. So, for most natural systems where lots of interactions are involved, the external history is determined by definite equations which do predict what happens. When lots of interaction occurs, the quantum view is just like the classical—overall history is determined.

So, if the reader finds the following section difficult to digest, worry not; its concepts will be rarely invoked as we progress up the hierarchy of matter to the new-physics description of atoms and living systems. Though it does have a role to play in the origin process as we shall see later.

Collapse of the Wavefunction

The connection between probability and what actually happens is called the "collapse of the wavefunction." In the slit experiment, we ended with a precise method of calculating the probabilities of the detectors firing in a slit experiment. But the slit experiment does not deal in probabilities; it deals with detectors firing. We take the final step of describing what determines which detector will actually fire.

Here is one of the most greatly unexpected concepts in the quantum view of the world: that there is absolutely nothing whatsoever that determines what happens in this final step.

Shoot a solitary electron through a 2-slit apparatus. We know how to calculate the probability amplitudes involved so can calculate the probability of a detector firing; but for this solitary electron we would like to know which detector will fire.

It turns out that this is an unknowable for it is firmly established in quantum physics that it is impossible to predict which detector will fire. Put another way, experiment insists that there is no natural law that determines which history the electron actually follows. In the new physics the laws all work on the internal level: there is no law governing the external.

Quantum Random Operator

Describing such random choice is difficult in mathematics, almost by definition. This inability is concealed somewhat in the official-sounding quantum random operator.

Experiment insists that nothing determines which path the electron will actually follow: which one it will "choose." The behavior of an electron is totally indeterministic—sometimes it will "choose" to go A, sometimes B, and nothing can predict which one. One path is chosen and the other is relegated to the realm of the might-have-been. This is why we can re-label the collapse of the wavefunction as "autonomy of choice" from the particle's point-of-view.

The math of the random operator's somewhat disreputable roots lies in gambling (sorry, gaming) theory—and is simplest to discuss in terms of "equal-probability and proportional representation." It sounds worse than it is. If you have ever tossed a coin for gain, you already know all you need know.

"Each aggregate describing all possible outcomes ... would be called a sample space .... In general, a sample space of a random experiment is a set of elements such that any outcome of the experiment is represented by one, and only one, element of the set."[40]

The operation that picks one of these from a set in game-theory mathematics goes under the name random choice generator. That's basically it—the quantum random choice operator. The math description of the choices made during the wavefunction collapse is just that simple and uninformative.

What the operator-description of the collapse of the wavefunction means is probably the most contentious issue in the philosophy of quantum physics. This debate of meaning, by the way, is almost irrelevant to most working scientists: the math of quantum mechanics has performed flawlessly under the most challenging of experiments. The worst fate that can befall the quantum description is, like Newton's, to be found to be an approximation of a more sophisticated reality. But wrong: never. So, the lack of consensus on meaning is not troubling while churning out the correct answers.

"The orthodox Copenhagen interpretation of quantum theory is silent on the question of the collapse of the wavefunction. The field is therefore wide open .... Any suggestion, no matter how strange, is acceptable provided that it does not produce a theory inconsistent with the predictions of quantum theory known to have been so far upheld by experiment. Our choice is a matter of personal taste."[41]

An extreme example of the random choice operator in action is that involved in the decay of a uranium atom by emitting an alpha particle. The probability that an alpha particle will make it out of the nucleus is extremely low. The nucleus, you see, behaves just like a liquid drop with a surface tension. A little ball of mercury escaping from a broken thermometer used to be a common example of this, the surface tension pulls the liquid metal into almost a perfect sphere. Now the surface tension of the nuclear 'fluid' is trillions of times that of mercury—it is not easy for a tiny drop of fluid (the alpha particle) to overcome this inward pull—there is a potential energy "wall" bounding the nucleus. The alpha particle collides with this wall trillions of times a second and usually bounces right back.

It takes, on average, hundreds of billions of years for the alpha to teleport across the wall and escape. Each bounce at the wall—each attempt to escape—is an event. Each event involves the random choice operator. In the equal-probability, proportional representation description of this operator, it is picking with equal probability an item from a huge set of probable histories that looks something like this:

1 : 1,000,000,000,000,000,000,000,000,000,000,000

While this seems odd and somewhat simplistic, the math does not get much more sophisticated than this—the key characteristic of absolute randomness, after all, is that it cannot be described by an equation. The math is telling us there is no describable process happening there—randomness is the absence of something defined.

Unlike everything else in the new physics, the random choice operator is difficult to describe in mathematics. It is impossible, for instance, to formulate an equation that has a random solution. This is why randomness is difficult to model on computers even though it would seem, at first thought, to be simple. The problem is that the generation of random numbers is not handled at all well by computers—programming a computer to come up with a truly random sequence of numbers is impossible. The best consequence of classical true-or-false logic in which "random" can only be approximated—pseudo-random numbers, as these best approximations are called.

Apparent Determinism

Indeterminism has given many a theoretician difficulties, one of which is, where did determinism go? This challenge to the quantum perspective is the reverse of the old: Why do so many systems in nature seem to have no freedom? Why are material systems so predictable and apparently ruled by law?

Even on a fundamental level, apparently deterministic laws such as "light travels in a straight line" and "light travels at the speed of light" are now understood as an artifact: photons actually have a probability amplitude to travel faster than the "official" speed limit and to veer off that way rather than toe the line, but these tendencies are canceled out by the tendency to go slower and veer this way—the only tendency that does not cancel out is the tendency to travel in a straight line at the speed of light. "The amplitudes for these possibilities are very small compared to the contribution from speed c; in fact, they cancel out when light travels over long distances. However, when the distances are short ... these other possibilities become vitally important and must be considered."[42]

If, at a fundamental level, nature is indeterministic, where does the apparent determinism come from? Colloquially put, how can "diamonds be forever" if the electrons and quarks they are made of are free to do their own thing?

"Einstein ... spent much of the rest of his life looking for the deterministic clockwork that he thought must lie beneath the apparently haphazard world of quantum physics. The clockwork has not been found. It seems that God does play dice."[43] Our everyday experience, however, is that "God doesn't play dice"—things in nature seem very ordered and predictable.

This is probably just as well for the development of science, for the fact is that most simple systems, the ones studied by physicists and chemists, are not at all stochastic; they are quite predictable and amenable to being described by simple laws. This convenient state of affairs arises for reasons that can be roughly classified into "multiple choice" and "no choice."

Multiple Choice

Systems made of electrons can seem to be deterministic and predictable because the number of electrons involved is so large and they are all in similar states. The Law of Large Numbers (LoLN) applies and what happens externally will absolutely reflect the probabilities.

A simple example of this as it applies to autonomous electrons is the deterministic behavior of electric current that "obeys" the deterministic law:

voltage = current x resistance

This simple relationship underlies much of our civilization. The reason for such absolute predictability is that a current of even a few micro-amps involves quintillions of electrons in very similar states all moving in a QPF called a conduction band. Even though the path a single electron takes through a metal cannot be predicted, the LoLN ensures that the behavior of the whole swarm accurately expresses the probabilities.

The indeterminacy of the single electron is lost in the predictability of statistics. (This, of course, applies to the most sophisticated of systems—the success of insurance and mass-marketing institutions is an indication that even we humans are ruled by probability and are more statistically predictable than we like to admit.)

No Choice

Even on the individual level, such as an electron participating in an atom, the behavior seems to be deterministic. This occurs because, while the electron is free, it has a strictly limited set of probable paths to choose from, all of which happen to lie within the atom. The set of states is bounded.

In the simplest situation, if all the possibilities have a zero probability except for one that has 100% probability, then autonomy has no choice: it has to pick the path with 100% probability. This, for instance, is the case for high-energy photons, electrons and the like: they have a 100% probability of plowing ahead in a straight line.

The way to make an electron do what you want, therefore, is not to look for some external law which will force it to your will, but to arrange things so that the probability of it doing what you want is 100% and the probability of it doing anything else is 0%.

An example of this is the ubiquitous TV tube. Here high-energy electrons are manipulated such that the probability of them illuminating the correct pixel is 100% for all intents and purposes. Of course, a lot of electrons are used, so the above multiple choice situation is exploited by the TV designer.

Electrons in stable structures, such as a helium atom, retain their self-determination, but the choice is limited to being close to the proton. All the possibilities that lead the electron away from the proton have zero probability, and the electron never goes that way and helium is stable.

From this lack of choice comes the stability and structural consistency of the world around us. Electrons do not get to exercise their autonomy in a chaotic fashion, not because they have lost their autonomy, but rather because the probabilities are constrained by the electron's environment. In this way, the autonomy of particles is constrained and complex stable structures can be constructed out of them.

The great chemical differences—and the structure of elderly stars—turn out to be an example of freedom-but-no-choice, for the probability of two electrons being in exactly the same state is zero; it is impossible and never happens.

The reason why the 96 electrons in a uranium atom don't bump into each other is simple: they have a zero final probability amplitude to do such a thing and thus, with zero probability, don't do it. Electrons don't zip around—and thus perhaps bump into another—they jump around without traversing what lies in-between and, the key point, they have a zero probability of landing anywhere close to another electron—the probability amplitudes just add up this way.

It is these restrictions that the electrons impose upon one another that make the solution of the wavefunction equations so difficult in atoms other than hydrogen (where there are no other electrons to take into account) and currently impossible for atoms such as uranium.

Time and Probability

There is a useful relationship between the probability of something happening and the time involved in it actually happening—simply put, probable things happen sooner than improbable ones. In a few situations, the theoretical structure developed in a science is mature enough to directly calculate the probable future of a system. Such is the case, for instance, with the electron.

In other situations, the probable future can only be determined by observation—but probability, of course, is not directly accessible; it has to be measured. A useful measure of probability takes advantage of the LoLN and the larger the number of attempts, the closer the results will reflect the probabilities. And if the attempts are spread out in time, the more time there is involved, the more attempts will be made. The actual history over sufficient time will accurately reflect the probabilities.

So, what is sufficient time? In radioactive studies, this period is called the half-life, the time it takes for 50% of a large number of radioactive atoms to decay. This gives a very useful measure of probability in terms of time—the period of time in which the system has a 50% chance of changing.

A neutron, for instance, is a triplet of quarks in a certain state and one of them can change. When it does, the neutron becomes a proton. A collection of 4 quintillion neutrons has only 2 quintillion left after about 12 minutes, 1 quintillion after 24 minutes, etc.—its half-life for the change, rounded up, is 12 minutes. The probability of a neutron changing into a proton over this period of is 50%.

As the word "lifetime" is unsuitable as a general measure of the probability of change in state—it connotes breakdown—we will use the more general term "transition time" and this is related to the probability that the transition will occur. As the transition time increases, the probability decreases, so the probability is inversely proportional to the transition time. This allows us to use the transition time as a measure of probability.

Estimating Probability

There are two situations in which the actual history fully expresses the probable history: a large number of systems in the same state or a single system iterates the same state. This is the Law of Large Numbers at work.

When there is just one system being observed, the LoLN can give no help, but an estimate of abstract probability by concrete measurement is still possible. In this case, there is only one measurement, the time it actually takes for the change to occur. For instance, although the half-life of a neutron is about 12 minutes, there is a probability that it could go for a whole year without decaying—and at the end of the year there would still be a 50% chance of the neutron falling apart in the next 12 minutes.

Such longevity, however, is extremely unlikely. Statistical theory asserts that it is highly probable that the single measurement will fall within a certain distance of the "mean lifetime" which can be considered the probable time period in which the system will change. Each of the italicized phrases has a defined meaning in probability theory—for instance, highly probable is usually set at 95, 99 or 99.9 percent of the time, depending on the situation.

The two measures have a simple connection derived by the math of probability statistics:

mean lifetime = transition time x 1.44

This relationship between the probability of the change in state (measured by the transition time) and the observed time of the change will prove useful in the discussion of "origins" when only one, or just a few, systems are involved. If only one measurement of an event is made, a good estimate—to the accuracy given by probability theory—of the probability of the change in state is

transition time = observed time/1.44

A provocative use of this relationship, which we will mention here and return to later, is that current estimates of the time taken to go from an abiotic earth to the emergence of the triplet code and the prokaryote-level of life forms was about 100 million years. A good measure, then, of the probability of this occurring is that the transition time is about 70 million years.

This contrasts dramatically with calculations based on classical concepts of random aggregation which give the probabilities of even moderately-complicated proteins emerging—let alone the sophisticated constructs necessary for the functioning of the triplet code—in times vastly in excess of the 15-billion-year age of the universe. Something is clearly missing in these calculations based on classical random chance-and-accident principles.

Arrow of Time

The movement through time has always been a bit of a puzzle in science. Our common sense division of time into past, present and future has no basis in classical science. The laws are reversible; they apply equally well to time "running backwards"—moving in either direction along the world-line is possible according to the classical perception. So where does the one-way nature of time come from?

"The laws of science do not distinguish between the forward time and backwards direction of time. However, there are at least three arrows of time that do distinguish the past from the future. They are the thermodynamic arrow, the direction of time in which disorder increases; the physiological arrow, the direction of time in which we remember the past and not the future; and the cosmological arrow, the direction of time in which the universe expands rather than contacts."[44]

As human beings, we are mortally aware of the difference between the past and the future:

The Moving Finger writes; and having writ, / Moves on: nor all thy Piety nor Wit Shall lure it back to cancel half a line, / Nor all thy tears wash out a Word of it.
[45]

The problem with classical physics was that it could not establish this arrow of time—the equations of motion work just as well for time running backwards.

The past state and the future state look the same to the classical equations which determine what happens to the external extension of a system. You can run this "movie" of reality backwards and it still makes sense.

The laws of nature are all, fundamentally, time-reversible. This symmetry is easily illustrated by replacing the present with a mirror—reflecting either the past or future in the mirror does not change the look of things—the world-line is symmetric to time reversal.

Another way of looking at this is that if time were to run backwards the same laws would be apply. This is true for all natural laws: both the externally-acting laws of classical science and internally-acting laws of quantum physics.

The laws of the new physics are time-reversible—they apply equally well to the internal extension whether time is running forward or backwards—it's the same action equation after all.

But once we bite the bullet and accept the random choice operator, we have a loss of time reversibility. The random choice operator makes no sense running in reverse: how can you add an actual history back into a set of probable histories? It makes no sense in reverse. The description is no longer time-reversible—you can't run the random operator in reverse—unlike the other operators which do make sense in reverse.

The full quantum picture of the past, present and future is not symmetrical—the reflections in the mirror no longer look the same.

The past-to-present is external and singular—the path the system just chose, while the future is internal and plural—the paths it has a probability of following. The present is the time when the random choice operator does its work. This does not look the same when a mirror is inserted.

This is the fundamental difference between the past and future implicit in the new physics. This aspect of quantum physics actually appeals to our basic sense of the difference between past and future—the moving finger has a single past to repent of what it has writ, but there are multiple possibilities for what it might write in the future.

This arrow of time appears naturally in the quantum view for, while both classical and quantum views have deterministic natural laws, the new physics has these laws acting on the internal extension, which can be multiple, and not the external extension, which is singular.

The only reason that this in-built arrow of time has not been fully recognized is that thinkers have usually taken the route of thinking of the internal extension as not really "real"—as a mathematical tool, but nothing more. If the internal extension is just a mathematical fiction then, of course, the in-built asymmetry in reality of past, present and future is no longer apparent. We have avoided a lot of difficulties and taken the path of least resistance in accepting that, if science insists on including an internal extension in its description of the bricks and mortar of reality, then we might as well take things at face value and accept that the internal is just as real as the external.

Simple Schrödinger

Now, when someone mentions Schrödinger's Equation at a dinner party (I wish), you can say:

"Oh yes, pixels and probability. So simple, really, that the negative steepness to the gradient has to fit into itself. I'm just fascinated by those little 'h' squares of chess jumps to existence. And I just 'sigh' over the wavefunction, don't you. So enigmatic. All those little cupid arrows pointing in not-spacetime. Sounds like my minister giving sermons about sin as "hooks" that stick out in spirit world and catch on more sin. Wonder if they are pointing in the same abstract space. And that final conjugation of the wavefunction into rough and ready probability is so romantic, somehow. Where would we be without that cute little 1s orbital, I ask you? Nowhere in a body, that's for sure!"

$$ -\frac{d^2\psi}{dx^2} \;=\; \frac{2m}{\mathsf{h}^2}\big(\;E - V(x)\;\big)\psi $$

That should lead to calls for "More Wine!" around the table.

By the way, just in case asked, the rest of the equation is boring stuff from high school: like mass and distance from the center, and electromagnetic energy penduluming from kinetic to potential and back.

See, once you know how to translate appropriately, math is not so hieroglyphic after all.

We looked at this equation in detail for one simple reason; it is basically as far as the quantum revolution has reached in terms of a rock solid mathematical equation. And it can barely be solved—not for quantum reasons, but because a precise classical-aspect calculation of the penduluming from kinetic to potential energy in multi-electron atoms is beyond current techniques.

You may breath a psi of relief; there are no more quantum equation-hieroglyphs to deconstruct.

Absolute Power

To my way of thinking, the most remarkable concept established by the new science is the awesome power-over-matter ascribed to the wavefunction, the supremacy of quantum probability over common sense notions.

The concept of probability having power over matter is gobbledygook in classical science where probability it is considered a lowly result of a lack of knowledge, if anything.

This is not so in the new physics where quantum probability plays a central role in the conceptual framework of the new physics. Thus, the quantum probability of two electrons being in the same state is zero. Now this is not the almost zero of everyday life, or even the calculus; but exactly zero. The power of probability is so absolute that all scientists are quite confident that not a single electron in the entire span of the universe, in the entire past and future of the universe, has ever been in the same state as another electron. Never has; never will; Verboten.

The power of this quantum impossibility—which in math-speak is almost as simple to calculate as one minus one equals zero—is so profound that it can hold up an entire star without any assistance. So, in a billion years or so, when our sun runs out of hydrogen to burn, it will collapse under its own weight until it has shrunk a million-fold. At this point, however, the electrons will be on the verge of being forced on top of each other, to share the same state. As this has a zero probability of happening, the sun will abruptly stop shrinking and become a stable white dwarf. All that is holding it up against the lash of a billion gravities is the power of quantum probability. An exhibition of Power that even Superman might marvel at.

In the following discussion is important to keep in mind that quantum probability is fundamentally not at all the same as classical probability—the coin toss-MegaMillions variety we are more familiar with. Rather, it involves sophisticated concepts such as complex numbers, probability amplitudes, orbitals and wavefunctions. As we are dealing with broad strokes, however, we need not delve into the details.

There are two major differences between the two concepts that have to be kept in mind during the following discussion (as the classical concepts so laboriously learned will reassert themselves on the unwary):

Quantum probability is causal and measured with complex numbers • Classical probability is resultant and measured with real numbers.

Quantum probability forms are discrete and relatively few in number • Classical forms are continuous and multitudinous in number.

Quantum Sharing

Classical and quantum theory both basically agree on the external aspect of what happens when systems interact: they trade, exchange and share bits of themselves with other systems.

In the last section on the structure of systems, we focused on the subsystems that were tightly held. Now we will deal with those with a tendency to stray.

"In the physical realm, operations arising from the interplay of four forces are transmitted by messenger particles.... In the biological realm, operations... are transmitted by messenger molecules.... This correspondence reveals a fundamental program of nature...."[46]

There are three things that have to be taken into account in this trading or coupling with subsystems. We have established that a system has:

1. Subsystems in quantum probability forms with intense gradients that have no tendency to stray. If things "fall into" probability then these QPF are deep and with steep walls. In the atom, examples of these are the protons, neutrons and inner, contented electrons of the massive atoms.

2. Subsystems in quantum probability forms (QPF) that have a tendency to leave the system. In an atom, these are the outer, valence electrons.

3. Unoccupied QPF that have a tendency to offer occupancy to passing stray bits that have taken off from other systems.

Only the last two categories are significant in interaction.

Coupling subsystems are those occupying a QPF with a probability of taking off from the system. This is a subset of all the subsystem QPF.

In an atom, for instance, this tendency to lose electrons is called the electropositive valence. The unpaired, single electron in the atom of lithium is a good example. This lone electron is easily lost when interacting with other atoms. On the other hand, the lithium atom has a very low tendency to take in extra electrons.

The empty QPF with a significant probability of taking up a passing subsystem into the structure of the system is a set of taking-in tendencies, the negative coupling capacity of the system.

The overall capacity for interaction by both giving and receiving subsystems is the combination of these two, a subset of the system's QPF.

This coupling capacity determines what we can call the sophistication of a system, the ways in which it can interact.

Correlation

This tendency to interact has consequences if the system is not an isolated one; if there are other systems around to actually interact with. The environment in which the system finds itself, its milieu, also has a tendency to interact. A system and its milieu, considered as a single system, have the same basic structure of occupied and unoccupied QPF, some of which are open to trading and exchange.

Both system and its milieu have their particular tendencies to interact. From this internal QPF comes the probability of subsystems shuttling about. This can happen in two ways:

The system has a tendency to give out a subsystem that is matched with a tendency of the environment to take that subsystem.

The system has a tendency to give matched with a tendency of the environment to receive.

The give-out QPF of one system merges with the take-in QPF the other creating a path of least action for the subsystem to slip along. Both directions taken together give the overall tendency to interact, the QPF for the interaction.

This is the interaction quantum probability field, a QPF.

Intensity of interaction

From this internal interaction QPF comes the external probability of the subsystem actually skipping on over rather than doing any of the other things it could probably do. The number of coupling subsystems making it over per unit time, the intensity of the interaction, will be proportional to this probability. The more probable the trade, the more intense is the trading.

While the random choice operator will have its say here, the law of large numbers is usually there to cancel its influence over the long run.

In the long run, the external 'form' or intensity of the interaction will reflect the internal form of the QPF.

It is not as complicated as it all looks as an example will show. Take two atoms, one of lithium, one of fluorine. In our step diagram, the QPF orbitals, some occupied, are:

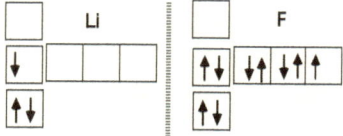

Their coupling capacities are quite different:

Lithium

a high probability of the singleton electron leaving

a zero probability of taking in another electron

Fluorine

a zero probability of any electron leaving

a high probability of taking in an electron

The correlation of fluorine giving out and lithium taking in is 0. This route makes no contribution to the interaction.

The correlation of lithium giving out and fluorine taking in gets the extra quantum boost when adding probabilities we encountered before.

Just how far the electron jumps as the two atoms approach is not recorded, but it's fast. The electron leaps from the lithium to the fluorine and everyone is happy; all the electrons are now in contented pairs. The result is two ions clinging to each other, the so-called ionic bond in chemistry.

Carbon and hydrogen are not as extreme as these two; they have about equal tendencies to give out and take in. They end up sharing contented electrons in pair-bonds. These bonds are the sticks holding the balls together in classical models of molecules.

Fundamental Interactions

That interaction involves coupling was not that obvious at the lowest levels of the material world. Rather, classical physics described interaction in terms of "forces" acting on material systems, some by direct contact like balls colliding, some at a distance like gravity.

In 19th-century physics, systems were thought to interact through the mediation of abstract, intangible fields of force, such as gravity or the electromagnetic field. In the 20th-century view, these fields were understood in terms of probability of coupling with subsystems. The four fundamental interactions of physics involve coupling with particles. The field equations of modern physics describe the probability that the coupling with subsystems will occur. First, we will take a look at the phenomenon of coupling and then at why coupling creates what classical science calls forces.

Experiments reveal that the universe contains just two types of fundamental particles—called the fermions and the bosons—which have been likened to the "bricks and mortar" out of which everything is constructed. While both virtual fermions and virtual bosons are to be found in the description of the quantum foam of the last chapter, it is the presence of the bosons that has the greatest implication for interaction.

The bricks of the material world are the fermions, the "bits of matter" such as the electrons and quarks. These fermions have the rather odd characteristic property—called "spin half-integral"—of needing a rotation of 720° to return to the same orientation. This otherwise rather mysterious property has been interpreted as support for Superstring Theory in that it is topologically equivalent to behavior on a Moebius strip[47]—a twisted surface on which it takes two circuits to return to the original orientation.

The mortar is the interaction between these fermions which involves an exchange of bosons—particles with the more familiar property, called "spin integral," of needing just a 360° rotation to return to their original orientation. It is the exchange, or "coupling," with bosons that unites the fermions together into composite structures.

Basic Interactions

Such coupling with exchange particles lies at the heart of the four basic interactions (or classical forces) known to physics: gravity, electromagnetism, the "strong" and the "weak" nuclear interactions. The best-characterized of these is the electromagnetic interaction where the bosons are the photons (particles of light) and the fermions are the electrons and quarks, both of which have "electric charge."

Electromagnetic Interaction

In classical physics, electric charge was something a particle had and the electromagnetic interaction was described as an action at a distance through electromagnetic fields.

One of the many reversals that occurred in the development of quantum physics was the realization that electric charge is not something a particle has but rather something a particle does—charge is simply the tendency of a particle to absorb or emit photons. To say that a particle has an electric charge means exactly the same as saying that it has a distinct tendency to absorb or emit photons—it "couples" to photons. This coupling is not an electromagnetic interaction—the photon itself has no charge—and exactly what is going on as an electron and photon merge or separate is a mystery since the structures of both of them are unknown.

Particles with "charge" emit and absorb "virtual" photons. Virtual photons do not suffer the time restrictions on virtual electrons since they do not experience, so to speak, the passage of time. Einstein's Special Relativity revealed that the faster you travel the slower time passes until it stops altogether at the speed of light. While from <u>our</u> reference frame, it takes a photon of light 20 billion years or so to cross the visible universe, in the photon's reference frame it takes no time at all.

So, during the brief existence of a virtual photon—which, of course, travels at the speed of light—it can actually travel an infinite distance. On its travels, the virtual photon can be absorbed by other electrons or quarks—coupling the particle that emitted it with the particle that absorbed it—giving the electromagnetic "force" an effectively infinite range. While it is a geometrical requirement that the "density" of these virtual particles falls off with distance, it is never, no matter how far the distance, exactly zero. Interaction is usually described in terms of fields. The intensity of the electromagnetic field depends on the probability of encountering virtual photons at a location.

Plus and Minus Charge

In non-quantum terms, the electron has a "negative" charge and the proton a "positive" charge, in the convention established by Benjamin Franklin. The difference between these charges does not reside in the capacity for the emission and absorption of photons—for either charge, the tendency to emit a photon is always exactly equal to the tendency to absorb one. In fact, when two particles interact electromagnetically, the uncertainty inherent in subatomic systems makes it impossible to know which system did the emitting and which did the absorbing; all that can be said is that a photon was exchanged.

The tendency of a plus or minus charged particle to couple with a photon is called its "coupling constant" and it has a value of about 1/137 for the charge on the electron. The difference between plus and minus charge is related to a type of polarization. While the exchange of virtual photons does not transfer energy between particles, it does transfer momentum. For particles with the same "charge," such as two electrons, the polarization of this transfer results in the electrons moving away from each other, just as it does when two positive charges couple. For particles with opposite charge, such as a proton and electron, this transfer moves them towards each other. We will return to this point when we, eventually, get to the discussion of the consequences of coupling with subsystems.

QED

The electron and proton in an atom interact by coupling with a prodigious number of photons—classically speaking, there is a powerful electromagnetic force between them.

The theory that describes this exchange is Quantum Electro Dynamics (QED). For all the staggering complexity of the actual calculations, the underlying structure of the QED equations is simply an iteration of two tendencies:

1) the tendency of a charged particle to absorb or to emit a photon.

2) the tendency of a photon to move from one place to another.

In modern physics, these tendencies—or, more technically, probability amplitudes or, more simply, internal probabilities—are at the root of all phenomena involving light and electrons—which embraces just about everything except gravity and the structure of the nucleus.

The electromagnetic force as a classical "force at a distance" working through an abstract force field is not part of modern science; instead, it is now understood as a substantial exchange of particles.

It is true that we usually do not think of an electron as having photons inside itself—after all, photons are huge compared to the size of the electron (or atom, for that matter). Yet, photons do definitely emerge from electrons as well as disappear into them. The vacuum foam of virtual particles obviously has to be included in the list of what a system is composed of. So, ignoring the outrage engendered by the thought that systems can contain things bigger than themselves, we shall include the photon—as well as all the virtual particles we will shortly encounter—in the substructure of the electron (and other particles).

The three other "fundamental" forces are described in the same way as the electromagnetic.

The weak nuclear force

Particles that "feel" this force (they are said to have a "weak charge") couple with particles called the W and Z intermediate vector bosons. These were predicted by theory, then detected and are now 'factory-produced' by a team of European scientists. They are now being produced in quantity in at least two high-energy facilities.[48] These weak bosons, like photons, are emitted in a virtual form, travel to other particles and are absorbed—the effects of this exchange being what we call the weak nuclear force. Unlike the massless photon, however, the bosons are massive and, traveling way below the speed of light, are all-too-mortal, falling apart in time measured in trillionths of a second. This ponderous mortality severely limits the scope and effect of the weak force and also accounts for its name.

The weak interaction plays a role in changes within the atomic nucleus. It has little to do with everyday life except for the essential role it plays in moderating the first step in the fusion of hydrogen to helium that powers the sun and, ultimately, all life on this planet.

The strong nuclear force

Particles that "feel" this force couple with gluons. While gluons and quarks cannot be isolated—they are "confined"—and can only be detected indirectly, both types of particles are considered firmly established in quantum physics. The confinement of the gluons limits the effect of the strong interaction to within the nucleus, but there it is hundreds of times stronger than the electromagnetic interaction.

The strong interaction is described by "Quantum Chromo Dynamics" which mirrors the equations of QED: the quarks making up the protons and neutrons in the nucleus couple to "gluons" which bind the quarks together. Analogously to electromagnetism, quarks are said to have a "color" charge, though now there are three types of polarization whimsically called red, blue and green.

Gluons, unlike the electrically-uncharged photons of the electromagnetic force, couple to themselves; they have "color charge." This is one of the reasons why QPF is much more complicated than QED—another being that there are three charges and the coupling constant is close to unity.

During the period when physicists were taking a good look at the strong nuclear force, high-energy colliders produced hundreds of particles. They were real but unstable. Many of these had lifetimes so short that they could not be easily detected directly and were instead observed to be "a resonance" spike in a graph of the results.

Note, that while these lifetimes are brief indeed—on the order of 10^{-24} of a second before disintegrating—the atomic nucleus is so small that these ephemeral particles, moving at a fraction of the speed of light, live quite long enough to make a few circuits of the nucleus. Thus, as virtual particles, they all had to be taken into account if the strong force that holds the nucleus together is to be properly understood. The coupling substructure of the proton and neutron thus include a large number of these ephemeral resonance-particles and thus have to be included in the list of subsystems available for coupling with.

Thus, while the positive protons in a large nucleus are copiously exchanging photons and experiencing an intense repulsion as a consequence of the momentum exchange they, along with the neutrons, are also exchanging copious numbers of all sorts of virtual particles—a resonance which exerts an exactly opposite effect and pulls the nucleus together. The "strong force" between protons and neutrons is based on the more fundamental strong force between the quarks—just as chemical bonds are based on the more fundamental electromagnetic force between electrons and protons.

Our earlier analogy of Yankee Stadium is no longer large enough. An excellent overview of the inner secrets of the quarks puts the quark size scale in perspective:

In the magnified analogy … with a human reaching the stars, atoms the size of the Earth, and [the protons and neutrons in the nucleus] fitting inside a playing field … we can say that a bare quark must be smaller than 10 centimeters, or about two-and-a-half inches, across."[49]

Electrons are about the same size. So, in a hydrogen atom, we have one electron and three quarks about the size of baseballs in a volume the size of the Earth. Clearly, an atom is a lot of orbital, a little bit of stuff doing the 'filling in.'

Gravity

Even though a quantum theory of gravity is not established, the gravitational interaction is also thought to involve coupling with hypothetical particles called gravitons. This concept, however, introduces a schizophrenia into modern physics since Einstein established the phenomenon of gravity as a curving of space-time, a bending that is mild in the vicinity of a star such as our sun but can be intense enough to "pinch off" a piece of space-time—as happens in the formation of a black hole.[50]

There is a growing consensus that at very high energy—such as abounded in the moments after the Big Bang—the differences between the particles—electrons, quarks, photons, weak bosons, etc.—disappear. These are the Grand Unification Theories and Theories of Everything that are at the cutting edge of modern physics. One of the more successful of these, the Superstring Theory, suggests that all particles, both fermions and bosons, are the result of an extremely intense curvature of extra exotic dimensions, and that regular gravity and gravitons are just a pale echo of this more fundamental level of reality. It is this sort of convergence of ideas that gives many theoreticians hope that quantum mechanics and gravity can be formulated in a consistent way.

Inertia

One of the enigmas associated with gravity is the link between inertial mass and gravitational mass. Mass is the measure of a system's capacity to gravitationally attract other systems; inertia is the measure of a system's reluctance to change velocity. There is (currently) no compelling reason why these should be the same, yet they are, to the accuracy of the best measurements.

Einstein avoided the problem for linear motion by removing the concept of absolute motion in the Special Theory and by postulating the equivalence of mass and inertia as a foundation for the General Theory.

Unlike the validity of absolute linear motion, the validity of absolute rotational motion is still under debate. There is a protracted and confused debate, which continues to this day, as to whether Einstein's General Theory of Relativity does or does not incorporate Mach's Principle: That there is no such thing as absolute rotational motion, only relative rotation. (Similar to Einstein's position on linear motion.) The implication of this, however, is that rotational inertia ("centrifugal force") is caused by the presence of the rest of the universe. Would the earth bulge at the equator if it rotated alone in the universe? No one really knows, but as linear inertia involves the Higgs—see below—we can guess that the answer will also involve them as well.

For all these questions surrounding gravity, both mass and inertia are an expression of graviton coupling. The most useful scientific measure of graviton coupling is actually not mass. While classical and quantum physics have quite different perspectives on what the mass of a system is and how it changes, both agree in placing momentum in a central role. Momentum combines both the gravitational and the inertial into a quantity that is neither created nor destroyed: rather, it is a measure of graviton coupling that is rigorously conserved.

Momentum is the product of mass times velocity. Mass is the measure of graviton coupling. Velocity is change of position in the inertial frame with time. This movement through the combined gravitational field of all the matter is a measure of the inertial gravitational coupling with gravitons.

Momentum, the product of mass times velocity, is thus a conserved measure of the gravitational and inertial interaction of systems with gravitons.

Higgs

In order to bring us up to the cutting edge of modern physics we will briefly mention the Higgs mechanism. In all of the interactions discussed so far, theory implies that the properties of the couplers should be very similar—clearly wrong as, while the photon and gluon are massless, the weak bosons are not. A mechanism has to be introduced to explain this symmetry breaking, and this mechanism involves particles coupling to the "vacuum" with Higgs bosons, massive particles (i.e., of very short range) that the super collider in Texas was supposed to look for.

It is this coupling to the vacuum that is thought to give each particle—bosons and fermions—its characteristic rest mass which is also somehow involved with gravity and gravitons. One of the key differences involves the quantum concept of spin: fermions have half-integer spins, bosons such as the photon, gluon and W have unit spins; gravitons are predicted to have a spin of two while the Higgs is expected to have spin zero. All this will hopefully become clear when a quantum theory of gravity becomes established and science moves down the hierarchy of matter.

We have included the subsystems in the vacuum foam in the substructure of particles such as the electron and quark. One puzzle is: if all particles have the same vacuum foam around them, why do particles have different abilities to couple with them? The vacuum foam of the quarks contains virtual photons, W bosons, gravitons, gluons and it can couple with all four of them. Presumably, the same stuff envelops the electron, but for some reason, it fails to couple with gluons; it does not feel the strong force. What happened to the virtual gluons in the electron's substructure? The neutrino, embedded in the same vacuum foam, does not utilize the photons we would expect to be there—the neutrino does not feel the electric or the strong force. Where did the virtual photons and gluons go?

See the Appendix and Volume Two for some speculations on the Higgs.

Bits of self

In all of the fundamental interactions just described, we see that the coupling subsystems are drawn from the substructure of the interacting systems—the quantum foam structure of the vacuum in the most basic cases.

Interactions involve them coupling with subsystems. Things interact by exchanging bits of themselves in quantum science.

Just as the "fundamental" particles interact by coupling with the subsystems from their substructure, systems at every level do exactly the same thing: they exchange bits of themselves and thus interact with each other.

Systems can interact by coupling (both sharing and exchanging) with any of its subsystems, though they do not necessarily have a significant tendency to couple with all of them. Here, "significant" implies that the tendency has to be taken into account if the behavior of the system is to be understood. These active couplers are a subset of the external hierarchy of stuff filling in the QPF.

The coupling substructure does vary somewhat with situation—for example, water at −200°C has a different set of tendencies to couple with its subsystems than when it's at +200°C. The following general discussion assumes everyday standard-temperature-and-pressure situations.

The list of active couplers is a qualitative description of the external aspect to interaction, delimiting the types of interaction a system can get involved in.

This is what systems do. In a nutshell, they interact by exchanging and sharing bits of themselves with others via QPF. What classical science describes as force is a consequence of the quantum probability of the exchange happening in the new physics.

The Path not Taken

There are consequences to the inherent contingency in open histories. Sometimes, taking one path and not the other can have historical consequences. A potentially good example is our current understanding of why all life uses only the left-form of amino-acids and right-form of nucleotides.

As far as we know, however, there is no reason to think that right amino acids or left nucleotides would not be just as good at working together. One explanation is that there is no good reason why our L-R set-up emerged—it was a contingent step along the way and the random choice operator picked one path from the possibilities. Based on this one event, a whole tree of possibilities opens up that ended up with us.

The other combinations never made it to this step, if they did, we out-competed and extinguished them for they have left no trace. We can diagram this with a simple wavefunction: the random choice operator "picks" the path at each node ending with the left-right connection event. They are the paths not taken in Earth's history. Once this L–R situation became established, it is theorized, it preempted all the resources and prevented any attempt to establish any other chiral combination.

If we could do the calculations (as far as current knowledge seems to predict) there is no reason why a probability that life would develop on the right amino acids and left nucleic bases is also there. a probability, however, is not an actuality. It seems the explanation is contingent; the left-right system appeared first and preempted the stage leaving no probability that another system could develop—the 'contingent evolution' promoted by evolutionist S. J. Gould.

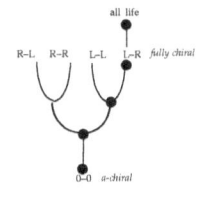

Or, then again, perhaps it will turn out—when we know and can solve the equations that describe all the internal systems and action equations—that a system such as a primate ape is almost certain to emerge over a period of 20 billion years after a Big Bang.

Hierarchical couplers

A hierarchy of interactions is quite simple to list: each level inherits what came before and adds the capacity to couple with its primary subsystems: at each level, the level below contributes an emergent interaction to be added to the inherited ones. Not all of these capacities, of course, are expressed at each level

SYSTEMS	COUPLERS	EXAMPLE
electrons	gravitons, photons	charge
quarks	" & gluons	color charge
nucleons	" & quarks as pions	nuclear force
nuclei	gravitons, photons	weight
atoms	" & electrons	valence
molecules	" & atoms	H bond
macromolecules	" & molecules	water structuring
organelles	" & macromolecules	protein action
cells	" & organelles	fertilization
organs	" & cells	immunity
etc.		

We have now discussed the second concept that the quantum and classical views agree upon—systems interact by coupling with subsystems. Systems are composed of interacting subsystems coupling with sub-subsystems. For example, a water molecule is composed of interacting H and O atoms which couple with electrons and photons.

Primary and secondary

A distinction that will be significant when we get to discussing origins is that of primary and secondary interactions. Put simply, while a system can couple with its primary subsystems, an isolated primary subsystem can only couple with secondary subsystems. The origin of systems deals with the fact that history is filled with examples of times where a system is absent followed by times where it is present. A simple example is the moment thousands of atom-less years after the Big Bang that saw the appearance of atoms. This marked the appearance of a new interaction on the cosmic stage. An atom can couple with electrons, a primary subsystem. Electrons, however, do not couple with electrons, they do not have atomic valence. Only the combination as an atom has valence.

Valence emerges, so to speak, with the formation of the first atom. Valence, involving the primary subsystem is a primary interaction. (Note, that in no way is this to be interpreted as most significant.) Atoms also interact by coupling with photons and gravitons, but these are secondary in that they are inherited from the particles.

We will call the primary interaction the emergent, and the secondary ones the inherited capacities for coupling.

The atom, for example, inherits the electromagnetic interaction from its constituent electrons and protons. On the other hand, the atom does not inherit the color charge of its constituent quarks, so chemists do fine without including color charge in their atomic and molecular action equations (such as they have).

As systems congregate together as a supersystem, the supersystem can start to couple with those very systems doing the congregating. We will return to this point when we get to the Origin of a system after we have figured out what makes the systems do the congregating in the first place.

The atom can couple with electrons—the realm of chemistry. But electrons and quarks do not have electrons as subsystems and so cannot couple with them—they do not have valence.

The valence interaction is what is called an "emergent property" of atoms—it only happens when electrons, protons and neutrons have assembled into atoms.

Valence coupling with electrons, then, is an emergent interaction of atoms while electromagnetic coupling with photons is an inherited one.

Bottom of hierarchy

Where does this hierarchical structure root itself—if a photon is a system, then from the above it must have its interacting subsystems, or coupling sub-subsystems. The suggestion in Superstring theory is that particles are self-sustaining vibrations, or solitons, in curled-up multi-exotic dimensions, and such "strings are not, of course, visible … impossible to detect by any means known to science today; they are mathematical curves."[51]

Is this pundit saying that the coupling sub-subsystems of particles are of the same stuff as mathematics? Perhaps not, but we really have not come across another suggestion.

While this rooting of the material hierarchy in such abstract stuff seems to be verging on metaphysics, it would tie up one loose end: if the root systems of the material hierarchy are really the same as—or even just similar to—"mathematical curves," then Wigner's "unreasonable effectiveness of mathematics" will no longer seem remarkable or needing of any further explanation.

Currently, the simplest suggestion of what our universe started off as is a 11-dimensional featureless sphere made of whatever it is that dimensions are made of.

The first thing of note that happened was the inflationary era in which the four time-space dimensions part company with the others and the four forces differentiate out from each other. The best mathematical description of this is currently group theory: "Towards the end of the last century, many physicists felt that the mathematical description of physics was getting ever more complicated. Instead the mathematics involved has become ever more abstract, rather than more complicated. The mind of God appears to be abstract but not complicated. he also appears to like group theory."[52]

The next phase is the conversion of inflationary energy into particle-pairs and the era of particle interactions which is currently described by complex numbers, Hilbert spaces, etc.

This is about as far as the "hard" sciences get (fully described mathematically) but the same principle applies. In a similar way, all the following developments—atoms, molecules, … bacteria, …primates etc. of the hierarchical structure—are all a result of the function working on the previous level.

We have already encountered the concept in Superstring Theory that the "stuff" out of which the "fundamental particles are made is more mathematical stuff than material stuff.

"This mathematical stuff is then processed by the natural law function such that "the entire sequence of events that unfold ...—the stars, the planets, the molecules, and the 'people'—are all just mathematical states ... a vast web of mathematical deductions spanning out from the starting state.... This speculative line of reasoning turns the Platonic position inside out. We no longer need to think of mathematical entities as abstractions that our material minds are battling to make contact with in some peculiar way. We exist in the Platonic realm itself."[53]

Made of Math: Run by Math: Described by Math. No wonder math has been called the Queen of the Sciences.

Coupling and forces

Before we move on we will mention here why classical physics does not describe interaction as exchange of subsystems but rather as forces, acting at a distance, that bodily move things around—obvious examples being the gravitational force, the electric force and the magnetic force. This is understandable when we realize that we can expect there to be consequences when systems couple with each by exchanging bits of themselves.

This is quite apparent in contemporary understanding of why the exchange of virtual photons in the electromagnetic interaction creates an apparent and measurable electric or magnetic force that moves things around. The explanation is quite simple—virtual particles can carry momentum along with them as they couple, and momentum—that key mix of: gravity and inertia—determines how mass moves through space. A change in momentum is a change in the way the mass of the system moves through space, it appears to be moved around by "forces" (in the classical sense).

Photons have momentum so that particles emitting and absorbing virtual photons will experience a change in momentum.

An electron coupling to another electron with photons can exchange momentum—its mass-through-space—in such a way that the change is such that the electrons move away from each other—there is a repulsive "force" between them.

It is this exchange of momentum via the virtual photons and the resulting effect on the history of the electron that is the classically-described "electric force" acting at a distance between charged particles.

This situation of subsystems being exchanged carrying their capacity for interaction is clearly a general one.

The other fundamental "forces" of boson coupling exert their influence on the fermion-bits-of-matter in a similar way

Gluons have the coupling capacities of momentum, spin, electric and color charge and transfer these from quark to quark. Unlike the other bosons, gluons have a strong tendency to emit and absorb gluons themselves—they couple strongly to themselves. This is just one of the reasons why the strong force is so difficult to describe mathematically.

W-bosons are like photons in that they carry momentum, they can also carry charge along.

An electron coupling the valence interaction between atoms, for instance, carries along with it its capacity to couple photons, its charge. But atoms inherit their ability to couple photons from the electron—charge is a secondary, inherited interaction in atoms and such a transfer of interaction capacity will clearly alter its interactions with other systems.

Similarly, a molecule coupling with a H atom in the H-bond—a chemical "force"—can expect that valance capacity is going to be transferred along with the H atom.

Types of coupling

Every system interacts in some way—the neutrino, the helium atom and the putative Dark Matter albeit rather minimally—for even if there were such a thing as a system that did not interact, we would have absolutely no way of ever knowing anything about it. Hedging just a little, then, we can categorically state that all known systems interact—they have a tendency to couple with at least some of their subsystems.

In this section we will see how this tendency of a system to couple with its subsystems is, like its overall form derived from a wavefunction—the internal aspect of interaction. The external density of interaction will be derived from this by the hopefully-by-now-familiar random quantum operator assuming there is sufficient time available so that the random choice operator can be ignored.

The primary interactions of the system are not inherited from the constituent subsystems. All the other interactions—the secondary interactions—are. The valence interaction of atoms with is a primary interaction while the electromagnetic ability is "inherited" from the electron and quark subsystems. Only the valence interaction is novel to the atomic level, the electromagnetic and gravitational capacities are inherited from the electron's and proton's charge and mass. In the discussion we need only consider the primary interactions of a system—the interaction capacity it does not inherit from its subsystems. To describe secondary interactions later on all we will need is a frame shift.

The tendency of a system to couple with its subsystems is a reflection of the tendency of some, if not all, of its subsystems to disassociate from the system in some way—the subsystems are not monolithically integrated but are somewhat loosely associated. Another way of saying this is that there is a tendency for such a subsystem to "escape" from one system and gets "captured" by another system in some way. These labels are from the subsystem's point-of-view but it is all relative; the system's frame of reference these migrations are emission and absorption, they are coupling.

While all interaction wavefunctions basically the same, for purposes of exposition we have three possibilities for how two systems might couple with their subsystems. In practice, many interactions are a mix of them as they lie on a spectrum ranging from sharing through exchange to at-a-distance.

The simplest situation is that of exchange—the center of the spectrum. Crudely put, the wavefunctions of the two systems come into contact in some way and a subsystem hops from one system to the other. An example of this is the formation of sodium chloride, common salt, from sodium and chlorine atoms.

The exchange interaction occurs when the systems are in contact. As its name implies, interaction-at-a-distance, involves separation between the two systems.

Here the subsystem hops out of the system—as in exchange—but then has to make it across the separation before it has the chance to hop into the other system and consummate the coupling. Our illustration of this will be the four fundamental interaction of physics in terms of charge—tendency to emit and absorb—and fields—the probability of making it across the separation.

The third possibility for coupling is the most interesting in its implications for it leads to stable structures, to links between systems, to the chemistry of atoms. The sharing wavefunction leads to subsystems being a part of two systems—or more—at the same time—the two systems are stuck together by a bond. The other interactions do not involve such implicit commitment. Our example of this will be the covalent chemical bond that links atoms into the molecules of life. An impressive example of this is a DNA molecule in which billions of atoms are linked by covalent bonds into a single, stable structure.

Clearly these categories are not that distinct: exchange interaction blends into indirect interaction as the separation increases, and into sharing in the other direction with the intimacy of sharing in a bond. Each of these paths of the subsystem that leads to coupling will have a probability amplitude, a little arrow pointing in an internal direction.

In the following we shall show that all three ways of interacting involve a correlation wavefunction, a constructive interference between one system's tendency to take in and the other system to give out subsystems.

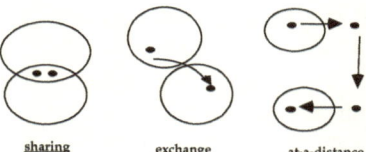

sharing   exchange   at-a-distance

The remainder of this chapter is devoted to describing the correlation wavefunction for each of these three varieties of coupling—they are basically very similar. Once we understand the correlation wavefunction, the rest is simple. The familiar step of a wavefunction becoming an actual density.

Always allowing sufficient time for the law of large numbers to counteract the unpredictable random aspect, the actual density of the coupling will be that of a probability density derived from the collapsed correlation wavefunction.

This density-of-coupling over time is the intensity of the interaction. We will return to this point in the next chapter when we look at the consequence of interaction—the higher the intensity, the more are the consequences. But first the internal aspect of interaction in the perspective of the new physics.

Valence

In our discussion of form we restrained the discussion to that of isolated, stable systems that was not involved in gaining or loosing primary subsystems.

The isolated, lithium atom, for instance, has no tendency to lose its outer, solitary 2s electron. Such a lithium atom is, however, not a happy one in the sense of being in a state of least resistance. There are two things that are paths of high resistance: for the system—the electronic state is not that of a noble gas, and the outer electron is not paired. We have already seen how little arrows explain why the pairing of electron spins in an orbital is a high-probability state. The noble gas state is similar in that it is the state where all the orbitals in a shell—the main quantum number, n,—are filled with paired electrons. This is such a low resistance state that almost all the chemistry of atoms can be explained by the impulse to inhabit this blissful—I mean, low resistance—state.

The coupling capacity of atoms is very significant in chemistry. The coupling capacity of an atom for electron exchange called its electro-valence. The overall tendency to take in an electron is called the electropositive character of an atom while the overall tendency to give one out is its electronegative character. An atom is usually characterized by which of these tendencies is the stronger though some, like hydrogen are equally capable in both directions.

The coupling capacity is measured by interacting atoms together to give a relative measure of such tendencies. Thus a current definition: "Electronegativity is the <u>relative</u> tendency of an atom to acquire negative charge.... [for example the] relative scale in which the most electronegative, fluorine, has a value of F: 4.0... are: O: 3.5, N: 3.0, C: 2.5, B: 2.0, Be: 1.5 and Li: 1.0."[54]

This is simple exchange. Exchange involves the correlation between the positive tendency of one system with the negative tendency of the other.

Not all atoms are so eager to participate in exchange interactions. If both positive and negative coupling capacities are zero the system has no tendency to lose or gain a subsystem. This is the situation for an interaction-indifferent system such as the helium atom.

Nuclear forces

Another basic example of interaction in contact is that of the strong force that holds the atomic nucleus together. The protons and neutron exchange virtual pions when very close to each other—a derivative of the strong color force that holds the quarks together inside the nucleon. The consequence of this is a massive transfer of momentum that pull the nucleons together with a fierce force—the quark degeneracy pressure making sure they don't get too close. It is this attraction that holds the nucleus together. It has to be strong because the positive protons that are right on top of each other have an intense electromagnetic repulsion that has to be overcome.

It is a balance between the pion exchange pulling the nucleons together and the photon exchange pushing them apart. The balance is such that two protons will not stick together by themselves—there is no helium nucleus with just two protons. This is just as well, actually, for if not so all the hydrogen atoms—single protons—in the sun would rapidly combine and its 10 billion years worth of energy would be released rapidly in a titanic explosion that would wipe out the solar neighborhood.

With just one neutron added to the mix, however, the balance is radically shifted—a helium-three nucleus—one neutron and the two protons—is a very low resistance state—energy is given off in its formation from the free nucleons. The neutron indulges avidly in the attractive pion coupling but not in the repulsive photon coupling. In the sun the only way two protons can stick together is if one of them changes into a neutron first. Then they can embrace with pion coupling—no disruptive photon coupling—with great release of energy. This is hydrogen-2, the deuteron, the first stage in the nuclear burning of hydrogen in the sun. The trick is getting a proton to change into a neutron—the reverse of neutron decay—and this involves the weak force. Being weak, it takes billions of years, on average, to flip a proton into a neutron and thus the essentially-slow rate of burning at the center of the sun.

The pion coupling does have one limitation—it depends on the virtual pions. And pions are quite massive—about a half the mass of a proton. Such a massive virtual particle has a very short lifetime—such disobedience of the law of the conservation of energy cannot last long enough to create a quantum of action. So brief is the allowed lifetime of these virtual pions that, even moving close to the speed of light, they can only cross distances about the size of the nucleon. It is a very short range force—even though it is very strong, its influence is severely limited to the size of the nucleon. This is why nucleons separated by more than their diameter do not attract each other by pion exchange. This is also why the weak force is weak, its victual particle is super massive and has a correspondingly tiny sphere of influence.

The repulsive photon exchange, however, has no such limitations. The virtual photons have zero mass which gives them infinite range. Thus in a massive nucleus a protons is only attracted by the nucleons in its immediate vicinity while it is being repelled by all the other protons in the nucleus. Eventually this accumulative repulsion overwhelms the non-accumulative strong force and the nucleus becomes unstable. By the time we get to uranium with 96 protons squished in the tiny nucleus the balance swings over to the repulsion and the nucleus is unstable, it tends to break up, it is radioactive.

Sharing

We now have a basic picture of interaction by exchange. Next we will look at interaction by sharing. The two are very similar in that sharing can be thought of as partial exchange,

The simplest example of this is the hydrogen atom. It is in a doubly high-resistance state—it has a singleton electron and it is one short of the desirable helium-like state. There is a certain tendency to lose an electron and a similar tendency to take one in. This equal matching of tendencies precludes either one of two hydrogen atoms gaining total control of both electron. Rather, the constructive interference between the correlations creates correlation wavefunction in the tendencies in either directions are the same. The correlation wavefunction is just like that for the exchange except that all four coupling capacities appear—not just a plus of one and the minus of the other.

Both directions are important. From this QPF comes the actual density of the coupling, the probability density, or intensity, of sharing.

This is all the theory we need to understand the nature of the covalent chemical bond.

Bonds

When two hydrogen atoms share their electrons the two high-resistance states disappear. There are no longer two unpaired electrons, there is a single, low-resistance pair. And both atoms can now lay claim to the helium-like structure—they have a filled main orbital even if its a shared one. One electron from each atom inhabits the bonding orbital. This is such a low-resistance thing to do that this sharing holds the two atoms together in a covalent bond. Such a pair of electrons inhabiting a shared orbital is usually symbolized by a single line joining the atoms.

A simple picture of this bond is that it is a resonance—there are two exchange interactions going on at the same time: each hydrogen atom has the electron pair

50% of the time. This picture makes apparent the involvement of all four coupling capacities in the correlation. The bond is a resonance of two forms where they "alternate" being in the helium-like state of being by having the pair of electrons. There should be no problem, at this point, in understanding how a wavefunction can be a mix of opposite states. Each system being helium-like 50% of the time has lower resistance than both of them being singletons 100% of the time. Being 50-50, the bond is not at all polarized—the hydrogen molecule does not, for instance, participate in the hydrogen bond

Resonance is commonplace in chemistry—many bonds are best understood as resonances of more elementary wavefunctions.

In chemistry, the correlation between two 1s orbitals are hybrid "molecular orbitals." It is the filling in of these by electrons that is the covalent chemical bond. There are two ways in which the 1s orbitals hybridize—one low resistance, the other not. The bonding orbital is the low-resistance state for a paired set of electrons. The anti-bonding orbital is a high-resistance state for one, let alone two, electrons to be in.

Helium molecule—not

Exactly the same thing holds for two helium atoms, each with two electrons in the 1s orbital. When they are in proximity, the sharing orbitals can be filled up. In such a case, however, while two electrons pair up in the bonding orbital, the other two would have to inhabit the anti-bonding orbital.

This is such a high-resistance state of affairs that helium atoms do not form a chemical bond with each other and helium molecules never form. Helium atoms are so self-satisfied that do not like to get to close together and, consequently, helium gas only reluctantly turns into a liquid when the temperature is almost absolute zero so they have no kinetic energy to get away from each.

In the terms we have just established, helium has a zero coupling capacity for valence, both plus and minus components are zero. So the correlation is also zero.

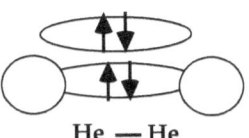

Carbon bonds

By far the most significant example of such equal sharing is the bonding ability of carbon. We have already discussed how the carbon atom is exactly half-full of the electrons it needs to complete its 2 shell. The s-orbital and three p-orbitals hybridize into four equivalent SP3 orbitals, one at each corner of a tetrahedron.

When two carbons are in proximity two hybrid orbital open up—a bonding and an anti-bonding orbital. Each carbon atom contributes one electron to the pair, the carbon-carbon bond that is, without exaggeration, the basis upon which life is built.

In our simple description with quantum operators, the bond is a filled in wavefunction with probability density.

This is the single carbon bond. The other three orbitals make bonds in exactly the same way. This is such a satisfying state of affairs—such a low-resistance state—

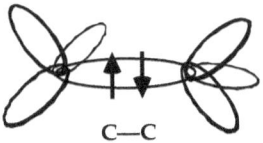

that carbon is very reluctant to break this bond. It is this reluctance that makes diamond—each crystal a single molecule in which every carbon atom is singly-bonded to four others.

Carbon likes its company so much that it will form double bonds—the unsaturated in fat—and even, on occasion, triple bonds. Making four bonds, however, is distorting things to much and are anti-bonding.

Carbon also bonds well by equally sharing with hydrogen, and just about every element for that matter.

C=C

The singleton 1s orbital of a hydrogen hybridizes with a SP3 of the carbon—their correlation wavefunction—and each contributes an electron to the low-resistance pair that inhabit the bonding orbital. Four such bonds satisfy both the hydrogen and the carbon in the methane molecule.

The measure of the sharing tendency of atoms—its valence—is somewhat different to the electrovalence for complete exchange. For instance, fluorine, which has a distinct tendency to rip electrons from others is tamed by carbon's sharing tendency to such an extent that the fluorocarbons are amongst the most stable of molecules, famous for their non-stick aspect so self-satisfied are they.

Unequal sharing

Where different atoms are concerned we do not always see such fair sharing. For instance, the bond between oxygen and hydrogen in the water molecule. Both atoms contribute one electron each to filling in the correlation. We can think of the bond as a resonant form where the oxygen has both pairs 60% of the time—which puts it into the desirable argon-like configuration—and the hydrogens have a pair each just 40% of the time putting it in the helium-like state.

As the electrons spend more time with the oxygen it has a relative negative charge compared to the hydrogens, and this is the basis for the all-important hydrogen bind. The polarity indicates that the positive and negative coupling capacities for valence are not equal—that the tendency for the oxygen to take in is greater than the tendency of the hydrogen to take in. It is this polarity that accounts for the ability to hydrogen bond.

There is a distinct probability of a molecule of water splitting into a hydrogen ion and a hydroxyl ion. Although the hydrogen ion is only a bare proton, it does not behave as an "elementary" particle in physics, rather it behaves as an atom with all its orbitals empty. All this bare atom needs to be helium,-like, however, is a pair of electrons. It finds these by latching onto another oxygen on a neighboring water molecule—in the sharing the oxygen contributes one its electron pairs not already involved in bonds. These outer pairs are called lone pairs.—nicely filled orbitals with nothing blocking access to them. The three hydrogen bonds to the oxygen are all equivalent—the positive charge gets smeared out over the three hydrogens—a quite low-resistance state.

The hydrogen ion hybridizes its empty orbitals with those of the lone pair orbitals on the oxygen. A filled bonding orbital results, binding the hydrogen ion into a hydroxonium ion. As water molecules are everywhere, this is how the hydrogen ion always is in solution. Here we see the significance of empty orbitals—they are just as real as the inhabited ones.

The proton can easily skip from one water molecule to another and this is how the hydrogen ion moves through liquid water. Directed and controlled, such proton transport is of fundamental importance to living organisms where "proton pumps" generate almost all the useful energy for a cell.

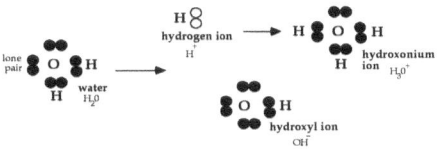

It is the sharing interaction that leads to linkage of systems into stable configurations. Such interaction is clearly going to play a significant role in subsystems hooking up as stable systems and systems hooking up together as supersystems. We saw this example where the consequence of the sharing interaction of hydrogen atoms and oxygen atoms is a water molecule. This molecule can interact in ways the atoms cannot—it can H-bond for a start. A system higher in the hierarchy of coupling capacities has emerged from systems lower in the hierarchy.

Where the sharing interaction can be expected to be pre-eminent in big-picture system building, it is not alone. We see an example of system -building by exchange at-a-distance where electrons and protons electromagnetically interact. As the electrons fill in the orbitals provided by the nucleus; an atom emerges and valence is on the scene. Natural law will determine the path of least resistance for valence and it will contribute to the internal wavefunction; this will determine the probability of what the atom will do. in the usual way

At-a-distance Interaction

To conclude this chapter we will look at interaction at a distance where separation between systems is involved.

The process has three basic steps from the subsystem's point of view: It escapes from one system. It then travels as a free system. It is captured in the vicinity of the another system.

From the systems' point of view the process is similar. One of them emits a subsystem. The subsystem makes it across the separation. The subsystem is absorbed by the second

system. The first and third steps in coupling at a distance involve subsystems leaving and entering a system. This is very similar to the coupling capacity we have already established for the exchange interaction; except hat here the giving out and taking in does not depend on there being another system in the vicinity.

Charge

Coupling at a distance can only occur if one system has a tendency to loose one of its subsystems and the other has a tendency to gain it. There is a positive coupling capacity to give out a subsystem and a negative coupling capacity to take one in. Our example of interaction-at-a-distance is the electromagnetic—the coupling with photons. A virtual photon in the quantum-foam structure of the electron has a probability amplitude to be emitted with little consequence for the structure and stability of the electron. In the new physics, it is this probability of gaining or losing a photon that is the measure of the electric charge off the electron.

This probability of emitting and coupling subsystems is called the "charge" of a system in basic physics. The symmetry of natural laws—acting on the internal aspect in the new physics—ensures that these two tendencies will be equal—at least in the sense of being conjugates of each other. When natural law is providing the wavefunctions, the plus and minus directions are always equal in probability. Note the caveat "when natural law is providing," for later we will deal with situations where natural law is not the direct provider of the wavefunction and the two directions are no longer necessarily symmetrical.

In basic physics, the probability of a system emitting and absorbing a coupling subsystem is called its charge. The well-characterized electromagnetic interaction is coupling with virtual photons. The probability of a particle emitting or absorbing a virtual photon is called its electric charge and is measured by what is called the fine structure constant or coupling constant. For a with unit charge this probability is given by the collapse of the coupling capacity. The actual density will equal the probability density over time.

One can think of this in a simple way: given 137 opportunities to emit or absorb a photon, the electron will do it once.

The reason why the probability is the same for the positive and the negative directions is the reversibility of the natural laws that govern the internal realm. In fact, as the amount of energy tied up in each individual coupling event with a virtual photon is very small, its extension in time is so fuzzy and impossible to pin down that it is impossible to say which charged particle give out the photon and which one did the taking in—all that is certain is that the exchange did take place.

As we are really quite ignorant of the inner structure of the electron, the fact that the fine structure constant is this has to be "added by hand" into current theories as it cannot, as yet, be calculated from first principles.

It is this number that is the proper measure of electric "charge" in modern physics. Incidentally, the same number measures the magnetic "charge" as well which, as it turns out, is simply another consequence of things not routinely traveling at the speed of light. Traveling at low speed, we interact with virtual photons in two seemingly distinct ways. The description of this effect is obtained by combining Maxwell's classical field equations with Einstein's relativistic ones.

Nuclear forces

The weak force is remarkably similar to the electromagnetic—so much so that they are often referred to collectively as the electro-weak force. The probability to absorb or emit a weak boson is exactly that as for the photon. The big difference is that the weak bosons are massive and cannot get very far—the mobility playing a determining role here. The color interaction of quarks is much "stronger"—hence its name—than the electro-weak interaction.

This is because the coupling constant for gluons—the probability that a quark will absorb or emit a gluon is essentially unity:

Given the opportunity to emit or absorb a gluon, the quark will always do so.

To make things even more complex, gluons themselves have color charge, they also couple with gluons—unlike the photons which have zero tendency to absorb or emit other photons. And these sub-coupling gluons, o to peak, also couple with each other. It all gets very messy as all of this has to be taken into account to "solve" the equations that we describe the color interaction. It recently took an IBM supercomputer almost year to process all the terms that have to be taken into account just to figure out the interaction of the three quarks making up a nucleon. Apparently, almost 25% of the "mass" of a proton is actually the energy tied up in gluons coupling with each other. In comparison, the "mass-energy" of the electromagnetic field coupling the proton and electron in the hydrogen atom can be ignored for all but the mot accurate of computations.

To even things up a little, however, because of all this promiscuous coupling, gluons don't get very far; just over distances commensurate with the size of a nucleon. Both gravity and electromagnetism do not suffer from this limitations of scale

Gravity

On the largest of scales, electromagnetism's tendency to cancel out its effects—because of the overall balance of positive and negative charges in nature—leave the largest of scales to be ruled by the unimaginably-weaker force of gravity; graviton coupling.

The classic illustration of this disparity in strengths is that the gravitational force of the proton in the hydrogen atom on the atomic electron is equal to the electromagnetic force of a proton on an electron at a distance of a star in our neighborhood, about 100,000,000,000,000 miles away.

On the largest of scales, of course, even these minuscule forces start to amount to something. Gravity rules by dint of the absence of a negative type of mass/energy that could cancel out the attraction of regular mass/energy for itself. It is gravity that rules the structure of planets, stars, and so on up to superclusters and the cosmic level. The "gravity charge" or mass of a system is based on the probability it will absorb or emit a graviton:

Fields

We now have a measure—in the coupling capacity—for the first and last step in the three stages involved in interaction at a distance:

(a) a system looses a subsystem,

(b) the subsystem moves,

(c) the subsystem is taken in by another system.

Now we will deal with the intermediate step, the mobility of the coupling subsystem. While step one and three involve the systems, the second step does not. The freed subsystem is now an independent system. As established, the such a system has a probable future, and the autonomy of the system will pick one of these. This is an open-ended history and it will involve an open-ended wavefunction as discussed earlier.

Unbound

All the other paths are those infamous not-taken ones.

This wavefunction gives the probability of finding the system at a particular location as the systems moves through it exhibiting its random nature. This spread out wavefunction is called a field. If a lot of systems are involved, the field is the cumulative probability of all of them. So, for example, the overall probability of finding a virtual photon at a location is called the value of the electromagnetic field at that point. If just one system is involved, the random choice operator will have to be taken unto account but, when large numbers are involved, it can be ignored.

The electromagnetic field always involves lots of photons so we can ignore the random aspect.

Naturally, we could measure this probability density by adding up all the little arrows, a tedious and time-consuming method, but one that gives the correct answer. A much simpler method, and the one used throughout physics, is the use of a field equation. This is similar to the way that the Schrödinger equation simplified calculating the electron orbitals of hydrogen. In fact, Schrödinger's equation is a field equation, one that treats the electron field. In the form of systems, the fields dealt with the probability density of structural subsystems; for interaction, the fields deal with the probability density of coupling subsystems.

So much so of fundamental physics can be expressed as field equations, in fact, that some physicists have gone so far to declare that objective reality is fields, and field alone. We have not taken this route, in our point of view the fundamental reality is systems of interacting subsystems. Mathematically, however, they are equivalent.

A field equation simply allows one to calculates the density of coupling subsystems at any point one is interested in. For instance, they are capable of calculating the quantity we were just discussing, the probability density of the mobile subsystems making it from system 1 to system 2.

In our perspective we can say that field equations give a measure of the probability density of the coupling subsystems at any location.

All the field equations of modern physic—and they are daunting in their details—can be thought of in this way. A simple way of thinking about this is that the system "throws" out this field of influence based on its ability to couple subsystems. One point too note is that, echoing the way that mathematicians recognize the "null set" or set-without-members as a significant entity, physicists accept that the field is still there even when its value is zero. The definition of the vacuum is that all fields have a zero value

The field is theoretically measurable by a "minimal test particle," a particle that, while it couples with the field, does not itself alter the field in any way. To measure the field, the test particle is placed at a location and the amount of coupling is noted, a measure of the field at that location. (Such a measure will be a measure of the consequence of the coupling—such as a force—the topic of the next chapter.)

Travel

The key to interaction is the probability that the coupling subsystem will make it across the separation between the two interacting systems. As always, this probability is given by the collapse of a wavefunction, in this case, the field.

Interaction at a distance has three steps: emission, travel across, and absorption. The wavefunctions for these three steps are the positive capacity of the first system, the field of the coupling subsystem, and the negative capacity of the second system. The correlation between the two systems will be the constructive interference between these three. The intensity of the interaction will be the collapse of this correlation wavefunction.

This is a somewhat hybrid expression as it combines attributes of bound subsystems—the positive and negative coupling capacities of the system—with attributes of unbound subsystems—a QPF, the field wavefunction.

Exactly the same holds for coupling in the opposite direction. For all the fundamental; interactions where the tendency to emit is the same as the tendency to absorb, the correlation will involve both directions.

Electromagnetic field

A good example of a very successful field theory is the electromagnetic influence of charge at a distance. The electromagnetic interaction is carried, as the physicists say, by virtual photons. The classical field equations of electromagnetism give the probability density of the virtual photons at a distance from the "charged" particle. The entirety of this over all space is the "electromagnetic field" generated by the particle. As mentioned, the electromagnetic influence spreads far indeed in that the electromagnetic influence of an electron and proton separated by interstellar distances equates with their gravitational influence at atomic distances.

The electromagnetic field at a location is nothing more than the probability of finding virtual photons there to couple with.

In our general discussion of emission we spoke of a subsystems escaping from the system This process is not understood. Just what happens at the start of a photon's journey—or at its end—is not understood. But leave and enter they do; photons begin and end on electrons. The situation is made even more hazy by the relative spatial extension of an electron—which does the absorbing and emitting—and a photon. While the spatial extension of an electron is less one millionth of a nanometer, the spatial extension of visible light is huge in comparison on the order of millions of nanometers.

Whatever happens, the initially-released virtual photons take of at the speed of light—they spread out symmetrically—they have no preferred direction—and expand into space. The field equations describe this very simply, the electromagnetic field, the density of coupling subsystems, falls off as the square of the distance.

The field equations do not take their inspiration from sound waves in organ pipes, rather they are modeled on the density of fluid flow.

Our example of indirect coupling will be the electromagnetic interaction of the electron and proton which is basically:

1. the electron (or proton) emits a virtual photon—its charge

2. the photon travels from place to place—the electromagnetic field

3. the proton (or electron) absorbs it—its charge

The discussion is applicable to all four fundamental interactions as they are all similar though the terms used are somewhat varied.

Getting across

Unless the "background" over which coupling-at-a-distance occurs is very inert, it can have a great influence on the probability that a subsystem will make it from one system to another. It could be absorbed and never make it, for example, or be retarded by being absorbed and then emitted along the way.

In the electromagnetic interaction coupled by virtual photons, for instance, the measure of how they are influenced by the intervening space the photons are traversing is called the dielectric constant, a measure of the ability of the virtual photons to traverse whatever it is that separates the interacting

Interaction	*emit/absorb*	*"charge"*	*subsystem movement*
electromagnetic	γ	electric	electromagnetic field
gravity	gr	mass	spacetime curvature
weak	W, Z	weak	weak boson current
strong	gl	color	gluon field

systems. Some systems, like iron atoms, enhance the mobility factor in the electromagnetic interaction but even the "nothingness" of the quantum vacuum foam has a slight, and measurable, effect in retarding photons as they pas by.

As mentioned, it is the mobility of the coupling subsystems, rather than the tendency to couple, that gives the weak interaction its moniker: "...the amplitude for a particle to emit a W is really no smaller than the amplitude for the particle to emit a photon, but the W is so massive that the probability amplitude for it to pass from one particle to another is very small—it gets so 'tired' that it's prone to turn right back. This [explains] why the weak interaction is so much weaker [than the electromagnetic one]."[55]

Inflation

On the other hand, as far as we are aware, nothing seems to influence the mobility of gravitons, the gravitational interaction is oblivious to what lies in between systems. In the very early history of the universe—about 10^{-34} of a second[56] after the Big Bang—it is widely held that an exponential expansion occurred, an abrupt inflation of atomic-size extension to galactic supercluster dimensions. When this inflation abruptly stops, the shock energy of this change then kicks off the "classical" hot Big Bang about a trillionth, trillionth of a second after the true beginning point.

The inflation is driven by a cosmic negative pressure field which is like negative gravity—it is intensely repulsive in its effects. Conditions were such that graviton coupling was powerful and expansive, unlike its pale descendant today which is feeble and contractile.

In many, if not most, cases, the subsystem mobility is such that the interaction decreases with increasing distance, usually described by an inverse square-of-the-distance law which simply reflects the geometric realities of volume with distance.

This is not always so, however; the interaction of quarks via gluons is at a minimum when they are close together but rapidly increases in intensity as they move apart—the 'infrared slavery' that further complicates our ability to fully describe color charge. It can also be very complex, as it is in cells coupling with hormones and other factors where the transportation by blood is involved.

Hydrogen bond field

Field formulations are a useful perspective for more complex at-a-distance interactions. For instance, it is useful to think of hydrogen boding in terms of fields. The exemplar of this capacity is water in bulk. The electromagnetic field is the probability of finding a virtual photon at a location; the H-bond field is the probability of the orientation of a water molecule at a location. water tends to structure the water around it, to attain the low-resistance state of the ice-like mesh The molecule structures the water around it, and this field can stretch an appreciable distance before it is overcome by random thermal motion or the imposition of the field of another. Water molecules are equally matched, they move each other around.

This equality does not hold when massive molecules are involved. Many molecules with oxygen (and nitrogen) in them are good at hydrogen bonding as are almost all of the molecules of life. As they are massive they move the water around much more than the water moves the massive molecule around. The molecule imposes its H-bond field on the surrounding water. In the formation of macromolecules, however, the cumulative push and shoving of many water molecules H-bond fields is very significant in moving the molecule around. An example is the spontaneous folding of an amino-acid chain into an active protein enzyme, a process driven by the combined desire of the macromolecule and the multitude of water molecules to structure into a state of least resistance. The process clearly involves wavefunctions with steep gradients in them for a "denatured" protein can refold into the active form in milliseconds.

This attempt by the molecule to structure the water around it will impinge upon the attempts of other molecules to structure the water around themselves. This would be coupling through the H-bond field. The form of biological molecules in water is not just that of the atoms it is composed of, it also includes the structure it imposes on surrounding water. The surrounding water molecules have to be included in the structure of the molecule.

Water is a somewhat polar molecule. While the bonds between the hydrogens and the oxygen atoms are predominantly sharing they also have quite a bit the nature of exchange as well. The oxygen takes more than its fair share, it pulls the electron pair it shares with the hydrogens close to it, making it relatively negatively charged, leaving the hydrogen somewhat positively charged. In comparison, the bond between carbon and hydrogen is scrupulously fair and there is zero polarity and thus no hydrogen bonding. The negative oxygen of one water molecule can attract the positive hydrogen of another molecule, this is the hydrogen bond.

These bonds are directional and the molecules have a sticky tendency to mesh with each other. When thermal energy is low the stickiness of these bonds is sufficient to hold the molecules in place and we get the open mesh structure of ice—it floats because of this open mesh structure.

Steam results when the thermal energy is much greater than the stickiness and the molecules fly free of any bonding. Between the two is the magic zone that allows for life. When the thermal energy is similar to the stickiness energy the alignments are temporary—they form and are then disrupted—and

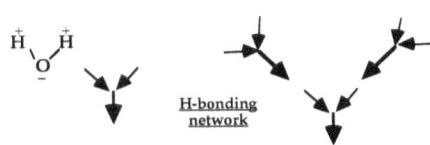

we have liquid water. There is alignment as in ice but it is only temporary as thermal motion tends to break it apart. Keeping in mind that we are really speaking about paths of least resistance, we can crudely characterize the hydrogen bond as the "desire" of water to take up the ice mesh structure in the same way that chemistry can be crudely described as the "desire "of atoms to take up the noble gas electronic configuration—filled paired shells.

The tendency to hydrogen bond is carried outwards in the structuring and polarization of surrounding molecules. We can think of the water surrounding a biological molecule as a field of structured water—a wave of alignment—and all the interesting interference effects that can be described by complex numbers. Just as the electromagnetic field is a simple description of the probability of absorbing a virtual photon at each location so the hydrogen bond field is a simple description of the probability of a water molecule having a particular configuration at each location.

Change in history

We will now deal with the simplest, and most common, consequence of interaction where one system influences the history of another system. We are still basically restricting the discussion to peer interactions—systems interacting with systems on the same level in a hierarchy. Two systems influence each other's history as a general consequences of the fact that subsystems take their capacities along with them as they change allegiances during the interaction. In the most general sense, the consequences will depend on how much coupling capacity is carried along by each subsystem and how many subsystems are being coupled.

The consequences will be proportional to the intensity of the interaction—the collapsed correlation wavefunction—and to the coupling capacity carried along by each subsystem.

The amount of consequences will depend on the intensity of the interaction—the amount each system carries along with it times the number of them making the trip. These consequences of interaction can be roughly equated with the forces that appear in classical science descriptions.

At this point the discussion bites its own tail, so to speak. Very much earlier we spoke of the source of the probability amplitudes that have informed our discussion of modern physics. We spoke of natural laws described by action equations. The action equations take into account all the contributions of each interaction. The item that actually appears in the equation is what we have been calling the consequence of interaction. When this changes, the wavefunction changes. The system now has a new wavefunction with a new collapsed probability density. The system will follow one of the probable histories in this new set-up subject to the vagaries of the random choice operator, The sequence describing simple change is: 1. internal correlation 2, external filling in 3. transfer of coupling capacity 4. change in wavefunction 5. change in history.

Leaving out all the system labels for simplicity's sake we have a simple sequence of wavefunctions. The subsystems carry their capacity to interact along with them on their travels. The capacity to interact that is being carried along by the transfer of subsystems are the secondary interactions of the system itself.

We established that a system has an overall capacity to interact, internal system, that was the composite of two qualitatively-different types of interactive capacity.

(a) a. primary not inherited coupling with primary subsystems, unique to the system itself, not a capacity possessed by any of the system's subsystems

(b) b. secondary inherited coupling with secondary subsystems, a capacity possessed by primary subsystems; includes the tertiary and on down as similar.

The capacity of the system for secondary interactions is inherited from the primary subsystems so when those subsystems are transferred they take the secondary interactions of the system along with them.

Contingency

The concept of contingent history that pops up throughout the sequence is just the simple requirement that there be interaction for there to be change If there is no interaction there is no change. But systems can only interact with each other if they are in the vicinity of each other (or at least close enough for coupling at a distance to be significant. For simple systems, in the vicinity can be equated with being close by each other. We are no longer talking probabilities here, the two systems have to be in the same place and the same time—a particular set of histories. As we have established, a particular course of history involves the random choice operator. The random choice operator of both systems must pick the same place at the same time—all interaction occurs in the present—so that they end up on the scene together—ripe, so to speak, for interaction. This is the contingent side of history and it very much involves randomness and is to be avoided if possible. And, as noted, possible permutations of even a small number of possibilities involve large numbers.

This would be an impossible situation if an infinite number of possibilities—a continuum—was involved as in classical physics. Luckily, the way wavefunctions interfere does not involve an uncountable infinity of states, not even a countable infinity of them, but just the combinations of a small set of small numbers. Even better, nature almost always involves large numbers of systems on the scene at the same time. Even tiny-probabilities can (relatively) quickly appear on the scene—in fact, sooner or later, any not-exactly-zero probability must appear on the scene. In this way, the influence of the random contingent aspect of history is somewhat nullified

Contingency does rule in the actual origin event, however, as the random operator comes into play to get them there. At some point in time the subsystems were on the scene, they did fill in the system wavefunction, an the system emerged on the scene.

Contingency also enters as the larger environment intrudes:. An example would be the results of a slit experiment performed when the nuclear pile next door goes critical and explodes. We will deal with interaction with the environment after we have dealt with two-system interactions.

This contingent history has an internal and external component: the systems have to be in the neighborhood and they have to have a significant correlation. The contingent prerequisite for interaction to occur is the systems must be in a situation—a configuration—such that there is a non-zero correlation between them:

internal correlation of systems
external configuration of systems
The consequence of interaction also has an internal and external aspect.
internal change in wavefunction and probable future
external contingent history actually followed

Movement

We will now look at examples of this somewhat general discussion. One example is the electromagnetic interaction where the exchanged virtual photons carry momentum—the gravitational interaction—along with them as they shuttle between the interacting systems. It is this exchange of momentum that is the electromagnetic force of classical science.

force of interaction (external consequences) =
intensity X amount carried by the coupling subsystems

The capacity for coupling transferred by this flux of photons involves just one, the capacity to couple with gravitons. Early in the discussion we saw that graviton coupling had two aspects: gravitational mass, the ability to couple, and inertial mass, involving a change in coupling. Virtual photons do not transfer gravitational mass/energy—being virtual, this is to be expected. (Real photons, on the other hand, do transfer real energy.)   Virtual photons do, however, transfer the inertial aspect of graviton coupling. This inertial aspect is measured by momentum, a measure that is as well-defined in quantum physics as it is in classical physics—unlike "energy" which, as we have seen, can be somewhat fuzzy over time. In classical mechanics, momentum is the product of mass times velocity:

When virtual photons are exchanged they transfer momentum between the coupling systems. The input of the electromagnetic interaction to the electron is the transfer of external photons and internal momentum.

Electromagnetic force

The transfer of momentum carried by the virtual photons is such that the electrons move apart, their inertia is altered by the coupling. The mass/energy of the electron, on the other hand, remains constant as the photons do not transfer it. It is this moving apart that classical physics calls the "electromagnetic force" pushing them apart. This bodily movement of the electron is external, and it is a reflection of what is happening on the internal, wavefunction of little arrows.

Momentum transfer is important at every level in the hierarchy of matter for almost all the higher capacities involve subsystems with real mass and real energy—so, unlike the virtual photons, they transfer mass and momentum along with them. Much of the movement of matter derives from this transfer.

The movement of the system, as a consequence of the coupling, can influence the correlation. Our example are two electrons interacting and, as noted, they move away from each other. As they separate the intensity of the interaction falls off. Less photons is less momentum transfer. Less momentum transfer decreases the "force" pushing them apart, the acceleration apart decreases with time. This is a negative feedback, the interaction, and hence its consequences, decreases over time.

Lipids, on the other hand have a positive feed back to the movement towards each other, the closer they get, the easier it is to displace discontented water molecules and the hasten together.

While virtual photons only carry momentum, real photons carry both energy and momentum, both of the aspects of graviton coupling. So a slow-moving electron that absorbs a high-energy gamma photon has both its momentum and energy changed, it becomes a high-energy electron zipping along at high speed.

When an atom absorbs a real photon, one of its electron moves to a higher energy state. Such "excited" atoms (or molecules) often have a quite different tendency to interact compared to their "ground" state. It is this phenomenon that underlies the photosynthetic powering of almost all life on earth: a photon-excited electron in a chlorophyll molecule is whisked away down a metabolic pathway;

the energy in the ensuing charge separation is then used to power a cascade of chemical transformations that ultimately turns carbon dioxide and water into carbohydrate and oxygen.

Proteins

Unlike the electron and proton which basically only couple in one way, proteins have multiple ways of coupling. Proteins are remarkable for the versatility of their interactions and do most of the "doing" in a cell. Proteins, for one thing, are marvelous organic chemists and are capable of many chemical syntheses impossible for the man in the lab. It is truly remarkable what just twenty-odd amino-acids can do when they are linearly liked in their hundreds. All these interactions will contribute to the external and internal input to a protein.

Almost all of the important interactions of proteins involve the spatial pattern of the interactive capacities on the extended system. Some of these important "patches" of interactive capacity on a protein "surface" are: ± H-bond ordering, ± charge, lone pairs, empty orbitals, metal ion interactions, aromatic ring resonances, etc.

Lipids

The ordering about of water by H-bonding capacity is one of the main contributors to moving large molecules around into their active structures.. All of life's molecules are in an environment of water molecules and have to deal with water's determination to minimize its resistance by forming oriented fields of H-bonding. The interaction of a single water molecule with a macromolecule has consequences for both of them—by the reversibility of natural law these will be equal and opposite. The tiny water molecule is drastically altered while the huge macromolecule gets a tiny tug. But there are a lot of water molecules around and the tiny pushes and pulls can add up to significant imbalance which the macromolecules bodily moves to correct. The movement stops when all the pushes in one direction are balanced by the pushes in the other. This is just how a massive amino acid chain folds into its active form just from myriad tiny tugs of water molecules.

A molecule that in has no capacity to H-bond will have around it a shell of very unhappy (high resistance state) water molecules. A excellent of example of this is a lipid (fat) molecule that has, as its main bulk, a long hydrocarbon chain in which the hydrogen and carbon fairly shares their shared electron pair—the molecule is non-polar as it is the greedy tendency of the oxygen atom to hog the electrons that polarizes the water molecule and sets the stage for H bonding.

This shell of water molecules is highly imbalanced—on the lipid side each molecule is unable to form a H-bond while on the bulk water side it is H-bonding. This unequal state is surface tension and its consequence is repulsion. This is a strong "force"—a small bead of water will lift itself up against gravity as it beads on a waved surface. This time it is the lipid that is repelled—it moves away from the water and into itself. "Oil and water do not mix" is a significant principle in the structure of the macromolecules of life..

Unlike the multi-talented proteins, this is about it for lipid interactions except for a slight stickiness most molecules feel for each other—think Post-it-Notes—called the Van der Walls attraction. This is why as any gas cools it eventually turns into a liquid—the stickiness and the kinetic energy of motion are similar. This is residual electromagnetic force based on the fact that the negative electrons in the atom are so spread out compared to the point-like positive nucleus that perfect cancellation of charge is not possible, the positive charge is not totally shielded by the electrons even in the neutral state. Helium atoms with their very stable electron pairs all very tight around each nucleus have the least capacity for Van der Walls attraction, the positive charge is very effectively shielded, as they say, by the tight skin of electrons. But even they will condense into a liquid when the temperature gets close enough to absolute zero. They have so little energy of motion that the not-quite zero imbalance is sticky enough to match it. Helium is not only the least reactive of the elements, it is the hardest gas to liquefy. But, given ridiculously-low temperatures, helium will liquefy. But them, in extremes, even a helium atom can be forced to give up its electrons. An encounter with an iron atom totally stripped of all its electrons (as could happen in a supernova explosion) will result in an exchange interaction—more a rape, really—dominated by the avidity of the ionized iron to take up electrons. The result is the helium atom is stripped of its electrons which plunge into the inner, empty orbitals of the Fe^{+56} ion. This ion, for that matter, is quite capable of stripping a fluorine atom— this is extreme chemistry; super-valence run amok. Back to the regular world of water moving molecules around.

Lipids take up a structure that minimizes the surface tension of water. A very important class of lipids are those with a highly polar end group attached. One very stable configuration of these is the lipid bilayer. All the long hydrocarbon chains are in the center and all the polar end groups are on the outside interacting readily with water.

These bilayers are very important in isolating compartments in living systems. This is a sophisticated example of simple change—the movement and change in history. It all follows the dictates of the internal wavefunctions and their combinations and collapse.

Pattern matching

Hydrogen bonding is also important in genetics, the complementary matching of base pairs in DNA/RNA. Here the bases do have the capacity to H-bond. But, just like the lipid scenario, when complementary patterns of H-bonding couple with water molecules they also eliminate water molecules and move together. Almost all the basic mechanics of genetics is based on the pairing-preferences of the four "bases" which are linearly strung in their millions and billions as the nucleic acids. (Yes, it is little confusing that linking millions of bases creates an acid, but that's the terminology.) Each base has a pattern of H-bonding-capable patches that complement those on just one of the other bases. Nucleic acids form duplexes—two strands lying side-by-side—when each base on one strand finds its complement opposite it on the other; their H-bond patterns zip together like a mini zipper being closed.

A similar, if much more versatile, movement together underlies much of the work of proteins. For instance, the H-bonding of a protein enzyme and its substrate is such as to eliminate water between them and unite—setting the stage for the substrate to change and, no longer fitting so well, be released.

Again, what is calling the shots is not so much the external form of the system but the patter of internal capacities. It is the patterns that are important in biochemistry and genetics, not so much the molecules on which they are being expressed.

The patterns flow from storage in nucleic acids to proteins and back to influencing the patterns being retrieved from the nucleic acids. This can be likened to music which can pattern grooves in vinyl, dots on CD's, radio waves from TV antennas, surges of electrons in amplifiers, movement of loudspeaker membranes and pressure waves impinging on the ear and on as patterns of neuron firing to who knows what in the brain. The external is not of primary significance—though necessary as carrier—it is rather the pattern being passed along. We will return to all this later.

What Are Things Made Of?

So, in brief, the answer that quantum science gives to the question, "What are things made of?" is that they are stuff filling in quantum probability forms.

Quantum science has an equally-brief answer for, "What do things do?" They interact by exchanging and sharing bits of themselves with other things. Just as before, the probability of this sharing is a reflection of a quantum probability form or field.

Forces are resultant things in the new science; forces are a consequence of actually sharing bits of self with another. The classical magnetic force, for example, is a result of the quantum probability of absorbing and emitting virtual photons (phantom bits of light that flit beneath the pixilation of reality and thus do not 'officially' exist.

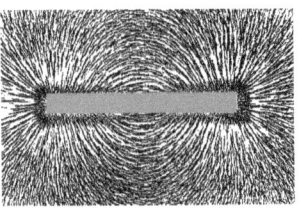

The simple form to a quantum probability field, the exchange of virtual photons that is magnetism, can be simply seen, just sprinkling iron filings around a magnet.

T-shirt Slogan

So, if classical science can be epitomized for T-shirts as "all is matter in motion manipulated by forces," then quantum science can be aphorized as "all is matter in external motion manipulated by internal probability fields and forms."

All the sciences would actually like to be 'modern' and manipulate quantum, not classical concepts. Physics, of course, is thoroughly modern. Chemistry with its quantum orbitals is as well. Biology, genetics, evolution, neurology, etc. are decidedly not modern. Biochemistry is currently straddling the fence as the quantum revolution slowly makes its way up the scientific edifice.

It is actually very difficult to switch from the classical way of thinking (probability a result) to the new quantum concepts of causal probability—even Einstein refused to accept the implications of the new physics, in the end, and he helped found it.

So, physics and chemistry now tell us that material objects are made of stuff and probability forms.

Here is another difference between the classical view and the modern: in the new physics, anything that is not forbidden is compulsory, it will happen. Something has either zero probability, or it has a non-zero probability. And even very small probabilities can be significant as they very occasionally get picked.

Let us assume that the same holds for all the other sciences—which are founded on physics and chemistry after all—before returning to our example of applying the concept to protein folding.

Quantum Probability Forms

We have seen the power of quantum probability—holding up aged stars by 1–1=0 alone. Now we are going to look at quantum probability on a more subtle level. For, as we shall see, it is the sophisticated manipulation of quantum probability that underlies—in an internal sense—the marvelous phenomenon of life.

But before we get to living systems, we have to start at the very bottom and work our way up to it as is appropriate in the scientific, bottom-up approach to deconstructing our universe.

The power of quantum probability underlies the less dramatic, but essential, exclusion that gives the elements such as carbon, oxygen and gold their different chemical properties.

For instance, a hydrogen atom is not just an electron and proton near each other. What makes all the difference is the 1s orbital, an intangible quantum probability field with a ball-like form. Quantum mechanics calls this aspect of the hydrogen atom an "internal extension" to distinguish it from the more familiar external extensions in space and time.

The 1s orbital is what gives the hydrogen atom all its character—it is a quantum probability form that is reflected in the overall form to the history of the atomic electron—what the electron does. And chemistry is all about what electrons in atoms do.

All the great difference between the remarkable chemistry that hydrogen atoms participate in—think water—and the null set of helium's relationships is a simple consequence of the fact that hydrogen has a "dissatisfied" singlet electron, while helium has a highly satisfied set of paired electrons. Two electrons in one orbital: one fitting this way, the other fitting that way. And, while the probability of two electrons being in the same state is 0%, the probability of being in the paired state is almost 100%. For a helium atom at room temperature the probability is exactly 100%—helium is totally indifferent to chemical sharing of electrons. Only being totaled in a violent collision can smash the electrons away and this takes a very high temperature, such as in the sun's furnace where even helium is fully ionized.

What are Little Things?

Significantly different from any classical concept is that the total-empty orbitals are just as significant in quantum chemistry as the occupied ones are.

For, even though an empty intangible quantum probability form (QPF) might seem to not belong in considerations of material objects, they are just as much a part of objective reality as the filled ones are. Just ask a chemist if empty orbitals play a role in the behavior of a hydrogen ion or the iron atom at the center of blood-red hemoglobin.

Furthermore, the "size" of an atom (those little colored balls that get tinker-toyed in chemistry) reflects the orbital's sphere-of-influence, not that of electrons and protons. Consider the atom scaled up enormously. The 1s orbital is now the size of a dark and empty Yankee Stadium. The proton has inflated to the size of a baseball at center field. The electron is as a brilliant, but tiny, firefly leaping from spot to spot so much faster-than-the-eye-can-see that the bowl of the stadium is filled with a misty glow, very bright near the baseball but hardly noticeable at the cheapest seats.

If the electron-firefly leaves the stadium, the remaining hydrogen ion is as a dark stadium with a baseball in the middle. But that emptiness is permeated by a quantum probability field, and this is what gives acid its kick.

So classical and quantum physics give different answers to the question: what is a hydrogen atom made of?

The classical answer is: an electron and a proton.

The quantum answer is: ditto, plus a set of quantum probability forms. Some of these QPF are full, some are half-full, and the rest are empty.

This holds for all the elements: they are composed of electrons, nuclei and quantum probability forms. The same holds for molecules in quantum chemistry—which involves a molecular wavefunction—and macromolecules in quantum biochemistry.

Orbitals are perfectly described by complex numbers and, if you have ever seen the Mandelbrot set you have seen the form-making capacity of complex numbers at work.

Providing QPF

We have already rejected the 'lock and key' concept of molecular binding and have embraced the quantum concept of things leaping in and out of quantum probability forms. It would be interesting to know just how close a substrate has to come to its enzyme before it teleports into the highly-probable bound state.

Consider again our hydrogen role-model. One way that we can translate that fearsome-looking quantum equation of the atom is to say that the proton provides a quantum probability form for the electron to fall into. It is an enabler.

The electron, on the other hand, controls the probability of what the nucleus will do. For, when a helium atom collides head on with another atom, it bounces off of it just like a solid billiard ball as in the classical picture.

The quantum view is a little more sophisticated: the attempt by the electrons of the target to enter the filled orbital of the helium atom is repelled with absolute rejection, by the power of the utter impossibility of this ever happening, a power of rejection that is the sole support of elderly stars. No wonder people considered atoms as little tiny bits of impervious solid stuff for such a long time; and did very well with the concept as it is a good approximation in simple circumstances.

Newton's insight still holds—equal and opposite reaction. The helium electrons also recoil in horror at the thought. At room temperature, the probability that the helium nucleus will follow along with these retreating electrons is 100%—the nucleus is constrained by the quantum probabilities provided by the electrons, just as much as the electrons are by the nucleus-provided orbitals.

This, in our example, is as if the baseball conjures up an empty Yankee Stadium; and if a pheromone attracts the fireflies, the whole stadium-baseball follows diligently along. The annals of quantum physics are filled with such odd-to-the classical mind phenomena.

Chemistry is all about providing quantum probability forms for other systems.

Smooth and bumpy

When free hydrogen atoms meet free oxygen atoms there is nothing to prevent their almost instant embrace. They slide right down the path of least resistance—least free energy in chemical parlance—to bonding as a water molecule.

We can mix hydrogen and oxygen molecules at room temperature, however, and nothing will happen. The gas mixture is quite stable—no water is formed. Even though a water molecule is by far the state of lowest resistance, the path to that state is not a path of least resistance. For the molecules are in a quite contented state. There are no unpaired electrons and all four atoms are in the noble gas configuration. Before the atoms can interact, they have to separate from each other—chemical bonds have to break so they can reform.

The path to this intermediate state is one of very high resistance—very low probability. There is a big bump in the road so the molecules stay intact and the gas mixture is stable. One way over this hill is heat; the hot molecules now have enough kinetic energy to smash each other into atoms. The atoms can now avidly combine. The excess energy is released and heats the gas even more; more smashing and rearranging; more heat released etc.—a runaway chain reaction. Spark a mix of hydrogen and oxygen and you will get an explosion.

In terms of probability, room temperature molecules of oxygen and hydrogen gas have an almost zero probability of making it over the barrier. The situation is just like that of the spontaneous decay transformation of a uranium atom by emitting an alpha particle to a state of much lower free energy. But the path to freedom has a big bump in it. The alpha particle moving through the center of the nucleus interacts with the other nucleons and is strongly attracted to them all. As it is in the center and surrounded, however, the mighty pull in one direction is balanced by an equally mighty tug in the opposite direction. The titanic forces are totally balanced all around and the alpha particle sails on through unimpeded.

At the edge of the nucleus, however, this balance comes to an abrupt halt: the alpha is still being pulled mightily backwards but there is no longer any pulling in the opposite direction. There is a surface tension, similar, if vastly greater, to the force that beads water on wax paper. This is the barrier, the bump in the road to a more stable state. So low is the probability of escape—so tiny is the wavefunction just outside the barrier—that the alpha has to hit the barrier trillions upon trillions of times before it has an appreciable chance that the random operator will smile in beneficent fortune and pick escape for once. While we have drastically simplified the complexities of both atomic and nuclear rearrangements, these general concepts will be sufficient for our purposes.

When the dust settles, the hydrogen and oxygen atoms are in water molecules—they made it through the high-energy intermediate phase riding the crest of the explosion. This is one way over a hump preventing systems from following the path of least resistance. Later we will discuss other, less explosive ways of overcoming such barriers to systems rearranging into states of low resistance. I mention it now only because the straight-downhill interaction of free atoms is somewhat of a rarity in nature; most of the interesting big-picture interactions involve bumps in the path of least resistance.

Catalysis by Provision

One of the key differences between living and non-living systems is that, while the wavefunctions involved in the structure of non-living systems are relatively static, living systems are anything but static.

We will start off with the simple concept of systems manipulating other systems by providing wavefunctions—paths of least resistance—for them to follow. The manipulated system is no longer directly dependent on natural law to provide a wavefunction. The system doing the manipulating is the generator.

In both cases, of course, the final step is the same, the collapsed wavefunction—be it natural or provided—has a probability density that will be the actual density given sufficient time and numbers involved.

Nature, of course, has the ability to "do organic chemistry"—molecules get manipulated in their interactions with others. High-energy processes—a spark in the experiment, lightning and solar ultra-violet in the primordial environment—initiated condensations of simple molecules such methane, ammonia and hydrogen and formed a whole mix of organic molecules including simple aminoacids. Today, of course, any products of natural metabolism are quickly swept up by living systems or destroyed by the omnipresent oxygen. But in the pre-biotic world, this would not have been so, and nature-in-the-raw is expected to have populated the early world with a wide variety of simple organic compounds.

It is only relatively recently that chemists have realized just how complex a "metabolism" natural law alone is capable of generating. The pre-biotic history of the earth could have provided many of the components of life—such as simple sugar and aminoacids—along with molecules with the ability to energize transformations such as high-energy pyrophosphates, iron-sulfide compounds, and thioesters. All of which are still to be found at the core of life's current metabolic activity. We can also expect that chemical catalysis was also involved in smoothing the way for these chemical changes to occur.

Catalysis involves providing a wavefunction so reactants can change into products. In the molecular realm, the measure of resistance is called the Gibbs free energy and chemical change follows the path of minimum free energy.

We have already noted that a bump in this path to least resistance occurs when the chemical change involves an intermediate. The block occurs when this intermediary stage has a higher free energy than either reactants or products.

One way around this block is to raise the energy of the reactants—heat or radiant energy are a few of the possible ways. Heat accelerates chemical interactions but is seldom used in living systems. Light, like heat, is capable of energizing many chemical transformations. While visible light energy is used for a lift in photosynthetic systems, this is a sophisticated level of organization.

Only ultraviolet light has much impact on non-living systems and that influence is usually disruptive. Iron atoms, however, can absorb UV and enter a relatively stable excited state—activated ferric ion—that can drive many chemical interactions such as the metabolically-significant high-energy thioesters. These are still to be found at the core of metabolism, and they are thought to have been the first systems that could drive the formation of ATP.

Thioesters breaking up is one of the few chemical transformations whose free energy release is greater than that for ATP breakup—thioester breakup can drive the synthesis of ATP from ADP. Such availability of thioesters provided by acti-vated iron can be expected to play a role in the early proto-metabolism of massive china clay beds fed by both by a black smoker and the surrounding sea water. Most black smokers are in the deep ocean where plates are pulling apart from each other with magma welling up such as all along the mid-Atlantic ridge today.

Such driving of chemical transformations by ATP or thioester breakup is very common in living systems and is well-documented in current science. The energiz-ing system plunges down a path of least resistance and is coupled to pushing the other system up a path of least resistance—making it go in the opposite direction. As noted, paths of least resistance are described by internal natural laws, and in-ternal laws are always reversible (it is the random collapse that is the irreversible step that clicks time ahead).

This coupling of two systems—one going down and the other going up—in-volves external interaction—there is a physical connection between the two. In this sense, it is an external phenomenon—which is why classical science handles this aspect of living systems very well. It is still a vertical phenomenon in that it can power interactions on many different levels of sophistication. ATP breakdown, for instance, powers all sorts of interactions in the uphill direction on many different levels in the material hierarchy—ions, molecules, macromolecules, spindle con-struction, cell division, etc.

In classical science, a surface is well-defined as solid boundary. A complex catalytic surface is such a well-defined solid boundary. Unfortunately for this sim-ple view, the new physics says that there is no solid boundary—what we used to think of as the surface of atoms, molecules, clay, etc. is actually tiny electrons te-leporting around in vastly larger extended orbitals. The surface is not really a solid boundary at all. For all that, filled orbitals can be roughly equated with classical surfaces. In catalysis, the filled orbitals that participate in providing a path of least resistance for others can be equated with classical catalytic surface that are not well-defined and somewhat fuzzily located. The providing of empty orbitals in catalysis, however, has no classical analog. In classical science, empty cannot be "real." It can, as I hope you remember from our slit set-up where "nothing" stopped projectiles from reaching their targets—and not just photons and elec-trons, but "solid" atoms as well.

So providing wavefunctions in catalysis has two basic aspects, only one of which has a classical approximation—proving filled-in wavefunction "surfaces" and providing empty wavefunctions with no classical analog.

See the Appendix for a look at the catalytic ability of clay and its role in the origin of life.

History of sophistication

This is a diagram of our current understanding[57] of life's history in terms of when each level of sophistication was established and set the stage for the emer-gence of the next level up.

Notes: The horizontal endosymbiosis is the internalization of 'bacteria' that became the ancestors of today's mitochondria and chloroplasts. The eubacteria are the familiar ones; the archae-type are a bunch of oddities that live in the most unlikely places like boiling water and volcano vents. The abiotic Earth was molten at first and without life. With cooling and the advent of the oceans, things quickly got going as natural QPF after QPF was sequentially filled in by the 'calcium effect'.

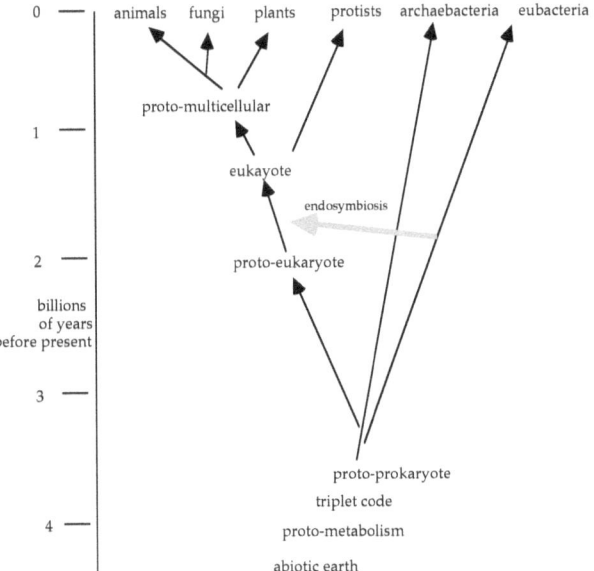

The most difficult step along this history seems, against intuition, to be that of getting the eukaryote pattern fixed as a player on the scene. Some of this difficulty could have involved many "extinction" events such as mar the continuity of the later fossil record. Extinction events, by definition, are those catastrophes—comets and asteroids being the prime suspects—that radically alter life on earth in relative geological instants. More, of these events in fact, can be expected to have occurred during the first three quarters of history. In general, bacteria are far better at surviving in extreme environments than are higher animals, so we can expect that many promising lineages were extinguished along the way to the one that was established.

"After the appearance of the first endosymbiont-containing protists, evolution once again settled into a relatively static mode, engaging mostly in diversification—endless variations on the same basic themes.... Then some eukaryotes "discovered" the advantages of getting together and pooling efforts. Why it took them so long to make this discovery is not clear. An enhanced interest in sex could be, at least, part, of the answer....[58] This is an example of horizontal exploration followed by an advance in sophistication. We shall encounter this later in the evolution of the operating systems.

Cambrian explosion

With the maturation of the eukaryote system and the exploration of multicellular possibilities, evolution apparently shifted into high gear in the final quarter of life's history.

The first multicellular organisms appear about 800 million years-before-the-present (MYBP), and, after a period of maturation, the floodgates of innovation opened about 600 MYBP.

In the next 200 million years, all the different phyla of life developed in a tremendous period of development known as the Cambrian explosion.

"About 570 million years ago, virtually all modern phyla of animals made their first appearance in an episode called 'the Cambrian explosion' to honor its geological rapidity. The [fossil record in the] Burgess Shale dates from a time just afterwards and offers our only insight into the true range of diversity generated by this most prolific of all evolutionary events. ... the fossils from this one small quarry in British Columbia exceed, in anatomical diversity, all modern organisms in the world's oceans today. Some fifteen to twenty Burgess creatures cannot be placed into any modern phylum and represent unique forms of life, failed experiments in metazoan design. Within known groups, the Burgess range far exceeds what prevails today. ... The history of life is a tale of winnowing and stabilization of a few surviving anatomies, not a story of steady expansion and progress."[59]

As might be expected, the more complex a system, the more possible varieties might be possible. This expectation is borne out—the mechanisms that emerged at this time capable of managing and duplicating systems at the level of sophistication controlling organs as a unified body opened the floodgates of exploration and innovation. Thus, the great difference in the number of species found in simple (pre-Cambrian) life as compared to the number of species found after the Cambrian explosion. The possibilities of organism structure seem to increase dramatically with sophistication, as seen in this chart:[60].

Compared to the billion it took for mature eukaryotes to emerge from the prokaryotes, the Cambrian period involved radical changes in systems taking place over periods of tens of millions of years—an explosion indeed in terms of speciation.

Kingdom	Features	Number of Species
Monera	prokaryote	4,000
Protista	eukaryote	20,000
Fungi	multinucleate	80,000
Plantae	plant	300,000
Animalia	animal	2,000,000

The rest of life's history has been variations on the themes initiated during that period and, it is only with the advent of the brain capacity that a radically new level of sophistication can be said to have emerged.

Generating QPF

The good news is that I am now abandoning my anthropomorphic way of describing QPF and the impulse to follow the path of least action. The bad news is that I am adopting a new analogy, that of the computer.

Quantum science states that the objective world is stuff filling in quantum probability forms.

A bacterium certainly fulfils this expectation. There are millions of protein-catalysts in the bacteria. Each of the 10,000 varieties is providing a QPF to the overall QPF that is the bacteria. The stuff of the bacteria flows through these QPF, like pipes and pumps in a chemical factory.

The stuff fills in this composite QPF and we see bacteria. While the stuff is constantly changing the bacterial form, reflecting the QPF in the usual way, remains constant.

We are now going to liken bacteria to a quantum-style computer.

First is the fundamental difference between a computer operating system and the programs that run 'on' the operating system.

I am currently running Mac OS X v. 10.3.8 on my PowerBook. My word processor, MS Word, is running on top of this, as well as many, many other different programs.

For bacteria, the operating system is the triplet-code RNA mechanism of protein synthesis. The programs are linear RNA written in triplet code.

In this section, I am going to deal with the programming side of living systems. In the next section, we shall deal with the operating systems of life and their origins.

The triplet code method of protein synthesis has been extensively described elsewhere. In essence, a digital RNA code is translated into an analogue protein effect contributing a QPF to the composite whole.

Digital: Linear RNA code

Translation into linear aminoacids, folding

Analog: Protein providing QPF for stuff to fall into.

This is the basic triplet code and the aminoacid "machine level" processes they call up. Most of the 'desires' of the aminoacids will be satisfied in the folding—which we will get back to by the end of the book—with a few left out as the active site, the QPF being contributed to the composite QPF. The code is "degenerate" in that different codons translate into the same aminoacid; the chart just lists one and the number of degenerates.

Why just 20 aminoacids, why just 20 'machine codes'. This is like asking why our alphabet has 26 letters. The best answer to this perhaps is that they suffice—the dozen or so phonemes of speech can be covered sufficiently well. For there are only a dozen or so elementary chemical reactions important to life's needs.

Designer codes

As an interesting aside: As this is being written, scientists are beginning to experiment with designer triplet codes—codes that are translated into aminoacids provided by the experimenter that are not found in nature. Two experiments involve bacteria with altered triplet codons that thrive on fluorotryptophan—a deadly metabolic poison to all universal-code users such as bacteria and us.

"One of these two bacteria with the designation "HR15" grew happily on it. Not only did HR15 thrive on fluorotryptophan, it was poisoned by tryptophan. HR15 is not just a picky eater, but an entirely new type of life, [researcher] Ellington says. [Researcher] Wong agrees, 'HR15 does represent a new form of life because the genetic code is the most basic attribute of living systems' he says. He calls the alteration of the genetic code, 'the ultimate test-tube evolution... we are altering the whole organism.' The public has nothing to fear from these artificial organisms, says [researcher] Schultz... Any bacteria that escaped the lab would starve without the researchers feeding them the unusual aminoacids. Bioethicist Caplan dismisses any charges that the researchers are playing God. He says that the scientists are 'playing man' and doing what people do best—creating new things. 'There's nothing wrong, morally, with inventing things,' he adds."[61]

I must say, I do like that delicate correction in the above—not playing God, playing man! Like Father, like Son.

Basic Process

There is also amplification: one mRNA can be transcribed into many copies of a protein, contributing many QPF to the composite.

In essence, though, we can describe the basic process of life as

A linear program is run on the operating system.

Quantum probability forms are generated.

There are differences of scale, of course. My Mac has one processor running the DOS while a bacterium has hundreds of thousands of ribosome processors all running at the same time.

So, while each ribosome runs just one program at a time, hundreds of thousands of them are all at work simultaneously. Massive processing of relatively few programs. Such massive, coordinated parallel processing is a major goal of computing science but with little success so far.

So, in quantum science, what is the basic description of a bacterium? It is a Mandelbrot Set-like concatenation of millions of quantum probability forms. The collapsed form to this internal aspect is the form we call a bacteria as stuff rapidly pours through the probability gradients.

The numbers involved are roughly 10,000 genes, 100,000 mRNAs, 1,000,000 ribosomes, and 10,000,000,000 proteins. Each protein is contributing a QPF to the composite final probability amplitude that makes a bacterium so probable.

Condensing this description even further with our computer metaphor, we can say that a bacterium is basically ten thousand linear programs running on one million operating systems generating ten billion analog quantum probability forms. Then there is, of course, the stuff flowing through the probability gradients, like electrons in orbitals of molecules.

So where did all those programs come from? We know pretty well how they are passed down and multiply-copied down the ages—some of our housekeeping genes are almost identical to those in use by bacteria, evidence of a common ancestor.

But where do the programs come from in the first place, what was their origin in the first place. "Who wrote the program?" is the first question asked when a Windows virus spreads like the plague—exactly like plague with email playing the role of carrier rats.

This brings us into the thickets and battles-royal of evolution. What are the origins of the one operating system and the ten thousand programs.

Generalized Schrödinger

We are now going to make some drastic generalizations about the nature of the well-formed QPF found in nature. Well-formed, as noted earlier, in that they are relatively long-term and stable forms.

First supposition: The Schrödinger equation that describes atomic orbitals is a member of a much larger class of equations that we will call the Generalized Schrödinger Equation, GSE. All well-formed QPF on any level of sophistication have a form that is accurately described by a Generalized Schrödinger relationship.

We will now dissect the Schrödinger equation, that intimidating hieroglyphic that, believe me, accurately describes the orbitals of the atoms. (Solving it, however, is another question entirely.)

A few points to remember: The proton (for hydrogen) or the nucleus, in the appropriate reference frame, is unmoving. The nucleus, in our time frame, is an unmoving, unchanging generator of QPF orbitals for electrons to fill-in. (See the sections on catalysis and enzymes for more sophisticated generators of QPF for others.

Generators of QPF are always unchanging compared to the transient nature of the stuff filling in the QPF. The Law of Large Numbers insists on it. Otherwise, the probabilities would never get a chance to be expressed by the stuff before the probability changed again. Certain francium atoms, for instance, can never be observed simply because the nucleus flips so quickly to another element that the 90 odd electrons only get to make a few jumps in the francium QPF before the nucleus decays and a new QPF is generated.

In the general scheme of things, note for later that the atomic nucleus plays exactly the same role as catalysts, proteins, and RNA programs running on a real and in a virtual operating system. In this sense, understanding the atomic nucleus from top-down is equivalent to dePrograming it.

Here is the monster we wish to tame.

The reason it looks so formidable is because it is written in the mathematical equivalent of assembly code. This is what the assembly code, the last step in software before its expression in hardware, for adding two registers together might look like:

$$-\frac{d^2\psi}{dx^2} = \frac{2m}{\hbar^2}\left(E - V(x) \right)\psi$$

1010011101000111001001000101000101001010001000010100

Or in hex shorthand, the command to insert the letter 'I' I just typed into my unsaved Word document in memory might look like:

564FA36EE765BA1000FFFFF22237656.

It is clearly impossible to code any but the simplest of programs in either form.

The top level language running on my Mac OSX, on the other hand, probably looks something like this:

CHECK KEYBOARD, CHECK NETWORK, RUN PROGRAM THREADS,
RUN HOUSEKEEPING THREADS, REPEAT.

This is the type of language we will be able to tame Schrödinger with. To give you hope, we will end up with a simple relation such as: s = I − qp

Thou Shalt Not

We are first going to do a simple algebra by dividing both sides by the same thing, the wavefunction, or that funny looking psi.

This of course is only allowable if the wavefunction is never exactly zero, which it never is. For, remarkable as it may seem, the 1s orbital of a hydrogen atom in your body actually has a non-zero value on the Moon. It is an extremely small probability and is essentially zero, but it never gets to exactly zero. It is like the zero limit in calculus.

For while the calculus zero is essentially zero, but it exactly. As any careful calculus textbook will say somewhere, when talking about the limit of one-over-infinity equals zero: "But note that it never becomes exactly zero."

At the very heart of calculus is the assertion that: "Nevertheless, as the difference between exactly zero and our essentially zero can be made as small as desired it can be ignored."

The only real difference is that, while you are allowed to divide by the calculus zero, you are not allowed to divide by exactly zero. Ever. Under any circumstance whatsoever. To do such a thing is to declare yourself a non-mathematician and your theories worthy of ignoring from henceforth.

This is why it is important that the wavefunction never be exactly zero, anywhere. Otherwise, our division would be disallowed.

Luckily, unlike almost everything else we have discussed in physics so far, the wavefunction is not discontinuous, it is not pixilated. It can shrink exponentially and infinitely without ever getting to exactly zero.

Let's get ridiculous for a moment to illustrate this seemingly trivial point. As big numbers are more impressive than the small, first we need to define a really, really truly-enormous number. We start with a big number, the familiar google-plex, 10 to the 10 to the 100, or $10^{10^{100}}$

Call this big number, g. Now raise g to the gth power in a tower of stories g high, $g^{g^{g^{\cdots}}}$. This is a really, really big number; call it G.

Now build another tower of G exponents, this time G stories high. $G^{G^{G^{\cdots}}}$. This is our really, really truly-enormous number; call it G.

Now flip it, calculate 1/ G. This is a really, really truly-infinitesimal number that any well-respected calculus major would be happy to call essentially zero, but would happily divide by it if need to. Call this essentially zero, o.

Now you might think that there would <u>not</u> be much room between o and 0. But you would be wrong. For it is proven that there is an infinity of locations even closer to the true zero. And not just a countable infinity, but an uncountable[62] infinity of points between this essentially zero and exactly zero. Infinitesimally close, believe it or not, still has an infinity of infinity of numbers between it and exactly zero.

All this implies that, while the probability of all your atoms deciding to be on the Moon might be 1/ G, it is not exactly zero. This can be considered a challenge to advanced technology: to manipulate and magnify such probability of teleportation before all the oil runs out.

Unlike almost everything else in the universe, the wavefunction is not pixilated, it is absolutely and smoothly continuous creating smooth probability gradients even across Planck pixels of space time. The value associated with that pixel then being the average of the gradient across the pixel. This means that the wavefunction can get arbitrarily close to exactly zero even at the far distant reaches of the universe.

In our stadium illustration of the relative sizes within atoms, quantum mechanics tells us that the firefly spends 99% of its time near the baseball, and 99.999999% of its time in the stadium. Yet it also has a non zero probability of appearing on Mars or Alpha Centauri or Andromeda, just for a Planck or two, before reappearing, unwearied by travel, back in your body on Earth.

Quantum mechanics can explain and justify such oddities with hands tied behind its back.

Dissecting the equation

All that was to justify dividing both sides of Schrödinger by the wavefunction, psi. As the wavefunction is never exactly zero, we are allowed to divide by it, and we get:

We are now going to slice this monster into segments, boil each part down to its essentials, then combine them back together. This will take a while, but it will be

$$- \frac{d^2 \psi}{dx^2} \frac{}{\psi} = \frac{2m}{\hbar^2} \left(E - V(x) \right)$$

worth the effort for our final result is a T-shirt equation that should not intimidate any but the truly math phobic.

The Twist Tensor

We shall start with the real-scary looking expression:

$$- \frac{d^2 \psi}{dx^2} \frac{}{\psi}$$

This is actually not as bad as it looks. For Newton's classical formula, F = ma, connecting force to inertial mass and acceleration, can also be expressed in this elegant, but complicated way.:

$$\frac{2m}{\hbar^2}$$

$$\left(E - V(x) \right)$$

$$- \frac{d^2 \psi}{dx^2} \frac{}{\psi}$$

$$\text{F/m} = d^2x/dt^2 = dv/dx = a$$

In words Newton's declaration is that the force, F, divided by the mass, m, equals either the rate of change of the rate of change in position with time, d^2x/dt^2, or the rate of change of velocity, dv/dt, or the acceleration, a.

So, what Schrödinger is describing on the left is the internal 'acceleration' in the form of the orbital, the rate of change of the rate of change in the form of the QPF wavefunction at any point.

And then this 'acceleration in the form' is divided by the value of the wavefunction at that point. We now have a value for the 'acceleration in QPF' per QPF. This value is then negated, it is rotated by 180° on the complex plane.

We shall call this final value a measure of the 'quantum twist' in the QPF.

For the 1s orbital, this twist about as simple as it gets—perfect, spherical symmetry and with no nodes, even at the nucleus. (A node is where the quantum probability is exactly zero.) In music, the 1s orbital would be called the fundamental waveform that fits and fills the degrees of freedom available.

This simplest, most basic filling-in corresponds to the simplest of programs running on a newly emerged OS.

The twist to the 5f orbital, on the other hand, is complex with multi-nodes, and an accurate description of such a convoluted twist is fiendishly complex in the extreme.

And the twists in a simple QPF such as the molecular wavefunction of a water molecule is intricate to describe.

Luckily, as Einstein discovered to his delight when looking for the simplest way to describe his General Relativity, math has these delightful things called tensors (a sort of sophisticated vector involving matrices). And tensors can describe the twists of even the most convoluted and complex of forms.

So, Einstein used to tensors to great effect to describe the way gravitational mass distorts and twists spacetime. See any good book for more info on this.

Now tensors look deceptively simple: they are just a letter with lots of little subscripts and superscript indices that have to be carefully kept track of in detailed calculations; e.g.: $T^{a}{}_{a}{}^{b}{}_{b}\cdots$. We can simplify by letting i stand for all the indices: $T^{a}{}_{a}{}^{b}{}_{b}\cdots = T^i{}_i$.

Tensors, and matrices of tensors, etc. are quite capable of handling even the most complicated twists to any QPF. This comes from an excellent introduction to tensors (translated from the Russian!):

"There are quantities of a more complicated structure than [real or complex numbers], called tensors... whose specification requires more than knowledge of a magnitude and a direction."[63]

So, we can now do a radical simplification. We will define the Quantum Twist Tensor or QTT, q, as the tensor array that accurately describes the quantum twist in a QPF. This does a nice job of simplification for us, just like the a in Newton's formula.

And, as we have no need to actually calculate the twists in various QPF, that's all we need to know about tensors.

$$q = Q^i{}_i = -\frac{d^2\psi}{dx^2}\,\frac{1}{\psi}$$

Schrödinger now looks a little less forbidding using the QTT instead of that double-integral mess.

We are dealing with two systems that are playing very different roles: the proton the generator, g, providing an orbital for an electron, the filler-in, f,

$$q = \frac{2m}{\hbar^2}(E - V(x))$$

to jump into and 'flesh out' over time by the LoLN.
If necessary, we can keep track of what belongs to who with the appropriate indices. I shall not generate clutter with this. If I did, the q would have a little g index while the m would have an f index.

Penchant and Passion

Now for the right-hand side of the genius-monster equation. First, we need another rearrangement, using the simple distributive law of algebra, to get a Planck's Constant, h-bar, inside the bracketed term. Then we also shift the 2 inside the brackets.

The right-hand side of Schrödinger now deconstructs into two fragments that we can treat separately.

The first fragment involves mass—which is Einstein-proved to be equivalent to energy.

$$m/\bar{h} \quad \& \quad 2E/\bar{h} - 2V(x)/\bar{h}$$

The second describes the pendulum-like balance between the maximum potential energy stored in the electromagnetic field, a constant, and the kinetic energy of motion which varies with position.

The kinetic energy is at maximum, the electron is moving very, very fast, at the proton-heart of the H-atom, the tiny baseball at the center of Yankee Stadium. On the other hand, the kinetic energy, the velocity, is essentially zero at the far 'edge' of the atom, the worst seats at the very top.

This is why, in classical terms, the firefly-electron does not just get to sit on the baseball-proton no matter how hard it tries to land. For, when it gets to the very center of the field, it is moving so fast it zooms right past the proton and is way up in the bleachers before it can slow down, turn around, and make another lunge to get to center. As electrons never learn or get bored, they can keep this up forever.

As the kinetic energy is a maximum at the center, the PE–KE expression in Schrödinger will be at a minimum at the very center of the 1s orbital. This is the Principle of Least Action appearing in a simple disguise.

Planck and Time

The little-h under each term is, as earlier discussed, just the conversion factor into the natural units of nature's pixels. This is Planck's Constant over $2\pi \lceil \rceil \pi$. As this is the ratio of the radius and circumference of a circle, h appears as a length, a radius, in Schrödinger. In other equations, the pixilation factor appears as an enclosure, a circumference, as an uncrowned h.

So "inertial mass over h-bar" is describing the inertia per pixel-radius. Same for other two fragments: the energy terms are per pixel radius.

The Generalized Planck's Constant is then the conversion factor, appropriate to the pixilation and timeframe, involved at any level of QPF under discussion.

Now, as mentioned earlier, action is the basic measure of existence. Its pixilation size is given h, Planck's Constant.

For the most fundamental level of reality, then, we can say that the time-frame for particle existence is of the order Planck seconds. So, this is the appropriate scale to use for the electron.

What about an atom of hydrogen? Is the Planck time still the appropriate conversion factor for atomic existence? I think not.

For an atom of hydrogen can only be said to 'exist' in objective reality over time periods measured in pico-seconds. And, while very, very short period of time, it is as an eon compared to the Planck time.

To make this clear, examine closely this computer-enhanced photo of an electron and a proton—taken with a Planck-time, freeze-frame flash camera—and then answer the question that follows.

Q. Is this a hydrogen atom? Or is it not a hydrogen atom?

A. Yes and No—it's a trick question :-)

The above is actually the supposition of two photos—you can see that the registration is not quite perfect for the two electrons at upper right.

One photo is of a hydrogen atom bonding together a G to a C nucleotide in a 50,000,000-year old sample of dinosaur DNA. This is clearly an atom that can be said to exist.

The other photo is of the inside of a TV tube displaying the "Golden Girls." The electron, at 500,000 mph, is moving horizontally from the cathode towards a red pixel on Blanche's heaving bosom. The proton is a tertiary cosmic ray moving vertically downwards at 10% the speed of light. In a picosecond, they will be miles apart. This second photo is clearly not of a situation that can be said to be an atom that exist.

The point is, at Planck time resolution, the two situations are indistinguishable. The concept 'atom' has no meaning over time pixels commensurate with Planck. A movie of either situation, taken at one frame a Planck, and screened at 100 frames a second, would take millions of years before there was any discernible movement in either electron or proton.

So, the conversion factor in the generalized Schrödinger will be different at different levels. As we are not going to actually do any calculations, the technical question as to exactly what values these factors can be left for now.

As a general rule, however, the timescale must allow for a QPF to be filled in a few score times. With this in mind, we can suggest some approximate time frames for systems with a lots of QPF that have to be substantially filled-in.

PARTICLES: Planck-secs	ATOMS: pico-secs
RIBOSOMES: mille-secs	CELLS: seconds
ORGANS: minutes	BODIES: hours
FAMILIES: days	NATIONS: years
SPECIES: decades	GENERA: centuries

Planck's Constant is not just about timeframes, it is a product of time and inertial mass. The inertial mass of a system is a measure of the system's reluctance to alter its current state of motion. We will discuss such reluctance-to-change, in general terms, in the following section.

So, the conversion factor in a generalized Schrödinger will also take into account the scale of the resistance-to-change.

That is all we need to note about the fact that the fragments of Schrödinger we are currently considering involve the 'radius' of a level-appropriate time and reluctance, conversion-to-pixels factor.

Reluctance to Passion

First, the simplest of the Schrödinger fragments (using $\hbar$ for h-bar as Word prefers it): m / $\hbar$

The inertial mass, m, measured in appropriate units, of the electron is a measure of its reluctance to alter its state of motion when tugged at by classical forces (actually moving in quantum probability gradients, of course). We can think of this simply as the tendency of the filling-in system to keep doing its own thing.

In the useful classical terms of the "movement" of the electron in the electric field, this inertial mass is a measure of the tendency of the electron <u>not</u> to respond to the electric force. So, the mass of an electron can be thought of as its reluctance, or resistance, to moving so as to fill-in the QPF orbital.

All systems can be expected to put up some resistance to filling-in a QPF, and this can be called the generalized inertia of that subsystem.

We shall define the quantum inertial reluctance of a subsystem, r, to be this pixilated, generalized inertia. This is the measure reluctance to move as the QPF dictates, not as the free system would if left alone. So, for an electron moving in the QPF of a proton: $r = m / \hbar$

We now invert this and, as the inverse of reluctance is passion, perhaps penchant, we now define the Quantum Pixilated Passion, the QPP of the electron, p, as: $p = 1 / r = \hbar / m$.

Substituting this into Schrödinger, then multiplying both sides by p, we end up with the much simpler-looking equation. Is it not great how math can hide a lot of detail with a few simple symbols! $q p = 2 E / \hbar - 2 V(x) / \hbar$

We can take this process of well-defined generalizing even further and simplify $E-V(x)$. This expression is notoriously difficult to solve explicitly. Luckily, we do not wish to calculate it, just understand what the expression is telling us about the way the internal world of the QPF is ordered.

Pendulum and Phase space

All we will need for this discussion is the simple pendulum, a favorite gadget in the elementary physics lab.

At point E the bob is momentarily at rest, it is not moving. All the energy of the interaction between bob and earth, via graviton exchange, is in the intensity of the interaction. A physicist would say that all the energy of interaction is in the field at this point. This we define as a measure of the intensity of the interaction, I. At point E, all the energy is in I, the potential energy of the interaction.

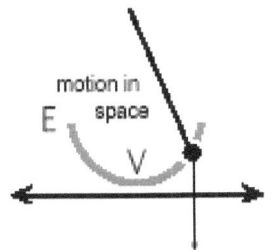

At point V, all the energy is now in the velocity of the bob. All the energy of interaction is now in the velocity of the bob. No energy is in the potential field, it is all in the velocity of the bob. Note that at V, the velocity is horizontal while the force of gravity is at right angles to it. At point V, and at V alone, the bob's velocity is not influenced by the intensity of the interaction. The bob is in totally-free movement and unencumbered by force. This we define as a measure of the bob's satisfaction during the interaction, s. At point V, all the energy is in s, the kinetic energy of the bob in free and full motion.

At the point where the bob is at in the diagram, at x away from V, some of the energy will be in the intensity of the field exchange particles and the balance will be the kinetic energy of the bob's motion (with horizontal and vertical components) of its attempt to attain complete satisfaction at V again. The bob 'wants' to stay at V, but it is moving way to fast to stay there. Life's like that.

This balance is described an expression that is familiar from Schrödinger. E-V(x) A better way to describe this back and fore motion (which involves sines and cosines) is as circular motion in a phase space.

The two axes are potential energy and kinetic energy. The movement of the back-and-fore pendulum at variable speed is now a point in this phase space moving at a constant speed in a perfect circle (assuming no friction, which is common). Even at this simple level, it is clear that constant motion of a point in a circle is easier to deal with mathematically than variable side-to side-motion of the actual bob.

Even though a pendu-lum bob is composed of quintillions of particles each with its own phase space, they all combine into the simple two-dimensional phase space that is suffi-cient to describe the behavior of the pendulum.

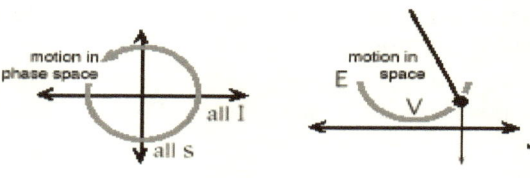

We can now apply this concept to the electron in the 1s orbital ground state, isolated H-atom.

When the electron is at the very edge of the atom, its velocity is zero and all the energy is in the intense electromagnetic field as the potential energy. This value appears in the E of Schrödinger, a constant. This is I, the intensity of the electro-magnetic interaction.

At the nucleus, the reverse is true. All the energy is now kinetic and in the motion of the electron, E=V at this point. The electron wants to stay there at the center, but when it gets there it is moving way too fast to stay there. This is the maximum satisfaction of the relationship and at this point, when s=I.

Next, draw a line connecting every point on the surface of the 1s orbital through the center to the point on the opposite edge. There will be a continuous infinity of such lines. Now consider the electron to be moving back-and-fore along every one of these lines at the same time (this only sounds impossible because we are using classical concepts (in a quite valid way) in the description).

As the 1s orbital has circular symmetry with no nodes except at infinity, its motion will be a perfect circle in an infinite-dimension phase space (a common creature in physics calculations). The other orbitals are, like Ptolemy's heavens, combinations of many such circular motions, or epicycles,in phase space. In phase space, the energy oscillations are a multidimensional circular motion.

The maximum potential energy for this situation, a constant, is the E in Schrö-dinger. The amount of this energy in the kinetic form, that varies with position from the center, is the V(x) in Schrödinger.

In the helium atom, all we have to do is consider two electrons at each end of the lines through the center. They oscillate happily together. When one is at the end of a line, the other is at the other end. When both are at the center, all the energy is in their mutual motion past each other. Whoosh. Maximum, internally-amplified, satisfaction. Over and over again.

Substituting our more general symbols for intensity and cyclical satisfaction, we have: $E - V(x) = I - s$

SIMPLE SCHRÖDINGER

Putting this back into our deconstructed Schrödinger, we end with the simple, general form that applies to any and all QPF:

$$q = (I - s) r$$
$$= (I - s) / p$$

In words, the twist tensor in any interaction equals the reluctance times intensity minus satisfaction, or equivalently, the intensity minus satisfaction over the passion.

This sounds like philosophy and theology, but it is not. Each term has been precisely defined mathematically.

When applied to the hydrogen atom, this general equation gives us back the highly-specialized original:

Now lets do a little algebra of the internal realm and rearrange the generalized Schrödinger equation to different forms:

$$- \frac{d^2 \psi}{dx^2} = \frac{2m}{\hbar^2} (E - V(x)) \psi$$

$$q = r (I - s)$$
$$r = q / (I - s)$$
$$I = p q + s$$
$$s = I - p q$$

If you put these into words, you will find a lot of wisdom. What is maximum satisfaction when two fill in a QPF together (a zero is when $I = pq$)? What is a minus twist tensor (when s is greater than I and r is non-zero). What situations generate maximum reluctance (inertial mass) and what is negative reluctance (when s is greater that I)? etc.

PROGRAM EVOLUTION

In the history of Earth, we find that species have distinct beginnings and endings. As Niles Eldridge puts it:

"If the fossil record has anything at all to tell us about the history of life, it is that species of 600 million years ago, or 400 million, 200 and 100 million years ago, are all different from the ones we have on earth today ..... we must further conclude that species undergo a 'birthing' process as well as a 'death'—or extinction—process."[64]

"Nothing in biology makes sense except in the light of evolution,"[65] a sentiment echoed by most biologists.

This 'keystone' element in the edifice of biological thought has gone through its own evolution and development resulting in what is known as the "Modern Synthesis," the 'received view' of contemporary scientists.

Although many natural philosophers such as the early Greek thinkers and Jean Baptiste Lamarck (1744-1829) had toyed with the idea of evolution—the idea that all life descended from a common ancestor—it is Charles Darwin (1809-1882) who, along with Alfred Wallace (1823-1913), is credited with responsibility for the foundations of our contemporary understanding of the evolutionary process. Darwin developed a comprehensive theory that he presented in "On the Origin of Species by Means of Natural Selection" published in 1859.

The theory he presented coalesced many of the disparate facts already catalogued by the exploratory science of that day. "Darwin was able to weave together an interlocking set of hypotheses explaining resemblance among organisms, their patterns of distribution, and their fossil records—a set of hypotheses making sense and providing coherence to a wide body of observation and experience that had accumulated by the mid-nineteenth century."[66]

Darwin's seminal work is the foundation of the modern synthesis. Other elements of the edifice were contributed by Mendel and other workers in genetics and, relatively recently, the explosive expansion of our comprehension of molecular biochemistry.

The disparate elements were first combined to create the modern synthesis by Theodore Dobzhansky in his "Genetics and the Origin of Species" published in 1937.[67]

This is a succinct description of the modern synthesis from a textbook on evolution: "New species usually arise through the accumulation of different genes within reproductively isolated populations of some parent species. These populations become so different that they cannot breed back to the parental population and thus can be recognized as distinct species."[68]

The contemporary view has answered many of the questions about evolution—it has many accomplishments to its credit. There are also, however, some questions that are not answered satisfactorily in the modern synthesis. There continue to be many challenges to the received view.

Well within the scientific mainstream are challenges that arise from recent developments in molecular and population genetics and in paleontology, the study of fossils. A recent review of "The Evolution of Darwinism" described some of these challenges: "One is a proposal that a kind of molecular determinism, rather than pure chance, impels the development of variation in DNA. The other is a contrasting claim, known as the neutral theory, that chance governs not only the initial appearance of genetic variations but also their subsequent establishment in a population. A different kind of challenge, based on new interpretations of the fossil record ... known as punctuated equilibrium ... holds that evolution proceeds not at a steady pace but irregularly, in fits and starts."[69]

In the 1940s, Ernest Mayr proposed that transspecific development might occur at a different tempo than subspecific development,[70] a proposal that was developed into the "punctuated equilibrium" theory of Gould and Eldredge.[71] This theory proposed that speciation occurred in small populations and very (geologically speaking) rapidly, an idea that has received a great deal of empirical support.[72]

One of the key concepts in the modern synthesis is that "evolutionary change must be dominantly continuous and descendants must be linked to ancestors by a long chain of smoothly intermediate phenotypes."[73] This idea was challenged by the extreme saltationist view that development proceeded by large jumps through the appearance of fortunate macro-mutation, the "hopeful monster."[74]

Although this idea was not well received by the scientific community at that time, recently it has:

"Been reborn as a product of the transposition of small regulatory elements of DNA, or by the translocation of large chunks of genome, leading in either case to major changes in gene expression by means of which, according to a flight of fantasy indulged by W. Doolittle, a toad might evolve into a princess with a minimum of intervening millennia."[75] The evidence for this 'quantum speciation'[76], its possible mechanisms[77] and saltationist models of evolutionary processes[78] are now the subject of debate in the scientific literature.

> Saltationist views have gained ground in the scientific community promoting the comments: "Quantum speciation of any sort was rejected ... in retrospect, it seems that Goldschmidt deserves posthumous accolades for his steps in the right direction."[79] And, "Quantum speciation entails no major elements not recognized within the Modern Synthesis of evolution. The new view simply differs in its emphasis on particular elements and in its implications for large-scale evolution."[80] There are those who have pointed out that Darwin himself can be considered a "punctuationist"[81] and, naturally enough, there is also a spirited defense of the Modern Synthesis or neo-Darwinian thought as is.[82]

The modern synthetic view predicts that, if the fossil record were exact enough, as the paleontologists dug and sifted through geological time, they would see a gradual drift, the transformation of a species into another.

Niles Eldridge, curator at the American Museum of Natural History and one of the developers of the theory of punctuated equilibrium, recalls his first experience of the difference between his expectations based on the modern synthesis and what he actually found in his explorations of the fossil record of the trilobite Phacops rana:

> "But that's not what's there ... in the entire 8 million years ... the greatest (though not the sole) amount of modification wrought by evolution in the Phacops rana stock was the net reduction from 18 to 15 columns of lenses. Hardly prodigious, this degree of anatomical retooling falls well within the normal bounds of 'micro-evolution'... We see something out of whack with prevailing expectations... as we climb up those rocks and check those samples, over what must be, in sum total, a 3-or-4 million year period, we see some oscillation, some variation, back and forth ... but no real net change at all ... This is the first element: simple lack of change. Stability, or stasis as [Stephen Jay] Gould and I began to call it. And the second element in this pattern is the apparent suddenness of the change: when it does come, evolutionary modification seems to be abrupt, an all-or-nothing sort of affair."[83]

The quantum nature of the fossil record had been quite apparent from the very beginnings of modern paleontology.[84] Darwin asserted, and this view has been incorporated into the modern synthesis, that this was an artifact. That, because the fossil record is incomplete, we gain the impression that a quantum change has occurred when in fact, if a temporally-complete selection of remains had been preserved we could then see the actual, gradual transformation occurring.

Darwin was very clear on this point:

"I have attempted to show that the geologic record is extremely imperfect; [a long list of reasons why]. All these causes taken conjointly, must have tended to make the geological record extremely imperfect, and will explain, to a large extent, why we do not find interminable varieties connecting together all the extinct and existing forms of life by the finest graduated steps. He who rejects these views on the nature of the geological record will rightly reject my whole theory."[85]

This is why there is almost a sense of relief in the paleontology world that the record is not complete. In the 1950s, one book on evolution clearly expressed it: Thank goodness, the fossil record is not complete![86] Why would such a strange sentiment exist—gratitude that the experimental data was incomplete? The simple reason is that, as already noted, for science to progress there has to be the ability to order and classify the complexity of nature. If the differences between individuals created a continuum it would be impossible to consign individuals into larger groupings, the neatly ordered, set of inter-nested boxes labeled with Latin binomials, into which we have been able sort individual organisms since the time of Linnaeus.[87]

As Darwin noted, the fossil record most definitely did not show the gradual transformation of one species into another:

"Why then is not every geological formation and stratum full of such intermediate links? Geology surely does not reveal any such fine graduated organic chain; and this, perhaps, is the most obvious and gravest objection which can be urged against my theory. The explanation lies, as I believe, in the extreme imperfection of the fossil record."[88]

There have always been, however, since the very start of the debate, certain paleontologists who have disagreed with this 'incomplete' interpretation of the quantum nature of the fossil record. They are known collectively as 'saltationists.'[89] Although in many ways their ideas can be very different, basically, they all maintain that evolution proceeds by leaps, sudden jumps from one state to another. This classification of scientists is broad, encompassing the early eighteenth century catastrophism of Georges Cuvier and the 'hopeful monsters' of geneticist Richard Goldschmidt of the 1940s.

dePrograming

The differences between species of bacteria are not to be found in the operating system for they are all identical (we use exactly the same OS ourselves in our organelles). It is the programs running on the OS that are different.

Evolution, then, must involve the evolution of programs—the evolution of programs that generate quantum probability forms when run on the OS. The task of the scientist, then, is to deconstruct the programming of living systems, or as I like to call it, dePrograming nature.

So, the struggle for existence happens first on the internal programming level, second in the external world. Moreover, the criteria for success in the internal world are quite different from those required for success in the external world.

Classical pictures of evolution have focused, of course, on the external stuff. What the material, the atoms, the molecules are doing. It has no concepts that can deal with why this flow of stuff is so probable that it happens all the time.

The inherent improbability, according to classical science, of many of the processes known to have occurred in evolutionary history has troubled many workers in the field.

One provocative book, in its attempt to solve this problem, in 1981 created a tremendous stir in scientific circles. In Britain, the reviews of the book *A New Science of Life: The Hypothesis of Formative Causation* covered both extremes:

Nature, the preeminent international science journal declared it "the best candidate for burning there has been for many years" while the New Scientist, a feisty news magazine, stated, "It is quite clear that one is dealing here with an important scientific inquiry into the nature of biological and physical reality."

The reason why the book created such a stir was that, after listing in "Some Unsolved Problems of Biology" the inability of classical theory to deal with questions of morphogenesis, evolution and behavior, Dr. Sheldrake introduces a new causal factor into the scientific picture of how the world works. He postulated a morphogenetic field: a non-energetic template or blueprint that guides physical, chemical and biological systems so that only one result occurs out of the many that are equally possible energetically. He uses the following analogy:

"In order to construct a house, bricks and other building materials are necessary; so are the builders who put the materials into place; and so is the architectural plan which determines the form of the house. The same builders doing the same total amount of work using the same quantity of building materials could produce a house of different form based on a different plan. Thus, the plan can be regarded as a cause of the specific form of the house, although of course it is not the only cause: it could never be realized without the building materials and the activity of the builders. Similarly, a specific morphogenetic field is a cause of the specific form taken up by a system, although it cannot act without suitable 'building blocks' and without the energy necessary to move them into place."[90]

It would seem that the 'morphogenetic field' proposed by Dr. Sheldrake is already a part of modern science—it is equivalent to the composite quantum probability form.

External Evolution

In classically-based evolution, evolution is left up to 'chance and accident.' Yet bacterial life appeared on the cooling earth not long after the oceans were finally stable and established.

We have already gone to great pains to show that such classical concepts of 'probability' have been totally repudiated in physics and chemistry.

Classical evolution then is definitely assailed from below by the quantum probability revolution in physics. It is also under attack from above, for classical concepts lead us to expect life to be highly, highly improbable. In classical science, a construct as sophisticated as a living cell is highly unlikely. As Fred Hoyle put it, the emergence of living systems is about as likely in classical physics as a hurricane sweeping through a junk yard assembling a fully-functional Jumbo Jet from the bits and pieces scattered around there.

Just as the chance of junk colliding in just the right way to form a fuselage is small, so, in classical physics, is the chance that atoms and molecules will congregate in just the right way to form cells, organelles, tissues, etc.

This view of evolution permeates all of biology, all of genetics, all of the brain sciences.

"Nothing in biology makes sense except in the light of evolution,"[91] is a sentiment embraced by most biologists. The classical science system put such great emphasis on fighting off a teleological explanation of evolution—that there is a purpose and a plan behind the origin of species—that biologists have gone to the opposite extreme and adopted the concept that there is no underlying organizing factor to evolution.

All scientists believe, of course, that the phenomena of Nature can be understood and that there are still many things that our science has yet to figure out—they would be foolish indeed to do "research" if they didn't believe there was anything left to discover.

So, it is strange that, while no one is saying that electrons and protons behave in a totally random chance-and-accident manner in the formation of simple atoms, many biologists are stating this to be the case for the much more complex rearrangement of the genetic molecules that occurs in the historical development of species, genera, etc., during evolution.

"Biologists think it essential to avoid asserting anything vitalistic. The only way to do this is to deny any vestige of entailment in evolutionary processes at all. By doing so we turn evolution, and hence biology, into a collection of pure historical chronicles, like the tables of random numbers, or stock exchange quotations."[92]

The reality of evolution, of course, is no longer a point of debate. There is such clear and abundant evidence that all life is lineally connected that it can be accepted as an established fact. Life is lineage; we are all connected through our ancestors.

Looking back some million years ago our lineages merge with those of the great apes, further back, with the primates, the mammals, the reptiles etc. This vast, interconnected lineage took its time developing: a few hundred million years or so after the molten Earth cooled off for basic bacteria-like organisms to develop from simple chemicals; another billion years or so for the development of complex cells; another billion for multicellular plants and animals to emerge; just a few tens-of-millions more for all the current diversity of living systems to be established; and the last half billion or so for the emergence of creatures aware enough to wonder about how it all happened.

The chance of even one specific protein being formed out of free aminoacids is of the order 1 in 10^{300}, while the odds of proteins etc. coming together to randomly form a simple bacterium are on the order of order 1 in $10^{34,000,000}$. Events with such odds against them could never be expected to happen in our universe that is only 10^{17} seconds old.

Whatever events occurred during evolutionary history, it is clear that such odds were never encountered at any step on the road from bacteria to man—certainly not a series of such highly improbable events. Rather:

"One can assume that life arose through an enormous number of small steps, almost each of which, given the conditions of the time, had a very high probability of happening ... a multiple-step process that relies on one improbable event's following another is sure to abort sooner or later."[93]

This quote is from Vital Dust, the best book I have come across that conveys the Big Picture. It covers everything known about what actually happened, from molten earth to human culture.

The perspective is classically-based, however, so cannot deal with its own conclusion that each of the many steps along the way "had a very high probability of happening."

The inclusion of quantum probability in the conceptual armory allows us to approach this central question. Classical biology is as incapable of dealing with why life "had a very high probability of happening" as classical physics found itself incapable of dealing with the two open windows as bulletproof as steel in our execution illustration of the slit experiment.

Just in case the above makes a theologian smirk of satisfaction in this swing in favor of God!, we should note that the new physics is just as inimical to one of many a religion's favorite axioms: God is in Control and knows what is going to happen.

For the new science asserts that it is impossible for even God to know which slit an electron will choose to go through. He can know the probability to the nth decimal place, but He cannot, in principle, know which slit will be picked by the autonomous electron. While God might manipulate probabilities in history like a divine psychohistorian, God is not going to know exactly what humans are going to choose to do. We have creative freedom and generate our own probabilities.

So, if there is blame to be laid for the historical misery of humanity's history it is either that God failed to make the probabilities of success great enough or that humans made very unwise, highly unlikely choices. "God is in Control" of what happens is totally incompatible with the quantum view of the world and must be ejected along with the other classical concepts we discarded earlier.

Both science and religion aim to describe the 'truth' about the reality we jointly inhabit. So eventually, if both get better at the task, they are going to end up converging. Right now, however, they are often so far apart, the concepts so black-and-white, so utterly contradictory, that it makes sense, in all current cultures, to ask acquaintances, "Do you believe in science? Or religion?"

I hope, I assume, that in some culture to come, this question will be as silly as asking, "Do you believe in physics? Or chemistry?"

Making it as black and white as it gets: religion insists that evolution is determined by fiat; classical science says that it's all chance and accident. The new physics suggests they are both wrong: natural law determines the probability of things happening; the rest is up to time and things filling in the probabilities.

Back to bacteria who, to misquote, God must love dearly because there are so darn many of them to make us look like afterthoughts.

Internal Survival

We shall leave the origin of the One Basic Operating System of life, the triplet code method of protein syntheses, until the next chapter. Here we shall focus on the origin of the evolution of the programs.

What do programmers do when they write new programs? Really efficient ones reuse the same code over, tweaked for different purposes.

So, the origin of new programs in bacteria can be expected to involve mixing, matching, and a lot of duplicated code with slight differences.

Both processes are well-documented in bacteria. Proteins fall into distinct line- ages with ancestral connections. They also mix their DNA stored programs with other bacteria in a simple form of sex. So, we can envision new programs as start- ing with new combinations of subprograms already in use.

But new raw code is not sufficient, it has to pass certain internal criteria.

For a program to do well in the environment provided by the operating system it will have to follow simple rules. We can illustrate this with my recollections of programming with MS Basic on the Mac XL.

The correct grammar must be followed for a program to run:

10 GOTO 20 10 GOTO 20

The correct syntax must be followed for a program to run. The program

10 GOTO 20 10 GOTO GOTO

Will crash as the syntax is incorrect. There must be a start—easy, and there must be an end—tricky, as endless loops are all too easy:

10 GOTO 20

20 GOTO 10

There must be nothing that crashes the OS:

10 DIV 1 BY 1+1 is OK while 10 DIV 1 BY 1-1 in not OK and will crash a computer

It must be elegant and use code efficiently. This is a higher level of program- ming success. You do not recode a wheel each time you need one; you do it once, then you call it up as a subprogram. When I type on my Mac, something like the following program runs:

CALL keystroke detected

CALL letter typed

CALL send ASCII to screen RAM

WRITE pixel pattern to screen

CALL keystroke detected

When a subprogram gets called a lot by 'higher' programs, it becomes rela- tively unchangeable and fixed down the generations. Most of our housekeeping genes, for instance, haven't changed much in a billion generations since the mud days. Changing them would be like changing the ASCII code for 'e'—impossible.

Only at the very top levels is program experimentation allowable.

We have already noted the fundamental principle that, in quantum science, what things are and what things do is all matter in motion in quantum probability forms.

This holds for bacteria. The myriads of proteins in a typical bacterium are each contributing a QPF to the mix. The overall composite of these, when filled in, is what one could call a healthy bacteria.

Metabolism is not a static thing; rather, foodstuff, etc. flows through the composite bacterial QPF. Each metabolic step can be likened to pipes in a chemical factory leading from reservoirs to reactors to another pipe; the width of a particular pipe reflecting just how many of that particular QPF are in the composite QPF. The ATP reservoir is small, for example, but it has a huge inflow pipe and thousands of small output pipes.

This is basic metabolism. But there is another level of control adjusting the size of the pipes and thus regulating the chemical transformations. This involves the regulation of transcription of housekeeping genes. Then there is a level that regulates this.

At the top is a program running that we can call 'life is good.' This Program is running when bacteria have food and are growing and dividing.

There is one more level of programming control. If the food disappears, the bacterium flips its state and becomes a spore, tough and resistant and awaiting better days.

The bacterium has allowed itself to be programmed by its environment—the environment is the final programmer of a successful bacteria. Survival of the fittest can be rephrased as survival of programs capable of being programmed by the environment.

The best analogy to all this is my computer. It sits there running dozens of threads doing who-knows-what until I hit a key. This simple input causes a cascade of changes to the RAM, to the video memory, to the pixel patterns on the screen.

At the bottom is the basic code that actually runs the machine. This is almost impossible to write a program with. So higher languages, such as C++ and on up, are used. At the top are the main programs, such as the MS Word virtual environment I am writing in. This sits right on top (for most of the time, as discussed later in Sleep).

In one of these 'higher' languages, we can say our little bacteria is running a simple program right at the top of the hierarchy:

WHILE input **IS** good

RUN life is good

ELSE RUN batten down the hatches

When famine strikes, in a very short time the composite Quantum Probability Form that is the thriving bacterium that is being generated by the "life is good" program switches to a composite form that is the spore. The stuff automatically falls into the new probability gradients and a spore results. Just three lines of code are sufficient for the trick.

The environment and such a bacterium are in a relationship just like my Mac and me. To the programs running in the Mac, I stand in the position of User. The Mac housekeeping programs are all busily running, layer upon layer, busily shifting stuff around, until I hit a key, tap, and Notice Must Be Taken and the book progresses.

The bacterial programs are the same, happily humming along until the environment goes, tap, and Notice Must Be Taken, and the BECOME A SPORE program starts to run. This state continues until a, tap, from the environment, and the BECOME A NOT-SPORE program starts to run. The environment is in the role of User to the bacterial program.

This motif holds throughout genetics. When, during development, a cell differentiates into a liver cell, say, it is because it received a tap from an organ program, which, to the cell program, stands in the position of User. To the organ program, of course, the liver cell program is just another trusty subprogram it makes many calls to.

Syntax Checker

Modern bacteria are so sleek, their programs so optimized and elegant, that it is not implausible that the final perfection of the bacterial form involved an efficient way of weeding out programs before they ever had the chance to run on a real operating system.

This is like a virtual reality. In this virtual reality, the novel programs could be virtually run with virtual consequences. Only programs that do not commit a faux pas are released to the real world and to get run on a real operating system such as a ribosome.

Lots of novel programs and combinations of them with the old can be released into the virtual OS and run. Only those that are deemed "well-formed" are allowed to graduate, while the failures are ruthlessly recycled. Only well-formed programs get to be tested in the rough and tumble Darwinism of the external world.

Not Random!

Note the generation of new subprograms and combinations of them is not a random event. See how difficult it is to shake oneself free of classical concepts. No, there will be relatively few combinations that have a Quantum Probability Form for them to fall into. To start, these QPF would have been provided by Nature just as molecular and atomic orbitals are provided by a beneficent Nature.

Later, there will be higher programs running that generate the QPF for the mixing and matching. This is exactly what our highly sophisticated immune system does in us. The immune system is capable of creating antibodies to millions of molecules only found in the chemists' test tubes. It is capable of generating antibodies to all but the simplest of the trillions of different molecules found in nature.

The immune system generates trillions of different such programs, each carried by a lymphocyte generated in the bone marrow.

Before it gets to run in the real world of the bloodstream looking for its complement to clutch and destroy, it first gets to run in a virtual environment created by a program running in the thymus. Only well-formed programmed lymphocytes are allowed back into the bloodstream as the mature T-cells of front-page fame. (The B-cells get tested in a different VR generated by an abdominal region called, for bird-related historical reasons, the bursa.) This is all well-documented and I am sure there was a great SciAm article about the it all recently.[94]

The virtual reality generated by the thymus is very simple. In essence, it just tests for one highly significant thing. The sequence goes like this

RUN a lymphocyte's linear program in the virtual reality

COMPLEMENT the QPF generated

WITH the QPF generated by all housekeeping programs also running

IF true

RUN destroy and recycle lymphocyte

ELSE

RUN activate and release on patrol.

Failures of this virtual testing is thought to result in the release of a lymphocyte that is programmed to attack a part of the body; arthritis is suspected to be such a malfunction. The lymphocyte is stimulated when it comes across cartilage, complements with it, and starts dividing, making lots of copies of the cartilage eating program. The resultant horde does its thing to the joints to great discomfort.

The Tetraplex

An even better example of testing in a virtual reality is the process of recombination between four strands of DNA in the mixing and matching and chromosomal rearrangement. This is closer in sophistication to the bacteria but, unfortunately, so little is known about speciation events that recombination is generally thought to be random even though there are well-known hot spots, as well as places strictly left alone.

We now propose that programs are run, and quantum probability forms get generated and tested in a virtual reality generated by the tetraplex stage.

While regular DNA involves just two strands of DNA, the tetraplex involves four strands. Surrounding these four intertwined condensed strands are a shroud of RNA and protein.

This internal testing of programs before they are released for external testing as new species explains why we just don't see lots of malformed individuals around. The external testing is the tip of the iceberg; most of the work has already been done internally.

What are the implications for bacteria having developed a simple program that generates a virtual reality for program testing? Such a program would be just like Windows running on

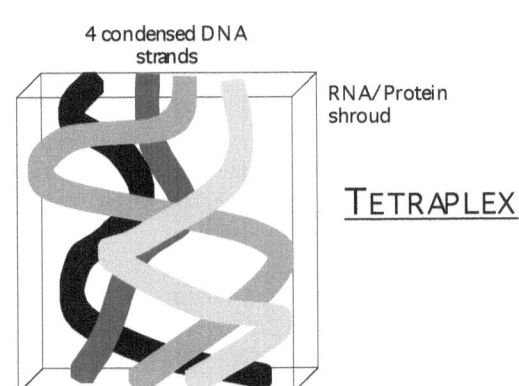

4 condensed DNA strands

RNA/Protein shroud

TETRAPLEX

my Mac: while Windows thinks it is running on a real Intel chip, it's actually running in a virtual reality generated by a program, VirtualPC, running on Mac OS X.

We are talking about a simple and primitive VR generation, of course, for bacteria were perfected billions of years ago and have changed little since.

Such a virtual reality as a pre-testing environment could be expected to be useful in the current day. Do bacteria just blithely accept any old DNA that is passed to them by a kind stranger? I bet they don't. First, the programs carried by an incoming DNA gets run in the virtual reality to see how it fits in with the home team. If incompatible, the DNA is fragmented so that it's program is destroyed.

Bacterial viruses run programs that subvert this testing and so get to take over the bacteria. We can expect that the way a bacterium treats an incoming DNA from a pal, and the way a viral DNA bullies its way in, should delineate just what the virtual reality program looks like.

Top down, bottom up

Biology is currently using the bottom-up approach to understanding how life works. This is like describing TV as "light stimulates a sensor array to transmit a series of electrical impulses down wires to an antenna where they are imposed on a radio wave which is picked up and causes phosphors to sparkle in patterns on a screen."

This is all very true—it's even more complicated—but it gives no insight into the Super Bowl phenomenon, the "State of the Union" or even "Lucy."

Classical science, knowing nothing of probability forms, has to take the bottom-up approach. This is good. Complementing this, however, we need the top-down approach, deconstructing the programs that are running in living systems. I choose to call this approach dePrograming.

Multiplication

Then comes the external struggle for existence in the real world. How well does the program do when it gets a chance to actually run.

This is the Darwinian decimation that has been studied and documented so well by classical science. So, I will say no more about it.

Science and religion should take a time-out to really digest these new concepts of quantum probability before returning to the fray.

So we have now a simple picture that seems not unlikely: clay beds hoisting an early metabolism controlled indirectly by clay supersystems and their interactions.

All of this will depend on clay macromolecules being on the scene, of course. In the scenario outlined so far, we depend on there being a variety of clays around each with a specific set of capacities to provide the wavefunctions for carbohydrate manipulation.

So far we have imagined that each clay macromolecule assembled out of its monomers under the direct control of a internal system and the indirect control natural law setting the rules—in essence, the same way atoms and molecules originate out of their subsystems and then control them.

The problem here is that even if one particularly useful clay molecule emerges on the scene, just one system is not going to make much impact. And we might have to wait a long time for natural law to assemble that particular clay again as we can expect the internal systems of clay to be multitudinous and for many, many varieties to be equally possible.

Template and complement

The solution to this problem is remarkably simple. Understanding this takes no new concepts, thankfully. We have already seen how one system (such as clay) can provide the wavefunction for other systems (such as carbohydrate metabolism). All we have to postulate is that clay discovered what biologists might call the "alternation of generations."

We have already encountered the significance of patterns on the surfaces of macromolecules and how they interact. Clay has pattern-making capacity par excellence—they have plus and minus charge, H-bonding capacity, etc.

In exactly the same fashion that a clay surface can provide the wavefunctions for carbohydrate transformation, this pattern can provide the wavefunction for assembling a clay molecule with an

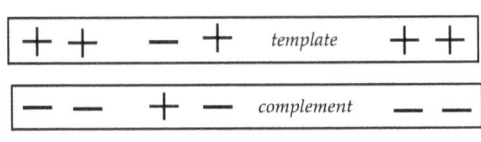

exactly complementary pattern on its surface. A biochemist would say that the first clay molecule acts as a template to produce its complement sequence.

Exponential growth

After the template and complement separate we have two possibilities:

(a)   the complement provides the wavefunction for a new template. We now have two template patterns on the scene—the template has multiplied.

(a)   the first template makes another complement—there are now two of them, complement multiplication.

The two of each can now repeat the cycle giving four of each, unlimited multiplication. And the power of exponential growth—which is what this doubling each cycle amounts to—is not to be underestimated. Any system which stumbles upon this simple method of multiplication via complements is clearly going to do well and become a major player. With just 300 such cycles of template-complement, for instance, a system could multiply to more that the number of particles in the known universe. Other factors of course—such as subsystem shortage—preclude such multiplication—but the possibilities are there.

Conservative multiplication

We can expect that, for any pattern-based system that has mastered the template-complement of multiplication, on Origin event is sufficient to make the system a major player. Once the very first of a qualitatively-different system emerges in an Origin event it can be rapidly multiplied by the template-complement process. The implication of this upward-compatibility process is that the template-complement process will be conserved up the hierarchy—its not going to change very much in history.

The providing of wavefunctions for running things exhibits enormous variation and "depth" of wavefunction hierarchies involved. The manipulative hierarchy involved in running reproduction—basically complex multiplication—remains remarkably unchanged all the way up. An example: The basic way we multiply human beings is exactly the same way single-cell plants and animals in pond water multiply: two haploid cells fuse, diploid cells multiply, diploid cell makes haploid cells, repeat each generation. The long-term world-line of the "human system" in history is just such a simple alternation of generations—the lineage diploid, haploid, diploid, haploid, etc. All the rest of the male-female dance is just a temporary housing that is built anew each generation.

As multiplication is so conservative we will only occasionally have to deal with any vertical movement up a hierarchy. The Origin of a new manipulative level of multiplication is, in almost all respects, the same as the Origins in the much more adventurous realm of manipulating stuff.

While I personally do not tend to the view that clay multiplication was that significant—I tend to think clay made buried lagoons of nucleotide-activated amino acids and nucleotides for the hell of it—this motif of multiplication via complementary patterns will appear over and over again.

Here we see a distinct quantitative difference between simple systems such a atoms which can only emerge by an origin process and pattern-based systems (such as clay perhaps) which can multiply via the complement process.

For atoms, the emergence of the first has little or no influence on the emergence of others. The rate at which a cooling plasma of electrons and protons forms hydrogen atoms is not influenced by the emergence of the first hydrogen—all of the atoms form under the direct control of internal systems provided by the indirect control of natural law.

For a system such as clay this is not true. While the emergence of the first is controlled by natural internal systems, the formation of more of the same can be controlled by the fixed internal systems, and the emergence of the first can have radical implications for the emergence of others.

Thus a clay molecule emerging originating in a rich supply of clay monomers could mop up the supply by multiplication—preventing any other clay molecule from originating. For the contingent stage of origins depends on there being subsystems around. Without them, the process stalls.

Fixation of a internal system through multiplication accomplishes two important things that are highly significant for the long-term survival of a system such as clay:

1. by fixing its internal system in its complement, the clay system frees itself from dependence on natural law to provide extra copies of itself Multiple copies do not need multiple origins. Just an Origin even suffices.

2. when multiplication replaces origins as the agent of change, the number of identical systems can increase exponentially. If one becomes two, then two become four, and four become eight, etc., then multiplication can rapidly dominate the scene. For instance, consider a clay molecule that can multiply once a day. While this might seem a slow rate of reproduction, this process could not go unchecked for even a year for, if each generation multiplied unchecked, the number of clay molecules by the end of the year would be more than the particles in the universe.

Horizontal providing

We can now outline what wavefunctions are involved in multiplication. We will consider a pattern-based system able to manipulate its subsystems for long term stability and also to multiply itself—the providing of wavefunctions for the manipulation of subsystems and the providing of wavefunctions in the template-complement "alternation of generations."

The template and complement is the kind of change we have labeled horizontal, there is no movement in a hierarchy, the template and complement are on the same level. Multiplication is a horizontal, back and fore, mutual providing of wavefunctions.

In terms of somatic activity—basically body-building—most organisms have a "sense" strand template that is actively translated into protein and an "anti-sense" complement that is usually not translated. There are good mathematical reasons for this; and there are also many interesting exceptions.

Both template and complement are both capable of providing wavefunctions for other systems, they are, after all, basically the same—they are on the same horizontal level of sophistication of structure. But the wavefunctions they provide can be expected to be quite, quite different. For example, an all-plus clay molecule will have quite different catalytic activity to its all-minus complement pattern. The all-plus clay will excel at providing paths for minus-charged molecules—like amino acids—while the all-minus clay would excel at manipulating positive charged molecules like nucleotide bases. One way in which an all plus template could separate from its all minus complement after assembly is by allowing activated nucleotides and amino acids to assemble between the two strands, pushing them apart. As noted, separation is an essential, if sometimes neglected, aspect of multiplication. Perhaps clay supersystems fed by black smokers discovered that making activated amino acids and nucleotides was a Darwinian asset to being a long-term player.

Here the putative two strands of the clay are playing similar roles. The patterns that are being multiplied are basically the same as the wavefunctions being provided for structural advantage. This is a static, non-living type of situation we can suggest for something like clay. not the case in living systems which, as we will shortly see, does not involve static providing of wavefunctions; rather, it involves a flow of wavefunctions—one of the steps being a translation of a coded wavefunction into a provided wavefunction.

This step can be likened to what goes on at the lowest levels of the computer I am writing this thesis upon—currently a woefully out-of-date Mac. There the CPU is translating machine code—long strings of binary ones and zeros—into instructions which are executed. There is a limited repertoire of machine instructions—the instruction set for that chip—that just get run over and over again—millions of times a second. A computer program is a long, linear sequence of ones and zeros. Microsoft Word on this hard drive, for instance, is a string of them seven million from end to end. A machine code is equivalent to a triplet code in life. The sequence of bits is translated into a machine code which is executed. This is equivalent to the triplet code being translated into an amino acid which, as part of a protein, will be "executed" as it provides wavefunctions for molecular manipulations. While machine code is "strict"—each instruction is a specific sequence of bits—the triplet code is "degenerate" in that many triplet codes, up to six, can be translated as the same amino acid.

For all that, the translation process in both computers and life is very unforgiving. Even changing one item in the instruction set can have major impact of what comes out the other end—crashes and sickle cell disease are both unwelcome.

At the start of a computer program we might find a simple instruction that is translated into an instruction set such as "run the loading program at location 111111." This instruction is a string of binary digits, or bits, say: . The inverse of this is obtained by flipping all the ones into zeros and all the zeros into ones; another, quite different sequence of bits: . This is so different that one thing is certain: it is not an instruction, when translated, to "run the loading program at location 111111." It could be the instruction to "add zero: repeat."

So you would surely stump any programmer with the request for a word processor program whose inverse was a picture editing program. We are asking a programmer to be so clever that when we invert the seven million bits of Microsoft Word we get the seven million string that, when clicked upon, is Adobe Photoshop. Impossible! Perhaps something much, much simpler, is possible—but if so, I have never come across it in my reading.

This is why, in general, only one of the template-complement pair—the sense strand—does the work of coding for amino acids. The anti-sense strand provides the wavefunction for multiplication but usually has no other role to play.

An inert anti-sense is not always the case however; some viruses have taken compacting to such extremes that they have accomplished dual coding—both template and complement are translated into functional proteins. While they are only relatively simple proteins, this is extraordinarily difficult to imagine. But this is equivalent to what those minimalist viruses have done: the template encodes, say, a DNA ligase while the complement encodes, a protein coat.

Even our own species has exceptions to the sense-translated, anti-sense-not translated. "[The m-proteins were translated] genes that were embedded within an intron of the [NF- protein] gene. ... These 'genes within genes' carry their coding information on the DNA strand that is the anti-sense strand of the NF gene."[95] This is not quite as clever as the virus—the intron gets discarded before translation of the NF gene occurs—it is still quite unusual.

But, in general, only the sense strand is usually translated in living systems

Finally we come to the actual process by which the template-complements assemble upon, and then separate from, each other. The actual density of this in history, as always, will reflect probability density of collapsed wavefunctions. These are provided by natural law—one of our quantum operators—and any systems around in the environment that are capable of providing wavefunctions. While a clay molecule cannot influence the contribution of natural law, it can alter the systems in the environment.

To my mind, the ultimate in clay based metabolism might be the appearance of peptides that could manipulate fat metabolism. This would be difficult for clay itself, being so full of charge, as it cannot provide wavefunction for hydrophobic molecules.

With fats came the possibility of compartmentalization and a new type of combinatorial exploration. Certainty clay supersystems that could array themselves in lipids might have an advantage of stability to local water flow—a water-repellent raincoat perhaps. This could keep out the rough environment and allow delicate patterns time to assemble their complements. A somewhat porous coat, of course, as getting totally cut off is not a good idea.

While I doubt that such sophistication was attained by clay, the general pattern applies to all living systems. Just as there is a hierarchy of manipulation of stuff, there is a hierarchy of manipulation in multiplication. The manipulation hierarchy of multiplication is, however, much simpler; there are only a few levels. There is conservation and upwards compatibility. Human multiplication involves just three basic levels of manipulative ability

1. Manipulation and integration of DNA multiplication: Separation of DNA double helix. Assembly of complement on each strand to form double helix. Result; two helixes. Alternation of template complement. Only one strand is (usually) involved in providing wavefunctions for the structural side of things; the other strand is not transcribed. This is what bacteria basically do.

2. Manipulation and integration of cell multiplication. Extended chromosomes are condensed, a process that halts the provision of wavefunctions by the central genetic system. The cytoplasm is now running on auto-pilot and will do so until the chromosomes are unpacked and un-condensed at the end of the whole process. The wavefunctions provided by the reawakened genetic system quickly bring things back under their beneficent control.   The condensed chromosomes are paired up on the mitotic spindle and duplicated. There is now a foursome. Two of them are pulled one way by contraction of the mitotic spindle, the other two are pulled the other way.   Each ends up with a chromosome pair, two daughter cells just like the parent cell.   Here we have an alternation between the paired and un-paired states. The condensed two-some and foursome states have the "complement" role here, they only have a brief time on the scene. Most of the history of chromosomes is in terms of the unpaired, opened-up state—the active, "template" role.

This is a sophisticated set-up unknown in bacteria and the difficulty in getting it up-and-running probably accounts for much of the billion years it took to get from bacteria to ameba. Once it was established it, just like the triplet code, took over the world: the mitotic spindles of animals, plants and fungi are essentially exactly the same.

In discussing the template-complement process we have neglected to take into account that, where wavefunctions are involved, we are always dealing with probabilities, probabilities that are rarely absolute certainties. In the process of multiplication via complements we have to take into account the probability of making mistakes in copying the pattern—alternating plus and minus forms—down the generations. While a template might arrange things so that the probability of its exact complement is very high, occasionally a low probability thing will happen—a sort of Murphy's Law of multiplication. This is an echo of the autonomy possessed by the subsystems. Remembering the inherent autonomy of systems to make choices, we should not be surprised if, occasionally, an unlikely pattern emerges that is not the perfect complement of the template.

Thus in our biogenetic clay Garden of Eden we can expect that the clay populations would be diverse in the extreme, but also related. As surface patterns can also be connect to the catalytic capacity of surfaces, we will see a corresponding variation in the metabolism thy indirectly control.

This can be compared with the champion multiplier, the DNA-based system that can make a complement with just one error in ten billion in general and much more in replicating crucial regions. Useful variation in higher organisms, useful in the sense of positive Darwinism, is rarely derived from mistakes in DNA. Such mistakes are rarely useful. Rather the variation so necessary to Darwinism is generated far up the hierarchy of genetic control in the process of recombination and chromosome manipulation.

A biologist would graph this variation as broad and wide for clay and narrow and sharp for a similar number of DNA generations of multiplication.

> "The particular type of organization that exists in the dynamic interplay of the molecular parts of an organism, which I have called a morphogenetic or a developmental field [a internal system], is always engaged in making and remaking itself in life cycles and exploring its potential for generating new wholes." [96]

This is a sort of micro-origin, a small part of the system goes through an origin process while the rest is participating in a multiplication process. This is why the variation in a population of systems arising by mutation is called micro-evolution, a sort of hybrid multiplication-origin process.

Once a clay internal system has been fixed, natural law disappears from the picture—the displaced internal system of the catalyst indirectly controlling metabolism has been fixed. This is a primitive example of what we can consider a proto-genetic system involving both displacement and fixation—the emergence of a clay "gene."

As Dr. Cairns-Smith concluded:

> "Clearly there are further observational and experimental clarification to be made of the big question: Do mineral crystal genes exist? At this point I can only answer 'Quite possibly' and go on to the next question: Could mineral crystal genes evolve? The answer to this, it seems to me, is 'Yes, they could hardly help it.'"[97].

The choice of clay is not that significant. Other suggestions are proteinoid, iron sulfide, lipids, etc. Whatever it was, it must have had the basic attributes we find attractive in clay:

Natural abundance—the contingent factor in origin history
Catalytic activity capable of performing basic metabolism
Plausible multiplication by the template-complement process.

We could extrapolate this perspective to the concept of clay supersystems controlling metabolic systems but it is difficult to see how clay supersystems could multiply by the template-complement process. Clay supersystems would be the original blob, just growing bigger and spreading their influence. Occasionally, some environmental upset might break a piece of which set up shop in some other location, but this is a pseudo-multiplication akin to crystals fragmenting—natural rules, not pattern rules are the ones in control.

Clearly multiplication does not increase the sophistication of the system—natural law is still just one step away—but it does enable the system to be a player, to be on the scene and explore the possibilities of positive Darwinism.

The ability of clay to multiply would have resulted in multiple copies of useful surfaces being on the scene—each with its entrained "useful" metabolism. Having multiple copies allows for rapid horizontal exploration of the possibilities of forming clay supersystems—the more systems doing the exploring the less time it takes for not-so-probable aggregations to check each other out.

A respectable mutation rate would provide for wide variation and for the clay systems to horizontally explore the possibilities of micro-origin.

Letting go

As always, the question of survival looms. There is what a biologist would call selection pressure, in an environment where many templates are competing for subsystems to make complements out of, any edge discovered by one will allow it to prosper. For positive Darwinism to occur we need a not-unlikely scenario for positive Darwinism to link keto-acid metabolic ability with the multiplication of clay. One plausible selection pressure on clay was what we might call the ability to let go.

Earlier on in the discussion we set the stage for our picture of multiplication by considering what happens when the complement forms on the template and then they separate. We neglected, at that point, to consider what would happen if the complement form but they did not separate. Clearly, not much. Separation is a key point in multiplication, for if the template and complement find each other's embrace so low-resistance that they never separate, the whole concept of multiplication stalls before it starts.

If clay complements tend to be sticky—the template and complement don't separate easily—this would be a bump-in-the-road for clay multiplication.

This is where keto-acids and simple sugars might have played a role. Much of clay pattern-bonding involves hydrogen bonds and, if sugars disrupt these bonds then they might facilitate separation of template and complement.

Such a multiplication-enhanced clay combo could multiply its components and their catalytic activity and monopolize the monomer resources provided by a beneficent natural metabolism. They would take over the clay bed eventually!

This is the most basic example of a living system imaginable: a clay supersystem with a metabolism supporting its multiplication—a example of a genetic-metabolic system that can multiply through time and space.

Another plausible suggestion is that the sugars helped the clay supersystems integrate, a glue to hold them together. And certainly polysaccharides are very sticky.

At this point we will just have to assume that there was a considerable advantage to keto-acid metabolism because, lacking a clay supersystem multiplication process, we are dependent on natural law to provide their internal systems, and this is not necessarily that likely. So we depend on the fact that once the capacity was established, it ensured the clay supersystems longevity for, on its continuance, depends the emergence of the amino acids and nucleotide bases. And we cannot realistically expect this clay process to be replaced until some sort of proteins and nucleic acids have appropriated the role of clay. The genetic takeover scenario championed by Cairns-Smith.

Sequential metabolism

Once we can envision a not-unlikely scenario involving clay supersystems—and a similar will hold for any suggested starting surface—multiplying and manipulating simple carbohydrates—we can suggest a similar patter for the origin of clay supersystems with the capacity introduce the nitrogen atom and manipulate amino acids. We can expect that the pace of clay metabolism to be relatively slow and not that specific.

Here the horizontal exploration of the clay supersystems is being played out as a metabolic construct creating carbohydrates and amino acids and peptides. Peptides formed from amino acids have a wide range of properties and could reasonable be expected to have occasional helpful roles to play in clay life.

Organic molecules have many properties that would be useful to the survival and propagation of clay systems.[98] In his section "Organic chemistry without enzymes,"[99] Cairns provides a provocative view of how clay systems could develop the ability to manipulate organic. It is also established that the catalytic activity of metal ions can be made very specific in a structured environment.[100]

A similar thing can be envision for nucleotide base metabolism. The great benefit conferred by this ability could be expected in the manipulation the energy of phosphate bonds—the centrality of ATP today being the "fossil" remains—and polynucleotides to perhaps find a useful role as storage repositories of nucleotides.

Having monomers on the scene opens up the contingent possibility of stringing monomers into polymers. This is just another catalytic activity we could expect in clay—the ability, at some late date, to create relatively primitive examples of proteins and RNA. With primitive proteins and nucleic acids as products of proto-metabolism we have the contingent requirements for life, as we know it to emerge.

Perhaps the best argument against clay as a proto-metabolism is that clay is not involved at all and has left no fossil remnants except perhaps for the ubiquity of metal ions in protein and nucleic acid interactions. For the two founding macromolecules of current life are the proteins and the nucleic acids, not clays.

In the clay proto-metabolic-genetic system we allowed that clays have both a catalytic capacity and a template-complement capacity. Clay does both. In our kind of life, on the other hand, we see a division of labor:

a. Proteins excel in catalytic activity, in providing surfaces to setting the rules for molecular and macro-molecule transformations. The changes wrought by proteins are extraordinarily diverse and seemingly unlimited. Proteins, on the other hand, have zero capacity to multiply by the template-complement process.

b. Nucleic acids excel at the template-complement process. At its extreme, honed by selection pressures, its accuracy is such that errors are kept to below one-in-ten billion (admittedly with the help f many proteins). Nucleic acids, on the other hand make miserable catalysts (though they do have a small capacity to manipulate other nucleic acids).

Whatever it was that marked the transition from proto-metabolism to primitive life must have involved these two macromolecules discovering how to make up for the other's deficiencies; to be able to do together what clay can do alone, control metabolism and multiply.

The fact that RNA does have some catalytic function has prompted some to speculate that the pre-life was RNA without protein (let alone clay), and that all the catalytic manipulations were being performed by the RNA. "... it is possible under different reaction conditions to entice this [RNA] to act either as an RNA polymerase, endonuclease, ligase, kinase, acid phosphatase, or phosphotransferase. Thus many processes related to reproduction of the genetic information in a prebiotic RNA world could have been catalyzed by self-splicing [RNA]."[101]

OPERATING SYSTEM
OF LIFE

We have now dealt, in broad terms, with the evolution of the 10,000 or so programs in a typical bacterium.

All of this, however, depends on the one basic operating system, the ribosome, to actually run the programs and allow them to generate quantum probability forms to contribute to the composite. The question now becomes, How did the operating systems originate? How did they evolve?

There are remarkably few operating systems for us to consider—for the triplet code, RNA system is just the first of the few we need to look at.

The evolution of operating systems is Macroevolution. For when a new operating system emerges on the scene, a whole new realm of possibilities opens up and exploration is exponential. This period of top-level experimentation ends, however, when they become subprograms and get called a lot by other programs. Change quickly becomes impossible, and that stage of evolutionary exploration is ended.

There are just four macro-evolutionary events. The chart also gives examples of when this is the 'top level' of programming sophistication. Otherwise, the level becomes a relatively invariable subprogram for a higher level.

Operating system	Linear program	Virtual reality
Basic OS	triplet code RNA	bacteria
Cell OS	spindle RNA	yeast
Organ OS	virish RNA	plant
Nervous OS	glial RNA	Fish mind
Emotional OS	basal RNA	reptile mind
Symbolic OS	bellum RNA	ape mind
Spirit OS	dual RNA	human mind

Virtual Reality

Each of these OS at the peak of innovation and exploration can be expected to develop programs that generate a virtual reality. A VR in which other programs can be run virtually; they can be tested internally before they are let loose in the external world. Much of the following is pure speculation, but I hope you enjoy it.

Again, this is like my Mac. The OS X runs many programs at the same time in threads that get a certain amount of CPU time. In the following screen grab, there is a program called "null" taking up a lot of time. This is the Mac OS 9.3, a virtual environment in which I can run my old programs. To OS 9, the virtual environment is a virtual CPU that it is running. If OS 9 gets crashed, OS X will burp, inform me, and blithely continue running the real CPU. On a real machine, such an OS 9 crash would necessitate a total restart.

Basic OS virtual reality—Virtual ribosomes on which programs can be run virtually. Foreign DNA gets tested first.

Cell OS virtual reality—Virtual DNA in the nucleus mixed and matched. Recombination patterns tested first here, then expressed on real chromosomes.

Process ID	Process Name	% CPU	# Threads	▼ Real Memory	Virtual Memory
311	Safari	0.40	6	54.96 MB	244.98 MB
175	WindowServer	3.40	2	53.79 MB	217.07 MB
585	(null)	91.10	14	39.79 MB	1.13 GB
587	Backup	0.00	4	20.24 MB	148.99 MB
281	MenuStrip	0.40	5	15.15 MB	158.53 MB
578	Activity Monitor	4.40	2	13.87 MB	165.23 MB
177	ATSServer	0.00	2	12.36 MB	87.39 MB
471	Stickies	0.00	1	12.36 MB	153.82 MB
588	BackupHelper	0.00	1	10.50 MB	36.73 MB
205	Finder	0.00	1	10.42 MB	157.11 MB
582	Grab	0.40	3	8.71 MB	161.20 MB
581	Windows Media Player	0.90	5	6.47 MB	152.18 MB
204	SystemUIServer	0.00	1	6.42 MB	148.99 MB
184	loginwindow	0.00	4	4.73 MB	123.52 MB
274	Mirror Agent	0.00	9	3.70 MB	145.43 MB
203	Dock	0.00	2	3.49 MB	138.98 MB
279	iCalAlarmScheduler	0.00	1	2.93 MB	133.51 MB
199	pbs	0.00	2	1.61 MB	44.41 MB
280	iTunes Helper	0.00	1	1.30 MB	124.53 MB

Organ OS virtual reality—Virtual chromosomes in the nucleus mixed and matched. Recombination patterns tested first here, then expressed on real chromosomes.

Nervous OS virtual reality—Virtual neuronal patterns tested first in the virtual reality generated by the glial cells. Well-formed programs are passed to real neurons.

Emotional OS virtual reality—Virtual emotional patterns tested first in the virtual reality generated by the basal ganglia. Well-formed programs are passed to the upper brain for real action.

Symbolic OS virtual reality—Virtual symbolic programs tested first in the virtual reality generated by the cerebellum before being passed to the front brain for real action.

Human mind OS virtual reality—Virtual symbolic programs run in a virtual reality. I can only suppose that this virtual reality is the one that I inhabit inside my head. Similar to the VR where, in your mind, dear reader, you are tossing around these words and ideas to see if they make any sense.

Basic OS 1.0

Now we will look at the various operating systems on which the RNA programs are running, starting with the simplest, and best characterized.

The most basic is the bacterial-organelle operating system. This is the basic triplet code method of protein synthesis that is being thoroughly explored. We will refer to it as the basic operating system or BOS.

Ever since BOS 1.0 was released, it is has remained virtually constant for billions of years (unlike the Mac operating systems I have followed from 1.0 to OS X.) The operating system in the microbe that turns milk into yogurt is exactly the same as the one running inside my mitochondria.

Next, a program and an operating system need a processor to run on. My PowerBook has one; my office G4 has two. A processor can only run one program at a time—my Mac OSX evades this bottleneck by running dozens of 'threads' but it can only deal with them one at a time so it spends milliseconds running each one in turn.

The bacterial equivalent of the processor is the ribosome; this is the BOS for the bacteria/organelle programs to run on. The number of processors in a single bacterium, however, runs to the hundreds of thousands.

As anyone who works with computers will attest, changing operating systems is a major hassle. Nothing old runs anymore; new versions have to be purchased. Living systems are fortunate not to be cursed with this "new, improved" burden.

At last, we have the first stage of quantum life science, the Basic OS.

BASIC OS

RNA linear program runs via protein

Generates Quantum Probability Form

Stuff falls into collapsed probability form

Cell OS 1.0

The operating system that runs cells—plants and animals—has elements similar to the bacterial level—there is a more sophisticated version of the ribosome, for instance.

Many programs, however, seem to be also running on the cytoskeleton network which is capable of remarkable expressions when properly programmed. The most dramatic example of this is the spindle of cell division that generates probability forms for the chromosomes to fall into. The units seem to be a dozen-or-so proteins, such as the actins, that pop together like legos.

Yet, another is the spliceosome system that snips all the non-triplet code out of DNA-to-RNA transcripts before being sent out of the nucleus. They are complexes of RNA and proteins, and an area of the nucleus, the nucleolus, seems to an element of the operating system here.

In the 1950s, the greatest advance in understanding the mechanisms of life was the elucidation of the triplet code in the DNA. This mapped the sequence of bases on the DNA to the aminoacid sequence in proteins. It only emerged much later that only a fraction of the information encoded in the DNA of complex organisms ever makes it out of the nucleus and gets translated by the ribosome into protein aminoacid sequences.

Molecular geneticists have found that, although the DNA is transcribed into mRNA, long lengths of the information are neatly and precisely excised from the mRNA before it is transported out of the nucleus. The DNA sequences that are excised are the "introns"—and they can be hundreds, even thousands, of bases long—while the remaining sequences are the "exons." It is only the exons that are spliced together by spliceosome complexes and transported out of the nucleus to direct the assembly of protein in the cytoplasm ribosomes. Having exons—roughly equivalent to protein domains—separated by introns has the advantage of being able to shuffle bits of proteins around and make new multi-functional proteins.

The mechanism separating the intron and exon material involves small complexes of RNA and a variety of proteins. They are called *small nuclear ribonucleoproteins* (snRNP).

"There are many different kinds of snRNP's, and functions have been assigned to only a few. ... They are the critical components of a sophisticated molecular assemblage called a spliceosome. As such, they take part in the splicing of mRNA ... a delicate operation that must be carried out with the utmost delicacy and precision.. Perhaps it is not surprising, then, that the snRNP's in spliceosomes specialize: each performs a different task during the splicing procedure. The picture of snRNP's working in concert in the spliceosome suggests nothing if not a well-oiled machine. ... One of the most intriguing aspects of spliceosome function is that the entire assemblage, rather than any individual component, seems to be responsible for the catalysis of the splicing reaction."[102]

"Sequence families similar to **Alu** are characteristic of mammals. They are not known to contribute to the survival of the organism. ... However, their presence does have important effects on mammalian evolution because interactions between **Alu** sequences at different sites may cause structural rearrangements of the chromosomes. The rapid evolution of chromosome structure in mammals may therefore be caused by the presence of **Alu**-like dispersed sequences. ... In the grasses, much of the DNA consists of short sequences, with copy numbers that may be greater than a million, arranged in tandem blocks, distributed over the chromosomes, but concentrated in certain regions. ... Highly repetitive DNA of this kind occurs throughout the animal and plant kingdom, but in varying amounts."[103]

While current science has decoded the triplet code—how the monotonous combinations of just four "letters" A, T, C, G codes for protein—it has yet to decode the patterns of this seemingly useless DNA.

From the perspective we have developed, we can expect that this DNA contains all sorts of "meaningful" codes

Genetic Programs

Life, it would seem, excels at the manipulation of quantum probabilities. What about a cell? The genetic system is running many, many programs that are generating a plethora of quantum probability forms that get filled in by all the stuff. A healthy cell is one where all the constituents are in a high-probability state. Disruption and disease moves things to an improbable state; healing occurs as the stuff falls back into the highly probable state.

One nice thing about this perspective is that it clears up a question I came upon in high school. The atoms in my body are all replaced in about a week or some shockingly-short period. What remains constant then? I wondered. Now, at least, I have a hint of answer—a quantum probability form that stays relatively constant.

Could development be a filling in of quantum probability forms as they are sequentially generated by an RNA-borne linear program? Interesting support for this is the quantum prediction of "fitting into a form" in two ways. While the great majority of people have their heart and lungs fit into the chest with the heart on the right, a few have everything reversed. But it all fits perfectly and such people have no problem. Very occasionally, a person has all the internal organs in the flipped position. Again, the fit is perfect. While such flexibility is built into the quantum view, it is alien to the classical perspective.

The closest classical analogy to a cell would be a computer running a factory producing computers programmed to run the computer factory. As this ascribes a Godly position to Bill Gates and Steve Jobs, and they don't need more elevation, we will pursue the analogy no further.

Selfish DNA

What is all the intron DNA used for if its information never gets to make protein? Most scientists have little to say about this strange surplus of DNA, although Richard Dawkins has been a little more inventive. He theorized that science has it all back to front: A body is actually only DNA's way of making more DNA and that much of the DNA had no function and in the triplet code was gibberish. This 'hanger-on' DNA Dawkins called 'selfish DNA.' Once selfish DNA had established itself, it just replicated itself down the generations along with the DNA doing the useful work[104].

In the genetic model used in the modern evolutionary theory, there is only one type of information in the DNA, and that is stored in exon DNA: aminoacid sequences written in the triplet code. We are proposing, however, that there is a great deal of information in DNA that has nothing to do with aminoacid sequence directly. The intron DNA provides a possible resting place for such information that is not translated by the triplet code, ribosomal mechanism.

"Because a direct function for this DNA is not readily apparent, it is often disregarded. However, a substantial portion of this excess DNA may specify genetic and structural partitions and may also provide essential recognition features that are important for orderly gene function. ...Indeed, excess DNA may be essential for the efficient 'compartmentalization' of genes at several hierarchical levels of organization."[105]

This concept that information is stored in the DNA in ways other than the Triplet Code is well supported by recent work on the effects that nucleic acids can have directly on other nucleic acid and their function.

One review of these developments stated that "Among the best studied ... are the self splicing introns ... [This self splicing intron] can act as either as a RNA polymerase, endonuclease, ligase, kinase, acid phosphatase, or phosphotransferase."[106]

A DNA sequence, it seems, has ways of controlling other DNA that does not involve it being transcribed into an aminoacid sequence. The authors of the same review think that this is good evidence for active pre-biotic RNA, a possibility discussed in Dr. Cairns' argument for the low-tech role of clay—see the Appendix.

It has already been shown that intron DNA has a specific (non-triplet code) vocabulary, "The presence of idiosyncratic words implies that the primary structure of introns is far from being random. We conclude that introns do carry some messages and, hence, should not be regarded as 'nonsense' DNA."[107]

Another possible mechanism has been raised by the recent work being done on the methylation of DNA. "Methylation of DNA is a ubiquitous phenomenon ... In eukaryotes, there are no established functions for DNA methylation, though recent evidence suggests that it may regulate gene expression.[108]

Another possibility is the actual structure of the nucleus itself and its control of the genetic information. For example, the nuclear matrix (the insoluble structural framework) has been shown to be involved in the splicing of introns and exons.[109]

Regulation mechanisms can be expected to be heavily involved in the process of evolution of the higher organisms. As one worker put it, "The most prominent evolutionary mechanisms in prokaryotes involve mutation and other genetic operations involving the sequence variability of DNA. The differences within wide taxonomic categories of metazoans are often regulatory ... The major adaptive radiations among these forms are likely to have been mediated by regulatory evolution."[110]

If it is correct to say that intron DNA has a host of regulatory functions, we can also conclude that the evolutionary development of the more sophisticated levels of regulation will involve intron DNA (regulation) rather than exon DNA (metabolism).

Whatever the details, however, the point has been made: There is plenty of room in the DNA for storing codes other than the storage of sequence information in the triplet code.

"It has been shown that the SMN protein is involved in spliceosome biogenesis and pre-mRNA splicing, there is increasing evidence indicating that SMN may also perform important functions in the nucleolus... These studies raise the possibility that SMN may serve a function in rRNA maturation/ribosome synthesis similar to its role in spliceosome biogenesis."[111]

Both operating systems and their codes are not well characterized and are currently under intense investigation. But it would clearly behoove geneticists to learn all about massively-parallel computer programming and do top-down studies to complement the well-established bottom-up approach currently doing so well.

The eukaryote ribosome and the protein-legos then are elements of the cell operating system. An active cell can have millions of eukaryote ribosomes in the cytoplasm (along with the myriads of prokaryote ribosomes in the mitochondria) and millions of the lego proteins. Massive processing running relatively few programs.

The ribosomes in our type of cell are similar to those in a bacteria, just a step up in size and sophistication. Much is still mysterious about how the cytoskeleton is organized. An organelle called the centriole seems to play a central role—it certainly does in cell division where its two aster-poles separate into the spindle which generates a quantum probability form that separates the chromosomes.

What could be programming the centriole and cytoskeleton? We need something that can carry a linear program to the centriole. It is messenger RNA (mRNA) that carries a program to the ribosomes.

Let's look at the usual suspects. Could it be RNA that is programming the centriole? I googled "centrioles and RNA" and right there at the top I found this:

"Evidence for a functional role of RNA in centrioles... We conclude first, that centrioles contain RNA which is required for initiation of aster formation, and second, that the centriole activity or ability to assemble a mitotic aster is separable from the basal body activity, or ability to serve directly as a template for microtubule growth."[112]

Sounds like a 'yes' to me. Spindle RNA (spinRNA) will not be carrying a program in the triplet code—the ASCII of the living world—but written in another code. This 'spinlet code' will be linear and, just like the triplet code and the ASCII, it will involve patterns.

Now it just so happens that our chromosomes—animals and plants alike—contain vast stretches of DNA patterns that are total nonsense in the triplet code. Like thousands of repeats of sequences of nucleotides such as **AAGGCCT** over and over again. This 95% or so of the genome has been called a variety of names, none polite, such as selfish and junk.

One reason for this disdain is that they apparently are not transcribed in RNA and shuttled out of the nucleus into the cytoplasm. I say apparently: There are hundreds of thousands of ribosomes and hundreds of thousands of tripRNAs pouring out of the nucleus to program them. But there is only one centriole. So, we might expect the proportion of spinRNA to mRNA to be about one in a million; probably below the resolution of current techniques.

Where are these few, non-triplet code patterns on the spinRNA coming from, I wonder. Perhaps sections of the junk DNA are transcribed into small amounts of spinRNA that conveys a program to run on the centriole.

As the spindle forms and separates in its majestic and stately fashion, we can expect that a maximum number of programs are being sent sequentially to the centriole/asters/spindle. So, this is where the maximum of junk transcription will occur if the suggestion is correct.

There is just one cell operation system in release—Cell OS 1.0—the ribosomes and mitotic spindles in my skin cells are identical to those in an oak's bark.

Just as a bacterium can jump from the active to the spore state, so do the much larger cells of our kind.

Unlike bacteria, cells have triplet code genes fragmented into exons, spaced apart by introns that are non-triplet code and are not translated into protein.

While all cells have such fragmentation, it appears to increase with overall sophistication. This also allows for a great deal of mixing and matching of the exon modules so that quite different proteins get made which generate quite different QPF.

A recent overview states it as:

> "In the {simplest cells} the splicing machinery can recognize only intronic sequences of fewer than 500 nucleotides, which works fine for yeast because it has very few introns, averaging just 270 nucleotides long. Bust as genomes expanded during evolution, their intronic stretches multiplied and grew, a cellular splicing machinery... [was] forced to switch... to a system that recognized short exons amid a sea of introns. The average human protein gene is 28,000 nucleotides long, with 8.8 exons separated by 7.8 introns. The exons are relatively short, usually about 120 nucleotides sequences [40 aminoacids when translated], whereas the introns can range from 100 to 100,000 nucleotides long.

> "The size and quantity of human introns—we have the highest number of introns per gene of any organism—raises an interesting issue. Introns are very expensive habits for us to maintain. A large fraction of the energy we consume every day is devoted to the maintenance and repair of introns in their DNA form, transcribing the pre-mRNA and removing the introns, and even to the breakdown down of the introns at the end of splicing reaction... By generating more

than one type of mRNA and, therefore, more than one protein per gene, alternative splicing certainly allows humans to manufacture more than 90,000 proteins without having to maintain 90,000 genes... Our genome already contains some 1.4 million ALU copies, and many of these ALU elements are continuing to multiply and insert themselves in new locations in the genome at a rate of about one new insertion per every 100 to 200 births. The ALUs were long considered nothing more than genomic garbage, but they began to get a little respect... Thus, ALU sequences have the potential to continue to greatly enrich the stock of meaningful genetic information available..."[113]

Sounds like encoded information is being passed down the generations.

Naturally, the classically-trained geneticists who wrote this consider the sprinkling of ALUs as random. We can expect, however, that the ALUs end up where they are because of a program-quantum probability form. Their positioning is specific and obeys a programming syntax.

This programming is clearly very important as, "Almost half the human genome is made up of transposable elements, ALUs being the most abundant."[114]

As we shall see, there is plenty of use that all this RNA can be involved in as layer upon layer of code is laid down.

Cell OS
RNA linear program runs
Generates Quantum Probability Form
Stuff falls into collapsed probability form **BASIC OS**

Organ OS 1.0

Next up the hierarchy of sophistication is the operation system of organs. How does our liver organize itself, how does a corn plant?

Let us assume that what we have seen so far still holds true on this level (nature is very conservative at the basic levels):

The organ is a consequence of a quantum probability form generated by programs running on a conserved operating system. The programs are linear and few; the processors multitudinous.

The obvious processors are the cells. Each of my organs has trillions of them in about 200 or so varieties. What could be carrying linear programs to the cell processors?

A clue is perhaps to be found in the AIDS epidemic. HIV is a strand of RNA wrapped in a protein coat. The RNA is not very long as such things go.

When this HIV RNA enters a T-cell (the T stands for thymus-trained, a B-cell is trained in the 'bursa' region) the HIV program does a remarkable thing: it flips the considerably more massive cell from its healthy state to that of being an HIV factory. One single strand of RNA can suborn a whole cell. All viruses, both the DNA and RNA variety, behave in this way. The calcium ion and the prions have a similar effect in their own provinces of action.

When HIV-RNA enters a T-cell, the first thing it does is get transcribed by a ribosome into an aminoacid chain, which folds, the assembly code runs, and generates a QPF with one special ability. It copies HIV-RNA into the human DNA that prup with a ribosome and get transcribed into

I have always thought it stupid of human cells to be so vulnerable—just one rogue strand of RNA, added to the millions already in there, causes a massive jump in state to occur. Why so defenseless?

Jumping to the conclusion: perhaps cells pass RNA to each other, this orgRNA carrying a linear program that runs in the cell. (I discount DNA, as all signs are that it is religiously segregated from the cytoplasm except during division.)

This is what the rogue HIV does, after all. Rather suggestively, our chromosomes are riddled with tens of thousands of genes for reverse transcriptase in various states of disrepair. And this is the very enzyme that allows the HIV to suborn a cell. All very indicative, you must admit.

This was second on the list when I googled "transfer of RNA between cells:" Evidence for transfer of macromolecular RNA between mammalian cells in culture. [115]

Unfortunately, that's all I could pull up, but it suggests that RNA might just be involved, yet again. (How many RNA codes are there? Clearly, we are still living in an RNA world!)

A cell in a developing organism receives a program on an orgRNA, obediently runs the received program, and generates a new QPF. The filling in of this new QPF is called differentiation. The sender of the RNA being in the position of User to the cell program. Just like the roles of bacterium and environment in sporulation as earlier discussed.

Just how does HIV suborn a cell? Clearly more than a simple protein or revamping of the cytoskeleton is involved. A new Master Program is being run on the cell OS and the infected cell turns into a virus factory.

We can expect that the code is not the triplet code, and is not the spinlet code: it can be called the organ-let code. One of the instructions will be to pass on some orgRNA to the cells around it. This can be modified to carry a result along with it. For in programming, counters are very important. Regular programming abounds in statements in the linear progression such as:

```
FOR N = 1 TO 10
        RUN program
        NEXT N
```

A changing count could easily be kept on RNA as it passes from cell to cell by something as simple as the number of a repeating sequence. The telomeres seem to keep a similar count of cell division.

This following might represent a simple counter:

```
FOR T = 5 TO 1
RUN NEXT T
```

Turning Worms

A similar thing apparently happened during evolution and we are living with the consequences. For all animals, the first stage of development is the formation of a hollow ball of cells with a hole in it. This first hole is the mouth; a second hole then forms as the anus. This pattern holds for a wide variety of "primitive" animals, including the hugely successful simple worms and complex insects.

In the lineage that leads to fish, frogs, dinosaurs, elephants and us, however, there is a sudden flip in the developmental process: the first-hole mouth-to-be flips to being the anus; the second-hole anus flips to become the mouth end.

While this is hard to reconcile with classical concepts, it is perhaps an example of there-are-always-two-ways-to-fit into a probability form. Something about this flip of the mouth to the second, and perhaps, more sophisticated hole opened up a world of possibilities including our big brains.

Development

In fact, it is clear that RNA mediated, HIV-like reverse-transcriptase transformation of the DNA has been important in the history of our evolution:

> "The commonest of all the [retrotranspons, long considered a genetic parasite] is a sequence of 'letters' known as LINE-1. This is a 'paragraph' of DNA between a thousand and six thousand 'letters' long, that includes a complete recipe for reverse transcriptase near the middle. LINE-1s are not only very common—there may be 100,000 copies of them in each copy of your genome—but they are also gregarious, so that the paragraph may n=be repeated several times in succession on the chromosome [a la Hox]. They account for a staggering 14.6% of the entire genome, that is, they are nearly five times as common as 'proper' [triplet code] genes The implications of this are terrifying. LINE-1s have their own return ticket. A single LINE-1 can get itself transcribed, make its own reverse transcriptase, use that reverse transcriptase to make a DNA copy of itself and insert that copy anywhere among the genes…

> "If LINE-1s are about, they too can be parasitized [it is supposed] by sequences that drop the reverse transcriptase gene and use the one in the LINE-1s. Even commoner are shorter 'paragraphs' called ALU. Each **ALU** contains between 180 and 280 'letters', and seems to be especially good at using other people's reverse transcriptase to get itself duplicated. The **ALU** text may be repeated a million times in the human genome—amounting to perhaps 10% of the 'book.'

> "For reasons that are not entirely clear, the typical **ALU** sequence bears a close resemblance to a real gene, the gene for a part of the… ribosome. This gene, usually, has an internal promoter, meaning that the message 'READ ME' is written in a sequence in the middle of the gene…

> "The genome is littered, one might almost say clogged, with the equivalent of computer viruses… Approximately thirty-five percent of the human DNA consists of various forms of [viral-like] DNA, which means that replicating our genes takes thirty-five percent more energy than it need. Our genome [and this we have to disagree with] badly needs worming.

> "There are sequences even shorter than **ALU** that also accumulate in vast, repetitive stutters… The 'word' can vary with the location [on the chromosome] and the individual, but it usually contains sentences of the same central 'letters': **GGGCAGGAXG**… The significance of this sequence is that it is very similar

to one that is used by bacteria to initiate the swapping of genes with other bacteria of the same species, and it seems to be involved in the encouragement of gene swapping between chromosomes in us as well…

"It turns out that the repeat number is so variable that everyone has a unique genetic fingerprint; a string of black marks looking just like a bar code."[116]

Sounds like info is being passed down a lineage to me.

Processors are the cells. Each is running a program depending on the RNA received from the neighbors. Running this program generates a QPF. Each cell in the organ contributes this to the composite whole, cell-level QPF.

Just like Mandelbrot, just like bacteria, this internal composite probability amplitude, when collapsed, has a quantum probability form to it that gets filled in by stuff. When filled in, this composite QPF is what we call a healthy cell. While the OS remains constant, the program running can change as easily as a ribosome can start translating another mRNA when finished with the first.

In a healthy organ, while each cell constantly receives a program to run, it is the same program that arrives. It keeps doing what it is doing.

When something like healing is called for, however, a new message arrives and the cell switches to an aggressive division mode appropriate for healing.

The arrival of a different RNA linear program can change many things at the analog end results: cell adhesion, cell division, cell differentiation, cell death, just a few dozen at most.

As the stuff fills in the quantum probability gradients being generated by running RNA programs, we observe the healthy functioning of a mature liver, or the rapid healing of a damaged lobe as the QPF change.

Cells falling into probability forms generated by programs running in the cell processor. A liver has a dozen or so types of cells—different processors—and trillions of copies of each. Truly massive parallelism.

There is also patterning on the DNA passed down a lineage. An example is methylation which adds lots of oily spots all over the DNA.

Organ OS	
RNA linear program runs	
Generates Quantum Probability Form	
Stuff falls into collapsed probability form <u>**Cell OS**</u>	
BASIC OS	

This is known, for instance, to signal if it was Mom or Dad who contributed that chromosome. Most methylation lies within transposons such as **ALU** and LINE-1.
117

"…the first [Hox] genes defined the head end of the fly and the last [Hox] genes made the rear end of the fly They were all laid out on order along the chromosome—without exception.

In mice, there it was again: almost the same 180-letter string—the homeobox. Not only that, the mouse turned out to have clusters of Hox genes (four of them, rather than one [in the fly]) and, in the same way as the fruit fly, the genes in the clusters were laid out end-to-end with the head genes first and the tail genes last... What was doubly strange was that the mouse genes were recognizably the same genes as the fruitfly genes... By having four [Hox cluster}, we and the mice have rather more subtle control over the development of our bodies than flies do with just one Hox cluster."[118]

EDEN, WOMBS AND BEDS

We have explored the internal aspect of quantum probability so far. But, as mentioned earlier, every quantum calculation ends with the collapse of the wave-function and the filling in of probability forms.

This is the contingent nature of history so well-explicated by the late Steven J. Gould. In brief, before stuff can fall into a quantum probability form, the stuff has to be already on the scene. In order for a quantum probability form to be filled, all the ingredients have to be present.

Our earlier example was of a protein embracing a calcium and jumping to fill a quite different QPF. In the absence of calcium, that QPF is empty, it is not filled in, and can play no role in the external world of interaction and the sharing-exchange of bits of self, the stuff filling in one's QPF. (Apparently, good car drivers extend their sense of self to embrace their car, sometimes even emotionally.

The great leaps in evolution each involve the emergence of a new, more sophisticated operating system. These leaps are few in number and took the longest times.

We can expect that the twin pinnacle of programming possibilities at each level of language involve:

First, the emergence of a VR in which programs can be virtually tested. This is the explosion of innovation each level went through as witnessed in the historical record.

Second, the elaboration of all the ingredients for the next level of programming language and a more sophisticated operating system for it to run on. When all the ingredients are together, they can jump to the new configuration of the QPF provided by Nature. We can call the last ingredient to appear on the scene, thus setting the scene for the quantum jump, the calcium factor. The jump is the same, just on a different scale.

We can take it that a small bacteria is about the maximum size over which natural quantum jumping occurs for it is rare for active components in living systems to be larger than this size.

Clay

Clay molecules are excellent catalysts, almost as versatile as platinum at providing paths of least resistance for other systems. The energy—the external enabler—is provided by nature in the form of the UV-driven iron cycle (which only shuts down with the advent of significant oxygen production by photosynthesis) and the smokers pumping out high-energy sulfides etc. (which they are still doing, and powering a bizarre ecology).

The clay provides the wavefunction down which these high-energy systems interact with each other. This is a vertical provision in that systems on different levels can be manipulated by clay: atoms, phosphate, carbohydrates, aminoacids, etc.

The clay provides a path of least resistance for reactants to turn into products. In classical terms, the catalyst is said to "stabilize" the intermediate by lowering its free energy. Now while this might seem like a rather complex way of looking at catalysis compared to the classical view of a lock-and-key where the reactants "fit" onto a surface and get stabilized, we shall see that this way has far more explanatory power when we get to more sophisticated levels.

Even today, all metabolic transformations involve catalysis by surfaces. Nowadays all these surfaces are provided by the endlessly versatile proteins, but biogenesis must have involved much simpler surfaces as proteins, themselves, can only realistically be created by a metabolism more capable that just nature-in-the-raw.

The systems that we know emerged with catalytic ability in the protometabolic era are many and various. Examples are the iron sulfides—can energize molecules from inorganic sources—and the clay molecules that are versatile in providing QPF for reactants to transform into product.

The black smokers under the oceans are prodigious providers of activated iron sulfides; and great beds of clay are a historical relic of this ancient period of time.

One reviewer of the current understanding of the origin of life concluded, "The most reasonable interpretation is that life did not start with RNA [DNA is not even under consideration as it is even more sophisticated than RNA]. The RNA world came into existence after many of the problems associated with prebiotic synthesis and template-directed replication of RNA had been solved. This implies that there was a simpler genetic system, or systems, that preceded RNA and that the evolutionary advances made by the ancestral system were somehow carried over to the RNA world."[119]

One of the most compelling suggestions as to what systems were involved in pre-life manipulation of molecules is the thesis, developed by Dr. Cairns-Smith, that primitive life first emerged in clay and clay structures.[120]

Dr. Cairns-Smith makes a good case in demonstrating that it is much more plausible to assume that simple systems, what he calls 'low tech,' emerged first and provided a foundation upon which more sophisticated 'high tech' systems could develop. He proposes that a 'low–tech' manipulation of molecules developed before the 'high–tech,' remarkably-complex manipulation of molecules by the gene/protein triplet code system.

This would be a forgotten sub-basement in the skyscraper of genetics.

While there is evidence that triplet code-based foundations of all life on Earth was established by 3.5 BYP. There is also evidence that life was making an impact on the earth even before this at almost 4 BYP[121]. The problem with both protein and nucleic acid polymers as ingredients of the earliest life has been called the

'Uroboros Puzzle' (the mythical serpent with its tail in its mouth): To make proteins, nucleic acids are required: to make nucleic acids, proteins are required—a chicken-and-egg type of can't-have-one-without-the-other conundrum. "This is the essence of the Uroboros problem."[122]

Dr. Cairns-Smith makes a compelling case for clay being the low-tech, pre-life provider of wavefunctions—he doesn't use this terminology, of course. He even goes so far as to imagine quite complex structures of multiplication and natural selection. On the other hand, clay might just as well have indulged in organic chemistry just for the hell of it, and living systems gradually took over the basic manipulation of molecules. The actual history was probably a mix of these two extremes.

This wavefunction provided by, say, clay—the classical catalytic "surface"—is a subset of the wavefunction of the clay system wavefunction. Earlier, we went to great pains to show that empty wavefunctions are as objectively-real as filled wavefunctions. Much of the catalytic activity of platinum, for instance, can be ascribed to the many empty orbitals just beneath its surface that provide a temporary home—a path of least resistance—for coupling electrons that are otherwise unable to get over a bump in the road.

Womb Eden 1

It took the bacteria and their colonies about a billion years to generate all the bits needed to fill in the QPF of the prokaryote type of cells.

When all the ingredients were in the right place, they all quickly popped into the now-highly probable configuration of the cell operating system.

The place was probably in a stromatolite and the eukaryote cell was rapidly perfected from this earlier prototype.

Stromatolites

The fossil evidence for the establishment of sophisticated bacterial colonies is striking in the billion-year old stromatolites. Similar to ones extant in a few exotic locations today, these are stratified layers of many different single-cell organisms. The strata reflect the history of their sequential evolution. These stromatolites are, "A dome-shaped structure about a foot high ... made of hundreds of wafer-thin layers of rock [whose] counterparts exist today ... in shallow water in restricted locations, such as the coast of Australia."[123] These are formed today by a primitive type of algae and the oldest of these structures found so far seem to be about 3,500 million years old.

The topmost level is the photo-synthesizers. The ability to use light as an energy source—forever liberating life from proto-metabolism—seems to have been discovered early on. It is apparently the result of a duplication of capacities.

First to be established was the ability of chlorophyll molecules—complex but not too difficult to make—to absorb light energy and raise hydrogen to an energy level where they can be used to create the two basic coenzymes needed for biosynthesis: reducing power in the form of NADPH and energy in the form of ATP.

While the formation of ATP by light energy is cyclical—the H-ion is returned back to the start—the formation of NADPH is not since the H is lost from the system and has to be replaced from somewhere. Unfortunately, there are few ready sources of H at the energy level required for this process to work. As shall see, this problem was quickly solved.

Chlorophyll works because the wavefunction inhabited by an unexcited electron is spatially very different from that inhabited by the excited electron. The excited electron is snatched up by a bucket brigade of small molecules and its energy can be used to drive NADPH and ATP synthesis.

With these two, the fixation of carbon dioxide becomes possible, an endless supply of carbon opened up. The basic reaction of this first step in photosynthesis is to create glucose out of carbon dioxide by reducing it with the NADPH driven by the energy released from ATP.

The problem of finding a source of the H atoms was solved in a simple manner by duplicating the photosynthetic apparatus. This second program-in-action specialized in taking the H atoms from water—at a low energy—and raising them up to the level required by the original system.

Only the hydrogen depleting synthesis of NADPH necessitates the release of oxygen. The formation of ATP, on the other hand, recycles its electrons in an endless loop.

Such photo-synthesizers make up the top level of the stromatolites and, as we shall see, prokaryotes capable of this double photosynthesis were the ancestors of plant chloroplasts.

Tapping into the endless source of water as a hydrogen source had a byproduct. The oxygen left behind is a waste product and escapes to the atmosphere. While much of this was absorbed by the inorganic world, eventually this byproduct appeared in appreciable quantities in the environment.

With a significant partial pressure of oxygen—and we can expect that this was higher within photosynthetic mats—a new possibility opened up. Duplicate the electron bucket brigade and disconnect one of them from the chlorophyll. Now run the electron cascade in reverse—combine glucose with oxygen to create ATP and NADPH.

Thus in a stromatolite, a layer of organisms could develop below the photosynthesizers—shaded from the light anyway—that lived off the droppings from the layer above, using their glucose and oxygen to make NADPH and ATP. This is the establishment of a simple food chain—the second layer feeding upon the upper layer. Such respiring prokaryotes were also, as we shall see, the ancestors of the mitochondria incorporated later into eukaryote cells.

Womb Eden 2

Organs were rapidly perfected and their forms are common to animals, fungi and plants.

It took a further billion years to generate the ingredients needed to fill in the empty QPF for the organ OS. This took place in the ocean somewhere, no doubt. Little is known about this step.

Womb Eden 3

It took a further billion years to generate the ingredients need to fill in the empty QPF for the animal OS. Again, this was in the ocean somewhere. Little is also known.

The result was the Cambrian explosion of experimental programming of the animal OS. As the programs developed in sophistication—filling in QPF that were previously empty in nature—experimentation was rife. Some were perfected, some fell by the wayside.

The evolution of the animal body plan along the branch leading to us involved fish-, amphibian-, reptile- and mammal-forms. The organ OS, as far as I can tell, has not changed. Just the programs getting more sophisticated.

It is in the nervous system where all the interesting stuff is happening, where new Operating Systems are emerging at each level.

The evolutionary pattern is the same as already established: When all the ingredients are present—the last 'calcium' system arrives in the mix, they all jump to fill-in an empty QPF provided by nature. This QPF, as always, will obey the generalized Schrödinger equation.

The emergence of programs to run on this new OS are few at first. The most obvious source of 'seed' RNA programs is the previous level. They will run poorly on the new OS at first and their limitations will be weeded out externally.

They have little competition, however, for they are the only inhabitants of this new level of sophistication and are in a womb-like eden where everything is provided. The multicellular life that flourished just before the Cambrian explosion, but died out, was probably an almost-ran but with something critical missing in its mix. A line of thought that did not lead to anywhere interesting, so to speak.

The evolutionary pace speeds up with the emergence of programs that create a virtual reality in which programs can be tested. This is similar to the emergence of the thymus as an organ that can virtually-test lymphocytes.

Internal and external evolution now work together and rapid progress can be made in perfecting programs to run as new types of animals in the external world. Shortly, we will discuss the Adam and Eve scenario, then assume is applies on every level where sex is involved, if in less sophisticated versions..

This burst of success rapidly perfects what is possible, given the language and its inherent limitations. In the computer realm, this is like CPM and Mac OSX. The levels of distinct mental operating systems are (at least):

The Survival or Basic Brain OS: We have inherited a basic, or fish-like sub-brain that runs all the lowest-level subprograms in a very simple RNA code-language. This involves the brain stem, the spinal column and the 'stomach' brain (the diffuse, but complex, abdominal ganglia that run our tummies). From top-down, this sub-brain is a collection of subroutines and subprograms that are called upon by higher levels and languages.

The Snake Brain OS: We have inherited a basic amphibian-reptile brain that wraps around our brain stem. Primary-color emotions—such as sex—are associated with this level of OS sophistication. The language is simple and, as it is a part of us, should be vaguely familiar. Rage and fear are others in this realm. Naturally, I am speaking of the actual experience of such emotions: Actually feeling, or worse, expressing a red-eyed, stab-and-rend, murderous Kali ferocity on a victim. Or feeling-expressing the bowel-releasing, prey-horror nausea of the living, human sacrifice.[124] There are nice things down there as well, too many to enumerate here. Talking about rage or fear, or thinking about the concepts as names, is of course, a function of the higher, human OS.

The Family Brain OS: We have inherited a basic mammal brain that is wrapped around the lower brains. Basic family-oriented emotions and concepts—without names—reside here.

The Tribal Brain OS: We have inherited a basic ape-hominid brain that is wrapped around the lower brains. Basic social skills involving many individuals. Names and actions, nouns and verbs, reside here as idea-of-sounds that are manipulated as a pidgin, both within the hominid VR-mind and without as sounds and gestures.

Womb Eden 4

Our brain, when fully functioning after about 18 years of development, generates a VR within which we, as the Main Program, run. We are capable of an internal language by which we can manipulate concepts into an infinity of possibilities. Some of them we actually do, with all the bother that physical work entails. The programs that are running in the VR, of course, think they are running on RNA in a real OS. But they are not. They are not on real RNA at all. They are divorced of any material.

When we are thinking hard, this implies, all the action is occurring in the non-material world of the VR. It is divorced from matter entirely. The neuron-firing patterns that the physiologists pick up with their scanners are just the program running that generates the VR in which we live.

The VR we live in day-to-day is generated, of course, by the physical brain. In Volume Two we will explore the possibility of the same programs running in a similar, if vastly larger, VR that is not generated by the physical brain.

It took until just 100,000 years BP to generate all the ingredients—which was the calcium factor, I wonder—needed to fill-in the empty QPF for the human mind and body.

All those different ingredients for the human OS were to be found in the different hominid races. Swimmers, climbers, upright-walking, long gestation, hairlessness, etc.—there were many ingredients that came together to jump as one to fill-in the so-far unoccupied human QPF. Simply put, miscegenation was essential to the human emergence.

When all these ingredients came together in the germ cell tetraplex—as we will shortly encounter—they jumped to the empty human QPF, then divided into male and female subforms.

The human capacity emerged from the primate recapitulation at about three when they started to create a true language out of the pidgin of their hominid forbears.

Quantum Adam & Eve

Where does a new body plan emerge? In the formation of the sperm, a tetraplex is formed.

The tetraplex jumps to a new configuration. Let us say it is the human. This is like the old Greek idea of man and woman being united at first, and then came separation.

Just so. This new tetraplex, filling in a previously unoccupied natural QPF, is the union of the male and female programs running in a virtual environment created by the recobinosomes (RNA again, no doubt.) What took so long was developing all the right ingredients in the various hominid races and then having them mix—human origins involved miscegenation.

We now imagine a multi-racial population of hominids. Through variation, these races explore the possibilities of the hominid system given the prevailing conditions. We propose during the differentiation of the hominid races there emerge morphemes that, while still being hominoid, we would recognize as one (or more) of the specialized morphemes making up a human. Such races are exploring the horizontal possibilities of morphemes such as upright posture, hairlessness, lowered voice box, enlarged brain, etc.

In the populations of the hominids, our direct pre-human ancestors, many different races will emerge by program development and innovation in the internal VR, then released for further testing in the real, external world.

Depending on the Darwinian environment, these variants will be selected for. Races will explore the possibilities of their inheritance. Some will explore hairlessness, some will explore the benefits of upright posture, some the benefits of larger brains and sophisticated vocalizations, others the benefits of opposable thumbs, etc. These can be developed in isolation but convergence is the key to the next step.

We can imagine a race that is at the center, overlapping many other races, in which all these components—a few dozen key ones suffice—mix together. We see hominids with all the human body characteristics but expressed in the hominid characteristic form.

Each of these racial adaptations, as they are called in classical science, can be expected to confer some advantages in an environment where there were many empty ecological niches to inhabit. In this sense, the environment is supportive of variation and exploration.

Speciation

In the tetraplex of this endowed hominid there of four copies of the hominid somatic chromosomes, c, two copies of the hominid female master program, x, and two copies of the tweak programs that flip some things the other way and result in a male, the y.

For, it is a well-established fact that the female is the default human form. The X chromosomes do all the work of providing certain high-level images for body development and functioning. The much smaller Y chromosome does little except tweak the impact of the hard-working X chromosome. The results of this tweaking are the difference between male and female, both primary and secondary. There is nothing that a man has that is not an exaggeration or reduction in something the female has.

From this tetraplex, he makes four hominid sperm, each haploid, and two of each 'sex.'

With all the ingredients present—the calcium has arrived, so to speak—there is a probability that two somatic and two of each sex will jump to a new configuration, the configuration that is the man-and-woman human composite.

This now separates into two, a haploid set of human chromosomes, C, and the human master program, X; and a haploid set of human chromosomes, C, and the human tweak program, Y.

As far as I am aware, the following proposal is quite novel. The two human sets are marked with a special "speciation event" imprinting. They are condensed and packed with a "do not open yet" pattern, of heavy methylation perhaps. A massive Barre body.

The tetraplex now forms two diploid sperm, each carrying a human program in a special locker along with a normal haploid set of hominid chromosomes. This is where, science tells theology, that God's creative input ar-

ENDOWED MALE HOMINID			FEMALE HOMINID	
tetraplex	sperm	speciation ✳	tetraplex	egg
hx	hx	H̶X̶	hx	hx
hx	hx	hx/hy	hx	
hy	hy	H̶Y̶	hy	3
hy	hy	hy/hx	hy	nurse cells

rives in the form of a previously empty QPF.

Speciation 1/2

The males and females of Generation 1/2 have the same father but can have different mothers.

The speciation-zygote created by their union does not unpack the specially-marked package. They are passed into the highly segregated germ cells and are probably deleted in the non-germ cell lineage, or kept as a massive, inert Barr body like the extra X in women.

> "The amount of phenotypic difference between two population systems is less significant than is the presence of reproductive isolation. In fact, pairs and sets of morphologically very similar but reproductively isolated species are known in many genera of insects, flowering plants, protozoans, and other groups. Such morphologically similar species are known as sibling species. For example, the malaria mosquito of Europe ... turned out on finer analysis to be a complex of six sibling species .... These sibling species, though reproductively isolated, are virtually indistinguishable ...."[125]

Both male and female are normal hominids with a deep secret.

Instead of a tetraplex being created in gamete formation, there is a focus on speciation. The program comes out of the virtual world and gets to be tested in the real world. In the last act of the speciation program, the hominid aspect is deleted and only the new program-hierarchy is sent out into the world.

In the male, hominid sperm carrying a now unpacked human program are generated. This is a generation 1/2 sperm, a hominid sperm with the human program as on a CD and ready to run.

In the female, the hominid program runs the nurse cells, now only two of them, and the hominid egg is endowed with the just-released human program.

MALE H, GEN. 1/2			FEM HOM, GEN. 1/2	
triplex	sperm	speciation *	triplex	egg
~~HY~~	~~HY~~		~~HX~~	HX
hx			hx	2 nurse cells
hy			hx	

MALE H, GEN. 1/2			FEM HOM, GEN. 1/2	
triplex	sperm	speciation *	triplex	egg
~~HX~~	HX		~~HY~~	HY
hx			hx	2 nurse cells
hy			hx	

Depending on what "Easter Egg" was passed down the germ cells, the possible combinations are listed in the table. The only real oddity predicted in this scheme of speciation is a hominid egg with a Y chromosome.

Ancestors

When these semi-sibling hominids mate, their hominid sperm and the hominid egg unite, as normal, to create a hominid zygote. The hominid program does a little tidying up, and then, as normal protocol dictates, turns the running of things over to the developmental program that is just starting to run.

The hominid zygote turns rapidly into a human embryo as the stuff falls into the human QPF being sequentially generated.

The Bible suggests they were not siblings, but as close as ribs are. For yes, this Generation 1/2 is nothing other than the parents of the human ancestors, Adam and Eve, at last on the scientific stage.

The basic program is: Turn cell into ball, make a hole, make another, flip them, and so on. The new programs that emerge then call on these earlier programs as subroutines. The programs pass overall control, the User effect, up the line as the higher operating systems start to kick in.

These are the first humans, born in a hominid womb, suckled at a hominid breast, raised in a thriving hominid tribe and destined to rule them all.

There is nothing to suggest that the first generation of humans number just two. A family load is quite probable. While these humans could not mate with the hominids, they could be semi-fertile with the Generation 1/2 as the Bible hints at, but their offspring would be sterile, just as with horses and donkeys.

If the first humans had not messed up, I imagine the hominids would have become human pet-servants—think dogs that do dishes—which would perhaps explain the almost universal desire for a personal servant-slave.

They did mess up—another story—and the hominids were eventually extermi-nated.   So, Adam and Eve did have navels, but their zygotes were created by hominid gametes and developed in a hominid womb and drank hominid food.

At four to six years old, they would have turned the pidgin hominidese into a real language, this first family of humans would have prospered and multiplied.

This sequence of internal speciation, a transition generation, and then external speciation is followed throughout the living world. I just picked on us as the ex-ample because it's such a hot topic that I hope I roundly denounced in as many media outlets as possible.

Conceptual conflict

Religion and science are offspring of the same impulse to understand what it's all about, but, like ill-matched siblings with incompatible characters, they can be at peace with each other when in separate rooms but easily brawl when sharing the same place.

Religion, at least when it's in a good mood, can be warm and supportive—giv-ing meaning and purpose to life in the grandest of terms, giving support and en-couragement, friendly and emotional. One of its character flaws, however, is that in its intermittent disputes with science, it has the most difficult time owning up when it is wrong. Just look at the retreat of religion into the petulant "He made it in six days to look as if it took ten billion years!" Perhaps this obduracy arises be-cause it's old and venerable and science is young and brash; perhaps it's a belief that love means never having to say you're sorry.

Science, for all its cold rationality, its rejection of purpose and meaning, it nit-picking passion for collecting facts, does not have this character flaw; it has no problem—at least when all the facts are assembled—in saying to religion, "Sorry, I was wrong."

Origins

One of the areas where they cannot avoid each other is origins: where did the universe come from? where did people come from? They have brawled over these two topics since science was kick-started back to life a few hundred years ago.

For a long time the bickering went something like this:

"The universe started suddenly with light!"—"Nonsense, it always existed!"

"The human race started suddenly with the first two people in one place!"—"Humbug, we came about as groups of humanoids all over the world gradually evolved into modern humans!"

Science has already gracefully conceded the first point: "Sorry, I was wrong, you were right! It did start suddenly, and light was the main event—I calculate the ratio as ten billion bits of light to each bit of matter."

Science is also coming around on the second point. It's not quite sure about it yet, but a great step in this direction appeared on page 31 of the January 1, 1987 issue of Nature, one of the most prestigious scientific journals in the world, under the heading "Mitochondrial DNA and Human Evolution." While the work was highly technical, its conclusions were starkly shocking:

"Mitochondrial DNAs from 147 people, drawn from five geographic regions, have been analyzed by restriction mapping. All of these mitochondrial DNAs stem from one woman who is postulated to have lived about 200,000 years ago...."

The authors, Rebecca L. Cann, Mark Stoneking and Allan C. Wilson, working at the University of California, Berkeley, had overcome a long and arduous course—not the least of their obstacles being the fulfillment of Nature's very strict standards—to stake their claim to a spot in the history books.

What it took to get to that point, and the reaction and rejection they received from the "old bones" paleontologists, has been documented in Michael H. Brown's *The Search for Eve: Have Scientists Found the Mother of Us All?* (Harper & Row, NY, 1990).

While this is not the place to get into details, we can at least lay down the general outline of what they accomplished.

Mitochondria

While most have a vague idea of what DNA is (or at least have heard about it), mitochondria probably need a little introduction.

Each of the trillions of cells that make up the body are divided up into compartments that allow incompatible processes to be kept apart. The practical wisdom of industry suggests why: a manufacturing complex—which is pretty much what a cell is—would have an overwhelming problem with quality control if duplicating computer programs onto floppy disks happened in the same quarters as burning coal to power an electric generator. Keeping such incompatible processes in separate areas makes a lot of sense

One of the great advances in the evolution of living systems occurred when a cell lineage stumbled on the great advantages of compartments and went on to become the common ancestor to all higher forms of life. The other lineages remained as simple bacteria who to this day do not have inner compartments and who, metaphorically, still duplicate their computer disks right next to the furnace.

The largest of these cell compartments is the nucleus, which is packed full of DNA. Industrially, the DNA is equivalent to hundreds of thousands of computer disks (genes) loaded with the instructions needed to program the industrial robots (proteins) that run all the myriads of processes in the industrial complex. The nucleus keeps the master disks safely stored away (chromosomes) and makes duplicates of them (messenger RNA) to send out to where they are needed in the running of the cell.

The mitochondria are usually the second largest compartment in the cell (some cells have one big one, most have lots of smaller ones). The mitochondria are the industrial equivalents of central power plants that burn fuel (glucose and fat) to generate power (ATP) for distribution to the other centers, including powering the computer-department labors of the nucleus.

All higher cells (eucaryotes) have these two compartments: the nucleus for information storage, duplication and dispersal, and the mitochondria for central power generation.

An idea that was shockingly revolutionary just a decade ago—but is now almost universally accepted—is that mitochondria are descendants of bacteria (procaryotes)—that the discovery of the advantages of keeping computer disks and coal is separate compartments involved a large simple cell (which was perhaps energetically inefficient) getting invaded by a smaller bacteria (which was energetically more efficient). While this infection was probably disruptive at first (even fatal), eventually the two learned to live together in mutual harmony—the big cell doing all the work of finding the fuel, the symbiotic bacteria, the proto-mitochondria, doing all the work of burning it up.

This insight caught on quickly because mitochondria are just like bacteria; they have their own little piece of DNA (only tens of disks-worth of information compared to the hundreds of thousands in the nucleus) and they multiply just as bacteria do: they get bigger and bigger, then split into two, with each "daughter" mitochondrion receiving its copy of the mitochondrial DNA. It is this which makes mitochondrial DNA so useful in the exploration of human lineage: its lineage is quite independent of that of the nuclear DNA.

Matrilineal Descent

The second point that makes mitochondrial DNA such a useful tool involves the way human beings are made—recall from Biology 101 that this involves the fusion of an egg cell from the mother with a sperm cell from the father.

The egg cell is huge; it has thousands of mitochondria and bulging fuel stocks all primed and ready to power the development of the new embryo. In cell terms, the egg is a big fat blimp floating lazily along, waiting for destiny to arrive.

If that destiny is not to be the flush of the menses, it will start with a single sperm piercing the egg and sparking the fabulously intricate process that ends up with a human being.

For the sperm cell, this moment of destiny does not come by waiting; the sperm has to take the gold—there is no prize for second place—in an Olympic marathon. As the run is equivalent to that from Moscow to Beijing via Mount Everest in competition with a hundred million others, the sperm can be no fat blimp; it is instead a stripped-down, sleek torpedo—just a head with its precious consignment of nuclear DNA from the father, and a powerful tail powered by massive mitochondria to push it ahead of the pack.

The single sperm that triumphs sends its head and tail to quite different destinies. The head merges with the egg and injects the father's nuclear DNA. Inside, this combines with the mother's and is packed away into the nucleus of the cell, now a zygote, ready to provide all the information needed in the construction of a human being.

The tail of the sperm, on the other hand, exhausted from its magnificent effort, drops away, its job done, and disintegrates. The result of this sacrificial effort is that none of the father's mitochondria gets into the egg—all the mitochondria in the zygote, and the human being it eventually turns into, come from the mother.

This also makes mitochondrial DNA very useful in studying lineage: all the DNA in the mitochondria in your cells—be you male or female—came from your mother. Furthermore, your mother's mitochondrial DNA all came from her mother—your grandmother—and hers from your great-grandmother, and hers from your great-great-grandmother, etc. All the way back into deepest time.

No Sex, Thank You

Yet another inducement for scientists to shift the study of human ancestry from fossilized bones to the DNA lab is that mitochondria don't indulge in sex.

Sex is the great mixer; it takes 50% of your dad's nuclear DNA and combines it with 50% of your mother's DNA to create a whole new 100% that is you. Then, in making your sex cells, it scrambles together (recombines) the contents of the dad's chromosomes with the same chromosome from the mom. That's why kids are different from their parents and their grandparents; sex keeps mixing things up in each generation.

This is the greatest thing about sex (from the lineage's point of view, at least): you get a totally different combination each generation. This blending of characters, however, is the worst thing about sex from the study-of-lineage point of view—tracing things back in time through the lineage is impossibly complicated after only a few generations.

Mitochondria don't do sex, so the copy of mitochondrial DNA which is passed on down the generations is an exact copy every time. Well, almost exact. Very, very occasionally (once in thousands of years, perhaps) a mistake is made in duplication and the DNA is changed. Most of the time, these mistakes foul things up and are quickly eliminated from the lineage. If the error is not disruptive (a neutral mutation) and happened in the formation of an egg cell, this little change can be passed on down the lineage from mother to daughter, in the matrilineal lineage.

It is these neutral changes that enable scientists to probe deep time.

Assuming that the rate of change, estimated to be 2 to 4 percent every million years, is constant—a tendentious assumption, but one that only alters the time scale—it is possible to calibrate a "molecular clock." For example, if two lineages differ by 0.3 percent, then their last common ancestor procreated roughly 100,000 years ago.

Search for Eve ...

The Berkeley group devised a technique to isolate large quantities of mitochondrial DNA from placentas (or afterbirths, the few big chunks of human flesh that are regularly chucked away) collected from a wide variety of women representing all the races. The changes in the mitochondrial DNA were identified by snipping them into little pieces with special bacterial enzymes that are very sensitive to DNA patterns—the "restriction mapping" technique.

The assumptions they made in interpreting their results were that a particular change only happened once in history (a very reasonable assumption based on what is known) and "that the giant tree that connects all human mitochondrial DNA mutations by the fewest number of events is most likely the correct one for sorting humans into groups related through a common female ancestry," as Dr. Cann put it in her excellent overview, "The Mitochondrial Eve," in the Natural Science section of *The World & I,* September 1987, p. 257.

From their data they constructed a lineage that could explain the global distribution of neutral mutations. Combining this with the molecular-clock estimates and with what is known about the timing of human migrations, they concluded that the best explanation of their data was that every human being can trace their lineage back to one woman who lived in Africa about 300,000-150,000 years ago, a woman quickly dubbed "the mitochondrial Eve."

As Dr. Cann is careful to point out, their data does not prove "that all humans stem from a single female ancestor," since the mitochondrial Eve is not necessarily the very first human ancestress. There is the "Smith" phenomenon to take into account, the one that plagues telephone-directory creators—one lineage can thrive at the expense of others (though, of course, this is a patrilineal phenomenon). There could have been a group of ancestral women, all of whose matrilineal lines died out except for one, the mitochondrial Eve whose DNA got passed down to every living human being living today—it only takes one all-sons generation to stop a matrilineage dead in its tracks just as an all-daughters one will end a family name.

But the research is certainly getting close to the original ancestress. Close enough, perhaps, for science to apologize to religion for deriding the Adam and Eve concept so scathingly in the past.

In the July 1997 issue of Scientific American, the work on mitochondrial DNA had progressed far enough for the presentation of a tentative map showing how human beings spread out to populate the planet as revealed by their DNA.

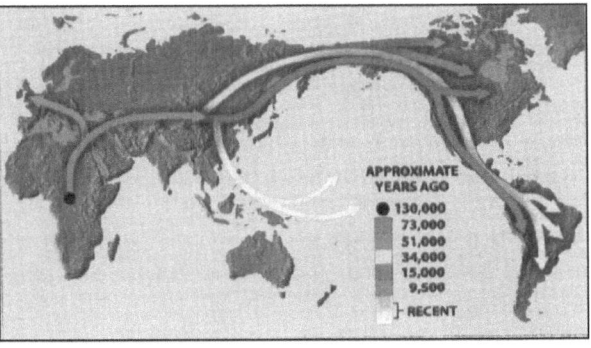

... and Adam

What about the men?

While there is no such thing as a mitochondrial Adam, there is another route. Sex determination—whether the zygote will develop into a boy or a girl—depends on what sex chromosome came from the father in his 50%: an X-chromosome will make a girl, a Y-chromosome a boy. Mothers always contribute an X chromosome: so girls are XX and boys are XY.

Boys get their Y from their dad, and he got his from his dad, and he got his from his dad, etc., etc., in a patrilineal lineage back in time.

Strangely enough, this sex chromosome doesn't get involved in sex. The X and Y that end up in a boy are so different that they don't scramble together the way the two X's do in girls. So, just like the matrilineal mitochondrial DNA in women, the Y-chromosome DNA in men is patrilineally passed on unchanged from generation to generation. Almost unchanged, that is, as it too can slowly collect neutral mutations which can be passed on. These are being studied and you can confidently expect this headline to appear one day: "Scientists find Y-chromosome Adam."

Surrogate Parents

It should be noticed that science's apology is conditional: while both now agree that there was an Adam and Eve, there is still a lot of debate and disagreement as to exactly how they got there—religion still has a very difficult time with the relationship to the great apes.

Religion is going to have to unbend, sooner or later, as the mitochondrial patterns found in chimps are closely related to the patterns of mutations found in humans, which implies that the zygote that developed into Eve got its mitochondria from a chimp-like ... what?

I hesitate to use the word "mother" here as it has the implication of like to like, equal to equal. As Eve is, by definition, the first human woman, this source of mitochondria cannot be human or a "mother" in the sense of equals. But, as this female-source-of-mitochondria stood in the position of a mother to Eve, the term "hominid mother-surrogate" is appropriate.

While this does not give the definitive answer in the theological debate on, "Did Adam have a navel?" it suggests, at least, that Eve had one.

The mitochondrial linkage suggests that Eve's hominid mother-surrogate and modern-day chimps had their last common ancestor a few million years ago. Research into this is currently a hot topic of investigation.

If Eve must have had a chimp-like mother-surrogate to get her mitochondria from, you can bet that Adam must have had a father-surrogate to get his Y chromosome from.

While I have yet to see any evidence collected on this subject, bets are that the father-surrogate to Adam was also a proto-human hominid like the mother-surrogate (though, in all likelihood, they came from different lineages, since same plus same generally produces same and Adam and Eve as the first humans were, by definition, different from their parent-surrogates).

While this is speculation beyond the bounds of where experiment has reached so far, it does give hope that one day science and religion will stop their bickering about how people originated and agree that they were both partially right and both partially wrong.

Nervous OS 1.0

The lowest levels of the nervous hierarchy is quite well understood, externally. The lowest level involves the pattern of ion flows across its membrane a neuron sends down its axon, a signal down its 'output' extension that influences other cells. These patterns of electrical signals influence other cells that the axon abuts onto—which can be tens-of-thousands of other neurons in some cases. Massive parallelism is in great evidence.

The best understood aspect of how the mind works is the sensory input—how information about the environment makes it to the level of 'awareness' which, in this discussion encompasses a dog seeing a cat and racing in for the kill.

The way the senses work is that a sensory neuron responds to a 'bit' of information about the environment such as red photons, a sound frequency, a pressure differential or a chemical concentration, etc.—the senses we call sight, hearing, touch, pain, smell and taste.

Some organisms also possess more than these four, such as a sense for magnetic and electrical interactions but we humans show little evidence for such sensitivities, perhaps made up for by the possession, if spiritual experiences can be taken into account, of the ability to perceive spirits such as Jesus and Mary, ghosts both benign and malignant.[126] But enough of such speculation, we shall stick to the senses we share with all mammals.

A sensory cell is rarely quiescent. It is usually firing off a series of electrical impulses down its output axon. On stimulation by a bit of information about the environment—such as a bunch of red photons—the pattern of firing changes, a different pattern of impulses is sent of down the axon.

This can be likened to the serial connection used in computers—a modem is a good example—where the pattern of bits is sent out one at a time.

This serial pattern-change might represent a minimal piece of information—a bit, a sensory pixel, so to speak—such as 'red detected.'

These sensory pixels are analogous to the particles at the bottom rung of the hierarchy of matter. This pattern change in the serial firing of the sensory neuron is at the very bottom of the sensory hierarchy.

The next level in the sensory hierarchy is also quite well understood. Sets of neurons—which, in the case of the eye, are not even in the brain but in the neural nets of the retina—allow these pixels of sensory information to interact with each other with all the possibilities of interference, both constructive and destructive, so well described by complex numbers. The super-systems created by these interactions are the sensory atoms, the next level of the sensory hierarchy. In the eye, for instance, these atoms of sense are items of information such as contrast changes, color gradients, etc.

This level of representation of the environment reads: A transition from deep red to light yellow was detected.

Further up the hierarchy of programmed processing are the nets of neurons that send parallel patterns of firing along their axons to other cells involved in the next level of processing. The hierarchy of visual processing is probably the best-described of the senses, at least in the bottom-up sense of looking at things.

In the brain, neural super-neural-nets, such as the retinal columns, allow the parallel input from such as the optic nerve to interact and form higher super-systems. The internal representations of these super-systems include shapes, such as a square.

Much of the early vision information processing—in massive parallel—involves simple logic such is found in regular computers. A simple example is AND: Are two inputs the same? Yes or no.

I recall reading somewhere, that all the basic logic functions can be accomplished by arrays of NOT-AND, or NAND, that is just the 'yes or no' of AND flipped to its opposite, to 'no or yes.' The primary levels of the visual cortex do something about as simple. Many such outputs are combined into the detection of lines or patches of the same color.

The 'you' doing the seeing thinks you are "seeing the outside world." But it's a virtual reality, it's a simulation. Just like my legal copy of Windows thinks it is running a real Intel Chip, it is being 'fooled' by a simulation, it is actually running in the virtual reality generated by Virtual PC running on Mac OSX running on … assembly code running a real Motorala chip.

You think you are "seeing reality" when you open your eyes. But it's a simulation, what is actually happening is intricately-pulsing neuron nets lighting up and fading. But the simulation sure looks real!

The visual cortex seems to be physically organized into columns of cells in which the sensory atoms integrate into more sophisticated entities. These columns of cells fire in correlated patterns when they 'perceive' things such as horizontal and vertical lines, areas of color, etc.

Sensory representations have been ascribed a process akin to the external Darwinism in classical evolutionary theory.

This so-called Neural Darwinism has gained supporters in recent years, notably Erdleman and his selection of neural representations by elimination. The law of survival in the sensory hierarchy is survival of the fittest representation. Here "fittest" implies "being a useful way of representing the reality" of the being doing the sensing—'useful' in the sense of the old biological mandate of survive to reproduce. A sensory image that indicates food, while the reality is a cliff is not at all useful then, in this sense.

This perspective is supported by what little is known about learning. The infant animal has its neurons in a way that can be characterized as "everyone is connected to everyone else." This plasticity is somewhat limited, of course, by the genetic constraints on the development of the brain. But there is not enough room in a trillion chromosomes—let alone the 23 of our species—to determine every one of the ways in which a quadrillion cells can connect with each other. In the totally plastic state, this number would be factorial-quadrillion which is so huge I have no idea how to calculate it.

Then there is the 'stuff' that falls into QPF in the nervous system seems to involve synchronized firing of neural nets. Are they also falling into quantum probability forms? And, if so, what might be generating the quantum probability forms for them to 'fall into'?

One possibility is the attendant, behind-the-throne glial cells that surround and embrace the well-understood neurons. As no other function except nourishment has been ascribed these mysterious "neuroglia, especially the astrocytes, oligodendroglia, and microglia" as Yahoo has it, we will not be stepping on anyone's toes.

Could RNA have a role in carrying the linear programs in the nervous system? Sure. Ten minutes with google and I came up with this:

Neural OS	
RNA linear program runs	
Generates Quantum Probability Form	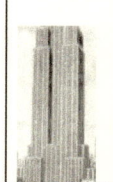
Stuff falls into collapsed probability form	
Organ OS	
Cell OS	
BASIC OS	

"At learning, a sequence of events leads to a fixation of memory: information-rich modulated frequencies, field changes, transcription into messenger RNA in both neuron and glial, synthesis of proteins in the neuron, give a biochemical differentiation of the neuron-glial unit in millions, a readiness to respond on a common type of stimulus.

"At retrieval, it is the simultaneous occurrence of the three variables: electrical patterns, the transfer of RNA from glial to neurons, and the presence of the unique proteins in the neuron, which decide whether the individual neuron will respond or not."[127]

"In neurons, localized RNAs have been identified in dendrites and axons; however, RNA transport in axons remains poorly understood... It is concluded that the specific delivery of RNA to spatially defined axonal target sites is a two-step process that requires the sequential participation of microtubules for long-range axial transport and of actin filaments for local radial transfer and focal accumulation in cortical domains."[128]

To My Mind

We will equate this with the emergence of the capacity to invent a grammar. This is a rather specialized aspect of language that adults do not have, inasmuch as they have "lost" it by the teenage years. Lost is probably not the correct expression, however. Rather, higher structures have come to depend on the constancy of grammar rather than its infant mutability. If this is correct, the "ontology recapitulates phylogeny" perspective suggests that this childhood stage is an echo of the Origin of Man.

That children temporarily have the faculty to invent grammar while adults do not became apparent when linguists investigated the origins of new languages in historical societies. The surprise was that only children are involved in the origin of real languages. This faculty is usually hidden since most children grow up immersed in the language of their parents: they learn that with remarkable facility and do not need to invent a new language with a new grammatical structure.

The rare exceptions to this—where children were not immersed in the grammar of their parent's culture—are where languages have been invented in recent times.

One example of language invention involved deaf children in a large institution in Central America. They were not immersed in the grammar of the adults, and so invented their own. They transformed the primitive pidgin signings of their few adult teachers into a true language with a fully-fledged sophisticated grammar.

As far as regular vocal language is concerned, there are many examples of true language invention in history when children developed in a culture in which a pidgin is spoken. Adults invent pidgins, they do not invent languages. A pidgin is not a true language in that it does not have a grammatical structure that can express any but the simplest noun-verb combos.

Pidgins have been invented by adults many times in history; they are quite common. When adults speaking many languages are forced to live together—as in port cities or slavery situations—they spontaneously develop a pidgin that allows for basic communication and economic interaction to occur.

A pidgin can convey basic information about things and actions; but not much more. In a pidgin, "John kill Jim," "kill John Jim," "john, Jim kill," etc. all associate a death with these two individuals; but it can convey no more. It is not possible to pass on a full description or understanding.

Nevertheless, a pidgin can be remarkably effective in allowing for basic social interchange in a polyglot population of adults. A pidgin is not capable of describing exactly who did what to whom; there is no grammar; there is no subject-verb-object structure to slot the words into. We suggest that pidgin was the highest linguistic ability of our ancestral hominids. They had sounds to represent objects and actions but no way of stringing them together into linear strings with a grammar structure.

Children developing in a pidgin-speaking environment are not exposed to a grammatical structure, they develop in a grammar-less world. They first learn all the sounds used around them; then they pick up all the pidgin words in use around them; and then they do the unexpected, they effortlessly invent a grammar; they organize the pidgin into a true language. They invent a true grammar and transform their parent's pidgin into a true language, a Creole. The Creole is the simplest type of true language—it matures by adding new words and speakers into a "regular" language. Adults do not have this innate capacity to improvise a grammar on the fly or effortlessly learn the language. Children exposed to a grammatical language use the innovative capacity to effortless language acquisition. In either role, the faculty is lost in later years. As adults, we can pick up new languages but only by strenuous effort and we can invent languages but it takes university-honed skills as a linguist to do it.

Grammar is like putting our thoughts in order. We think in language. I am sure that the deaf-language innovators thing in terms of sins.

The scientific worldview we will be constructing on the internal cause-of-probability of quantum physics has a certain resonance to a philosophical structure created by Karl Popper. In his classification, the objective reality studied by scientists—atoms, planets, cells, galaxies, brains, etc., etc.—corresponds to World One in his profound philosophical dissection of reality into three realms. [129]

Popper's World Three is the realm of the mind, what we have going on inside us. For example, the concepts and theories of science belong to this realm. This World is what goes on inside each person, the thoughts, theories, concepts, plans, emotions, passions, etc.

World Two is where World One and Three intersect as ideas are expressed in life and culture and, occasionally, in science. Expressions of scientific thoughts, plans and passions in the form of books, educational institutions, cyclotrons, conferences, etc., all belong in Popper's World Two. World Two is the expression of human thoughts, ideas, plans and passions in all that we see about us—buildings, washing machines, concerts, newspapers, dollar bills, interstates, etc.

One general way of interpreting this philosophical perspective is that human artifacts—which in terms of classical random chance-and-accident are highly unlikely aggregates of atoms—can only be comprehended if an internal influence is included in the discussion. The aspect of culture we call science, for example, can be thought of as scientific thoughts influencing what happens to scientific materials.

While I don't think Popper's concepts embraced the notion of probability being the fundamental link between his three worlds, the similarity between the two is apparent. If the similarity holds, modern science leads us to expect that ideas in the mind are linked to probability. Ideas in the mind are probability forms that manipulate matter: the idea for Mona Lisa provided the probability form for the oil paint to "fall into" in the painting process.

Does the human speech module, the human speech program involve programs encoded on RNA? I think so.

First, the program gets run in the virtual reality generated by the Main Program. To experience this directly, think this thought silently 'inside:

"What I think of as 'reading this sentence in my mind' is actually RNA programs running in a virtual OS environment. The 'I' doing all this is actually a Master Program."

Most of us have a limited success reprogramming this Master Program, but it is usually difficult. Just like the testing program running in the thymus, you can release these programs from the virtual reality to the real. To experience this, read this sentence silently until instructed to speak:

"What I think of as 'reading this sentence aloud' is actually RNA programs running in a virtual OS environment that I am releasing to the real OS where they are running and I found myself speaking this sentence fragment, 'releasing to the real OS to run as the speech fragment...'."

Did you start speaking on the first thought 'releasing?' If so, did you catch the "release" command you gave. It's kind of hard to do at first as we are so used to either reading silently or reading aloud so the switch in midstream is unpracticed. Try it a few times. Try whispering it.

The stored programs are probably in my cerebellum which seems to have a syntax checker as only well-formed programs are allowed into storage—'learning' or 'getting it.' Reading is a major program and takes a good while to assemble.

Learning, the, is the putting together of a program that runs. Once it runs properly, we have learnt it and it is stored in the "DNA" cerebellum for retrieval when called on. Are quantum probability forms involved in the human mind, I wonder? A simple experience of mine leads me to think so:

Ossining Longing

For over ten years, I commuted from Ossining to Manhattan on the MetroNorth train. Almost every day I was assaulted as I waited for my morning train with the strident computer-announcer insisting:

Attention! Attention at...!
OSSINING!

For 10 years, every day, this voice resonated inside my skull as I fretted, "What now!" I have now, as of this writing, lived in Mount Kisco for fourteen months. I still catch the MetroNorth. Almost every morning the same voice intones:

Attention! Attention at...!

And what is strange is that there is an "Ossining" shaped hole there in my head that, for a moment, is very perplexed when the Voice says:

... MOUNT KISCO!

For it does not fit. It's empty, a pixel of frustration. It happens to me every day and there is nothing I can do to stop that little glitch of surprise. Could such an expectation be an empty probability form? Is a program running in the front of my brain (cerebrum) and generating an "Ossining-shaped" empty probability hole with a 'desire' to be filled and not empty? A Pavloved-dog probably felt the same way when that darn bell rang but no food appeared.[130]

Now, if a part of my mind is made of probability forms, perhaps a lot more of it is; perhaps the whole shebang.

So, the QPF of the aminoacids are a distant cousin of the QPF for "Ossining" in my brain. Perhaps my anthropomorphic translation of quantum math into natural English has some justification after all.

In one of those delightful moments of synchronicity, as I was writing this a commentator on a WNYC spoke on studies of how familiar music runs in the head. A specialized part of the front brain (the auditory cortex) was highly active (running a program?) and remained active even when the catchy tune (The Pink Panther riff was the example) suddenly stopped midway. Something was there with no sound in it. When an unfamiliar tune was played, however, the activity stopped the moment the tune stopped.

Bacterial feelings

Actually this kind of feeling is not as sophisticated as we might imagine.

Consider a bacterium. As established, it is an internal composite of quintillions of QPF generated mainly by proteins. Of these, a fraction generates QPF for glucose, say. In a healthy, well fed bacterium, all of these trillions will be filled; they will be 'satisfied' in an internal way connected to the Path of Least Action.

But in a difficult environment, perhaps only 5% of these glucose QPF will be filled. What about the 95% of the quantum probability forms that are empty? All the ones with nothing in them. Does this void amount to anything?

Now, in the classical view, the concept of a bunch of nothing amounting to much is quite ludicrous. But, we know, from our weird execution, that a bunch of nothing can indeed amount to something very significant, like bulletproof vests made of nothing but a void.

So, what do these trillions upon trillions of empty QPF amount to as they clamor to be filled? Just like a simple aminoacid in a chain, just on a larger scale.

Could not this unhappy bunch be a primitive kind of feeling of hunger? The part of the overall composite that is 'not happy' with the way things are, is similar, indeed, to our own basic instincts, except in depth and scale, that I withdraw my earlier apology for using anthropomorphic analogies to describe the longings of aminoacids.

Bacterial autonomy

We earlier established that an electron is ascribed a simple autonomy in quantum physics. It is known in the labs that even God cannot know where the electron will land in a slit experiment.

We linked this autonomy to why it is impossible for a computer to create random numbers. The best computers can do is generate pseudorandom ones—"random numbers generated by a definite, nonrandom computational process" according to Yahoo. Clearly pseudorandom is not really random at all. Yet, you can reel off a string off random digits no difficulty.

Electrons, when faced with 'competing' probabilities—such as 50% go this a-way, 50% go that a-way—have a true autonomy.

We expect no less of bacteria. When faced with competing probabilities they also can be expected to have a true autonomy. The mind is mysterious, even in bacteria.

The Human OS

The mind is definitely to be found in layers, reptile, mammal, etc. But, simply put, we can add just one more level to the edifice.

For now we can think of these mental atoms in terms of what science has already established—or rather not estab- lished—about the "bind- ing problem." We can illustrate this with the visual system. The chal- lenge is that we know how the base of the visual system detects all sorts of

Human OS	
RNA linear program runs	
Generates Quantum Probability Form	
Stuff falls into collapsed probability form	
Neural OS	
Organ OS	
Cell OS	
BASIC OS	

"primitive" things about the world—edges, lines, and transitions in color, etc.—in many different areas of the brain. The challenge is to figure out even a good sug- gestion as to how the visual system integrates all these primitives together to "see" a distinct object such as a tree rather than just a lot of lines and colors.

We will look at this in terms of filling in of empty wavefunctions—coded im- ages corresponding to things in the environment. For the human capacity, we will further explore the idea that these images become "things in themselves"—they become the atoms at the foundations of some higher intelligence. We will equate this with the phenomenon of "naming"—being able to think of a "tree" without actually looking at one. The image has a reality that is quite independent of there being trees to look at.

Learning

It takes a human being about three years to assemble an "abstraction module" that can recognize an abstraction in the following story of a five-year-old's birth- day party.

A short story:

Richard was five. Aunt Marie and Uncle Willow were there, Aunt Betty (Peter, her husband had recently passed, so there was a little gloom about), and dad's spinster sister, Aunt Francis, and lots of friends.

When his mom asked him later about the party Richard, being a smarty pants, declared "there were three aunts here for three hours and I was so nice to them. But all I got was three shillings, and three kisses too. Yuck!"

We get so used to such counting skills, like riding a bike, that seems automatic and hence, common sense. For the little boy recognized that three aunts, three hours, three kisses and three pennies—which otherwise have nothing else in com- mon—do have something in common: the abstraction, three.

Such a familiar skill is difficult to view as special, but animals do not have an abstraction module, so see nothing in common between three prey and three predators.

It seems that the front of your brain—the cerebrum—is where you struggle to assemble new and sophisticated programs (concepts) while all your automatic ones, the ones that actually require effort to become conscious of, are in the back, the cerebellum. This is where the 'riding a bike' that "you never forget" is stored and can be called upon effortlessly as an RNA linear program is run.

All the learning takes place in the front, which is constantly trying to get the cerebellum to store it away. The problem is, the cerebellum is where programming rules are strictly applied and it only "accepts" what mathematicians call well-structured constructs. It is a syntax-nut, nit-picking module that will only run well-structured programs. This means, in essence, it will only accept programs that have already been proved to work in the VR up front.

The front, in the cerebrum, is where "you" speak English silently. The back-room cerebellum is where math-speak is spoken. The front can think $1 + 1 = 4$. The back will not accept it, it will reject the malformed program and wait for you to come up with $1 + 1 = 2$, a well-formed statement that is eagerly admitted into the math-speak program store. Somewhat like DNA.

There is nothing quite like the feeling of getting it, especially if you have been dumb enough not to get it while all the girls have already. But once you get it, you know from experience, you never forget because it's stored safely in back.

Sleep

My Mac has a lot of housekeeping chores it has do to do—update the clock, flush memory to disc, adjust the virtual memory, etc. It has lots of such programs to run. Being a 'threaded' CPU, each gets a few microseconds every second to run and do its thing. The chores are important; if they are not done, the big shot programs that get milliseconds will quickly grind to a halt.

Apparently, such rapid interpolation, such intimate mixing of necessary chores while running the Main Program is not possible when super-massive parallel processing is involved. For all animals with a brain have to sleep.

It looks like the chores have to be done at night; the Main Program—that's you, dear reader—is shut down and all the housekeeping programs come out of storage on neuralRNA and get to run in waves of shifts, such as REM sleep.

That 'refreshed' feeling that is so desirable when the alarm rouses the Main Program back into action is no ephemera. QPF that should be empty are empty. QPF that should not be are not. That moving finger Main Program that is "I am" starts to Run and generate probabilities that the day will be a good one as the external me interacts with others on another day on planet Earth.

Why is sleep so important? I would not suggest going without sleep for many days. For soon, no matter what kind of Main Program you are, you will start to experience "system crashes," things like not knowing who you are or why you are in this room that is unnervingly psychedelic. (Perhaps LSD triggers the running of a housekeeping program that, at least for me once, makes the Main Program believe in magic, and speaking to trees.)

This is a screen grab of my activity monitor as I type. The only two programs I am aware of using are Word and Finder. What are ATS and PBS doing in there with all those megabytes to sprawl around in?

If I were to somehow disable all these OS X chore programs, both Word and Finder would almost instantly on my scale of things grind to a miserable death, trust me.

Process Name	% CPU	Threads	▼ Real Memory	Virtual Memory
Word	11.40	2	40.36 MB	241.12 MB
WindowServer	2.40	2	39.37 MB	211.46 MB
Finder	0.00	2	16.78 MB	164.95 MB
MenuStrip	0.40	5	15.16 MB	158.57 MB
ATSServer	0.00	2	14.84 MB	87.43 MB
Activity Monitor	2.90	2	10.85 MB	161.69 MB
Grab	5.90	3	5.89 MB	149.58 MB
loginwindow	0.00	5	5.27 MB	124.21 MB
SystemUIServer	0.00	1	5.02 MB	148.00 MB
Database Daemon	0.00	1	4.43 MB	138.98 MB
Mirror Agent	0.00	3	3.43 MB	144.57 MB
Dock	0.00	2	3.42 MB	139.07 MB
iCalAlarmScheduler	0.00	1	2.81 MB	100.33 MB
pbs	0.00	2	2.23 MB	44.41 MB
iTunes Helper	0.00	1	1.13 MB	91.36 MB

Just so, without sleep, your Main Program consciousness will falter and crash badly, sooner or later. Getting a good night's sleep is as good as control-alt-delete is as in resurrecting a crashed Windows.

The brain's neural operating system keeps on working throughout the night—it's almost as active as when awake, just a very different kind of activity. But it is busy running all the many housekeeping chores; it is not running "you" anymore. The Main Program has been "written to disk" on virtual RAM, ready to commence running again in the cleaned-up real RAM in the morning.

In the old OS, a program crashing also crashed the operating system (which is mostly virtual) and I would have to pull the plug, count to three, and then restart it. Horribly time consuming.

Now, however, with OS X, a Force Quit program is (usually) available under the apple (thank you Steve, great job. Me and my 400 apples are grateful.)

This simple program must issue some very low-level instruction to OS X such as "terminate that program and wipe its data structures in its RAM partition." No matter that Word has totally frozen, and the hypnotic wheel is spinning endlessly. At most, I will have lost a paragraph, sometimes just a word, for I save compulsively from bitter experience.

I just calmly hit the right keys and this little window is generated. A click and all is new.

Going to sleep is just like that. Some low level timing program notes that it is time, and safe, to sleep. It runs the "Force Quit" program and clicks on Main Program. The Main Program instantly stops, the contents are written to disk on RNA, and you go to sleep.

As is well known, we cannot will ourselves to sleep. Going to sleep happens to us not by us. For this reason, it is not possible to actually experience "going to sleep," we just abruptly stop. There is no "you" there to notice what going to sleep feels like. The housekeeping programs get to run and start cleaning up and moving stuff around.

One of the side effects of these housekeeping programs is dreaming. This is probably bits of the day and other debris running momentarily in some module. The characteristics of this type of dreaming is that we are not in control, and in fact it's more like fragments of "I Am" running with a limited autonomy that is not

"I". Our bodies move a little; dogs scratch and sniff. The memory of such dreams rapidly fades within minutes. They can reveal you something about the fragment of yourself that was running, but little else.

There is another type of Dream, however, that is qualitatively and quantitatively quite different from 'animal' dreaming. This type of dream is very vivid, we are our whole selves, and we can speak. It is as real as the real world. In fact, on waking from such a vivid and real Dream it is the bedroom, the everyday world that seems unreal, dull and monochrome. The Dream is so real we expect it, not the bedroom, to continue.

I probably do not have to remind you that such Dreams can be nice, or they can be not nice. In fact, a real nightmare can be totally unsettling. When I was young, I had the same Nightmare many times: a hideous witch screaming after a terrified, running me. She never caught me. The dream was so utterly and vividly dreadfully real that I was shaken and miserable for weeks after each occurrence.

Then there are the glorious Dreams, where everything is perfectly and delightfully intense and wonderful. I have flown in such Dreams and once or twice flown in day-Dreams. If you have not had such a Dream, the best art that captures it for me is Disney's Peter Pan where they soar above Victorian London. The 'real' thing is much, much better. And it's so easy—the "up body" impulse is as natural as the "raise arm" one is.

'I Am', Fermat & Hilbert

The leap from the hominid to the human mind could be as simple as a programming language which goes from manipulating just real numbers to being able to manipulate complex numbers.

The first thing I do when I get a new calculator program is ask it to display the square root of minus one. Almost always, they say something like "not a number" and I sigh with frustration.

For it is safe to say that all the math used in science and technology involves complex numbers at one stage or another. So, I find this restriction of simple calculators to the real numbers hard to understand.

I am quite sure that it will be found that the human VR mind and 'I Am' Main Program all manipulate complex numbers, not just real ones.

Self Creation

Incidentally, you might be wondering where this "I Am" program that is you came from. It has been assembled, code by code, by yourself as you have lived your life. The human program is self-developing.

Naturally, much of the first layers of the "I Am" master program, the ultimate User, were laid down by your parents in the very early years. But as your sense of "I Am" my own person emerged, more and more it was you who assembled the code by the choices you made and things that happened to you. Naturally, the culture larger than the immediate family also plays an increasingly important role.

By 16 or so, it is a safe bet to say that you were making all the really important decisions by yourself, and a few 'best' friends and cultural role models (for both good and ill.).

So the current "I Am" program was added to by what yesterday's "I Am" experienced, which was added to by... etc. This is why we can be so complicated inside. We are a labyrinth of a program, layers upon layers upon layers, forgotten or forbidden-to-run places, etc. And, let's face it, some of the subprogram routines have bugs in them, some so awful they cause you to "crash" when the run in your mind like an old tape replaying yet again.

Unlike a simple computer, a massively parallel one such as the brain, can run two, or more, Main Programs at the same time. Perhaps the comment, "I have a divided mind..." can be taken literally; two slightly different "I Am" versions running at the same time. This is never a comfortable situation, and it can get really bad for some unfortunates.

Externally, I might be what I eat. But internally, in the realm of QPF generation where it really counts, the Main Program that is "I" was assembled by what I did. So the "I Am" program that runs everyday is a composite of all my life experience; some active, some quiescent but (waiting to spring to life, sometimes at very inconvenient moments).

This is the "I Am" program that is 'written to disk' every night when I fall asleep. Everything that is me is written onto an array of linear RNA programs.

My Mac OS calls this memory dump file that it creates when going to sleep, the VMfile. If a lot of programs were running when I closed the lid, this file can be many gigabytes in size on my hard drive.

So, that is where "I Am" is when I am asleep, stored in an organic VMfile in the form of RNA arrays of arrays. It will involve a truly-huge amount of linear information. This must involve a truly immense amount of information: whatever-bytes upon whatever-bytes of linear information all copied out of RAM and onto the RNA 'disk.' Each glial cell probably has just one "I Am"-RNA. There are a zillion of these, so a zillion RNA are involved; small if 'empty,' as big as necessary to store some complex experience.

The storage capacity of even milligrams of RNA, however, is really, really astoundingly huge, not one of our G numbers perhaps, but getting there: it's on the order of 4 to the power of Avogadro's Number, $4^{10^{23}}$. This expression is so huge that, just to reduce it to two stories high, 10^N, to write out N as 1,000, 000... I would fill the visible universe with them, even if I made each zero the size of a virus.

So, a milligram or so of RNA is more than adequate to store all that is the "I Am" program that is you or me. Everything. Nothing is edited out or deleted (a selective Erase module would be nice for deleting the bad stuff—but the 'moving finger' has no 'backspace-erase' function, unfortunately.

RNA Arrays

There is both good news and bad in this perspective for push-the-envelope brain scientists:

Good News: An entire human personality can be stored onto, and retrieved from, on a few milligrams of RNA[131]. Sub-programs, like riding a bike or speaking Korean as a second language, are also stored on even smaller amounts of RNA.

Bad News: Taking RNA as a pill is unlikely to work (at least with currently-feasible technologies). For the RNA molecules that are doing the storing are arranged in a precise array. The precise array corresponding to the positions, or 'addresses' of the glial cells. For the brain is not a jumble of cell, it has an intricate, if highly-repetitive, array of cells in layers and lattices.

Location is everything, as is well known. The hard drive on this Mac is divided into sectors that are sequentially numbered. The OS only deals with sectors, reading and writing to them as needed by calling up the position, the address, of the sector.

From the point-of-view of Mac OSX, a sector can hold a wide variety of things such as a bit of: data, a command, a word, a picture, a sound, etc.

From the point-of-view of the sector, it's all the same: a pattern of tiny N and S magnetic poles (or tiny pits and no-pits on a CD; really tiny bits and no-bits on a DVD, etc.)

So two things are important when you go to sleep—when the "I Am" program running in the VR is written to disk for the night's safe storage.

The subprogram written on the glial cell iAm-RNA.

The position of the glial cell in the multi-dimensional layered structure of the brain; its address.

Escher captured the idea:

A simple way to describe this is to use Hilbert multi-dimensional matrices. A simple 2-D, 2x2 matrix looks like:

The manipulation of matrices, with thousands of rows and columns, is a well-understood aspect of math and is used extensively in much of science and information technology.

$$\left\{ \begin{matrix} +A & +Bi \\ -Bi & -D \end{matrix} \right\}$$

High-end calculators, such as *Mathamatica*, can not only deal with complex numbers, they are proficient in matrix algebra as well.

A simple extension of this allows for a complete description of the RNA-glial cell array that stores the "I Am" as one goes to sleep.

First, the 2-D matrix is generalized to have an unlimited number of real dimensions, each at right angles to all the others. A real hypercube, a hypermatrix. At each location, there is a value. For the brain, this is the iAm-RNA program chunk stored in a particular glial cell during the night's rest.

All of these axis are real. This is like putting a line at a right-angle to another in 2-D space, then another at a right angle to both in 3-D space, then another at a right angle to them all in 4-D, etc.

The next step is to put, at a right angle to each real axis an imaginary axis. This is called a Hilbert space of complex dimensions.

So a 3-D Hilbert space actually has six axis, all at right angles to all the others—three that are real and three that are imaginary. Combinations of the two can be considered as a complex axis. Tensors are good at this kind of stuff.

Just to illustrate how useful such multidimensional Hilbert spaces are we shall take a short break to use them to solve a little—well, bigger than a margin—problem.

Fermat's Theorem

Fermat's conjecture involves this relationship between numbers:

$x^p + y^p = z^p$

Where x, y, z and p are positive real integers greater than zero, and p is a prime.

The hypothesis is that while there are an infinity of solutions for the solitary, even prime number, $p = 2$, there are absolutely no solutions for the denumerable infinity of odd primes starting $p = 3$, 5, 7, 11, etc.

The first point is that a real integer raised to a real integer is always a real integer, and this holds true all the way out to this side of infinity. The second point is that: If:

> The number x measures a line in 1D-space
> The number x^2 measures an area in 2D-space
> The number x^3 measures a volume in 3D-space
> The number x^4 measures a hyper-volume in 4D-space

Where each x-axis (or y and z) in the x-hypercube is at right angles to all the other x's in the cube. In the unit cube in n dimensions, there will be N points with a coordinate 1 and just one unique zero point. In an infinite-dimensional, the will be on infinite cloud of 1 around the one unique zero point. In an infinite-D Hilbert space there will also be an infinite could of points at I, but still one unique zero point.

Then:

The number x^p measures a hyper-volume in pD-space. Same for y and z.

So, Fermat's relationship is actually about adding hyper-volumes and then equating them to another hyper-volume. This is just a generalization of the way Pythagoras equates the area on the hypotenuse and the sum of the areas on the other two sides.

Translating all this into English, Fermat's equation states that the hyper-volume of the z-cube is measured by a real integer that is equal to the sum of the hyper-volumes of the x- and the y-cubes when added.

In order for the x and y cubes to simply add together as they do, they must be independent and distinct hyper-volumes with no overlap. For this to hold, every x-axis must be at 90° to every y-axis while also being at right angles to all the other x's. So, our hyper-volume inhabits a complex p-dimensional hyperspace. I believe the technical term for such is a Hilbert Space.

Hilbert Space

In Hilbert Space, the relationship between x and y is that of a real and an imaginary axis. This orthogonal requirement is satisfied in two simple ways. For every x there is a yi or, for every y there is an xi. It makes no difference, so we will use the first.

We can add this to our list above: The number $x^p + (iy)^p$ measures a hypercomplex-volume in pD-Hilbert space. (A pHD-space?)

We need just a few aspects of Hilbert space to make our point. All the axis, both real and imaginary, touch at just one point, exactly zero. All extensions in this space start at this common point (you cannot have an extension along any axis starting at −1/2 and ending at +1/2.

Most mathematicians keep track of which axis is which by using a cumbersome convention called the right hand rule (or is it the left thumb, I forget).

Much simpler is to accept the well-defined concepts of complex areas, complex volumes and complex hyper-volumes into the descriptive arsenal. I have already accepted complex 1-D lines, as complex number little arrows, so why stop there.

Consider the following four squares, each with sides 1 unit in length, and the following questions about it.

What is the area of the four squares if:

x is a real axis and y is a real axis?

x is an imaginary axis and y is an imaginary axis?

x is a real axis and y is an imaginary axis? Or the converse.

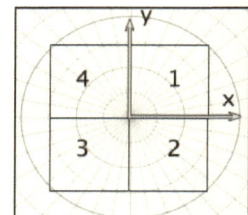

The answers are, with x going first:

Both x and y are real: Doing the usual math we have:

$$+1 \times +1 = +1$$
$$+1 \times -1 = -1$$
$$-1 \times -1 = +1$$
$$-1 \times +1 = -1$$

Yes, in Hilbert space, areas can be negative. As most quantum physicists find that a Hilbert space, not a real space, describes the way the world really works, this is no small matter

Both x and y are imaginary: Doing the familiar math we have.

$$+i \times +i = -1$$
$$+i \times -i = +1$$
$$-i \times -i = -1$$
$$-i \times +i = +1$$

The areas extended by two imaginary axis at right angles are also real, just with the signs reversed.

Real x, imaginary y—the symmetry showing why switching roles would have no effect on the outcome.

$$+1 \times +i = +i$$
$$+1 \times -i = -i$$
$$-1 \times -i = +i$$
$$-1 \times +i = -i$$

So the area extended by multiplying a real axis by an imaginary axis (technically, the Cartesian Product) gives us imaginary areas. What else?

As i is also known as the rotation operator, an imaginary area is at right angles to real areas.

This combination of a real and imaginary axis is the complex plain, so the regular complex plain is actually a 1-D Hilbert space.

Hyper-volumes

There is a subtle difference between a regular space and a Hilbert space that very is important for our discussion. This is because 50% of the Hilbert is composed of imaginary axles.

Construct the infinite real unit 'hypercube' in the following manner with each side of positive unit length: a 1-D line, a 2-D square, a 3-D cube, a 4-D hypercube... a countable-infinity-D hypercube.

The sequence of hyper-volume is:

1 1 x 1 1 x 1 x1 1 x 1 x 1 x 1 ...

Which is 1, 1, 1, ...

Now construct the same infinite hypercube in an infinite Hilbert space, but this time using all imaginary axes.

The sequence of hyper-volume is now:

i i x i i x i x i i x i x i x i ...

Which is i, −1, −i, +1, +i, −1 ...

The hyper-volume of the imaginary nD-unit cube is:

Imaginary for all n-cubes with an odd number of edges out to infinity (and beyond).

Real for all n-cubes with an even number of edges.

This essential distinction between volumes with an odd or even number of edges is not to be found in a regular space, only in a Hilbert space.

The hyper-volume of the imaginary nD-unit cube is rotating counterclockwise, the positive direction with a period of four through the sequence of volumes that are:

+ IM, − RE, − IM, + RE, + IM, − RE, − IM, + RE,

+ IM, − RE, − IM, + RE, + IM, − RE, − IM, + RE,

This is a very interesting cycle that seems to reflect a basic fact of life: the three dimensions of space and the one dimension of time.

Euclid used real numbers to describe spatial separation, length, area, volume, etc. Pythagoras codified the relationship between axis components and 'straight line' separation which, generalized for any number of orthogonal real dimensions, is:

$$s^2 = x^2 + y^2 + z^2 ...$$

Einstein added time to the familiar three of space, but he assigned it an imaginary axis to extend along. In the equations of general relativity, time appears as an extension along an imaginary fourth dimension. So time always appears as 'ti.'

The connection between components and separation in Einstein's unified spacetime is now:

$$s^2 = x^2 + y^2 + z^2 + (ti)^2$$
$$= x^2 + y^2 + z^2 - t^2$$

Plus-one and minus-one are distinctly different and easily distinguished. On the other hand, plus-i and minus-i are so identical that they can be switched with impunity. The distinction is not significant, it is just a convention. While there are two distinct real units, $+1$ and -1, there is only really a single imaginary unit, i.

The complex conjugate of a number is the same number, just with all the plus and minus signs to "i" flipped. This 'reflection' of a complex number in the real line is the one that appears in the collapse of the wavefunction when a complex numbers transforms into a real number.

Back to the four-cycle of hyper-volumes in a purely imaginary space.

Start with "i"—or NOT real—and call it time, an imaginary extension.

Rotate this unit imaginary extension by 90°—multiply it by i—and you have swept out a negative-real area in Hilbert space. Call this new orthogonal axis the 1st spatial dimension.

Rotate this by another 90°—multiply it by i—and you have swept out a negative-imaginary volume in Hilbert space. Call this new orthogonal axis the 2nd spatial dimension.

Rotate this by another I and you have swept out a positive real volume in Hilbert space. Call this new orthogonal axis the 3rd spatial dimension.

As positive-real is about as real as it gets. We seem to live in a reality that is just filled with things that can be described with real, positive numbers. And all the complexity of every wavefunction collapses, in the end, to a real, positive number.

We live in a veritable bubble of a real, positive universe. With just these four dimensions providing us with a "just right" positively real environment to inhabit.

From above, we conclude that the recipe for creating our universe—or at least the spacetime 'stage' on which all the interesting stuff can happen—might have been as simple as:

Take one i from Hilbert space. Cube it. Voila, a positively real spacetime. In Hilbert space, the relation between components and separation/distance is:

$$s^2 = (ti)^2 - x^2 - (yi)^2 + z^2$$
$$= -t^2 - x^2 + y^2 + z^2$$

So why did our universe extend to infinity into four dimensions, and only four, starting at the Big Bang? Why did just four do it after the Big Bang? The others clearly didn't, and those other dimensions remain un-inflated and Planck-sized to this very day—and there seem to be at least eight of them. These are curled-up, multidimensional strings and branes extending a tiny distance from every point in regular spacetime. (They are hardly noticeable but the Higgs finds them a fine place to deconstruct in, as we shall discuss in Volume Two.)

The answer probably goes something like this:

Eight dimensions would be positively real, but would be too confusing for simple folk, two directions in time would be particularly so. Twelve dimensions would also be real, but even worse with three time-like extensions (nightmares sometimes exhibit such a poly-dimensional time and space experience; and it's not particularly pleasant.)

One, two or three dimensions would not be positively real. Thus, four is the only one.

Fermat in Complex Hyperspace

If Fermat's is true in a real space, it will certainly be true in Hilbert space. If false in Hilbert space, it is most certainly false in real space.

With Fermat, we are actually adding two complex volumes and equating them another volume

All these volumes in Hilbert space have a common point at zero and they have to be non-intersecting. We can only accomplish this in arbitrarily-large dimensional hyperspace if one volume, say x, uses all real axes while the other uses all imaginary axes.

So, Hilbert Space, Fermat's relation is actually:

$$x^p + (iy)^p = z^p$$

where z^p is a hyper-volume measured by a real integer. For the solitary, <u>even</u> p, this reduces to the expectation that:

$$x^2 - y^2$$

has to be a real integer. As this is a trivial expectation, we pass it by without comment.

For all <u>odd</u> p, however, the expectation is not a trivial one. For the odd powers of i are either $+i$ or $-i$. We have established that we can make this simple substitution, setting $x^p = X$, an integer, and $y^p = Y$, an integer. This gives us the expectation that:

$$X \pm Yi$$

is a hyper-volume measured by a real integer. As this is only possible if Y is zero—which disobeys the requirement that y be an integer greater than zero—such a combo is not possible. This holds for all odd primes, p, out to this side of infinity.

Waking Up

End of detour and back to multidimensional matrices.

A matrix in Hilbert space is the next step in sophistication. These are the tools with which to describe brain function; words are quite inadequate.

So, when you wake up in the morning, the virtual file stored on RNA is read back into active memory and resumes running in the VR generated by the brain. The "I Am" wakes up and heads for the bathroom. When running in the VR generated by my brain, this is the conscious "I."

In Volume Two: Science in the Realm of Spirit, we will look at the possibility of the same, self-assembled, labyrinthine "I Am" program running on a VR not generated by the physical human brain.

Me and my Mac

We can summarize this discussion with current, trans-millennial, computer science. For there is an almost perfect, one-for-one analogy—with a few 'minor' differences—between Me and my Mac. Bear me out, this is not just extreme Mac-love speaking. The differences to keep in mind while appreciating the analogy are:

Parallelism:

Mac: When not in 'sleep' mode and functioning efficiently, hundreds of regular programs run at a time in my Mac's active memory. A large percentage of these are on "idle" until they are called upon to do something.

Brain: When "I Am" is awake and functioning efficiently, a zillion programs are running in my brain's active memory. A percentage (estimated by some to be as high as 95%) of these are on "pause until they called upon by another program.

Operating system:

Mac: All these hundred-or-so programs are multi-threaded on 1 Motorola PowerPCG4 chip—an intricate assemblage of doped silicon—at a speed of 1,200,000,000 cycles/sec. This chip is running one legal copy of MacOSX.

Brain: Each of the zillion programs is single-threaded on a glial cell—an assemblage of CHNO—at a speed of scores of cycles/sec. Each of these zillion glial cells is running one copy of the brain's operating systems. The probably-incomplete list of these glial-run operating subsystems is: humanOS, hominidOS, mammalOS, reptileOS, fishOS, wormOS.

Program copies:

Mac: There is just one copy of each program running in active memory—such as Finder, Word, Photoshop, n-kernel, pbs, etc.

Brain: Each glial cell has a RNA linear subprogram that is just slightly different to that of the two glial cells on either side of it along a linear mental axis (a concept which we will shortly make mathematically rigorous). A glial cell can be a component of many such mental axes which thus cross in that cell. The six-degrees-of-separation rule probably holds, as billions of connected-glial lie along, and define, each active mental axis. Every one of these mental axes crosses the zero mental axis which is generated during development. This is equivalent to the unique zero point at the corner of a multidimensional cube in hyperspace. Each of the glial subprograms associated with an axis can be called upon by other RNA subprograms, from another axis, also in that glial. Thus, while there are a zillion programs running in my brain right now, there are only thousands of basically-different ones at work—each running in billions of slightly-different versions along each mental axis.

Connecting Buses:

Mac: The serial connections between central modules—the buses—are 64- or 128-bits wide, each bit being distinctive.

Brain: The serial connections between major modules—the white matter—are a zillion-bits wide, each bit being very slightly different to its neighbor's.

Long-term storage:

Mac: The contents of my Mac's active-memory—intricate longitudinal patterns of $+^{ve}$ and $-^{ve}$ electric charges in RAM—are constantly written and read, to and from, the short- and long-term sub-types of storage-memory. The short-term version is just another variety of RAM. The long-term version is quite different, it is a linear pattern of tiny N^{th} and S^{th} magnetic poles on a well-organized hard drive. Unused storage space on the hard drive is either blank or old stuff that can be written over. From the perspective of active-memory, however, both types of storage are identical except that short-term memory takes a few cycles to access while the long-term takes many, many cycles. There is a single hard drive which stores absolutely everything in long-term storage when I close the lid and the "Put to Sleep" program runs. This hard drive can be removed, plugged into another Mac and the active memory brought back to life, the lid can be opened, in a totally-new Mac.

Brain: The contents of my brain's active-memory—intricate transverse patterns of Na^+/K^+ membrane depolarization patterns in neural nets——are constantly written and read, to and from, the short- and long-term sub-types of storage-memory. The short-term version is just another variety of neural-net firing. The long-term version is quite different, it is a linear pattern of tiny A, U, G & C nucleotides on a memoryRNA subprogram in a glial cell. Unused storage space is always a blank—there is no such thing as writing over old memory, it is just added to. The moving finger writes in memoryRNA code, and it never overwrites, never even backspace-deletes. From the perspective of active-memory, however, both types of storage are identical except that short-term memory takes a few cycles to access while the long-term takes many, many cycles. Each glial cell stores such a memoryRNA subprogram of each of the mental axis it is a member of. This program is added to when that mental axis is in action. Each glial cell has stored absolutely everything in long-term storage when I am lying down, thinking vague thoughts of Oprah and tensors, and the "Put to Sleep" program runs. "I Am" when asleep is entirely stored as a billion-D real matrix of glial, with a set of memoryRNA programs at each location. All that really needs to be stored here are the values needed to specify a QPF, just q, p, I and s. As any three determine the fourth, only three values need to be kept in storage. Along the axis, these values would be the ones that alter a pixel at a time along the mental axis. This matrix of memoryRNA could theoretically be removed, dropped into another brain and the active memory brought back to life, the smell of good coffee is a good trigger, in a totally-new brain. Unless the brain is somehow a blank, I do not think this is nice idea, but it is theoretically possible. Subprograms of "I Am," such as knowing Hilbert-matrix-tensor algebra (see below for an introduction), would be a much better commercial prospect.

Programs running on a real operating system/chip:

Mac: The hundred-or-so programs running in active memory generate, in general terms, a single QPF that directs the functioning of keyboard, screen, or memory, etc. In the bottom-up perspective terms used in this book, this external activity is an example of stuff filling-in the QPF provided by programs. When I type a key on the Mac active and short-term memory spring into feverish activity. When I choose "Save," the long-term memory also gets involved.

Brain: When a glial cell runs one of its memoryRNA programs it generates a QPF that strongly influences the behavior of the local neural net it "serves". Each glial cell running a program generates such a QPF. This QPF, of course, obeys the equation q=rI-rp As a mental axis is either active or inactive along its entire length, billions of similar programs, altering in pixel-by-pixel fashion the shape of QPF they generate along that axis. As thousands of higher programs, and billion sand billions of the lowest type, are all running—each on its own a mental axis—we end up with a zillion little QPF all interfering with each other. As the Mandelbrot set illustrated, in a very simple and 2-D, such mingling of complex numbers can result in interesting things. It is this final, massively-composite nervous system encompassing   mindQPF that actually directs the firing of neural nets (even though the local glial have the most. When I type a key on the Mac, active-, short- and long-term memory all spring into feverish activity in my brain. The composite mindQPF around the finger changes and my finger moves to fill it in, hitting the key. From the bottom-up perspective, this is a flood of nerve impulse arriving in the muscles, and a flood of calcium ions making my actin molecules flip to a shorter shape. However, this is effect, the cause was the body-spanning composite mindQPF. There is no "save" button in my mind brain to call upon: Every thought, to a lesser extent, and every experience to an absolute extent, is reflected in the active and short-term memory as well as the endlessly extending memoryRNA program in each and every glial cell (many are blank, of course). This is the permanent hard drive storage where that implacable moving finger stores its, so far, unchangeable record of you and your life. The programs in this memoryRNA are what you and I really are. Some might thing this demeaning; I think it's brilliant! I'm delighted to know what I am. (It's better, to my mind's sense of dignity, than the 'you're a piece-of-meat, brain secretion' the classical science perspective suggests I am.)

Mobility:

Mac: The long-term storage form and transport is highly variable. This applies to all types of storage but we will use a program as an example. I can drag a copy of Word:Mac2004 from my hard drive to a virtual drive, from there drag a copy to a CD, from there grad a copy to replace the original on my Mac's hard drive (do not attempt at home). Same program, different external form. If I had downloaded this latest version of Word from the web over my cable wide-are LAN connection it would have been the same program but gone through the following forms as patterns of:

Pits and hills on the master DVD. Laser pulses. Electric charge on a Microsoft server's high-speed virtual hard drive. AC current in a laser driver. Light pulses in a coaxial cable. AC current in my cable modem, Ethernet cable and Airport module. Radio waves between Airport and Mac. AC current in my Mac's antenna. Electric charge in my RAM active- and short-memory . Patterns of Nth and S^{th} on my hard drive

Brain: Glial-matrix-memoryRNA is the only storage form we know of in which the human "I Am" comes in. Radio waves seem out of the question, though the reported 'golden cord' connection in out-of-the-body experiences is suspiciously like a computer bus or the *corpus callosum*.

This completes our list in which the computer Mac is not the same as the brain computer. For all that, they are minor differences compared to the low-resolution, general analogy we will to use in discussing the mind.

Keeping the limitations of the analogy in mind, we can now discuss how my brain is just like my Mac. Perhaps that is why we get along so well.

Consider this hierarchy of programs running on my Mac—Word-for-Windows running on Windows NT running on VirtualPC running on Mac OSX running on the Unix shell calling Machine code controlling a Motarola chip.

Wake up!

I am typing away at the top, in Word-for-Windows, when I stop and start thinking about what I am writing. After one minute of inactivity, my Mac goes to sleep. It turns off the screen, writes a virtual image of the entire active- and short-term memory to the hard drive, halts the disk then turns everything off except a low-level little program that detects when a key is hit and calls the "wake up!" program. All this is to save battery power.

This file stores, all in one linear array of magnetic poles, an image of:

The state of Word-for-Windows + the state of WindowsNT + the state of VirtualPC + the state of Mac OSX + the state of the Unix shell + the state of the machine code

This is the Mac when it is asleep. We can call all this the magnetic image, m-image of the Mac in stored form.

Using this as an analogy, with all the limitations just listed, I can now describe myself at work on this book as:

The "I Am" User program running in a
  Virtual simulation of solid reality generated by the
  Virtualworld program running on the
   Brain OS running on the
   Module OS shell calling
    Neuron firing controlling a
    Body

While not as simple as the 'mind-and' concept, it is probably more accurate.

We can now use this to describe what happens when some low-level programs decides it is time to sleep and calls the "put to sleep" program from its RNA store along the glial on some mental axis.

There is no need to pause to "write to disk" as the glial have been doing this continuously all day. So active- and short-term memory can be just switched off, to clear them, and then filled with the nighttimes housekeeping chores program. Especially shoring up any new mental axes established during the day's experiences, like finding a long-elusive appreciation for rap music.

While I am asleep, the matrix of glial holds all of me in storage. This ultra-file stores, all in one 3-D array of memoryRNA, an image of:

The state of "I Am" + the state of Virtualworld + the state of the VR program + the state of the brain + the state of the modules + the state of the neurons

This is "I Am" when I am asleep. We can call all this the RNA image, the r-image of Me in stored form.

Waking up is so similar I will not discuss them separately. The image file is loaded back into active- and short-term memory, the neurons or RAM, and "I Am" and my Mac spring back to life to resume where we left of.

The human mind is actually programs running in a virtual reality. Philosophers talk about mental 'qualia,' such as 'red' or 'sweet." They are, of course, constructs of the VR. The program that generates the VR has a table that it compares the sensory input from the optic nerve, it stimulates the red mental axis appropriately. We only imagine we are experiencing a sweet red apple—it's really just an avalanche of neuron impulses. As we all inherited the same table from our distant ancestors, we all must 'see' the same red as the VR is the same for all of us. If this reminds you of The Matrix, you might have a point. (In mitigation, I believe that the purpose of it all is benign and beneficent, if given a chance, so Don't Panic.)

We stated that, if classical science is correct, this image of Me can only be contained in an r-image file. If most religions are correct, then we can assume that there is at least one other form that this Me image can be stored as.

Perhaps, like Airport and my Mac, they are as different as radio waves in a room and pits-on-a-DVD's silver surface are. We shall pick this up again in Volume Two.

Brain Science

Now for the concept of a mental axis and how it can be mathematically-described. Each concept we have, each thing to which we give a name, a word. Each concept has a mental axis. Each concept has a word—or a structure of words that we use for it both inside our mind and when we speak externally of it.

Take the word 'quantum' as an example. You knew the word before you read this book so there was a mental axis assigned it, a linear chain of glial cells starting at the zero point zero.

If you are a well-read non-scientist, this was a meager chain, with few connections or intersections with other words and concepts. If you are a specialist biologist, you started with a similar, if longer and better connected, mental axis for the word and concept "quantum."

If you are a quantum physicist, you will have started with a mega mental axis for the word with a mega number of intersections with other words and concepts each on its own mental axes.

If you have never come across the word quantum before, you started without a mental axis for it. If you have made it this far, however, a new mental axis will definitely have been extended and activated.

In all cases, hopefully, you will have made a lot of new connections and intersections between glial-stored mental axes.

We have already discussed all the math we need for basic brain science:

The glial cell array and its connections—the long-term storage mode—can be described This matrix is as a regular polygon in a multi-dimensional real space.

The active- and short term memory involves a QPF being generated at each entry, we have extension into complex space. This can be described with a multi-D Hilbert matrix with a q=rI-rp QPF at each entry. This matrix is as a regular polygon in a multi-dimensional Hilbert space and odd things can be expected to happen.

Popper's World

We can conclude that Popper's three worlds are actually remarkably similar when looked at the quantum way we are suggesting:

World One, the physical universe. The material world is a composite of myriads of interfering QPF generated by protons & nuclei, filled-in by a myriad electrons.

World Three, the mental universe. The mind is generated by a composite of myriads of interfering QPF generated by RNA-programed glial cells, filled-in by a myriad neurons firing. There are many ideas: a few are allowed to 'escape' from the virtual reality

World Two, the cultural universe. There are many ideas in every human mind: a few of these ideas are allowed to 'escape' from the virtual reality and are expressed, for good or ill, in the external world. A culture is generated by a composite of myriads of these expressed QPF generated by the minds of the citizens, filled-in by a myriad of things such as families, audiences, law-courts, paintings, crime, corporations, concrete, wars, etc.

We conclude that, at least in this book, all three worlds are, basically, massively composite QPF being filled in by stuff. A theoretically Unified Science, indeed—all we need is a testable prediction...

TELEPORTATION

Did we forget anything? Oh yes, protein folding. Linear aminoacid chains finding their correct configuration; and calcium jumping.

Hopefully by now you have been so fully impressed with the power of quantum probability that that teleportation of aminoacid chains will pose no intellectual problem for it at all.

Consider a set up of two boxes separated by a good distance such that the probability of the test object being found in either box is 50% and being in-between is 0%. The experimenters regularly check the boxes. Half the time the object is in one box, half the time it is in the other. Yet, even if the boxes are miles apart, they are never found in transit between them. Basic teleportation. (Actually, experimenters usually do the much simpler, if totally equivalent 'slit experiment,' but it's the same phenomenon.)

It takes a lot of ingenuity to demonstrate teleportation—the manipulation of quantum probability is still in its infancy—as things usually conspire to make the probability of moving incrementally exceedingly high and the probability of not doing so exceedingly small.

Starting with the simplest thing—the photon of light—experimentalists, over the years, have developed set-ups that have demonstrated the teleportation of electrons, atoms and even molecules—big ones.

I came across this with a google search:

The Vienna team, now jointly led by Zeilinger and Markus Arndt, has performed a new experiment on tetraphenyl-porphyrin molecules. These biological molecules are present in chlorophyll and hemoglobin. They have a diameter of about 2nm, which is over twice as big as a carbon-60 molecule.

These are big molecules the size of lipids, which is the polite term for fat and, as any dieter will attest, fat is decidedly flesh and "solid matter." Scientists are doing the equivalent of teleporting tiny bits of matter. The 'beam me up' days of commuter heaven will be here when these baby-step experiments are scaled-up somewhat. Don't burn your MetroCard yet, though, as the scale factor is about a trillion, trillion fold.

Now we can look at protein folding in terms of quantum probability forms instead of using the classical concept of lock-and-key.

Earlier we diagrammed the 'dissatisfied' aminoacids linked in the natal extended chain with their various needs and desires as the outies and innies of jigsaw pieces.

In the quantum view they are not 'solid' at all, the "bumps and holes" on them actually represent arrays of fully-filled, partially-filled and totally empty quantum probability forms. Now a 100-aminoacid chain is, admittedly, larger than a tetra-phenylporphyrin molecule. But now the scale factor is a much more reasonable 50-fold

So, it is not unreasonable to suggest that a folding aminoacid chain takes the short cut and quantum jumps from the very-low probability extended state to the very-high probability, active form without trudging through everything in-between. Almost as rapidly, the chain could make a series of jumps to the final state.

Teleporting Proteins

Most of the interesting biochemical changes take energy input to make them happen, usually by involving ATP in the process. They are all endothermic energy-absorbing changes. Without the energy input from ATP, they would never happen even with all the correct enzymes present. Protein folding is almost unique in that it is an ectothermic process—it gives off heat and happens quite spontaneously.

So there just might be a way to test this. The extended chain has a higher free energy than the folded state. In the classical picture, this energy given out gradually and appears a slight raise in temperature. In the quantum view, this energy will be released all in one lump in the quantum jump and will appear as a photon of a specific frequency.

As the energy difference is not that great, the photons will be in the invisible infrared and microwave regions. I googled "visible light emission during protein renaturing" and got null results.

The outline of an experiment to test this is conceptually simple. Two cells—one the protein deficient control, the other a dilute solution of a simple and very pure enzyme—watched by sensors that span a wide frequency range.

We heat the solutions to above the denaturation temperature and then let them cool quickly. The sensors measure the spectrum of frequencies and make their report.

The classical prediction is that the thermal radiation of the test will show a simple shift compared to the control.

The quantum view suggests that there will be no shift in thermal spectrum but rather a spike, or series of them.

Heat denatures the chain, cooling renatures it. If quantum jumps are involved, then the opposite effect should also be observed: a laser tuned to the emitted wavelength should pop a folded chain back into the denatured state.

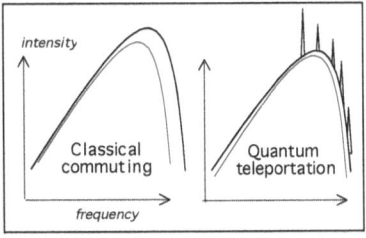

A few questions to conclude:

Can a tuned laser denature and renature enzymes?

Any entrepreneur have some spare venture capital to explore weightloss by teleport-o-liposuction?

Summary

In Science, there is recognition of the universal impulse to choose a history that maximizes Quantum Satisfaction—a generalized inverse of the Principle of Least Action.

In a Quantum Science, natural law determines the Path of Greatest Satisfaction and thus the Quantum Probability of choosing to do something, or nothing.

Quantum Probability is very, very different to the chance-and-accident probability of classical science (and the chance of leaving Vegas with a cent).

Quantum Probability is all-powerful—from supporting old stars to the glint a diamond building to a baby's smile—and rules our universe with an inexorable, iron hand.

In a Quantum Science—other that the probability weighting just mentioned—all systems, from the simplest to the most sophisticated have true autonomy of choice. The only mathematical description of this is the True Random Number generator/operator (which, as such does not exist, does not help much). Moreover, things occasionally choose to do things with a vanishingly small Quantum Probability. This firmly- and painfully- established principle is summarized in the popular aphorism: Even God doesn't know which slit the electron will pick. Let alone me!

Therefore, while in classical science, the principle is that things must follow the Path of Least Action; in Quantum Science, things tend to follow the Path of Greatest Satisfaction.

The autonomy of all things is trumped by the well-accepted Law of Large Numbers. All this insists on is that, given enough repetitions, the actual history will faithfully flesh-out the Quantum Probability.

What things are involves long-term, interacting subsystems moving in a Quantum Probability Form (QPF). Our exemplar is the 1s (pronounced one-ess) orbital. We will also deconstruct and generalize the Schrödinger Equation to a form suitable for dinner conversation and T-shirts (see info for agent).

Reality is pixilated. Long-lasting QPF on each level of science are distinct and relatively small in number. Unlike classical science, which allows a continuum of forms, QPF are discrete and, at most as in the Human OS generated Virtual Reality, a denumerable infinity, not a continuum, is involved. Well-formed ideas are distinct.

What things do involves them exchanging body parts, with others of complementary tastes, via a Quantum Probability Field (also QPF).

One system (the generator) can provide a QPF for another, perhaps quite different, system to move in. Given sufficient time, it is inevitable that the filler-in system's history will reflect the QPF. Clay is our primitive example, proteins our most sophisticated.

Protein folding involves teleportation of the whole linear chain.

A triplet-code program, written on RNA, running once on the Basic OS 1.0 of life, generates a single metabolic QPF.

The Basic OS is described and its Evolution discussed.

Evolution occurs first on the internal, programming level. Novel programs are generated in a way reminiscent of antibody generation in lymphocyte.

There is a Virtual Reality QPF, generated by programs as above, in which programs are first tested before release. Errors are ruthless eliminated. Our exemplar is the lymphocyte and its testing at the hands of the thymus, a harsh master who kills those who fail his requirements.

I call this Internal Survival of the Well-formed, and it all occurs in the virtual, internal QPF VR.

Released into the real world to run, the program now faces the well-documented gauntlet of External Survival of the Fittest. This is the second, and quite subsidiary, step in evolution: the Darwinian reaper.

Variation is not random, it's stuff filling in distinct QPF and, in the long term, its destiny is utterly determined. (That should mollify the theologians for that seemingly-demeaning quip about God and the electron through two slits. They can claim God set up the probabilities, made some subatomic particles, then sat back to wait for the stuff to fill into the QPF and humans, at last, appear. Fear Not! For God will only appear, like this, in the text when I have a sidebar comment to inflame the to-the-death blood feud between classical science and traditional religion. As both have been superseded, however, they should take a time out for a breather, perhaps to watch the coronation of the Last Pope, Benedict-N, I will probably see.)

A bacterium has thousands of programs stored on its DNA. They are programmatically-called upon, as RNA transcripts, and the programs run in massive parallelism and repetition on millions of copies of the Basic OS. Each little program-run adds another QPF to the overall composite QPF.

The composite of billions of these QPF—interfering as waves in a wavefunction, a grand final probability amplitude with a tinge of Mandelbrot—is the QPF we recognize as a healthy, hearty bacterium when it is filled in by the atom-stuff. The QPF is relatively constant; the atoms are not, they are constantly replaced.

Metabolism can be usefully considered as flows of atom-stuff pouring in to fill in the QPF generated by the bacterial RNA programs. Then pouring right back out again. The timescale is longer, but the same is as true for us.

In brief, the massively-composite bacterial/organelle QPF is generated by many linear triplet-code RNA programs running massively in parallel on multiple Basic OS 1.0. Filled in by over time by a flow of atom-stuff, this filled-in QPF is what we call a healthy bacterium

Damage is simply healed by regenerating the QPF and waiting for the stuff to fill it in.

The elementary aspects of feelings and desires are discussed and established here on the bacterial level.

We now apply this basic concept to each of the levels in life's sophistication. In each case, we discuss the nature of each OS and its development in a womb-eden-mom-clay environment. On each level, healing and damage control is handled as above. Same for emotions. So:

Cell level: The massively-composite Cell QPF is generated by many linear spindle-code RNA programs running, massively in parallel, on multiple Cell OS 1.0. Filled in by over time by a flow of organelle-stuff, this filled-in QPF is what we call, and recognize, as a healthy eukaryote cell.

Organ level: The massively-composite Plant/Organ QPF is generated by many linear virish-code RNA programs running, massively in parallel, on multiple Organ OS 1.0. Filled in by over time by a flow of cell-stuff, this filled-in QPF is what we call, and recognize, as a healthy organ.

Body level: The massively-composite Body QPF is generated by many linear, higher-language RNA programs running, massively in parallel, on multiple Body OS 1.0. Filled in by over time by a flow of cell-stuff, this filled-in QPF is what we call, and recognize, as a healthy organ.

Basic Mind: The massively-composite mind of a worm is generated by many linear glial-code RNA programs running, massively in parallel, on multiple Neural-1 OS 1.0. This is filled in by over time by a flow of patterns of neuron firing. This is expressed via the nerves in the well-established way. This filled-in QPF is what we call, and recognize, as a healthy and happy worm.

Fish Mind: The massively-composite mind of a fish is generated by many RNA programs running, massively in parallel, on multiple Fishy OS 1.0. This is filled in by over time by a flow of patterns of neural net firing. This filled-in QPF is what we call, and recognize, as a healthy and happy fish.

Amphibian Mind: The massively-composite mind of a fish is generated by many RNA programs running, massively in parallel, on multiple Fishy OS 1.0. This is filled in by over time by a flow of patterns of neural net firing. This filled-in QPF is what we call, and recognize, as a healthy and happy turtle.

Reptile Mind: The massively-composite mind of a reptile is generated by many RNA programs running, massively in parallel, on multiple Dino OS 1.0. This is filled in over time and is what we would have called, and recognized, as a healthy and happy dinosaur.

We conclude that, yes, proteins do teleport in an RNA world.

The task of science is to deconstruct all these RNA-bourn programs on RNA, or dePrograming as I like to call it.

Time Frame. The rule is that internal and external evolution of programs can only proceed until they become subprograms called by a higher language. An OS rarely changes, and then by just a tweak.

Basic OS 1.0 emerged 4.2 billion years ago in a black-smoker per fused China clay bed; the first eden-womb.

Cell OS 1.0 emerged some 3 billion years ago in a womb-eden stromatolite.

Organ OS 1.0 emerged in the Ocean womb-eden some billion years ago.

The animal body-basic brain or Fish OS emergence kicked off the Cambrian Explosion some half billion years ago. The VR testing place was perfected as the Basic Mind we also have as a part of our Mind.

Then the Amphibian OS 1.0 emergence and its subsequent VR development and perfection as the basic Amphibian Mind we also have in our composite Mind.

Same for the Reptile OS 1.0.

And the Mammal OS 1.0.

The Primate, Ape and Hominid VR sequentially emerged with increasing sophistication until, less than a 100,000 years ago, the VR was perfected as the Human OS, the Mind in which you, a Master Program are running and trying to make sense of all this. What you think of as thinking is actually a program running in a VR, which is, to my mind. Somewhat less demeaning than being told I'm just a bunch of neurons sparking.)

The emergence of any new sister species, and in particular, the first emergence of the Human Mind VR in Adam and Eve involves:

The tetraplex of meiosis and the little-understood recombination complex. It is in the tetraplex of the male that a previously unoccupied QPF gets filled-in with a quantum jump (or 'the Word of God' takes form, as the religionists would have it).

Four generations:

A pre-grandfather in whose testis the empty Human VR generating program is first filled in. His offspring are the:

Hominid pre-parents, who carry the Human Program as a massive, inactivated Barre Body in their germ cells. Parents of:

The first True Humans, born from a hominid womb with regular navels. The language instinct emerges about age four and culture begins.

The birth of humans in the regular way. A new species, but much more significantly, a new VR, has successfully evolved internally, and the become established externally.

Diploid sperm, triploid zygotes, four-ploid germ cells and other such oddities.

The nature of sleep, dreaming and Dreaming is discussed. This ends with a teaser for Volume Two, which deals with the Planck Mirror that separates the physical and spiritual realms.

VOLUME 2.
MATHEMATICS, PHYSICS AND CHEMISTRY

UNIFIED SCIENCE

This work is a presentation of the author's concept of a scientific worldview based on the Divine Principle (theological) and Unification Thought (philosophical) worldview as first taught by the Reverend Sun Myung Moon.

This worldview embraces three great realms that, together, make up the Cosmos:

1. The Abstract Realm. This is where God, the Natural Laws and the Mind reside.
2. The Physical Realm. The objective reality in which we have a physical mind and body. We are born, we grow up, and we die.
3. The Spiritual Realm. The objective reality in which we have the spiritual mind and body we developed while in the physical realm. We exist here without end.

Current scientific understanding of the Abstract Realm is restricted to mathematics. The exploration of the Physical Realm is well developed, and will be the main focus of this work. Current scientific understanding of the Spiritual Realm is almost zero although, as we shall see later, a start might just have been made in exploring this realm.

It is not our intention to prove current science 'wrong.' It is, however, our intention to show that current science is incomplete; that it is a subset of a larger picture. The Unification view as presented here has four important implications for current science:

1. That spacetime is a two-sided entity, and that the physical realm only inhabits only one side of it.
2. That natural law is a sophisticated, not simple, abstract construct that is mathematically-similar to spacetime.

3. That it is incorrect to ignore the quantum wavefunction aspect of physical systems in all the sciences except physics and chemistry.

4. That living systems involve the RNA-controlled interplay of analog form and digital memory. Their evolution history is driven by accumulated wisdom, not by chance-and accident.

Each of these four points will be explored in detail as the discussion progresses.

Science and religion take the opposite tack in their attempts to explain the world we live in. Religion is a 'top-down, big-picture' discipline; it starts with God and works downwards to dealing with simpler matters. Science is a bottom-up discipline; it starts at the bottom with the simplest things, and works upward towards comprehending the big-picture.

This work takes the scientific approach, so God will only appear towards the end of the discussion, not at the start. We will find it necessary, however, to replace the vague and simple concept of Natural Law with a much more sophisticated and mathematically well-defined entity we will call the Logos. Only towards the end of this work will we need to discuss how and why such a sophisticated abstract entity as the Logos came into existence. Only the natural law Logos which is immutable and unchanging will be necessary to understand the structure and history of the physical realm up to the origin of Man. We will not have to invoke any direct action of God, a view that is compatible with religious history and the obvious immense difficulty that God has in directly influencing events in the physical realm. In the final section we will briefly discuss the reasons for such a 'hands-off' approach to the creation and running of the physical realm.

We start with the very basic question of all: Why is there 'something' rather than 'nothing' and why we have to include an Abstract Realm if we want to aspire to a complete picture of the 'something' we find ourselves in.

THE ABSTRACT REALM

While there is probably a technical term in philosophy for the simplest of entities that inhabit the Abstract Realm, here I shall just use the term 'concept.'

We start with two concepts: that of 'absolutely nothing' and 'a set.'

The concept of a set and what it contains is a fundamental concept. It is commonplace in language where the word 'humans' is the set of all humans, 'cows' the set of all cows, 'integers' is the set of all counting numbers, etc. It could also be a set of simpler sets.

The contents of a set are usually placed inside two curly brackets and separated by commas; so the word 'cows' is the set {cow1, cow2... cowN} while the integers are the set {1, 2, 3, ...}—the first being a finite set while the second is an unbounded set as there is always a larger integer.

Linear Extension

'Absolutely nothing' is the concept we call the empty set, the null set and its symbol is {} and it is said to have zero, 0, members. Any set whose members can be put in a one-to-one correspondence (1-to-1) with this set is said to be of count, or size, zero, the integer (the counting numbers) symbolized by 0.

This null set is now made the member of a second set, a set with a single member, the null set. Any set that can be put into 1-to-1 with this is said to have one member, the integer 1.

The concept of absolutely nothing automatically leads to the concept of one, so we no longer have 'absolutely nothing', we automatically generate the concept 'one.'

As there is an unending supply of nothing we can make endless number of these sets of 1 and by combining them generate all the integers, one after the other in linear order, starting with 0 and going on without bound—which is often simply illustrated on the integer line, starting off at zero and stretching off to an infinity of integers.

$$\{\ \} \equiv 0$$
$$\{\{\ \}\} \equiv 1$$
$$\{\{\{\ \}\}\{\ \}\} \equiv 2$$
$$\{\{\{\ \}\}\{\ \}\{\ \}\} \equiv 3$$

The integers

This is the additive concept that generates the integers. Now while the structure of the abstract realm is vertically hierarchical, there are also many horizontal connections. The 'power set' is another approach to getting something, albeit an abstract concept, out of the concept of Absolutely Nothing.

Given a set of members, how many unique unordered subsets, called the power set, can be constructed. As each member of the set can either be in a subset or not in a subset, there are two possibilities for each member. The number of subsets is thus easily calculated by multiplying by 2 for each member. The power set of a three-membered set is going to be 2 x 2 x 2, symbolized by 2^3, which is eight.

You can always make two 'improper' subsets by either taking none of the members, the null set, or taking all of the members, the entire set. All the 'proper' subsets have at least one member in them. A simple set of three

$$\{A,B,C\} \xrightarrow{\ power\ set\ } \begin{cases} \{\ \},\{A,B,C\} \\ \{A\},\{B\},\{C\} \\ \{AB\},\{AC\},\{BC\} \end{cases}$$

improper subsets
proper subsets

members: A, B and C will serve to illustrate. Note that only different sets, ignoring ordering are considered, i.e. {B,C} = {C,B} so only counts once.

We start with the power set of the set with just one member, the null set, the first to appear out of nothing. We can either take this set or we cannot take this set. In either case we end up with the empty set. As we are only interested in different subsets when calculating the power set this is just one member of this power set.

The power set of a zero-member set is measured by the integer 1, a single set. This gives us $2^0 = 1$, an 'exponential' expression that has caused much confusion.

Now the 'set whose one member is the empty set' is not the same as the empty set, so the power set of a one-member is measured by the integer 2, and in exponential notation, $2^1 = 2$.

Next we take the power set of 2, and this has four members, $2^2 = 4$. We can iterate this process and create a hierarchy of power sets. These rapidly get very large: the power set of a 65,000-member set is an integer that is 19,000 digits long—just printing it here would take up all the room in this book.

A more gradual hierarchy of exponentials is given by the simple rule of adding exponents while multiplying. Just by iterating $2^{N+1} = 2 \times 2^N$ we create each level in turn.

$$2^2 = 2 \times 2$$
$$2^3 = 2 \times 2 \times 2$$
$$2^2 \times 2^3 = (2 \times 2) \times (2 \times 2 \times 2)$$
$$= 2^{2+3} = 2^5$$

Using this hierarchy we have the basis for the binary expression of the integers rather than our familiar decimal system involving adding powers of ten together, where the integer $132 = 1 \times 10^2 + 3 \times 10^1 + 2 \times 10^0$ in decimal and 10000100 in binary.

Even in the temperate binary sequence the size of the power set grows very quickly: for a ten-member set, the power set already has over 1,000 members.

$2^0 = 1$	$2^0 = 1$	
$2^1 = 2$	$2^1 = 2$	
$2^2 = 4$	$2^2 = 4$	
$2^4 = 16$	$2^3 = 8$	
$2^{16} = 65,536$	$2^4 = 16$	
$2^{65,536} \approx 2 \times 10^{19,728}$	$2^5 = 32$	

2^3	2^2	2^1	2^0	decimal
8	4	2	1	number
0	0	0	1	1
0	0	1	0	2
0	0	1	1	3
0	1	0	0	4
0	1	0	1	5
0	1	1	0	6
0	1	1	1	7
1	0	0	0	8

Gregor Cantor was the first to prove that for any set, be it finite or infinite, the power set is always has more members, it cannot be 1-to-1 with the it. This is a trivial observation for finite sets, but not so obvious when dealing with infinite sets, and this leads to some new insights into infinity.

We have already encountered infinity in the unbounded integers. This is called the countable infinity, and as it is the first of many, and this countable 'denumerable' infinity is called 'aleph-zero.' A simple but unintuitive fact is that the infinite set

$$\text{countable infinity} = \aleph_0$$

of even numbers can be put into a 1-to-1 match with the set of integers (which intuition suggests that, as this has all the odd as well all the even numbers, should be twice as big). This simple example illustrates how different infinity is to an integer in which multiplying by two always makes a greater integer while $2 \times \aleph_0 = \aleph_0$, and that a proper infinite subset can be the same size as an infinite set.

1	2	3	4	...	$\aleph_0$
↓	↓	↓	↓	↓	↓
2	4	6	8	...	$\aleph_0$

Passive and active

Starting with the concept of absolutely nothing, we inevitably ended up with the infinite integers. Next we have the concept of integers interacting with each other. The operation of addition is akin to placing two integers side by side and seeing what new integer they are equal to. In this 'placing them together' operation the integers play a passive role. Subtraction, the reverse of addition, leads to the negative integers which zoom off from zero in the opposite direction to the counting number line to negative infinity, and we will discuss them in more detail in a moment.

In multiplication, one of the integers plays an active role while the other plays a passive role. This difference can be seen in the sensible "five times four oranges" versus the nonsense of "five oranges times four oranges." The active role is "five" and the passive role is "four oranges." The active integer 'five' is stretching the passive integer 'four' and transforming it into the integer 'twenty.'

These two roles are not so apparent in the integers where multiplication is 'commutative' and the order is irrelevant. There are, however, important areas of mathematics which are distinctly non-commutative and the order of multiplication is crucial.

SUBJECT

OBJECT

LINEAR STRETCH

RESULT

1 2 3 4 5 6 7 8 9 10 11 12 13 14 15 16 17 18 19 20 21

$x \times y = y \times x$ commutative

$x \times y \neq y \times x$ noncommutative

	1	2	3	4	5	→	$\aleph_0$
1	1/1→2/1	3/1→4/1	5/1→...				
2	1/2 2/2	3/2 4/2	5/2 ...				
3	1/3 2/3	3/3 4/3	5/3 ...				
4	1/4 2/4	3/4 4/4	5/4 ...				
5	1/5 2/5	3/5 4/5	5/5 ...				
↓	:	:	:	:	:	⋱	
$\aleph_0$							

The squaring function is when an integer acts on itself; it takes both the active and the passive role.

The inverse of multiplication is an integer divided by another integer (except zero) which leads to the fractions, the 'rational numbers.' There are a countable infinity of these as can be seen in an array where the ratios of all the integers is to be found. Starting at the corner with 1/1, we can now count all of them in order along the diagonals (that some ratios, such as 1/2 and 2/4, are identical is irrelevant) with a complete 1-to-1 correspondence of the integers to the array of rational numbers.

$$\aleph_0 = \aleph_0 + \aleph_0$$
$$= \aleph_0 \times \aleph_0$$

Many infinite sets are countable and can be put into a 1-to-1 correspondence with the integers. It was a great surprise to the early Greeks to discover that there were numbers along the number line that could not be expressed as a ratio of two integers. A particularly simple example is the length of the diagonal of a square of one unit. Now Pythagoras had proved that in a flat geometry the square of the hypotenuse is equal to the sum of the squares of the other two sides, so the length of the diagonal, d, squared was equal to the sum of one squared plus one squared. The number, d, when squared resulted in the number 2, so d is the square root of 2.

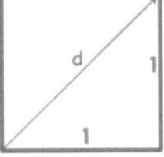

It did not take long before it was realized that there is no ratio of two integers that when squared, results in the integer two. The diagonal was not measured by a rational number, it was a number that was 'irrational.' A modern approach to proving this irrationality is used in

$d^2 = 1^2 + 1^2$
$= 2$
$d = \sqrt{2}$
$\neq n/m$
$2 \neq (n/m)^2$
$2m^2 \neq n^2$

the final expression: that there is no integer that when squared is equal to twice another integer squared. This approach involves another concept that emerges naturally when integers are multiplied and divided; the concept of the prime numbers.

Primes

Integers fall into three classes in terms of their behavior in multiplication and division.

$$1 \times n = n$$

1.   **The Defining integers**. There are just two of these, 0 and 1. Multiplication by 1 does   nothing at all, nothing changes; while division by 1 also has no effect. Multiplication of any integer, no matter how huge, by zero results in zero; while division of any integer by zero does not result in an integer (NAN-=Not A Number) but an unbounded, endless process heading off to infinity. This is why division by zero is not allowed.

$$\frac{n}{1} = n$$
$$0 \times n = 0$$
$$\frac{n}{0} = \text{NAN}$$
$$1 \times p = p$$

2.   **The Prime integers**. A prime number, p, can only be created by multiplying the prime by the unit, and it cannot be divided by another integer to give another integer.

$$\frac{p}{m} \neq n$$
$$c = p \times q$$

3.   **The Composite integers**. A composite number, c, can be created by multiplying a set of primes together, its 'prime factors,' and it can be evenly divided by these primes.

$$\frac{c}{q} = p$$

The 'fundamental theorem of arithmetic' says that every positive integer has a unique prime factorization. Although the proportion of primes falls off as the integers get larger, it was proved by the early Greeks that they are infinite in number; there is no largest prime.

The first few prime numbers are 2, 3, 5, 7, 11, 13, 17, ... All integers, except the defining integers 0 and 1, have a unique set of prime factors. This fact is called *The Fundamental Theorem of Arithmetic.*

$$n = \{p^a q^b\}$$

When an integer is squared, the number of primes in the result is just double that of the integer. It is simple to see, in exponential notation, that the prime factors of an integer-squared must come in even numbers. As zero is an even integer—it has a zero remainder when divided by 2—this also holds for primes that are not in the factorization, such as 2 in the integer 15.

$$n^2 = \{p^a q^b p^a q^b\}$$
$$= \{p^{2a} q^{2b}\}$$
$$15 = \{2^0 3^1 5^1\}$$
$$15^2 = \{2^{2 \times 0} 3^{2 \times 1} 5^{2 \times 1}\}$$
$$225 = 1 \times 3 \times 3 \times 5 \times 5$$

In particular regard to the existence of irrational numbers, this implies that the result of squaring any integer, n, will be an integer with an even power of 2s in its prime factorization. Multiplying the resultant integer by 2 will result in an integer with an odd number 2s in its prime factorization, as in the example.

$$2 \times 15 = \{2^0 3^1 5^1\}$$
$$2 \times 15^2 = \{2^1 3^2 5^2\}$$
$$450 = 2 \times 3 \times 3 \times 5 \times 5$$

Returning to the length of the diagonal of a unit square, assuming that it is the ratio of two integers leads to equating two integers, one with an even, and the other with an odd, power of 2s in its prime factorization. As this is not conceivable, the diagonal cannot be the ratio of two integers and it must have an irrational length. (The name given them only suggests the extreme discomfort the Greeks experienced by this discovery.)

$$d^2 = 1^2 + 1^2$$
$$d = \sqrt{2}$$
$$\sqrt{2} = \frac{n}{m}$$
$$2 = \frac{n^2}{m^2}$$
$$2m^2 = n^2$$

odd power = even power

Using this method, all the square roots of the integers that have an odd-power prime among their prime factors (which is trivially the case of the primes) can be shown to be irrational numbers.

Infinity of infinities

We have already noted that there are a countable infinity of fractions. We can, however, generate a countable infinity of the 'proper' fractions that lie between zero and unity by the reciprocal function, one divided by each of the integers.

The reciprocal function can pack all the integers from 1 out to the unbounded infinity of integers in order into the space between 0 and 1 without missing a single one. Just as infinity is not a number but a limit, so is zero in division (although it is treated as an honorary integer, it's not really, as hinted at by the edict against 'division by zero'). This is an emergent property of integers that are 'interacting' by division, the inverse of the multiply operation. And there is plenty of room for the fractions such as ⅔ that are not on this list. This is an illustration with a vengeance of the fact that a countably infinite set (all the ratios of integers) can have a countably infinite number of infinite proper subsets (the fractions between 0 &1, the fractions between 1& 2, the fractions between 2 & 3, etc.)

The rational numbers are everywhere dense, but somehow there are gaps between them into which the irrationals fit along the number line.

All rational numbers, both integers and fractions, can be expressed as decimals that have a repeating pattern (which can be all zeros) off to infinity (which is usually indicated by a bar over the repeat). Irrational numbers do not have any repeating pattern to them.

$\frac{1}{1}$	$1.000000000\overline{0}$	rational
$\frac{1}{2}$	$0.500000000\overline{0}$	
$\frac{1}{3}$	$0.33333333\overline{3}$	
$\frac{1}{7}$	$0.142857\overline{142857}$	
$\frac{2}{11}$	$0.1818181818\overline{18}$	
$\sqrt{2}$	$1.4142135623730...$	*irrational*
$\sqrt{3}$	$1.7320508075688...$	

It took almost 2,000 years before Georg Cantor proved that the infinity of irrationals was greater than the countable infinity of the rationals. The infinity of irrational decimals was not countable, it was the first example of an 'uncountable' infinity. He showed, by an ingenious diagonal argument involving choosing its nth digit to be different from the nth digit of the nth decimal, that it was always possible to construct a decimal that was not on any list. The irrationals could not be put in a 1-to-1 matching with the integers, there were always irrationals left over.

The counterintuitive conclusion is that, even though the rational numbers are everywhere dense along the number line, they leave space for a much greater infinity of irrational numbers along the continuous number line.

Cantor also showed that the power set of any set, be it finite or infinite, is always greater than the set itself. (A trivial concept for finite sets, but of great subtlety when dealing with infinite sets.) He thus established a hierarchy of infinities founded on the countable infinity of the integers, which he called 'aleph zero.' The power set of aleph-zero and the uncountable infinity of the irrationals is this 'infinity of the continuum' called aleph-one. The power set of aleph-one is aleph-two, an even greater infinity. The taking of power sets can be continued without bound, resulting in a countably-infinite hierarchy of uncountable (except for the first) infinities each greater than the previous.

$1, 2, 3, 4... = \aleph_0$	rationals
$2^{\aleph_0} = \aleph_1$	irrationals
$2^{\aleph_1} = \aleph_2$	shapes
$2^{\aleph_n} = \aleph_{n+1}$	
$... = 2^{\aleph_0}$	

This is quite something, starting with nothing we inevitably end up with the concept of an infinity of infinities. This is the nature of the Abstract Realm as higher levels with more and more sophisticated emergent properties emerge from lower levels without those properties.

Transcendental numbers

To complete our list of numbers along the linear number line we have to accept that some important irrational numbers are more irrational than others; they are transcendentally irrational compared to the algebraically irrational numbers just discussed. There are actually two classes of irrational numbers, numbers that cannot be expressed as the ratio of two integers or as a repeating decimal.

1.   Algebraic irrationals. These involve a finite set of operations on the integers—a finite number of additions, multiplications, taking roots, etc. More precisely, mathematicians say "are roots of a finite polynomial with integer coefficients." So $\sqrt{2}$ is an algebraic irrational; a single operation is involved, and it is a one root, or solution ($-\sqrt{2}$ is the other) to the quadratic polynomial, $x^2 - 2 = 0$

2.   Transcendental irrationals. These involve an infinite number of operations on the integers. They cannot be expressed in a finite form.

We will mention just two of the important transcendental numbers that play important roles in the structure of the Abstract Realm, one is called 'pi' and the other just 'e'. The number π (pi) can only be equated with an infinite number of operations. One such awkward looking equation involves adding up six times the reciprocals of all the integers-squared and taking the square root of the result.

$$\pi = \sqrt{\lim_{n \to +\infty} \frac{6}{1^2} + \frac{6}{2^2} + \frac{6}{3^2} \cdots \frac{6}{n^2}}$$
$$= 3.14159265358979323846264\ldots$$

While π appears in many unexpected places in the mathematical structure, its history is not so much connected with linear extension as with circular rotation (a topic we will soon deal with in detail).

The diameter of a unit circle (one with a radius of 1) cuts the circle into two halves that each have a length of π (while the straight line distance is $2\sqrt{2}$), while the area of the unit circle is also π.

A spot in circular motion on a circle of linear radius 1 making one 'period' travels a distance of 2π going once around the circle. This measure of rotation is called radians, so there are π radians in ½ a period, and ½π radians in ¼ of a period.

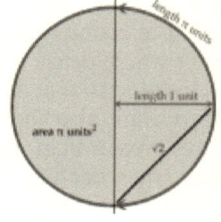

While π gives a relation between linear and angular motion, the number 'e'—the 'base of natural logarithms,'—deals with the balance between the unit and the infinite.

If you multiply the unit 1 an infinite number of times it remains 1. If you multiply 2 in this way, the result rapidly zooms off towards infinity.

For numbers only slightly larger than 1, the result depends on how many times you repeat the multiplication. Enough times, and any number even infinitesimally greater than 1 will eventually zoom off to infinity. If the number of repetitions is small, however, the result stays close to 1. As an example, we will use a number, n, that is just one millionth greater than 1. When multiplied just a hundred times, the result is also just barely greater than 1. Multiplied a hundred million times and the huge number that results has 43 digits before the decimal point.

There is a balance point between these two extremes, a Goldilocks point of being 'just right' we might say. If the number is raised to the power, n, that is the reciprocal of just how much the number is greater than 1+1/n, the result in the infinite limit ends up as the number 'e', which is between 2.5 and 3.0.

$$e = \lim_{n\to\infty}\left(1+\frac{1}{n}\right)^n$$

$$n = \left(1+\frac{1}{1,000,000}\right)$$

$$n^{100} = 1.000100005$$
$$n^{100,000,000} = 2.6\times10^{43}$$
$$n^{1,000,000} = 2.718280469$$
$$e = 2.718281828...$$

The number e is also the limit of adding an infinite number of fractions. The partial series with just 7 fractions to add yields e correct to three decimals. Both of these formulas are a lot more elegant than the one for pi.

$$e = \sum_{n=0}^{\infty}\frac{1}{n!}$$

$$= \tfrac{1}{0!}+\tfrac{1}{1!}+\tfrac{1}{2!}+\tfrac{1}{3!}+\tfrac{1}{4!}+\tfrac{1}{5!}+\tfrac{1}{6!}\cdots$$

$$= \tfrac{1}{1}+\tfrac{1}{1}+\tfrac{1}{2}+\tfrac{1}{6}+\tfrac{1}{24}+\tfrac{1}{120}+\tfrac{1}{720}\cdots$$

$$= 2.7180555$$

$$e^x = \sum_{n=0}^{\infty}\frac{x^n}{n!}$$

Many operations in mathematics involve exponential behavior (ever faster) or logarithmic behavior (ever slower) which are both connected to e. The number e is the basis for the exponential function which, with its inverse, the natural log function, appear throughout mathematics and science. (The term 'log' used to refer to a base of 10, and 'ln' was used for the natural log to the base e, but in most science and math 'log' implies the natural base.

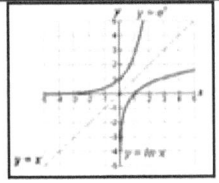

$$x = e^y$$
$$y = \log x$$
$$x = e^{\log x}$$

The structure to this foundational level of the Abstract Realm has many unexpected connections, such as the one between the transcendental number 'e' and the prime integers.

Distribution of the primes

Ever since they were discovered, people have been interested in just how the primes are distributed among the integers. The table of a few primes that we presented above illustrates some facts about the primes:

1. All the primes but the first are odd numbers. As 50% of the integers are even numbers, this eliminates 50% of the integers. Similarly, ⅓ of the integers are divisible by 3, ⅕ of them by 5, 1/7 of them by 7 and so on.
2. As the integers get larger, the primes get fewer in number.
3. The spacing between consecutive primes can be as small as 2 and varies in a seemingly random fashion.

In such studies the number of primes equal to, or less than a given number, n, is called the prime counting function and symbolized by $\pi(n)$, an example of overloading a symbol as the π function is not the same as the number π. The table lists $\pi(n)$ up to 11 (note the plateau between 7 and 11) and the graph gives its value for the first 140 integers.

$$\pi(2)=1$$
$$\pi(3)=2$$
$$\pi(4)=2$$
$$\pi(5)=3$$
$$\pi(6)=3$$
$$\pi(7)=4$$
$$\pi(8)=4$$
$$\pi(9)=4$$
$$\pi(10)=4$$
$$\pi(11)=5$$

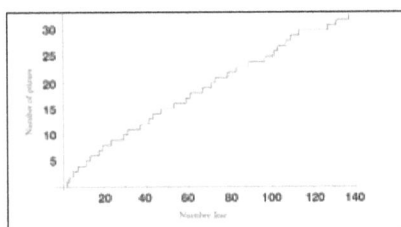

Just a few centuries ago, the proof of the Prime Number Theorem was established. It is this PNT theorem that connects the transcendental 'e' with the integer primes. The PNT states that the number of primes less than a number, n, is approximately that number divided by its natural logarithm, and that the relative 'error' in the count falls to zero as the number gets unboundedly large.

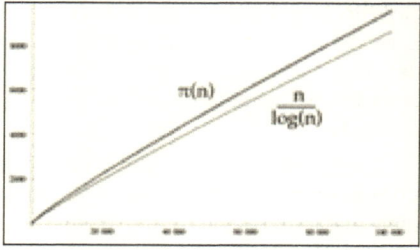

$$\pi(n) \approx \frac{n}{\log(n)}$$

$$\frac{\pi(n)}{\frac{n}{\log(n)}} = \xrightarrow{n \to \infty} 1$$

It will be seen from the graph of $\pi(n)$ up to 100,000 that while the absolute error is increasing (it is about 1,000 at 100,000) the relative error is decreasing (it is 1,000/100,000 or 0.01).

So there is some order to the appearance of primes amongst the integers. The focus then moved to the error in this estimate. A great improvement involved integrals of the logarithmic function, and currently a complete understanding of the distribution of the primes rests on a proof of the Riemann Hypothesis about the zeta function. We will return to this topic later, but it should be noted that we have already touched on yet another Abstract Realm connection in our formula for the transcendental number π. This formula was just a restatement of a formula for calculating the value of $\zeta(2)$, the zeta function of the integer 2.[132]

$$\zeta(2) = \frac{1}{1^2} + \frac{1}{2^2} + \frac{1}{3^2} + \frac{1}{4^2} + \frac{1}{5^2} \cdots$$

$$= \frac{\pi^2}{6}$$

****

So far, our discussion has been based on the concept of linear extension, or size. The image is that of the integers emerging from zero and extending ever outwards in a line. All the numbers we have discussed so far—integers, fractions, irrationals and transcendentals—are on this linear extension from zero.

Along with this fundamental concept of linear extension, there is another concept that is almost as fundamental, that of angular rotation. While this is also an unbounded, you never get anywhere but just go around and around in circular motion, which is the topic of the next section.

Circular Rotation

We have already encountered the unit circle in our discussion of the number π.

We start with a unit extension from zero (linear extension comes first). This establishes the 'real' axis with a rotation of zero. Using the origin as a fixed pivot, we then rotate the line a full circle until it comes back to where it started. This full circle of rotation is variously referred to as:

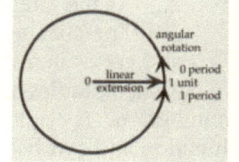

1 period = 2π radians = 4 right angles = i^4 = 360°.

For a constant circular motion, the period can be measured by time. An example is a circular motion that repeats 60 times a second, it has a frequency of 60/sec and the period is $1/60^{th}$ of a second.

In all our diagrams you might have noticed that all the circular motion is in an anticlockwise direction. This is the convention for the plus direction that is invariably used in discussing angles. Clockwise is just as possible, and is designated the minus direction.

Circular motion is found throughout the mathematical foundations of the Abstract Realm where it often appears as its linear components called sine and cosine.

The Sine function

While sines and other such 'trigonometric functions' are usually introduced via static right angle triangles, they are more intuitively grasped when thought of as the connection between circular and linear motion.

If we look directly at a constant circular motion, the end of the unit extension moves in a full circle, measuring off equal lengths per tick of time. If we look at the circular motion from the side, however, that end appears to just go up and down along a linear line in a movement that does not measure off equal lengths per tick of time. The illustration is for the ¼ of a period, the sine of the angle, a, sin(a). At 0 the sine is zero, it then increases rapidly at a decreasing rate until it slowly arrives at 1at ¼ period. The reverse happens in the second quadrant as the sine slowly decreases at an increasing rate until it arrives back at 0. In the third quadrant the opposite happens, the sine decreases to −1 at ¾ before returning back to 0 at a full period. This up and down movement repeats for each cycle of circular motion.

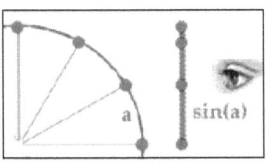

This 1-D view of the 2-D circular motion is called the sine motion. It is an example of 'down-sampling'. There are two things to note about the relationship between circular angular motion and sine linear motion:

1.   The circular motion is in two dimensions and involves a constant change in position
2.   The sine motion is in one dimension and involves a constantly changing change in position.

While the angle change is in regular steps and increases without limit, the sine change is in irregular steps and oscillates symmetrically between +1 and −1. The sine function compresses the two dimensional motion into a simpler one dimensional component of linear motion.

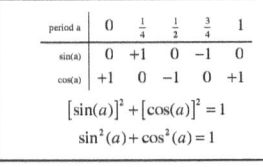

If we look sideways from the bottom a similar motion occurs that starts at +1 goes to zero at ¼ period, to −1 at ½ period, 0 at ¾ and back to 1 after 1 period. This is the cosine component of circular motion.

The sine and cosine movements are equal and opposite complements. When one is unity, the other is zero, for example, and the two functions are said to be out of phase by 90°. From the Pythagorean relation, the sum of the squares of the two components is unity (as in the first equality, which is invariably symbolized by the second, somewhat misleading, equality).

period a	0	$\frac{1}{4}$	$\frac{1}{2}$	$\frac{3}{4}$	1
sin(a)	0	+1	0	−1	0
cos(a)	+1	0	−1	0	+1

$$[\sin(a)]^2 + [\cos(a)]^2 = 1$$
$$\sin^2(a) + \cos^2(a) = 1$$

If the circular motion that is generating them flips from counter-clockwise (plus) to clockwise (minus) it has no effect on the 'symmetric' cosine function but reverses the 'asymmetric' sine function.

$$\cos(-a) = \cos(+a) \quad while \quad \sin(-a) = -\sin(+a)$$

Symmetry

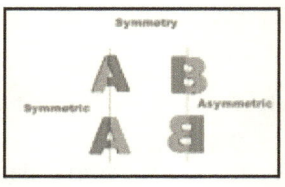

Symmetry plays an important role in the Abstract Realm. A simple example is reflection about a central axis. The letter A, for example, is symmetrical about the vertical center while the letter B is not. The 'B' turns into a 'ꓭ' as seen in a mirror. The letter 'A' is said to be symmetric while the 'B' is asymmetric.

So a single flip does not change a symmetric object. Flipping the letter A a second time also makes no change. This is similar to multiplying plus one by itself repeatedly; at each step you always have plus one.

If you flip the letter B a second time, you get back the original letter B. Flip it again, you have the inverted letter. This is similar to multiplying minus one by itself repeatedly; at each step you alternate between plus one and minus one; it has a period of 2.

Symmetrical, period 1, $(+1)^n$ = +1 +1 +1 +1 +1 +1

Asymmetrical, period 2, $(-1)^n$ = −1 +1 −1 +1 −1 +1

This is why a symmetrical object is said to have a positive (or even) symmetry, while an asymmetrical object has a negative (or odd) symmetry. Mathematical functions also have similar symmetry properties.

We have seen that the sine and cosine functions are identical except for a matter of phase; the cosine is a maximum at 0 or a ½ rotation, while the sine is a maximum at ¼ and ¾ of a turn about the circle.

It is this difference that makes the sine movement an asymmetrical function while the cosine function is a symmetrical function. This can be easily seen if the two curves are drawn on a graph where the zero axis of rotation is at the center and +½ and −½ of a period are placed at either end.

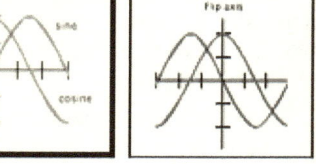

When the cosine (seen in red) flips about the vertical axis it stays exactly the same shape, just like the letter 'A'. The cosine view of circular motion is a symmetrical function.

Flipping the sine function turns it into a different shape, the mirror image, just like the letter 'B' — the sine function is an asymmetric function. There is one final difference between the sine and cosine function.

The sine wave has 3 nodes; one at either end and one at the very center, and 2 disconnected crest areas (or antinodes) in lobes on either side of the zero where all the amplitude of the wave is concentrated.

The cosine function is the opposite. Both the cosine and the flip-cosine wave have just 2 nodes where the wave is at 0 (at ±¼ rotations), and all the waviness is concentrated in just 1 antinode.

Waves

The circular motion is taking place on two of the three dimensions. If the circle now moves along the third dimension, this linear motion combines with the sine motion to generate a sine wave. The distance the wave travels in one period is called the 'wavelength.

In a similar way, the cosine movement combines with the linear motion to create a cosine waves. The sine wave and the cosine wave are identical except for being 90° out of phase. This phase difference makes the sine wave a 'closed' wave that is zero at either end of the cycle, and the cosine wave an 'open' wave that is at a maximum at either end of the cycle.

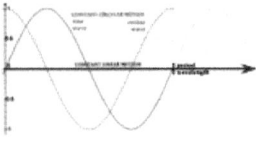

A flip of the circular motion generating the components from anticlockwise to clockwise has no effect on the cosine wave, while the sine wave flips and 'waves' in the complementary direction.

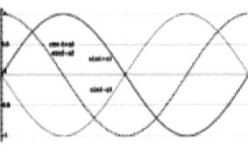

Waves appear throughout the Abstract Realm and the difference between open symmetrical waves and closed asymmetrical waves is often of significance.

The sine and cosine waves we have just discussed are components of circular motion on the unit disk. This creates a sine/cosine wave with an 'amplitude,' or size, that has a maximum of ±1.

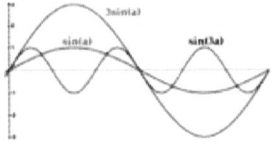

Larger and smaller amplitude waves can be created by just multiplying these 'unit' functions by a scaling factor. For example, the wave $3\sin(a)$ has an amplitude 3 times greater than a $\sin(a)$ wave, while the wave, $\sin(3a)$, has a frequency that is 3 times as great and a period that is ⅓.

One important property of waves is that two or more of them can seamlessly blend into a single wave. The three waves just mentioned combine into a wave with the unit period with four 'crests' in the amplitude per period. In the 1800s, Joseph Fourier showed how any wave could be broken down into a sum of simple waves, and that any wave, no matter how complex, could be created out of a sum of simpler waves. Some waves, such as the jagged sawtooth wave, are the limit of an infinite sum of waves.

The illustration shows how the abrupt sawtooth wave is the limit of a simple summation of a combination of sine waves of the form $1/n(\sin(na))$. The sum when n=1 to 3 and n=1 to 10 is illustrated in green, while n=1 to 100 is in red. Even this number of smooth waves is already getting close to the sawtooth wave. In the limit, as n goes to in-

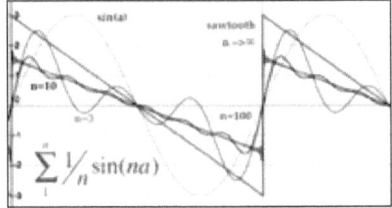

finity, the sharp sawtooth wave (in blue) emerges. We will now look at a few of the properties of waves that will be important when the discussion moves out of the Abstract Realm and into the Substantial Realms.

Wave intensity

We have already mentioned the squaring function, where a number operates on itself in both subject and object roles. This 'self' action is also important for waves. The square of a wave is called its 'intensity,' and the intensity of the minus half of the wave is equal to that of the positive half, by the rule of signs (which we will deal with shortly) where $-1^2 = +1^2 = +1$. A flip from anticlockwise to clockwise motion has no effect on the intensity of a sine wave.

The wave intensity of a (co)sine wave does not have the same shape as a sine wave, it is somewhat 'sharper' in shape and varies from 0 to +1. The intensity is a measure of the energy, or power, in a wave which is all in the 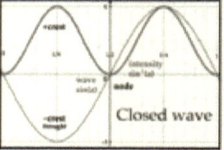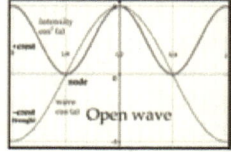 ±crests and zero at the nodes. A closed sine wave has zero energy at the boundary and at the center of the wave. All its energy is in two 'lobes' about the center node. An open cosine wave has all its energy at the boundary and at the center, and there are two internal nodes on either side of the center.

Standing waves

We have dealt so far with just single waves. A more sophisticated wave is one where the circular motion continues and the sine and cosine movements cycle endlessly as they pass a reference point. If such a "traveling wave" has no discernible beginning or end, it is impossible to tell whether it is a sine or a cosine wave. It is also impossible to tell if the related circular motion is clockwise or anticlockwise. The wave can be represented as a simple wave that repeats endlessly.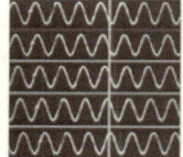

Unlike the traveling wave that is always changing its location and has global properties, a confined 'standing wave' stays in the same location. It is still an endless wave, but it is reflected endlessly between its boundaries. Standing waves are localized, confined waves. When dealing with waves and boundaries, an important measure is the wavenumber, the number of waves that fit into the space allotted.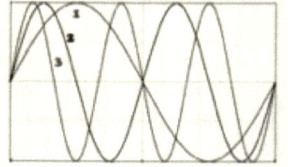

If a sine wave fits by waving once, another will fit by waving twice as will one waving three times. These waves are simply sin1x, sin2x and sin3x, as illustrated, or in general, "*sin nx*," where n is the integer number of waves that fit. We are now going to change the graph. Instead of going off the right side and appearing on the left, we now imagine that the wave is reflected by either end. Now ½ the period fits into the space.

The simplest example is one-half of a sine wave. When it is reflected at the boundary it turns into the other half of the sine wave. All the energy of the bounded wave is in the center of the wave. Examples of closed waves in the Physical Realm are 'closed' organ pipes, the photon of light, and the electron orbitals of atoms.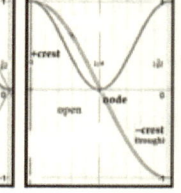

For an open wave, all the energy is in the boundary of the wave, and there is no energy at the center. Examples are the 'open' organ pipes and the gluons of the strong nuclear force.

For waves that are zero at the boundaries, we have a simple series of waves that will fit—rather like a string vibrating with fixed, zero-movement ends. Their periods are ½, 1, 3/2, 4/2… of the fundamental period, and the numbers in red are the wavenumbers, the counting integers, 1, 2, 3, 4, … n.

These numbers play a prominent role in describing confined standing waves. In music they are called the "harmonics" of a sound generator: e.g., the 1st (or fundamental) harmonic, and the 2nd, 3rd, 4th … harmonics of a violin string. In chemistry, they are called the "primary quantum number" of an atomic electron: e.g., the 1s (or ground) state, and the 2s, 3s, 4s … 'excited' states of an electron in a hydrogen atom.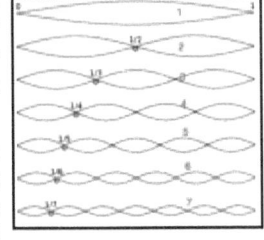

A similar discussion holds for waves that are open at the ends, the cosine wave, with the maximum energy at the boundaries. The energy, for example, of the 'gluon ball' that is the atomic nucleus is all in the surface and zero at the center where the quarks dance.

Such waves are called 'standing waves.' The standing wave is continuous circular motion confined within local bounds, unlike the traveling wave which is globally unbounded. The Greek letter ψ [sigh] is often used to symbolize a standing wave.

The waves so far have been either open or closed standing waves. Depending on the boundary, we can also have mixed waves, one with a node at one end and an antinode at the other end as in many organ pipes.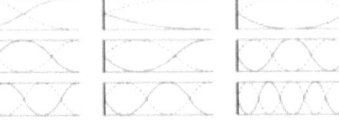

Paired waves

There is just one more important property of waves we need to establish. We have just finished a discussion based on flipping the sine and cosine waves about the vertical axis. This next property involves flipping the waves about the horizontal axis.

Now both functions change shape. The sine wave just flips into its mirror image as before, while the cosine function flips upside down.

This is a neat illustration of the adage that: "for every up there is a down." If a wave fits waving one way, it will also fit by waving the other way. "Waves that fit" come in pairs waving in complementary directions. Other than this, the paired waves are identical.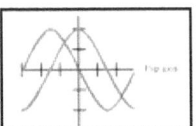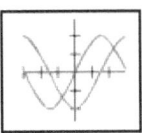

This property of waves plays a central role in the workings of the logos. For every wave that fits, there are two complementary forms—Where the one goes out, the other goes in; where the one is concave, the other is convex. All things that are given their form by waves come in complementary pairs that fit together.

We will now spend much time describing how all things—animal, vegetable, or mineral—are given form by waves, admittedly much more sophisticated ones than those just discussed. So it will take a while before we are able to properly describe how this wave property relates to the male and female form and its delightful way of fitting together.

For now, we will just note that the great difference between the intuitive behavior of the electromagnetic waves that structure the atom and the non-intuitive chromodynamic waves that structure the atomic nucleus is due to the first being a closed wave with a zero energy boundary (sine waves), while the second is an open wave with a high energy boundary (cosine waves).

Interference

So far we have discussed one wave that fit inside the resonator with an integer number of wavenumbers. When two or more waves occupy the same cavity they "interfere" with each other. This can be either positive or negative.

Two identical waves that fit simply add together at every point—the wave has twice the amplitude. The waves are said to be "in phase." While the amplitude of the wave doubles, the intensity of the wave is quadrupled; if the amplitude is tripled, the intensity is nine times as great, and so on. This is positive, constructive interference of the two waves and is energetically favorable.

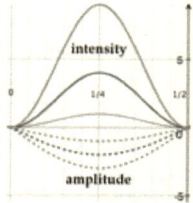

Wave amplitude:	1	2	3	4...
Wave intensity:	1	4	9	16...

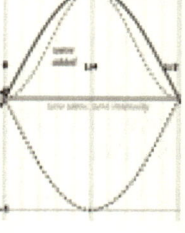

Negative, destructive interference of waves occurs when the 'up-ness' of one is cancelled by the down-ness of the other. Perfect cancellation occurs when a sine wave is added to the negative sine wave. The resulting wave is zero everywhere, it is no wave at all. The two waves have annihilated each other. This is destructive, negative interference. The two waves are totally "out of phase."

Adding sine and cosine waves of the same period together generates a new curve with a slightly greater amplitude than either. This wave also has a period that is larger, where the complete sine and cosine fit into a θ space, the composite wave has only completed ≈1⅓ wave-

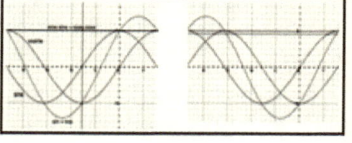

lengths as can be seen from the diagram. The composite wave is out of phase with both the sine and cosine waves. While waves that are in phase interfere constructively, waves that are out of phase interfere destructively.

Consider a sine wave starting off in a resonator which is 99/100th its wavelength, a non-integer number. The reflected wave is still the opposite of the incoming wave, but now the wave is not at zero when it hits the boundary: the reflected wave is now not the same, it is a different looking wave that is "out of phase" by a small amount with the incoming wave. Things get worse when the reflected wave is reflected back again, the wave is out of phase with the original. The wave is diminished rather than being strengthened.

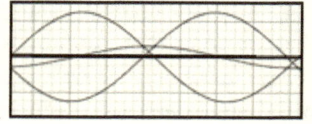

With constant reflection the waves are all out of phase and the result is zero—a wave that does not fit quickly dies away, the resonator will not resonate at that frequency. This is why all standing waves are waves that fit by ½ periods into their bounds.

Harmonious resonance

Next we are going to consider a sound generator that is creating a large set of sine waves of randomly varying frequency—noise as we call it. This is placed as input to a reflecting confined space which has an outlet.

Only those waves that fit within the confines as integer wavenumbers will survive, all of the energy of the random input ends up in the 1st, 2nd, 3rd etc. harmonics. The mix of harmonics in this 'standing chord' is different for different resonators, and is what makes a 'middle C' generated by an organ, a flute, a violin and two human singers sound so different and distinctive.

If the confining space is changed, the standing chord shifts to fit. The time it will take for this sorting out of waves will depend on the speed of the wave. For sound and light with finite speed, this change in the standing wave, the wave that fits, takes a finite time. As we shall later explore, changes in the 'state of the wavefunction' occur within a "pixel of time" no matter the spatial separation involved. As far as current technology is concerned, the change in the wavefunction-that-fits is instantaneous.

When two waves with wavelengths related by a simple fraction to each other combine many interesting phenomena occur.

Some very interesting patterns of constructive and destructive interference patterns occur even when the traveling waves are the same frequency but coming from slightly different sources, like passing through two slits.

This setup is called a "slit experiment" and it was used to discover the wave aspect of photons (a boson) and electrons (a fermion).

The following diagram shows two waves whose wavelengths are 5/6th different in length traveling separately and then together. They make a beautiful pattern of wave-within-wave. In the case of sound waves, this beautiful pattern is perceived as a melodious chord, a pleasant combination of two notes.

Only the small fractions are perceived as harmonious. The large fractions, such as 10/11 or 11/12 are dissonant and are perceived as unpleasant to most listeners. The small fractions relating wavelengths are considered harmonic and are given their own names in musical theory.

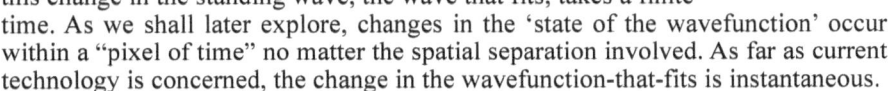

A "chord" in music is when two or more harmonious notes are sounded together. A 'major chord' involves a combination of three harmonious waves. So that all musical instruments can be in tune with each other, there are standard wavelengths, standard frequencies, each with its own name.

A symphony

In music, very intricate harmonious waves are created. A sophisticated example is a choral symphony being performed in a concert hall. This system involves a great number of subsystems. There are a relatively small number of 'wave generator' subsystems, the musical instruments and choir members who create the complex sound waves. There are also a relatively great number of 'wave resonators,' the air molecules whose history and form reflect that of the combined waves projected by the generator subsystems.

The generator subsystems are few, massive and barely move. The resonator subsystems are many, lightweight and mobile. It is the resonator subsystems that 'couple' the audience to the symphony performance. We will find this differentiation between the two roles of subsystems in a system most useful when dealing with the sophisticated waveforms we regularly encounter in everyday life.

SYSTEM OF	INTERACTING SUBSYSTEMS	
	WAVE GENERATOR	WAVE RESONATOR
CHARACTERISTICS	FEW MASSIVE IMMOBILE	MANY SLIGHT MOBILE
SYMPHONY	INSTRUMENTS	AIR MOLECULES

Complex numbers

While the sine and cosine functions are similar, they involve looking sideways at circular motion along different dimensions: the cosine is looking along the length side while the sine function is looking along the height side. The only point they have in common is the zero point. They involve different dimensions, and this difference must captured in the mathematical vocabulary.

The description of this difference involves the concept of operators. We have already seen that in multiplication there are two different roles, one the subject that is acting on the object. The technical way to express this is that an 'operator' is acting on an 'argument.' In the self-acting process of squaring, the number takes on both roles.

The real number line of the cosine axis is rotated anticlockwise by 90° to be the real number line of the sine axis.

The Rotation Operator

This operation "rotated by 90° from the real axis" is indicated by multiplying by letter, 'i.' This is the "rotation operator" that rotates anything it multiplies by 90°. The operator '–i' rotates everything by 90° in the opposite direction.

Rotating by 90° and then again by 90° in the same direction, i^2, ends up back on the real line, but in the opposite direction. This is -1, the negative operator that flips everything by 180°. We get to the same place if we do this in the opposite direction. We have

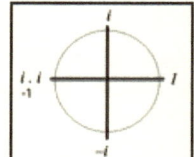

$$-1 \ = \ +i^2 \ = \ -i^2$$

which is why 'I' is also called the 'square root of minus one.'

Performing a third 90° rotation brings us to '–i' at ¾ around the circle, and this is just '–i'. A final and fourth 90° rotation brings us back to the starting point at +1. This is a cyclic group of 4 rotational operators related in the following way.

$$i^1 = i \qquad i^2 = -1 \qquad i^3 = -i \qquad i^4 = +1 \ : \quad i^5 = i$$

This is the group of orthogonal *rotation operators*:

+1	0°	*identity*	+i	90°	*anticlockwise*
–1	180°	*inverse*	–i	90°	*clockwise*

The sine wave occupies an orthogonal dimension to the start axis of the cosine wave, and this distinction result is symbolized as:

cos x   +   i sin x

In adding the two real numbers, the + signifies bringing the two numbers together and uniting them while it is assumed that both numbers are positive—the form x + y is just a simplification of (+x) + (+y) that conflates adding with direction.

This brings us to a discussion of one more type of number to add to our menagerie of concepts that emerge inevitably from the concept of Absolutely Nothing; the integers, rational, irrational and transcendental numbers of linear extension and angular rotation. These are the 'complex' numbers that combine both linear and angular motion into a single number.

Negative numbers

The number -1 is not so mysterious in terms of addition and subtraction, the number line just extends in the opposite direction to the counting integers. It is, however, when considered in the subject role of multiplication. On the real number line, the object number first shrinks to zero and then extends outwards from the other side. This works for even a huge number such as a google $-1 \times 10^{100} = -10^{100}$. As zero plays a unique role in multiplication, this is hard to comprehend.

But when considered as a rotation operator of 180° (π radians, ½ period), multiplying by -1 just rotates the number around and this problem of going to zero does not arise.

A number rotated by 180° has a zero rotation when rotated by another 180°. This is a much more intuitive understanding of the 'rule of signs' than the mnemonic, "Minus times minus is a plus, for reasons we will not discuss."

Note that this flip back and forth is indifferent to clockwise

and anti-clockwise direction (or mixtures of the two). It is only rotation by ½ that is sensitive to direction, and ½ plays a special role in the properties of waves.

Imaginary numbers

With the concept of numbers rotated by 180°, it is a simple step to numbers rotated by 90°, numbers that are on a number line that is orthogonal to the real line, and intersecting the real axis at zero. Multiplying the number 4 by i results in the number $4i$.

Such numbers are called 'imaginary numbers' (the name again suggesting the struggles the pioneers had in accepting such numbers). Multiplying a number rotated by 90° by a number also rotated by 90° ends up with a number rotated by 180°, a negative number, so the square roots of negative real numbers are not found on the real number line, they are imaginary numbers.

The real and imaginary numbers only have one

point in common, the integer 0. Rotation by 90° is sensitive to clockwise and anti-clockwise. Going in the clockwise direction is the operator $-i$. Doing this twice also ends up on -1, so $-i$ is as much the square root of -1 as is $+i$.

$$+i \times +i = -1 \qquad \& \qquad -i \times -i = -1 \qquad so \qquad \sqrt{-1} = \pm i$$

Complex numbers

It is then but a small step to numbers with any linear size, m, (its magnitude) and any rotation, a, (its amplitude). These are the complex numbers, and they not ordered points along a linear line but points on the 'complex plane'. It can be indicated by an arrow of length m with its origin at zero, making an angle a with the real positive axis, the zero for rotation. This is the 'polar' form for a complex number, (m,a).

Complex numbers can also be resolved into real and imaginary numbers components, the cosine and sine components times a scaling factor. This is the rectangular or Cartesian form of a complex number, $(x+yi)=n(\cos a+i \sin a)$ with the n as a scaling factor.

COMPLEX PLANE

The polar and rectangular forms of a complex number, z, are related by this set of relations (where the arctan function is "the angle whose tangent—the ratio of sine to cosine—has the value"):

The great mathematician Leonhard Euler used the infinite expression for 'e' recently mentioned to explore what happened when 'i' appeared in the exponent position. By regrouping, he ended up with the infinite expansions for the cosine and imaginary-sine functions. He established 'Euler's Formula when the angle, a, is in radians. The angle of ½ period, π radians, then gave him 'Euler's Identity' (which many think the most beautiful formula in math) linking the transcendental numbers 'e' and π to the integers.

$$z = m@a \qquad = m(cis\ a)$$
$$= x + yi \qquad = m(\cos a + i \sin a)$$
$$m = \sqrt{x^2 + y^2}$$
$$a = \arctan \frac{y}{x}$$

$$e^{ia} = \cos a + i \sin a$$
$$e^{i\pi} = -1$$

This leads to yet another way of expressing a complex number as an imaginary power of that remarkable transcendental number, e. This gives (at least) five different ways of expressing the same complex number, z.

$$z = me^{ia}$$
$$= m@a$$
$$= m(\cos a + i \sin a)$$
$$= mcisa$$
$$= x + yi$$

Each expression has its advantages and disadvantages, but is it always possible to move easily between expressions using the identities already discussed.

For instance, adding and subtracting complex numbers is very simple in the rectangular form of x and y, while multiplication and division are not. Conversely, it is simple to multiply and divide complex numbers in the polar form of m and a, while addition and subtraction are not. As the not-simple expressions are rather ugly, I have not included them in the table.

	rectangular	polar
$z_1 + z_2$	$(x_1 + x_2) + i(y_1 + y_2)$	
$2z$	$2x + i2y$	
$z_1 - z_2$	$(x_1 - x_2) + i(y_1 - y_2)$	
$z_1 \times z_2$		$m_1 \times m_2 @ a_1 + a_2$
z^2		$m^2 @ 2a$
z_1 / z_2		$m_1 / m_2 @ a_1 - a_2$

The complex numbers are 'complete' in that nothing you can do to them results in something that is no longer a complex number. The real numbers are not complete, for the square roots of all the negative numbers are not to be found on the real line. Complex numbers are found throughout the math level of the Abstract Realm, and they are extensively used in the modern scientific explanation of the workings of the Physical Realm.

The positive real numbers are 'honorary' complex numbers with zero rotation, so the complex numbers inherit all the emergent properties of the linear real numbers. The combination of linear and angular, however, gives the complex numbers a host of new emergent properties. One example is the Mandelbrot set of complex numbers.

Mandelbrot set

$$z_{n+1} \xleftarrow{\quad n \to \infty \quad} z_n^2 + z_0$$

Creating the Mandelbrot set involves iteration of squaring and adding a complex number, z_0, such as in the example shown, and seeing what happens with the result of the repetitions. If the result is bounded, the number, z_0, is in the Mandelbrot Set. If the result zooms off to infinity in any direction, the number, z_0, is not in the set.

In this illustration of the Mandelbrot Set (MS) of complex numbers on the complex plane, the members of the set are colored black, while the numbers not in the set are color-coded according to how rapidly they zoom off to infinity.

The set is symmetrical along the imaginary axis (as befits the equivalence of clockwise and anticlockwise) but is not on the real axis which goes roughly from −1.8 to +0.25.

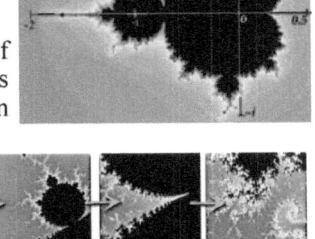

The fascination of the MS is the fractal, irregular boundary of the set. 'Fractal' is the property of having a structure on every level of magnification, while irregular is the property of never quite repeating the pattern.

Numbers that differ by billionths, or trillionths, etc., can be segregated by the boundary. In the illustration, a spot on the upper left of the MS is magnified so that the final one is a 100,000,000 times magnification of the first.

Harmonic Primes

Another of the fascinating properties of the complex numbers was discovered when they were combined with the 'harmonic series.'

This is the name that is given to the sequence of numbers created by adding up the reciprocals of the integers. While the rate of

$\frac{1}{1}$	1
$\frac{1}{1} + \frac{1}{2}$	1.5
$\frac{1}{1} + \frac{1}{2} + \frac{1}{3}$	$1.8\overline{3}$
$\frac{1}{1} + \frac{1}{2} + \frac{1}{3} + \frac{1}{4}$	$2.08\overline{3}$
$\frac{1}{1} + \frac{1}{2} + \frac{1}{3} + \frac{1}{4} + \frac{1}{5}$	$2.28\overline{3}$
$\frac{1}{1} + \frac{1}{2} + \frac{1}{3} + \frac{1}{4} + \frac{1}{5} + \frac{1}{6}$	2.45
$\frac{1}{1} + \frac{1}{2} + \frac{1}{3} + \frac{1}{4} + \frac{1}{5} + \frac{1}{6} + \frac{1}{7}...$	$2.59\overline{285714}$

increase is slow, it was proved that the series was not bounded, you could make the sum any size you pleased by taking the sequence out far enough.

It was also known that the reciprocals of the integers-squared created a bounded series, and the genius of Euler was to show that this limit involved π-squared as noted above. Nowadays this limit is called the zeta of the variable, s, so Euler's result is the "zeta of 2" and infinity is the zeta of 1. His method allowed the calculation of the zetas for all the positive even integers. For the odd integers, this method fails, however, and the only thing known is that they have an (unknown) limit.

$$\frac{1}{1^2} \qquad\qquad 1$$
$$\frac{1}{1^2} + \frac{1}{2^2} \qquad\qquad 1.25$$
$$\frac{1}{1^2} + \frac{1}{2^2} + \frac{1}{3^2} \qquad\qquad 1.36\overline{1}$$
$$\frac{1}{1^2} + \frac{1}{2^2} + \frac{1}{3^2} + \frac{1}{4^2} \qquad\qquad 1.4236\overline{1}$$
$$\frac{1}{1^2} + \frac{1}{2^2} + \frac{1}{3^2} + \frac{1}{4^2} + \frac{1}{5^2} \qquad\qquad 1.4636\overline{1}$$
$$\frac{1}{1^2} + \frac{1}{2^2} + \frac{1}{3^2} + \frac{1}{4^2} + \frac{1}{5^2} + \frac{1}{6^2} \qquad\qquad 1.4913\overline{8}$$
$$\frac{1}{1^2} + \frac{1}{2^2} + \frac{1}{3^2} + \frac{1}{4^2} + \frac{1}{5^2} + \frac{1}{6^2} + \frac{1}{7^2} \dots \qquad 1.51179\dots$$

$$\zeta(2) = \sum_1^\infty \frac{1}{n^2} \xrightarrow{\quad s=\infty\quad} \frac{\pi^2}{6}$$

$$\zeta(s) = \sum_1^\infty \frac{1}{n^s} \xrightarrow{\quad s=\infty\quad}$$

This 'sum of reciprocals' formula for calculating the zeta only works for real numbers greater than 1, but another method for calculating the zetas, established by Bernhard Riemann, allowed for the calculation of the zeta for any number, including the complex, and every number except the integer 1 was found to have a finite zeta (that might be a complex number).

$$\zeta(-2) = \zeta(-4)\dots = 0 \qquad \text{trivial zeros}$$

$$\zeta\left(\frac{1}{2} \pm iy\right) = 0 \qquad \text{zeta zeros}$$

There were certain numbers, that when plugged into this formula, everything cancelled out and the result was zero. These numbers are called the 'zeros of the Riemann zeta function'.

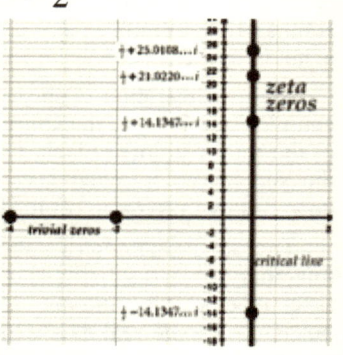

While the positive even integers have a 'closed form' for their zeta involving powers of π, the negative even integers are all zeros of the zeta function. This infinite set of negative even integers is called the 'trivial zeros.'

The non-trivial zeros, and is what is invariably meant by the 'zeta zeros,' is the infinite set of complex numbers that also give a zero when plugged into Riemann's extended zeta function. He proved that if x+iy was a zero, then so was x–iy, and his as-yet-unproved Riemann Hypothesis was that all infinity of these complex numbers had the form, ½±iy. While the trivial zeros are regularly spaced out on the negative real axis, the zeta zeros are irregularly spaced out on the x=+½ imaginary axis, called the 'critical line.' As the size of the imaginary component (ignoring signs) gets larger, the number of zeta zeros along this 'critical line' increase, the zeta zeros get more numerous. This is the opposite behavior to the irregular spacing of the primes which become less numerous as the size of the integers gets larger.

All this brings us back to the integers and the primes, for this was what Riemann was actually exploring, the error function in the prime counting theorem. What he discovered was that the increasing spacing of the primes along the real axis was intimately connected to the decreasing spacing of the complex zeta zeros along the imaginary axis. That the 'error function' in the Prime Number Theorem—which links the distribution of the primes to the transcendental number 'e' in its logarithmic form—

$$\pi(n) \sim \frac{n}{\log n} \qquad \text{the PNT}$$

$$= \frac{n}{\log n} + \text{error}$$

$$= \frac{n}{\log n} + f(\text{zeta zeros})$$

was a function of the zeta zeros. The more zeta zeros included, the more accurately was PNT adjusted.

We have already seen how jagged waves, such as the sawtooth, can emerge from a combination of smooth sine waves. In a similar way, the irregular, jagged graph of π(n) emerges as the contributions of more and more zeta zeros are added to the PNT.[133] In the limit, the prime counting function, π(n), emerges.

Even deeper and sophisticated mathematical connections

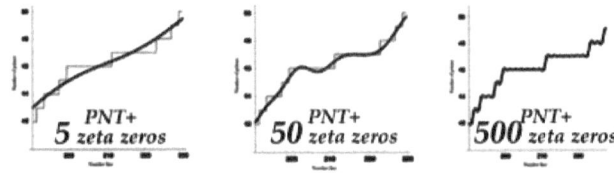

between the primes occur beyond the simple math I am using, so you will have to read his book to find out what prompted an author to exclaim in puzzlement: "What on earth does the distribution of prime numbers have to do with the behavior of subatomic particles?"[134]

Abstract Hierarchy

This concludes our brief look at the mathematical levels at the foundations of the Abstract Realm. Some thinkers do not accept that the Abstract Realm as an objective aspect of reality, and consider Mathematics as a creation of human beings while the Physical Realm is all there is to objective reality.

This is a rather provincial view, however, as human beings have only been around for <100,000 years, while the Universe has been around for ~13.5 billion years, according to current science. Eugene Wigner called it "the unreasonable efficacy of mathematics in the natural sciences." He said this because all of what are called the 'hard' sciences express their concepts in the language of mathematics. (Sciences, such as biology, that rely on fuzzy 'natural' language, such as English, to express their concepts aspire to hardness, but are in the meantime considered 'soft.') Furthermore, all scientists firmly believe, quite rightly, that the mathematical laws uncovered in this age are exactly the same as those at work throughout the billion-year history of the universe when no humans were around. The universe runs by mathematical concepts that preceded humans, and to think otherwise goes against the facts.

An attempt to read an advanced paper on fundamental physics should be sufficient to convince even a devout materialist that the writers consider their subject to be pure mathematics. Modern physics considers the 'fundamental particles' out of which atoms are constructed to be much more 'bits of pure math' than 'bits of pure matter.' There is, for example, an area of simple mathematics called 'group theory' (dealing with sets of things that transform into each other, but not discussed here) that successfully predicts what comes flying out of high-energy smash-ups in particle accelerators. This can only support the concept that entities in the Physical Realm are actually constructed out entities in the Abstract Realm.

The Physical Realm, we conclude, is made of math and run by math. Naturally, such a view precludes any worry about "the unreasonable efficacy of mathematics."

Starting with Absolute Nothing, we have the inevitable appearance of the integers and how they interact with each other (addition and multiplication) and upon themselves (squaring). The interactions at this level along with the concept of rotation leads to the inevitable emergence of higher levels in the hierarchy, the rational, irrational, transcendental, imaginary and complex numbers. Waves are the emergent properties of combinations of linear and angular motion.

On these foundations are the 'higher' mathematics based dealing with very sophisticated entities with many subtle emergent properties.

The basic tool in mathematics is the 'proof'—such as we saw in the proof that the square root of two or any prime number cannot be the ratio of two integers. We can restate as 'a universe in which the square root of two is a rational number is inconceivable.' Using such language, we can state that it is inconceivable that the universe does not have an Abstract Realm that is as real as any other realm. A realm that exists independent of any human exploration of its structure.

Emotional waves

We have seen in the zeta zeros that the hierarchical structure of the Abstract Realm can get very sophisticated. We will now look briefly at just how sophisticated the highest levels might get.

One indication that waves play an important role at very high levels of sophisticated emergent properties is to be found in music, which can convey a wide range of emotions. A punk-rock band pounding out their distaste for society, and a choir & orchestra affirming the joy of life in the climax of Beethoven's 9th Symphony, are equally loud, but the emotions conveyed to the audience are quite different.

Scientists have minutely scrutinized music to see what it is made of. The are by now throughly convinced that all music is pressure waves in air. Whatever an emotion is, it must be conveyed by the shape of the wave, as the 'air' aspect to music is immediately discarded by the listening ear and only the wave is passed on, transformed in turn, into the vibrations of a membrane, the quivering of tiny bones, another membrane, pressure waves in water. This last is so shaped that standing waves form with their antinodes of energy at different places, a 'Fourier Analysis' of the complicated wave at each instant into its component sine waves. Cells lining this shaped water have little flags, called cilia, on their surface. Such cilia located at an antinode wiggle about in the energy there, while cilia located at a node do not as there is no energy.

The amount of energy determines how vigorous the wiggle, call it x, and the cell responds by reporting a digital number up the neural hierarchy that for its set position, call it period a, the energy in the water pressure wave is $\sin^2(a)x$. Each patch of cells does this for all frequencies.

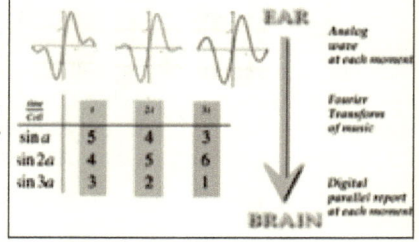

The ear converts the sound wave each moment into the sum of a series of sine-squared waves, and converts this into digital form that is sent, in parallel, up the auditory nerve for further processing. The details of what happens next is still being worked out, but the end result is that we 'hear' the analog wave not the digital numbers, so the wave must be recreated by a reverse transform from digital information to an analog wave.

A simple conclusion from all this that an emotion is the form of an intricately-structured wave. Simply put, an emotion is the property of a wave with a fractal structure akin to that of the Mandelbrot Set.

We can conclude that these sophisticated waveforms in the higher levels of the Abstract Realm have the emergent property we call 'emotion.'

Structure of Man

What about the very highest levels in the Abstract Hierarchy? Can we say anything meaningful about them. Modern science is well in advance of mathematics in answering this question as a result of (metaphorically) taking a human being to bits and seeing what it is made of.

The answer (and we will soon deal with this in detail) turns out to be remarkably simple as scientists found there were only two things:

1. There found a set of 'fundamental particles,' FP, which, as mentioned, are treated as mathematical entities in modern science. The properties of the FP are measured with real numbers, and divide the FP into two classes that obey a different math: The 'force' particles (bosons), and the 'matter' particles (fermions.)

2. They found a 'wavefunction' that was strictly determined by a mathematical law and could only be described with complex numbers. The wavefunction determined what happened to the FP of both kinds, but only indirectly. Rather than the wave directly determining what happened to the particles, it was the intensity of the wave that determined what the particles would do over time.

The conclusion was that if the wave had an intricate form then the particles would take up the mathematically-related form of the intensity of the wave. If the wave was a simple sine, the form of the particles over time would be sine-squared.

It is bounded waves that confine particles together long enough to be considered a 'system' and given a name (while the transitory encounters of traveling waves are usually not so dignified). The number of particles of both kinds involved in the make-up of a human being is enormous—trillions of trillions of trillions of them—and many of them are on their way in or on the way out of the human being. In order to keep track of this enormous number of particles we employ a hierarchy of names—such as organ, cell, molecule, atom, etc.—but they are all just various numbers of particles in a hierarchy of waves.

We have already discussed how simple waves readily combine into more complicated waves. On a human timescale, the zillions of particles making up the human body move as if they were confined and organized by a unified wave, rather than as a collection of isolated waves. Watching a ballerina pirouette or an athlete sailing over an obstacle, it is clear that a unified wave is moving all those zillions of particles in a unified manner.

So a human being is a unified wave, described by complex numbers, that confines and structures a zillion or so fundamental particles, described by real numbers. Nothing more, nothing less is there.

We have already established that emotions are complex wave forms. In everyday language we say that it is our mind that experiences emotions, that it is the "I Am" who is experiencing the emotion (usually along with a lot of other stuff going on in the mind as well).

We can tie all this together by equating the the human mind with the unified wave, and the human body with the intensity of this wave as expressed in particles.

The human mind, in this view, is a wave at the pinnacle of possible waves, with the emergent properties of "I Am" capable of intellect, emotion and will.

As this utterly sophisticated hierarchy of waves is also to be found at the pinnacle of the Abstract Realm, we conclude that the emergent properties at this level include 'I Am,' emotion, intellect and will. This is what is usually called God.

So, starting with the concept of Absolutely Nothing it is inconceivable that we do not end up with a God of heart, intellect and will.

We shall assume that this God initiated the creation of the Substantial Realms out of the Abstract Realm, and created an abstract entity to run the creation along a track to the fulfillment of a grand purpose. Questions about the purpose, how things got off the track, and what is being done to get things back on track, are best dealt with in theology. All I will say on this topic, for now, is the question, "If you were an eternal being with the capacity for love, wouldn't you want to spend eternity with those you love?"

We will call the abstract entity created to run the universe, the Logos. It is also hierarchical and, at its lower levels, embraces the mathematical principles that scientists call 'Natural Law.' We will explore what natural law is and its similarity to spacetime and physical systems in Book Two.

The Abstract Realm is founded on, to put it in the simplest mathematical form, the inevitability of Absolutely Nothing leading to the concept of one, and so on. The two substantial realms are founded on a different principle, that zero can be separated into two equal but opposite quantities (which, if allowed to combine would amount to nothing).

Two such possibilities are $0=+1-1$ and $0=+1+i-1-i$, the first creates a line while the second creates a unit complex plane. Using a multiple of this second method, the Logos first separated a zero of the Abstract realm into four orthogonal complex units. Then the Logos asymmetrically twisted apart this set of eight entities into two sets of four; the 's-metric' with one imaginary and three real axes, and the 'p-metric' with one real and three imaginary axes. The asymmetry was such that the s-metric had a twist to the right, R, while the p-metric had a twist to the left, L.

Under the direction of the Logos, these two metrics went their separate ways and expanded into the two Substantial realms, called the Physical realm and the Spiritual realm. In both metrics, there was a

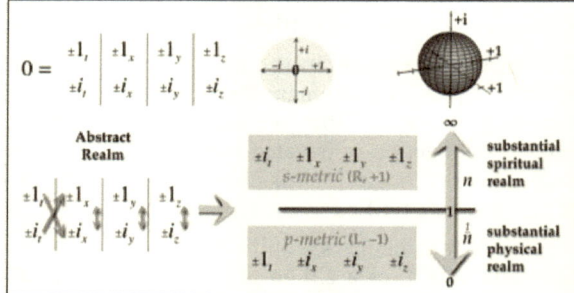

wave of period 1 and velocity 1. In the p-metric with the three imaginary axes, the intensity of the wave was −1, while the intensity of the wave in the s-metric with the three real axes was +1. This difference gave the two metrics entirely different histories under the direction of the Logos, with many properties of the two realms having complementary relationships such as +n and −n, or n and 1/n.

As the history of the p-metric is known in some detail, we will spend most of the rest of this work studying the structure and history of the Physical realm. We will only return to the s-metric towards the end of the work as it is currently almost unknown to the sciences. I say 'almost' since one aspect of it has recently emerged into scientific purview as the 'Dark Energy' that is accelerating the expansion of the visible universe.

THE PHYSICAL REALM

While the destiny of the p-metric was the substantial Physical realm, its origins were not at all 'substantial' but Abstract. At the moment of creation, the Physical realm was the p-metric with a wave of −1 intensity. The theory of Special Relativity defines a quality in this metric called 'separation' related to the four axis by the Pythagorean relationship.

This speck of −1 is called a the 'false vacuum' in modern physics, and it has what is called a 'negative pressure' in the theory of General Relativity.

metric	$\pm 1_t \pm i_x \pm i_y \pm i_z$
separation	$s = \sqrt{(\pm 1_t)^2 + (\pm i_x)^2 + (\pm i_y)^2 + (\pm i_z)^2}$
	$= \sqrt{1_t - 1_x - 1_y - 1_z}$

This negative pressure caused the p-metric to enter what is called "a period of inflation" in which the separation increased exponentially to very large values. This much enlarged 'four dimensional' construct then entered a 'period of braking' when much of the rate of expansion was converted into a set of wrinkles and twists in the four dimensions, called 'fundamental particles,' in a period called the hot Big Bang.

The 'fundamental particles' of modern physics are more accurately called the 'fundamental entities' of the substantial physical realm, for while they have many aspects of a particle (such as location) they have at least as many wave aspects to them that are not local.

No matter the complexity, size and sophistication of the systems studied in the Physical realm as it currently understood by current science—be they galaxies, stars, planets or dust; be they bacteria, plants, animals or human—they are all composed of just a few kinds of fundamental entities interacting by exchanging a few other kinds of fundamental entities.

In order to understand things, we need to understand the things they are composed of, and how these things come to be and how they behave. For this reason we will have to spend a lot of time discussing the fundamental insights of modern science into what matter is composed of.

First we will discuss the nature of the fundamental entities and their relationship to the Logos (natural law) and only then return to what caused the braking of the global expansion into local twists and wrinkles of spacetime in the hot Big Bang.

FUNDAMENTAL ENTITIES

The impulse of the global inflation was braked and turned into local wave-like deformations of the p-metric with an intensity, measured by real numbers, called the 'external energy' of these fundamental entities. They also retained an 'internal wavefunction' aspect, measured with complex numbers, that retained an imprint of the global wave of the inflationary period called 'entanglement'. The external energy is equivalent to what current science calls the 'particle' aspect of the fundamental entities, and the internal wavefunction gives rise to the 'wave' aspect.

To summarize the rest of the chapter which will deal with both aspects in detail: The wave tells the particle what to do; the interactions of the particle tell the wave how to change.

Classical science does not have the concept of an internal causal wave and focused solely on the external interactions. It was only with reluctance that experiment after experiment necessitated its inclusion during the gestation of current 'quantum' science. "In a sense, the difference between classical and quantum mechanics can be seen to be due to the fact that classical mechanics took too superficial a view of the world: it dealt with appearances. However, quantum mechanics accepts that appearances are the manifestation of a deeper structure ... and that all calculations must be carried out on this substructure."[135]

We will discuss the external energy first, as it has been explored by science for many centuries, and then discuss the internal wave aspect which has only been explored for less than a century.

External Particle

The external energy of a fundamental entity is tied up in the two types of p-metric deformation called 'boson' and 'fermion'. Colloquially, these are sometimes called the relatively evanescent 'bits of force' and the relatively immutable 'bits of matter.' One remarkable finding of recent physics is that the all the intricate complexity of the everyday world in the Physical realm is composed of just these two types of fundamental entities, albeit both in enormous numbers of them. It is as if we had discovered that the world was constructed out of just two types of tiny Lego building bricks.

To understand the difference between the external aspect of these two types of entities we need a little mathematics from an area called 'topology.'

Topology of spacetime

As described, the p-metric has a Left-handed structure involving one real dimension, labelled t, and three imaginary dimensions, labeled x, y and z. While the plus and minus directions along the 'spatial' imaginary axis are very similar, as illustrated by the similar behavior of $+i$ and $-i$, while the plus and minus directions along the real 'time' axis are not similar, as illustrated by the differences between

+1 and −1. At the moment of creation, the wave had a velocity of 1 in the spatial dimensions and a velocity of 0 in the t dimension, as did the local deformations created during the braking period.

The x, y and z spatial dimensions get twisted up with each other during the chaos of the braking period, and it is the two basic ways that dimensions can be topologically intermixed by twisting, generating the two types of fundamental entities.

The external particle aspect of a fundamental entity involves distinct 'twists' in the spatial dimensions along the time dimension. The two possible types of twists correspond to the two types of particles: A boson involves one or more "oriented" twists, while a fermion involves one or more "non-oriented' twists.

To simplify, we can approach this impossible-to-illustrate four-dimensional situation with a familiar three-dimensional example.

Consider a transparent plastic band with width along the ±x-axis and length along the ±y-axis, but with a zero extension in the z-axis. The x-y plane has two distinct sides to it, one facing in the +z direction while the other is facing the −z direction. (In the diagram, these two are shaded blue and pink solely so we can keep track of them.)

We now perform these three operations on this strip:

1.   Cut the strip across the x-axis, separating out a top edge and a bottom edge along the x axis.
2.   Rotate the bottom-x-axis in the z-z plane around the y-axis. This rotation can be in in either of two directions, +x to +z, or +x to −z. The direction illustrated is +x to +z and is designated a Right-up rotation about the y-axis. Rotation of the x-edge in the reverse direction about y is designated a Left-down rotation.

There are just two ways of aligning the rotated split x-axis together so that the third step can take place:

a.   The bottom-x-axis makes a 180° turn about the y-axis in the z-axis.
b.   The bottom-x-axis makes a 360° turn about the y-axis in the z-axis.

3. The top- and bottom-x-axes are sealed back together into a single x-axis.

The strip has now got a topological twist in it that can only be removed by performing the three operations in reverse order (and the reverse of cutting is sealing. These topologi-

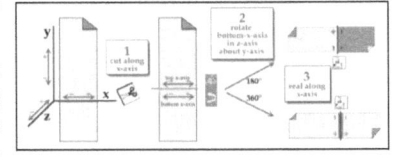

cal operations involve three orthogonal directions. In this simple example, the three directions involved are just the three spatial dimensions, but more complicated situations are possible:

I.   Cutting and sealing along the x-axis
II.   Rotation about the y-axis
III. Rotation in the z-axis

This topological operation in this simple example alters the three spatial dimensions in the following ways:

• Z-axis: This dimension of the strip is unaltered, it is just the same after the operation as it was before.
• X-axis: The x-dimension has either:

    a.   a discontinuity of + and – directions along it, or

    b.   a wave-like patch of disturbance along it.

• Y-axis: The effect on a directed x-segment (an arrow in the illustration) making a circular motion around the y-dimension in the z-dimension is either:

    a.   Flips the arrow by 180° into its opposite direction. This is called a 'non-oriented' y-axis, and the two-sided strip in the example has become a non-oriented, single-sided Moebius Strip. We can say that the y-axis has become 'disoriented' by the topological operation and has a 'spin' of ½. The spin can be Right or Left depending on the way the the x-axis was given its ½-twist. An

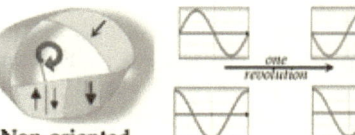

Non-oriented

asymmetrical sine wave, for example, traveling along the disordered y-axis would be altered to a minus-sine wave. A symmetrical cosine wave, on the other hand, would be unaffected by the revolution.

    b.   Does not flip the arrow, it just rotates it by a full 360° back to its original orientation. This is called an 'oriented' y-axis. Adding a second ½-twist to the Moebius strip (in the same direction as the first one, because a ½ twist in the opposite direction would just restore all three dimensions back to the original undisturbed state) results in a two-sided oriented band with a wave in it. While a (co)sine wave traveling along the y-axis will be rotated, it will be the same before and after. The oriented axis has a spin of 1, and this can be Right or Left.

Oriented

Topology of the metric

This simple introduction to Topology will suffice for an overview of the fundamental entities in the substantial entities in the Physical realm are constructed of. In the simple example in a 3-D space of real dimensions used above, the 'sidedness' of the real dimensions (the x-y plane) and the difference between Left and Right is dependent on the remaining real z dimension. For instance, while it is possible to create a non-oriented single-sided 2-D surface (illustrated) in 3-D space, the surface has to intersect itself (and is called a Klein bottle) and do not come in Right and Left forms.

This limitation is not the case for topological disturbances in the physical metric which, as described, is composed of components of four complex dimensions. A real (or imaginary) component of a complex axis has an inherent 'sidedness' (shown in pink and blue) that is provided by its orthogonal imaginary (or real) axis.

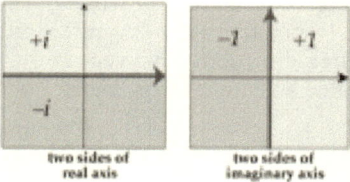

two sides of
real axis

two sides of
imaginary axis

The external 'particle' aspect of the fundamental entities is a topological disturbance created by a slicing of some or all of the three imaginary spatial dimensions of the p-metric, twisting them in the real time axis of the p-metric, followed by a resealing of the spatial dimensions. There are two topological results of this operation:

1.   A 180° twist-in-time results in a non-oriented topological defect. This has an asymmetrical and closed wave aspect, as exemplified by the sine wave, and these fundamental entities with a ½-spin are the set of fundamental 'fermions' each distinguished by the number of spatial dimensions involved in the slicing and sealing.

2.   A 360° twist-in-time results in an oriented topological defect. This has a symmetrical and closed wave aspect, as exemplified by the cosine wave, and these fundamental entities with a 1-spin are the set of fundamental 'bosons' each distinguished by the number of spatial dimensions involved in the slicing and sealing.

The fundamental entities have an external energy. One source is the amplitude of the wave which, as discussed, has an energy that is given by the intensity of the wave, its self-square. This energy in the disoriented, closed fermion wave is zero at the boundary and center of a full-period wave, while in a ½-period standing wave the energy is at the center and zero at the boundaries. The energy of the oriented, open boson wave is all at the boundary and center in a full-period wave, while in a ½-period standing wave it is all at the boundaries.

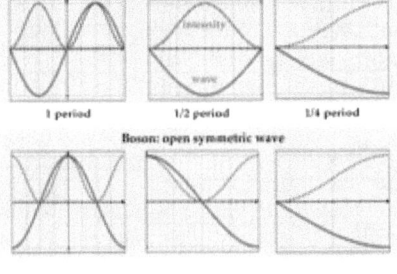

In discussing sound waves, we mentioned hybrid standing waves in organ pipes with one open and one closed boundary. Both fermion and boson standing waves can fit into such single node boundaries as ¼-period waves.

We will call this aspect to the energy of a fundamental entity its 'amplitude' energy, and, as intensity is the square of the amplitude, a wave of 2sin(a) has four-times the amplitude-energy of a sin(a) wave, all else being equal.

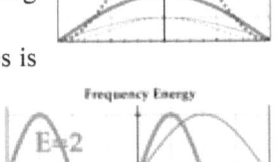

A second source of the energy in fundamental entities is in how quickly the wave is waving, so that a wave that makes two periods while another wave makes only one period has twice the energy, all else being equal. This 'frequency' energy of a fundamental particle is proportional to the frequency of the wave, so a sin(2a) wave has twice the energy of a sin(a) wave. The energy contributed by the wave is the sum of the amplitude energy and the frequency energy.

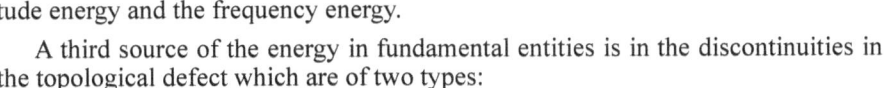

A third source of the energy in fundamental entities is in the discontinuities in the topological defect which are of two types:

The first kind of discontinuity occurs in non-orientated topologies. In the simple example we used of the Moebius strip, when sealing the twisted plastic band back together the + and − orientations along the x-axis were irrelevant. We can make our example 'polarity relevant' by making the x-axis a magnet with a North and South end. Sealing the x-axis now involves bringing a North and a South pole together, and this takes energy. The Moebius band now has energy locked up in the sealed x-

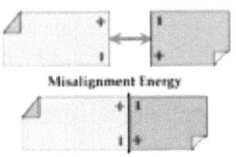

axis. In an analogous way, segments of the resealed p-metric that have an alignment mismatch also have an energy in them that we can call the 'misalignment' energy. The oriented topologies do not have any misalignment energy.

The discontinuity in the oriented topology arises from the open nature of the waves, there is an abrupt discontinuity in the p-metric from waving to not-waving at the boundary of the wave.

To summarize, all the fundamental entities have amplitude and frequency energy. In addition, the fermions have misalignment energy and the bosons have boundary energy. As there are three spatial dimensions in the p-metric in which the external topological twists in time can occur, there are only three types of fermions and three types of bosons, making six types in total.

We will just list the six entities here for now, and return to them in more detail after discussing their internal wave aspect.

The names given to the six types of funda-

Number of dimensions	Bosons Oriented, open	Fermions Non-oriented, closed
One	Weak-vector (W)	Neutrino (v)
Two	Photon (γ)	Electron (e)
Three	Gluon (g)	Quark (q)

mental entities are in no way systematic, and reflect the state of science at the time of their discovery. The customary symbols are also shown: the one-fermion is the Greek 'n' (nu), and the two-boson is the Greek 'g' (gamma); the rest are plain English.

Internal Wave

The topological defect (the particle) aspect of fundamental entities involves a deformation of the p-metric, while the wave (the wavefunction) aspect governs just what portion of the p-metric is involved in the defect, where the particle is located. The wavefunction is a complex wave that moves with a constant velocity of 1 through the p-metric.

This complex 'location' wave has a linear size (amplitude) of p, and an angular rotation (phase), of α, together comprising the complex number p@α. The amplitude and phase will vary at locations along the wave, and the intensity of the wavefunction at a particular location, the amplitude-squared, p^2, gives the probability of the particle being found at that location.

For example, a particle with a standing wave that has the form of a sine wave will have a standing 'probability density' that has the form of a sine-squared wave.

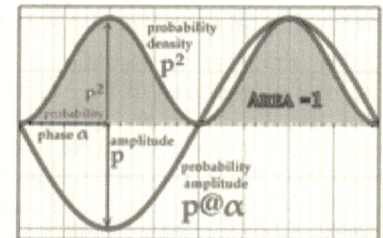

If the particle was always at a single location, the probability of it being there is unity, 100%. In the wavefunction, however, the probability is spread throughout the wave and all that can be said for a single particle in a standing wave is that the probabilities along the wave will all sum to one, it is somewhere to be found in the intensity wave, but we cannot say exactly where. In the realm of calculus (a level in the Abstract Realm we will briefly discuss later) this area under the intensity curve is unity, 100%. For the 1-period sine wave in the example, the technical term is that the 'integral' of the p-axis sine-squared wave over a single period along the α-axis is unity. The wave aspect has a complex-number value everywhere on the p-metric, and this can be zero at locations outside a bounded wave.

$$\int_{a=0}^{1} \left(\sin^2 p \right) d\alpha = 1$$

The physics of the internal wave reached its apotheosis in the "adding little arrows over-history' methodology perfected by Richard Feynman. (For some reason, he does not think the reader capable of grasping complex number, and calls his little arrows multiplying together 'shrink-and-turn,' shrinking because the amplitude is never greater than one.) This perspective is also called quantum electrodynamics (QED), the official name for the theory that describes the behavior of electrons and photons in terms of internal probability.

QED is extraordinarily successful and accurate. Feynman has modestly stated that: "The theory of quantum electrodynamics has now lasted more than fifty years and has been tested more and more accurately over a wider and wider range of conditions. At the present time, I can proudly say that there is no significant difference between experiment and theory! ... To give you a feeling for the accuracy [of the quantum description of the electron]: if you were to measure the distance from Los Angeles to New York to this accuracy, it would be exact to the thickness of a human hair. That's how delicately quantum electrodynamics has, in the last fifty years, been checked—both theoretically and experimentally."[136]

Probability Statistics

It would seem from the above discussion of the probability amplitude, given by the complex number p@α, that only the linear component, p, is an important factor in the probability, p^2, and that α is irrelevant. This is true in isolated cases, but when fundamental entities interact with each other (the topic of the next chapter) both have to be taken into account to understand the 'final probability amplitude' and the ending probability of the states after the interaction.

The operations involve the adding and multiplying of complex numbers (sometimes a great many of them) to calculate final complex number and its probability intensity. Complex arithmetic is more sophisticated, and it can successfully predict some quite unexpected results, as illustrated by the Mandelbrot set which is also the result of adding and multiplying complex numbers. (It is, perhaps, being uncomfortable with complex numbers that leads so many writers to speak of the "weird aspects of quantum physics," as if nature is at fault.)

Probability as it is usually understood (or 'regular' probability as we will call it), involves just two simple rules that, as usual, will be illustrated by coin tossing.

A regular coin has a 0.5 chance of landing heads, H, or tails. (This is simply a statement that, if the coin is thrown zillions of times, the probability density will be 50% H and 50% T to whatever accuracy you require).

There are just two simple rules for multiple events in regular probability:

- In '*AND*' situations, the probabilities are multiplied
- In '*OR*' situations, the probabilities are added

- In either case, the size of the result is the probability.

For example, the probability of throwing a head *AND* another head, giving HH, is obtained by <u>multiplying</u> ½ times ½ giving the probability of HH to be ¼. (i.e., if a pair of regular coins is thrown a zillion times, the the probability density of HH will be 25%. In the same way, the probability of TT is also 25%.)

The probability of throwing a two heads *OR* throwing two tails—the 'even' HH & TT throws as compared to the 'odd' throws HT & TH—is obtained by by <u>adding</u> the ¼ probability of each combination together, giving a probability of ½ for an even combination and ½ for an odd.

Much of the 'weirdness' that so many find in modern physics can be ascribed to the well-established fact that it is the probability *amplitudes* that are added and multiplied, not the probabilities, which only emerge afterwards in the final squaring step. Otherwise, the rules are almost identical:

There are just two simple rules for multiple events in complex probability:

- In '*AND*' situations, the probability amplitudes are multiplied
- In '*OR*' situations, the probabilities amplitudes are added
- In either case, the intensity of the result is the probability.

Even though the rules are the same, the switch from the real numbers of regular probability, to complex numbers can produce strange results. For instance, for probabilities greater than zero, adding two of them together to get a zero is impossible. The impossible, however, is possible with complex numbers. Two complex probabilities of the same size but with amplitudes that are 180° out of phase, α and $\alpha+\pi$, will add together to create a zero probability.

$$p + p \neq 0$$

$$p @ \alpha + p @ (\alpha + \pi) = 0$$

The complex wavefunctions of bosons combine to create what is called Bose probability, while the fermion wavefunctions combine to create Fermi probability rules; neither of which are as 'regular' probability.

The difference is quite clear in a thought experiment using three types of coins, each of which when tested individually, has a 50% chance of coming up heads. The difference is that when pairs are thrown, one type obeys the rules of regular probability, one type obeys the rules of Bose probability, and one type obeys the rules of Fermi probability.

Put simply, the symmetric waves of bosons add together, the waves combine constructively into one of twice the amplitude and four times the intensity. Bosons have a probability of 1 of entering the same state, so for a boson coin, while the probability of the first H is ½, the probability of the second H is 1.

Two familiar examples of such boson behavior are to be seen in the radio waves emitted and received by TV and radar antennae, and in the laser beams that adjust our eyesight and read our DVDs.

The opposite probability behavior holds for fermion waves. The asymmetric wave flips its sign, so two identical waves attempting to share the same state end up canceling out destructively to zero. This is why the probability of two fermions sharing the same state is zero, it is impossible, it is absolutely forbidden as quantum probability has absolute control over the Physical realm.

The only way two fermions can share the same state is if their waves are opposite, if one is clockwise while the other is anticlockwise. The flip of the asymmetric wave when two identical fermions of opposite spin attempt to share the same state leads to constructive interference into a wave with twice the amplitude and four times the intensity.

As attempting to add a third will inevitably clash with one of the two waves, this reduces to the rule that only two fermions of opposite spin can share the same state. This is called the 'Pauli Exclusion Principle' of natural law, which only allows two electrons of opposite spin in each quantum state. Two somewhat familiar examples of the enforcement power upholding this law can be found at both ends of the size scale: On the tiny level of the atom, this Exclusion Principle is cause of the hierarchical structure to the electrons in atoms, and the chemical interactions they are capable of. On the scale of our sun, all its enormous mass in very old age—in some 50 billion years or so—will be kept expanded in a sphere the size of the earth against the crushing force of gravity (which we will soon discuss) as a cold 'black dwarf' star. The physical universe at 13.5 billion is not yet old enough for there to be any of these to be around, but the transition stage of still-hot 'white dwarf' stars that are slowly cooling off are quite common in our stellar neighborhood.

Fermions have a probability of 1 of entering the opposite state, so for a fermion coin, while the probability of the first H is ½, the probability of a second H is 0 and the probability of T is 1.

Regular probability is based on 'independent assortment' in that the second toss is independent of the first toss. Neither bosons nor fermions show this independence as the second throw depends on the first.

	AND HH (TT)	AND HT (TH)	OR HH /TT	OR HT /TT
Regular	¼	¼	½	½
Bose	½	0	1	0
Fermi	0	½	0	1

This wave-derived behavior only adds to the opinion that quantum behavior is somehow 'weird.' This table compares the 'probability statistics' for a regular, Bose and Fermi coin. Note that that the regular probability of a coin composed of zillions of fermions and bosons is just the average of Fermi and Bose probabilities.

Probability Density

The internal wave of both types of fundamental entities can be a bounded wave or an unbounded wave. For an unbounded traveling wave, the intensity of the internal wave gives the simple probability of finding the particle at a location as the wave travels by.

If an internal wave is separated by some obstacle into two smaller waves, the external particle can appear to be in two places at the same time as there are now two probabilities moving in tandem, and the particle appears to jump between the two as if the external separation did not exist.

This 'entanglement' of a particle is perhaps the greatest cause of the unease that classical scientists feel towards the way the universe has been found to function.

The 'development of the wavefunction,' as such changes are called, is determined by natural law, and is the probability. This is the main difference between classical physics and modern physics:

- In classical physics, events were thought to be determined by natural law
- In modern physics, we know that the probability of events is determined by natural law.

The internal does not, however, determine exactly where a particle is in space at a particular time. This is very different from classical science which thought that the <u>exact</u> location was determined by natural law. It is this lack of external precision that has given rise to the false impression that quantum science is indeterminate.

For standing waves, however, which repeat themselves endlessly, this indeterminism fades away in importance, and the external probability density of a particle is completely determined by the internal wavefunction which, in turn, is completely determined by natural law.

This kind of determinism is called the Law of Large Numbers (LLN). This simply states that the percentage error, or difference, between the probability and the observed probability goes to zero as the number of 'trials' goes to infinity.

A more detailed treatment states that with N trials, the difference between the theoretical probability and what is actually observed almost always falls within the bounds (more technically, within one standard deviation) of plus or minus the square-root of N. So, when throwing a regular coin 100 times, it is quite normal to get 55 heads as this falls well within the bounds of $50\pm10=40$ and 60, and there is no reason to suspect the coin. If 10,000 throws end up with 5,500 heads, however, the coin is most suspect as this falls well outside the boundaries of $10,000\pm100=9,900$ and 10,100.

As the reciprocal gets smaller as the square root gets bigger, the indeterminism can be made as small as required by increasing the number of trials. The fractional difference goes to zero as N gets larger.

$$\frac{\sqrt{N}}{N} = \frac{1}{\sqrt{N}} \xrightarrow{N\to\infty} 0$$

N	10	100	1,000	1 million	10^{24}
$\sqrt{N}$	~3	10	~32	1,000	1 trillion
$\frac{1}{\sqrt{N}}$	31%	10%	3.1%	0.1%	0.000,000,000,1%

For an atomic electron in a bounded standing wave about the atomic nucleus (an orbital) that moves a trillion-trillion times a second (10^{24} trials), the difference between the intensity of the wave, the probability density, and the actual electron density is essentially zero. It is this aspect of the LLN that leads to such phrases as 'the electron is smeared out in an atom' even though the appropriate probes always find the particle, the topological twist, to be located at a precise point.

Given that it is not usual in nature for the number of trials to be proportional to the time that passes, we can restate this basic principle as: 'Given sufficient time, the form of the internal aspect determines the

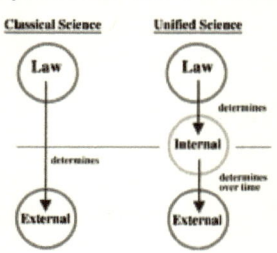

Classical Science — Unified Science

Law → determines → External

Law → determines → Internal → determines over time → External form of the external aspect.'

The electron density in an atom determines its physical and chemical attributes. So, on the scale of seconds, it is entirely appropriate to state that the physical and chemical attributes of atoms are determined by the natural law, and indeterminacy drops right out of the picture.

Entanglement

We have already seen in the 'entanglement' set-up of a slit experiment how an external particle can seem to be in two places at the same time. In classical science which does embrace the internal aspect of things, this is, of course, impossible to imagine let alone understand. But in a science in which the internal and external are united, it is the internal that takes precedence and the external just reflects the internal.

A node that has zero extension in the complex dimensions can have a sizable extension in the external spatial dimensions. For a simple bounded sine wave with two 'lobes' separated by a node, the node has essentially a zero extension and the two lobes touch at the central point. If the node has an extension, however, then the two lobes appear separated by a stretch of empty space. This is possible because, as mentioned, the wave in the p-metric is in *all* the p-metric, and if it just happens to cancel out to zero for a stretch and then no longer cancels out at another point, so be it.

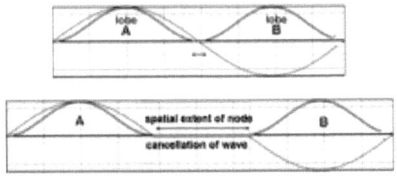

The particle aspect ignores any spatial separation, and its probability density continues to fill both lobes as usual. The particle appears to be both locations at the same time. As noted, however, interaction with an appropriate probe always finds the particle to be located at a precise point. The particle aspect of a fundamental entity appears to 'teleport' between location A and location B without ever having any presence in the space separating them. This phenomenon is called 'entanglement.' To make things even 'weirder' to the classical mind, the extent of the spatial entanglement is unbounded, it can stretch across the 13.5 billion light years of the visible universe without any problem, and for entangled particles that are now reaching us that went their separate ways right after the origin of the universe, say 12 billion years ago, the entanglement can stretch 24 billion light years into areas beyond the visible universe from which light has not yet had time to get to us.

Space Travel

As a cultural aside, it was science fiction writers who noted that human colonization of our galaxy, let alone all the distant other ones, would be sluggish and dull if transportation and communication was limited by the speed of light. To make things interesting, they had to invent things such as warp speed, hyperdrives and 'wormholes' through spacetime. The phenomenon of entanglement, when mastered, opens up such possibilities and the rapid colonization of the extraterrestrial planets that are found to be plentiful in our local neighborhood. Nature, as always, provides as there are many natural processes that send entangled particles off at essentially lightspeed in opposite directions, and we are daily showered with a plethora of entangled cosmic ray particles whose other lobe can be many, many light years distant.

As we will discuss in the next chapter, the interaction of a particle alters its wavefunction, and the entanglement is quickly lost once a particle starts interacting with other particles. So the cosmic rays—mainly protons—that penetrate the atmosphere to the earth's surface have lost any of the entanglement they had when entering the solar system. The Moon, however, does not have an atmosphere so it is constantly bombarded by entangled cosmic rays, and we can expect that research into exploiting this natural resource will only be accomplished by a colony on the Moon.

We have already encountered entangled particles on the everyday scale when an electron traveling wave passes through a setup called a 'slit experiment' and it appears to be in two different locations at the same time.

On a tiny scale, we see examples of entanglement in bounded waves. The electrons in an atom are entangled, and the standing waves, the atomic orbitals, can have many nodes in their probability density where the probability is zero. For example, the form of a simple standing wave, such as the 2p orbital, is akin to that of our simple 1-dimensional sine wave where the electron has a zero presence at the center; while the distinctly-

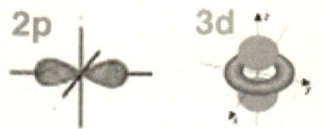

odd form of the 3d orbital can only be created by waves in multi-dimensions. As illustrated by the Mandelbrot set, complex numbers operating on each other are capable of creating a plethora of forms. The electron teleports between the two lobes of the 2p and the three lobes of the 3d as usual, and the probability density of the electron is the intensity of the wave.

We have now looked at entanglement, a property of the internal wave, on three different spatial scales to show that it is a mistake to consider it 'weird':

- Nanometer separations in the electron density of atoms
- Meter separations of an electron in slit experiments
- Gigameter separations of a cosmic ray proton.

Collapse of the wavefunction

Classical science, which insists that an electron has just one location, deals clumsily with the probabilistic aspects of matter by speaking of the 'collapse of the wavefunction.' When the electron moves to a new location and is observed, its extended wavefunction just disappears. This perspective, however, leads into such absurdities as Schrodinger's Cat which is 50% alive and 50% dead before it is observed.

The 'wave is real' perspective, however, has it that the wavefunction changes when the particle interacts, the wave of the particle just alters to a different wave. What happens is that when the particle interacts with the detector, its wavefunction alters and the entanglement is lost.

For example, an electron flipping from a traveling wave in the slit apparatus to a standing wave bound to an atom in a detector. The electron always has a wavefunction, it does not collapse and disappear on observation. This perspective does not lead to absurdities.

before   after
interaction

traveling wave   standing wave
in slit apparatus   in detector atom

Internal and External

We can now unify the internal and external aspects of the fundamental entities in our listing of them in the physical realm:

	Bosons	Fermions
Internal wave	Oriented open	Non-oriented closed
External defect *dimensions*	Weak-vector (W)	Neutrino (v)
One	Photon (γ)	Electron (e)
Two		
Three	Gluon (g)	Quark (q)

- Fundamental entities have an 'internal' aspect, a wavefunction that is described by complex numbers. This internal wave is determined by the Logos (natural law).

- Fundamental entities have an 'external' aspect, a particle described by real numbers that is a topological defect in space along time. The intensity of the internal wave determines the external probability density of the particle in spacetime.

Discrete Quanta

There is another difference between the math that applies to the internal level and the math that applies to the external particle. We earlier discussed the real number line and the difference between the countable infinity of the rational fractions and the uncountable infinity of the irrational continuum. It is this difference that applies to the internal and external.

1. The internal aspect of the p-metric measured by complex numbers is continuous, and it can have rational or irrational values.
2. The external aspects of the p-metric measured by real numbers is not continuous, and can only have integer rational values. The apparent continuity of the external aspects is actually an artifact of 'resolution.'

For example, a line of one meter that is composed of 1 trillion discrete segments each one a trillionth of a meter long would appear continuous to the naked eye, and it would be impossible to 'resolve' the individual segments. It would, however, be impossible to construct such a line that is exactly the square root of two in length, the best that can be done is a line that is a fragment of a trillionth larger or smaller, but never *exactly* the length of $\sqrt{2}$.

Color mathematics

Another more everyday example, which will prove most useful a little later on, involves the difference between positive color and negative color or, as it is called in the trade, additive colors and subtractive colors.

On the the screen of computer I am working on, my typing appears as smooth black letters on a pure white background. The colors in the diagram are just as smooth and vivid. When I turn the computer off, however, the screen is black. The smoothness and the colors are all artifacts of resolution.

With a magnifying glass, it will be seen that the computer is actually turning discrete of 'pixels' of Red, Blue and Green on and off. The pixels are so small—72 to the inch—that my eye cannot resolve them individually and I just see the overall impact on the three types of receptors in my eye.

Red, Blue and Green (RGB) are the positive, additive colors. When all the pixels are on, I see a smooth white, and when they are all off, I see a smooth black. When just the Red pixels are on, I see a smooth continuous red. Pixels of two colors add together into a color I also see as smooth and continuous. These are the rules for adding the positive colors to black to get the full gamut of colors we expect from computers and TVs:

+R+G = yellow, +R+B = magenta, +G+B = cyan.

For the subtractive colors, we look to the cover of a paperback book printed on heavy white card. Under a powerful magnifying glass—they will be at least 120 to the inch—the pixels reveal themselves individually to be either Cyan, Magenta, Yellow or Black (CMYK). The white paper starts off as 100% R, G and B. The RGB colors that affect the eye must now each be reduced, subtracted by an ink, to control the appearance of smooth color at a low resolution.

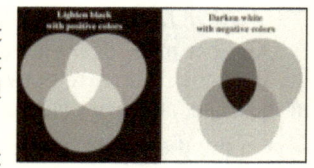

The ink that absorbs all the R and only lets the G and B from the paper through is a cyan color. The ink that absorbs all the G and only lets the R and B from the paper through is a magenta color. The ink that absorbs all the B and only lets the G and R from the paper through is a yellow color. As CMY just looks a dirty grey due to ink limitations, equal amounts of CMY are replaced by pixels of black ink, which absorbs everything.

Subtracting color from white   –CMYK

Pixels of two colors add together into a color I also see as smooth and continuous. These are the rules for subtracting the negative colors from white to get the full gamut of colors we expect from glossy magazines and book covers.

–C–Y = green, –C–M = blue, –M–Y = red.

It is found, in practice, that varying the intensity of the RGB and CMY pixels in 256 increments—more than 16 million combinations—is sufficient to give the impression of 'full color' on both computer screens and print jobs.

These colors will not all fit these relationships on the real number line but the color relationships do fit perfectly onto the complex plane. It can be seen in the mismatch of the magenta arrows that these three will not fit into the same three-fold symmetry, but they do fit nicely into a construct with a hexagonal, sixfold symmetry.

This can be simply illustrated using different modes in the image manipulator, Photoshop. In lighten mode, the overlap of three circles of positive color on a black background gives white and the three negative colors. In the darken mode, three overlapping circles of negative color gives black and the three positive colors.

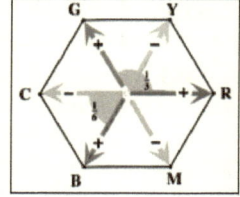

This a construct with a six-fold symmetry. It has each positive color and negative color along a plus-and-minus axis, and each axis has a rotation to the others of ⅓ a period, with plus and minus ends alternating at a period of ⅙. Each positive color is embraced by the appropriate two negative colors, and each negative color is embraced by the appropriate two positive colors.

As an example of the very "unreasonableness" of one area of mathematics applying to two very different levels of the physical hierarchy, this simple math of positive and negative colors with its hexagonal, alternating symmetry of three axes rotated by ⅓ a period and with ⅙ a period separating them will be all that we need to understand quarks, gluons and the structure of the atomic nucleus (where the colors are called ±R, ±B and ±G, and CMY is not used). Colorless is important in this area, and we should note that adding red and anti-red to black gives colorless white, while subtracting them from white gives a colorless black.

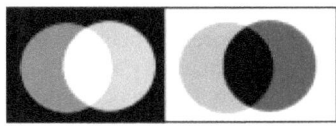

It is for this reason that I have gone into the subject of color resolution in some detail. When the pixels of color are small it is not possible to resolve the colors and the result is a colorless situation of white or black, or if all six colors are present in equal amounts, a colorless 50% grey.

Resonance

Resonance is phenomena that occurs with waves but not particles, and we shall encounter it a lot in the discussion. Waves that resonate together have a lower energy than when in isolation.

A relatively familiar example of paramount importance is the resonance of 4 different electron waves in the carbon atom resulting in the four 'hybrid' waves that account for carbon's chemical properties.

The four hybrid orbitals have less free energy than that of the single s-wave and three p-waves.

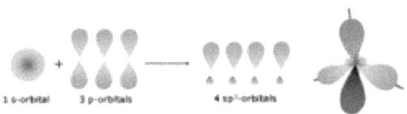

A Pixelated World

We will mention just three external aspects that only appear to be continuous because of resolution limitations:

Space

Space feels continuous, we certainly don't feel any 'grittiness' even when flying through the air. It is not, however, and it only appears continuous because our senses cannot resolve the grit. This 'pixel of space' is called the Planck Length, pL, and 'natural' units this size of pL is 1, while in human-scale units a pL is about a trillion, trillion, trillionth of a meter, 10^{-35}m. Our best experimental instruments have a resolution of only about 10^{-18}m, so space seems continuous even in the most delicate experiments—the pL emerges from theory and not from experiment.

Time

The sense of time passing continuously is even more ingrained, the concept of time being actually jerky is almost incomprehensible. As we pass through space (very slowly) and time (very rapidly), we cannot resolve the ultra-tiny 'pixels of time', called the Planck Time, pT, which is 1 in natural units and 10^{-44} seconds in human units.

Speed of Wave

We have discussed waves in the p-metric, and waves have a linear velocity as well as an angular phase. We shall now focus on the linear velocity of the wave and ignore its waviness. The velocity of waves in the p-metric is unchanging, it is a constant unity that never varies.

It was Einstein who first saw how velocity through time and space were connected in a 4-D world of spacetime that was an expression of the p-

$$c^2 = (t)^2 + (xi)^2 + (yi)^2 + (zi)^2$$
$$= t^2 - x^2 - y^2 - z^2$$

metric. The basic equation of Einstein's 'special relativity' is a simple extension of the Pythagorean Theorem. It states that sum of the squares of the velocities along the three imaginary axes (x, y and z) and one real axis (t) of the p-metric always sums to the square of lightspeed, which is unity, in natural units where c=1=1p-L/1pT.

As noted earlier, while +1 and −1 have distinctly different properties, there is only the 'point of view' difference between clockwise and anticlockwise to distinguish +i and −i. Moving in opposite directions along the complex time axis is the very significant difference between matter and antimatter. Moving in opposite directions along the imaginary spatial axes is the 'point of view' difference between going East and going West. This is why time is assigned the real axis and space the imaginary axes in special relativity nowadays rather than the space-is-real and time-is-imaginary of the early formulations.

To simplify the discussion it is no matter to orient the spatial dimensions so that the velocity is along a single axis, which we will call 'v'. In the physical universe, the constant velocity of the wave can be distributed differently between space and time. Almost all the entities found in nature fall into just two classes:

1. All space and no time. An entity with no inherent rest mass, such as a photon, the velocity through space is 1 (lightspeed) and the velocity through time is 0.

$$c^2 = t^2 - v^2$$
Zero restmass
$$0^2 - v^2 = c^2$$
$$vi = c$$
Non−zero restmass
$$t^2 - 0^2 = c^2$$
$$t = c$$

2. All time and no space. For entities with a rest mass, such as things composed of real electrons and quarks, the real energy (mass) has an inertial preference for moving through real time and reluctance to move through imaginary space (as will be discussed later). A velocity of a 1,000 mph is considered 'fast' in our reference frame, our resolution of space and time. In natural units, however, the velocity is one millionth and the square is a trillionth, essentially zero.

As humans belong to the second class, we customarily move at well below lightspeed. This is where another artifact of resolution influences our perception of the universe. The phrase 'quick as a wink' suggests that 1/10th of a second in considered a very short time, while normal events take place in brief seconds and 'just a few minutes' and long leisurely hours.

Things look very different when viewed through special relativity. In a wink of an eye, the movement through time is equivalent to 18,000 miles in space. The passage of a leisurely hour is equivalent to a voyage of 670,616,629 miles, of a trip past Mars to Jupiter.

Our experience of smooth time passing slowly by the second is an artifact of resolution as it is actually jerky and exceedingly fast.

Existence

While the notion of 'existence' is usually considered a philosophical question, modern science has a precise description of it (just as the words 'internal and external' have the precise descriptions, 'measured by continuous complex numbers' and 'measured by integer real numbers' respectively. Existence, as defined by science, is also pixelated, it comes in discrete 'quanta' called Planck's Constant, pC or h, that, unlike time and space, is within the resolution of experiment. The pC involves a measure of 'energy-in-time called 'the action.

The Action

Max Planck is considered to have started the second scientific revolution when he discovered that 'the action' came in discrete quanta that were all the same size (the first, the start of classical science, is ascribed to Isaac Newton).

The seemingly smooth continuity of existence is another example of the resolution as the pC is so very tiny. The pC has a value of 1 in natural units while in human units it has the value

4.136×10^{-15} eV secs. This applies to linear considerations of existence, for angular considerations the pixel of existence is $h/2\pi$ and 0.658×10^{-15} eV secs, respectively.

A 100 kilo (~4 x 10^{37} eV) man sitting and pondering existence for 10 seconds has ~4 x 10^{38} pC of existence, of pixels of existence, so it is no wonder he ends up thinking that reality is continuous as the jerkiness of existence is not resolvable, at least when unaided by machines.

The action is a measure of the Physical realm that plays a central role in modern physics. "Our search for physical understanding boils down to determining one formula. When physicists dream of writing down the entire theory of the physical universe on a cocktail napkin, they mean to write down the action of the universe. [The accompanying illustration is a contemporary action equation; 's' is the total action.] It would take a lot more room to write down all the equations of motion... The action, in short, embodies the structure of physical reality."[137] This is as true for quantum physics as it is for classical physics.

$$\mathbf{S} = \int d^4 x \sqrt{g} \left[\tfrac{1}{\varkappa} R + \tfrac{1}{g^2} F^2 + \bar{\psi} \not{D} \psi + (D\varphi)^2 + V(\varphi) + \bar{\psi} \varphi \psi \right]$$

Action and probability amplitude

It is an action equation that describes the combined influence of all the many interactions along an external 'history' of the entity to the changes to the internal wave at the end of that history. It is the action that connects changes in internal realm of the probability amplitude with the external interactions of an entity over a 'history' in time.

We have already encountered a touch of calculus when we discussed the area under a wave, the integral of the wavefunction intensity. The math connection between the internal and external is almost as simple. The line-integral of a wave is simply the linear-length of the wave if it was stretched out flat. For simple forms, such as a single cycle of radius r (a circle), the area-integral is πr^2 and the line-integral is $2\pi r$. For intricate forms, however, while the area-integral is a relatively simple to calculate, the line-integral usually takes a lot of sophisticated math; we will not pause to explore it in any detail.

It is the line-integral that gives the total action along a history, and it this action that determines the final probability amplitude and intensity for any history. "The fundamental law of quantum physics states that the probability amplitude of a given path being followed is determined by the action corresponding to that path."[138]

It can be just by inspection, that the line-integral length of a sine curve goes up as the frequency goes up. There is more action along a high frequency wave than there is along a low frequency wave, a point we will return to shortly.

The Uncertainty Principle

To be 'real' then, as least as far as external science is concerned, an entity has to have at least 1 pixel of existence, there has to be at least one 1 h of energy-seconds. Existence comes in integer units of 1h, 2h, 3h, 10^{38}h; never as ½h, 1⅓h , etc.

This is called "Heisenberg's Uncertainty Principle" when it is expressed in terms of observation and measurement: The uncertainty, Δ, of energy-in-time is the discrete Planck's Constant, the pixel of existence: $\Delta E . \Delta t = h$. Another relation is that of momentum, mv, and position, x, where $\Delta mv . \Delta x = h$. Attempting to resolve the physical world in greater detail, at say h/10 is futile, you can only get integer values of the pixels.

This is the insubstantial foundation upon which the Physical realm is constructed, as things are currently understood. The innocence and simplicity of matter-only classical science is gone, never to return, to the disappointment of those attached to classical science.

"Perhaps, someday, an experiment will be performed that contradicts quantum mechanics, launching physics into a new era, but it is highly unlikely that such an event would restore our classical version of reality. Remember that nobody, not even Einstein, could come up with a version of reality less strange than quantum mechanics, yet one that still explained all the existing data. If quantum mechanics is ever superseded, then it seems likely we would discover the world to be even stranger."[139]

Real and Virtual

Fundamental entities that have at least 1-pixel of existence are called 'real' particles. (The overloading of the word 'real' in math and science is not of my doing.) This requirement has different implications when applied to the two different basic types of fundamental entities: the bosons and the fermions.

Bosons

Bosons are easy to create. A +1-spin boson can be flipped away by a fermion which, by Newton's third law of Action and Reaction which holds in all sciences, flips by −1 into the negative state. A +½ rotating fermion that spits out +1 rotating boson will flip into the −½-rotation state—and the asymmetric wave is always doing this. When this flips out another +1 boson, it returns back to the +½ state (rotation add around the unit circle).

If this rotation is in both the complex wave of the p-metric as well as in the real topological defect, then the particle is a 'real' particle. In such a case, the fermion spin loses energy and the boson slips away with a pixel of existence. Fermions can rid themselves of surplus energy in the form of real bosons.

If the rotation of the fermion is only in the p-metric and not in the external space-time, the internal rotation of a fermion sends out 'virtual' bosons that only exist in the internal p-metric but not in the external spacetime. They do not have a pixel of existence, and the fermion does not lose any energy. Technically, a virtual boson does not have energy (which is always positive) but it does have momentum (which can be positive or negative). This is why a fermion can shed an unlimited quantity of virtual bosons—at a rate limited only by how fast it can rotate—sending them off in all directions in equal and opposite amounts to conserve momentum.

	FERMION ROTATING	BOSON EMITTED	
Internal	$+\frac{1}{2} \Rightarrow -\frac{1}{2}$	$\Rightarrow 1$	Real boson
External	$+\frac{1}{2} \Rightarrow -\frac{1}{2}$	$\Rightarrow 1$	
Internal	$+\frac{1}{2} \rightleftarrows -\frac{1}{2}$	$\Rightarrow 1, 1, 1...$	Virtual bosons
External			

A fermion is never in isolation, it is surrounded by a 'halo' of virtual particles at all times and places. It is actually a composite entity. This halo is the cause of what science calls the 'charge' on a fundamental entity, examples being the electric charge of an electron or the 'color' of a quark.

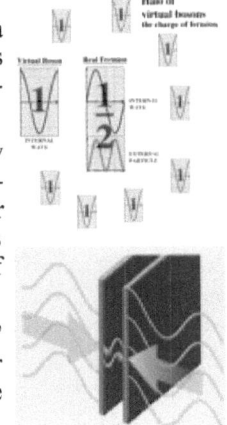

Virtual bosons, of course, being virtual and without any actual energy, are very hard to detect directly. They can, however, be indirectly detected because they can interfere either constructively or destructively, and the consequences of this are observed in the 'lines of electric and magnetic force' of classical science.

That the vacuum is awash with such "not really existing" virtual photons, is illustrated by the well-documented Casimir Effect of two silvered plates being pressed together by the 'vacuum fluctuations' when they get very close to each other.

This happens because the virtual photons in the vacuum can have any wavenumber while those between the two conducting plates must have wavenumbers that fit the space. So there are a lot more virtual photons on either side than there are between the two plates. It is this imbalance that pushes the plates together to create the Effect.

Fermions

Fermions are not as easy to create as the bosons, because they are limited by their non-oriented twists. By the Law of Action and Reaction, these can only be generated from integers in pairs whose twists are in complementary directions along all three possibilities.

If a fermion-pair involves both the internal and external aspect, the fermions have (at least) a pixel of existence and they both have a real energy. For example, a high energy 'gamma' photon (a 2-twist boson) is quite stable when traveling through the vacuum of space but if it hits a "disturbance in the Force"—such as the electric field about an atomic nucleus—the photon can fall apart. The boson turns into a pair of oppositely rotated 2-twist fermions; it creates an electron and an anti-electron (positron) that go their separate ways. This 'pair-production' of matter-antimatter out of gamma photons is the reverse of what happens when electron and positron merge their complementary non-oriented twists into integer-twist bosons, and gamma photons emerge in the mutual 'annihilation' of matter and antimatter.

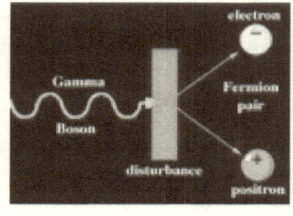

In exactly the same fashion, a disturbed ultra-high energy photon can fall apart into three quarks and three anti-quarks with equal and opposite color and electric charges, a 'nucleon-pair' such as a proton and an anti-proton. The reverse happens in annihilation.

Antimatter is extremely uncommon in the Physical realm; any positron that has the temerity to appear in the Universe is rapidly annihilated by an electron to photons. (We will discuss this asymmetry of matter and antimatter when we finish discussing fundamental entities and their interactions, and return to the discussion of the Hot Big Bang.) This is why matter is so long-lasting: there are no exactly-equal complementary twists to annihilate the electrons, which are 2-twist fermions, as the exactly-equal positive charge in the Universe is tied up in quarks, and they are 3-twist fermions. The quarks are eternal for the same reason.

This pair restriction means that virtual fermions always occur in matching pairs. An electron has virtual electron-pairs and bosons in its halo, but the pair effects are minimal because they do not have a pixel of existence. As the energy of a particle-pair is real, they can only appear briefly before annihilating back to nothing. For instance, the electron has a 'rest mass' of 511,000 eV so an electron/positron pair has a real energy of 1,022,000 eV. Such a virtual pair existing for 4×10^{-21} of a second would have a Planck's Constant of existence, which they don't, so they have to get back together before this time. Even at the speed of light, one cannot get very far on a there-and-back journey in such a brief moment. The halo of virtual electron-pairs about a real electron only extends for about 6×10^{-13} meters, and only the highest-resolution experiments are capable of detecting its influence.

The Vacuum

As noted earlier, zero is an integer, and the undisturbed p-metric, the 'vacuum', has a probability amplitude, albeit a small one, to fall apart into a a fermion-pair that separate, albeit briefly, before recombining back to the undisturbed state. The vacuum is speckled with brief sparks of electro-positron pairs. The vacuum also has a probability amplitude, albeit even smaller, to fall apart albeit even more briefly, into proton/antiproton, neutron/antineutron pairs as well as all the other particles discovered by high-energy physics.

$$0 \Rightarrow +\frac{1}{2} \ \& \ -\frac{1}{2} \Rightarrow 0$$

A problem arises because fermion-pairs appearing in the vacuum have a rest mass of a million eV while they flicker on and off. The probability of a pixel of spacetime turning into a particle-pair is very small, and they only flicker very, very briefly (for arguments sake we will take the probability of a pixel of vacuum turning into a million eV electron/positron pair in any second the very small one-over-a-google, 10^{-100}. The problem arises because even a very, very, very small number such as this can amount to a large number when multiplied by a truly gargantuan number such as a google-squared. The only exception is *exactly* zero. This conquers any number, no matter how large, such as a google-to-the-power of a google to the power google.

$$\frac{1}{10^{100}} \times 10^{197} = 10^{97}$$

$$10^{97} \times 10^{6} = 10^{103} \, eV \, / \, ly$$

$$0 \times 10^{10^{100^{10^{100^{10^{100}}}}}} = 0$$

If you do the math, the number of pixels in a cubic-lightyear-second is almost a google-squared at 10^{197}, so the energy each second in a cubic lightyear can be expected to be over a google eV. This is just for electron-pairs; all the other pairs contribute as well. Furthermore, there are a lot of cubic lightyears of vacuum out there—our Local Group of galaxies alone occupies about 10^{20} of them.

The gravitational effect of all this 'vacuum energy' in the virtual haze should compact space by a 'cosmological constant' factor of $\sim 10^{100}$. This expectation of quantum physics is not borne out by experiment which has that the observed cosmological constant is essentially unity, there is essentially a zero vacuum energy in the Universe.

This calculation is based on a consideration of the p-metric alone, however, and when the energy of the virtual particles in the s-metric are taken into account, the contributions of the two metrics cancel out almost completely. As we shall see in a later chapter, the energy of real particles in the s-metric is negative, the dark energy that was only recently discovered to be a major component of the Universe. Positive energy has a positive gravitation; it pulls together. Negative energy has a negative gravitation; it pushes apart. As plus a tiny number minus the same tiny number is zero, and zero conquers any number, no matter how large it is. The contributions of the equally-probable virtual 'supersymmetric particles,' as they are called, cancel out the contribution of the regular virtual particles, giving a dynamic cosmological constant at the observed size of unity and the observed zero vacuum energy.

$$\left(+\frac{1}{10^{100}} - \frac{1}{10^{100}} \right) \times 10^{197} = 0$$

$$0 \times 10^{6} = 0 \, eV \, / \, ly^{3}$$

Putting aside these global considerations for now, we will now return to the very local consideration of the real elementary entities out of which the Physical universe is constructed. We will take a closer look at each entity in turn, starting with the bosons, then the fermions, and then deal with their interactions with each other.

The Bosons

The bosons are the bits—more technically 'are the vectors'—of the fundamental forces that act on the fermions, the bits of matter. All things being equal, the symmetrical bosons of force are relatively ephemeral, and they easily appear and disappear. The asymmetrical fermions, in comparison, are relatively eternal, and only with difficulty appear, and then only in exactly opposite pairs of particles. The fundamental forces are:

1. The Weak force. The vector bosons are the W particles.
2. The Electromagnetic force. The vector bosons are the photons.
3. The Strong force. The vector bosons are the gluons.
4. Gravity. The graviton is thought by some to be the vector boson (an integer spin of 2) but, as we shall discuss in the next section, gravity is a consequence of wavefunctions being global, not local, on the p-metric.

We shall discuss the vector bosons in turn.

W-bosons

There are three W-bosons, the W⁺, the W⁻, and the W. This last is usually called the Z-boson, but as this obscures the basic similarity between the three bosons, we will not use it.

Name	1	2	3
W ⇅	1	0	0
W⁻	+½	−½	0
W⁺	−½	+½	0

The W-boson has 1 unit of quantum spin, an oriented open wave with an enormous boundary energy of 91,200,000,000 eV, or 91.2 GeV. This energy is reduced somewhat by transferring a ±½ twist to the second orthogonal direction. The result of this transfer is similar to an electron (or positron), as we shall see when we get to the fermions, but the magnetic spin is only ½ and it is going in the opposite direction. The boson resonates between the W and W± form, and this reduces the boundary energy of the W± boson to 80.4 GeV.

The real W-bosons are readily generated in high-energy particle accelerators, but their energy is so great that they quickly decay in fermion/anti-fermion pairs. The virtual W-bosons are a component of the halos of all the fermions, but their enormous energy only allows them to exist very briefly so the virtual halo does not extend very far. A virtual W of 91 GeV for 10^{-25} secs would have a pixel of existence, so its lifetime is less than that, and it cannot get far even at lightspeed, so the virtual halo reaches less than 10^{-17} meters. A virtual W⁺/W⁻ pair of 160 GeV has an even smaller halo.

These sizes are tiny, even on the scale of the atomic nucleus, and the fermions have to be extremely close, which is very unlikely, before their halos overlap and allow for interaction. It takes a long time, on a nuclear scale, for the weak force to play any effect. This is how the Weak nuclear force got its name. A neutron can decay, for example, into to a less-energy state of a proton, an electron and an anti-neutrino but this involves a W-boson intermediate so the 'half-life' of the neutron is ~11 minutes, which is akin to eons in nuclear-scale time.

The low probability of 'weak coupling' does play a very important role in everyday life as it this slowness that is responsible for the stately rate of thermonuclear reactions in the Sun over billions of years.

The half-life is a widely-used measure of probability when the probability of the change is constant over time. The half-life is the time it takes for 50% of a them to make the change. This is akin to measuring the probability of a regular coin by flipping it enough times so that, by the LLN, the desired accuracy is obtained.

Even though it is impossible to tell when an individual neutron will decay (there is a small probability of it lasting a second or a whole year, as there is of throwing 10 heads in a row), if unlimited numbers are available it is possible to measure the half-life that is precise as required. As large numbers are not a problem with neutrons, this measure of probability is known from experimental observation to be 611.0±1.0 seconds.

The Photon

The photon particle involves two oriented twists at right angles to each other. The first twist is called "magnetic" and the second "electric." These are two open waves that might be expected to have an enormous energy in the range of the single-twist W-boson.

Name	1	2	3
γ	1	1	0

Exactly the opposite is the case because of a resonance between the two waves. The first wave is constantly transforming into the second, and the second twist is transforming into the first. When the energy of one is high, the energy in the other is rapidly changing. The rate of change is called the derivative in calculus The derivative of a wave is the steepness, or slope of a wave is given by moving a dot along the x-axis by an infinitesimal length, dx, and seeing how the dot on the y-axis changes length, dy. The ratio of these lengths, dy/dx, is the derivative of the wave.

Closed sines from Open cosines

The energy of the open magnetic cosine drives the creation of the electric wave as its derivative, the change in the electric wave is maximal when the cosine is maximal. Similarly, the energy of the open electric cosine drives the creation of the magnetic wave as its derivative, the change in the electric wave is maximal when the cosine is maximal.

It can be seen from the illustration that a cosine wave is flat at its maximum, the derivative is 0 when the cosine is ±1. The cosine wave is rapidly changing as it passes through zero, its derivative is at a maximum of −1 (a downward slope when the cosine wave is passing through 0 from plus to minus) and a maximum of +1 when the cosine is going from negative to positive. A more detailed analysis shows that the derivative of the cosine wave is a nega-

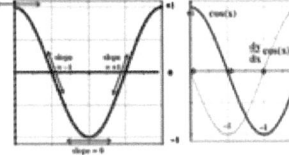

tive sine wave that is −90° out of phase with the cosine. So the energy of the open magnetic wave is driving a closed negative sine wave in the electric, and the electric open wave is driving a closed sine wave in the magnetic.

The resonant form of two such resonating open waves is two closed sine waves that are 90° out of phase with each other. There is no boundary energy in a photon, it has no 'rest mass.' A virtual photon is an electromagnetic wave with no inherent energy.

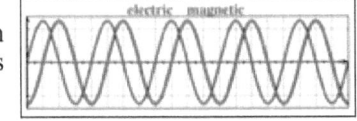

A real photon, however, has a pixel of action which is connected to the line-integral, and this goes up as the frequency goes up.

A photon whose wavelength is a light-year, has a very lazy wave with an ultra-low frequency, and has essentially no energy, the action is small energy and long time. A photon whose frequency is very high, on the other hand, will have a great deal of energy, the action is great energy and short time.

The magnetic and the electric are at right angles to both each other and the direction of travel. A photon that takes a long time, t, to cycle has little energy, E; a photon that takes a short time to cycle has a lot of energy. But the product of the two is always one pixel of existence, so for any photon, Et = h.

Being bosons in their behavior, photons readily link to each other creating what is called a ray of 'electromagnetic radiation.' Visible light is just 1-octave—when the frequency doubles— out of the 60 octaves of the entire spectrum. No matter what, though, each photon has exactly one pixel of existence, a Planck's Constant's worth.

With increasing time-period, below visible photons there are the infrared, the microwave and the radio photons which fade out until their action is all in time, and their energy content is so small as to be undetectable. With decreasing time-period, above visible photons are the ultraviolet photon, the X-ray photons and the gamma-ray photons. The limit in this direction would be a photon with a time period of the Planck Time, the Planck photon we might call it with all its action in energy and the minimum of it in time. It is thought that such photons were present in the ultra-high temperature of the first few moments of Creation.

While the difference between an ultra-low frequency photon being used in submarine communication and an ultra-high energy cosmic-ray photon shedding a shower of real particles as it tears into the earth's atmosphere is extreme, they are all identical in that they all have a single pixel of existence, a Planck's Constant of the action—it's just distributed differently. The table gives a few specific examples.

We have direct experience of these 'pixel of existence' photons. The light of a candle, for example, emits a wide variety of photons that peak in number at yellow light.

A yellow laser emits photons of identical frequency which, being bosons, all add together in lock step into waves of enormous size and power. All these photons are "real" in that they each amount to a pixel of existence.

PHOTON	TIME PERIOD	ENERGY	ACTION
Radio	4.1×10^{-7}	10^{-8}	4.1×10^{-15}
Infra red	1×10^{-12}	4.1×10^{-3}	4.1×10^{-15}
Yellow light	1.3×10^{-15}	3.2	4.1×10^{-15}
X-ray	1×10^{-17}	4.1×10^{2}	4.1×10^{-15}
Gamma ray	1.3×10^{-21}	3.2×10^{6}	4.1×10^{-15}
Planck	10^{-44}	4.1×10^{29}	4.1×10^{-15}

Gluons

Name	1	2	3
g	1	1	1

The gluons have three orthogonal oriented twists. If any two of these open waves try to put their boundary energy into driving closed sine waves in each other, the third one is going to be left out. This outlet for excess energy is blocked. This is where the section on the color math of positive and negative colors comes into play. While the gluons involve many complex dimensions, the math collapses down to the simple math on the color plane we studied. The gluon open-wave energy is used to twist the p-metric away from its fourfold symmetry of orthogonal axes into a hexagonal symmetry of three axes.

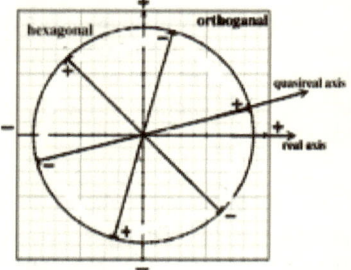

The amount of rotation is ⅙π. The diagram illustrates that, of the thee axes, one axis is the most similar to the original real axis and we can call it the 'quasireal axis. The other two axes have little in common with the original ones. The quasireal axis is pointing forward in time while the other two are pointing backwards. There are three spatial axes so there are three possible semi-real axes, a 'red' quasireal along the x, a 'green' quasireal along the y, or a 'blue' quasireal along the z axis.

In the hexagonal frame, the three non-oriented twists of a gluon fall along the three axes so that one is along the positive quasireal axis which moves forward in time and is the 'color' of the gluon, and the others fall along two of the negative axes, which combine as the 'anti-color' of the gluon.

| | QUASIREAL | | |
	R	G	B
YM	+R–R	+G–R	+B–R
CY	+R–G	+G–G	+B–G
CM	+R–B	+G–B	+B–B

This is where the color-math of three positive and three negative colors simplifies things, where Y and M going in the opposite direction combine as –R, the C and Y combine as –G, and the C and M combine as –B. The result is a gluon with color and anti-color ends whose center is colorless.

The possible combinations are listed in the chart, where the gluons with an overall color are distinguished from those that are grey and colorless over-

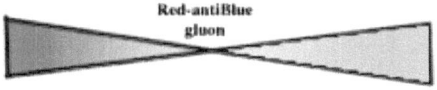
Red-antiBlue gluon

all. A gluon can be considered as an entity with two colored ends with a colorless patch them. This allows the three non-oriented twists to coexist, but it does not solve the boundary problem of open cosine waves. A more detailed calculation shows that a gluon removed from a proton would have infinite energy, that the attempt at removal would take an exponential increase in energy without any bound whatsoever. Isolated 'color' is quite impossible.

The gluons get around the open energy problem using the limits of resolution. If the ends of a ball of gluons are such that the colors cannot be resolved in a pixel, it has no overall color.

A ball of thousands of gluons aligned at their colorless centers into a sphere with the positive and negatives so mixed on the Planck scale that the surface is colorless has sufficiently reduced its energy so that can exist. The energy of all the open colored cosines is now in a colorless surface, a sphere of gluon-energy that has a colorless, energetic surface. All the energy of the three open cosines is now in this colorless surface. A colorless state can be accomplished in two ways:

1. All the colors and anti-colors are equally represented, a grey of triple color pairs.
2. There is only one color and its anti-color in equal amounts, a grey of a single color pair.

The illustration shows these two types of "glueballs" as they are called, and a cross-section of the second type, a glueball of +G–G (grey) gluons. The glueball has all its energy in the grey-of-triple color pairs

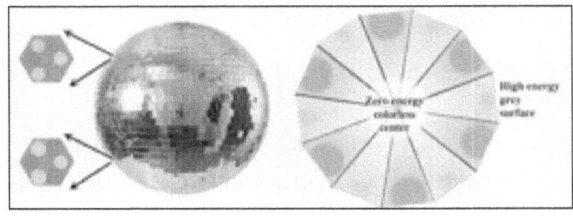

surface and none at the colorless center. The surface of the glueball has the equivalent of a surface tension, akin to but immensely greater than, the surface tension that makes water ball-up on an oily surface.

Plain glueballs have a high energy of hundreds of millions of eV, and they are unstable and their energy readily decays into a host of real fermion-pairs and bosons. The glueballs are, however, stable when there are colored quarks in the composite.

Three quarks of different positive color (the nucleons, e.g. the proton) are surrounded by the a grey-of-triple-color surface. There is tiny imbalance of color over anticolor, but not enough to matter. A quark/anti-quark pair (the mesons, e.g. the pion) are surrounded by the single grey surface with no tiny imbalance of color over anti-color.

The final thing that we shall mention that holds for all the bosons is that they are open cosine waves, and these are symmetric waves, so they look exactly the same going in either direction along complex time. The bosons do not come in matter and antimatter forms (discussed in the next section). The complement of a +R–B 'matter' gluon would be the –R+B 'antimatter' gluon, but this is already present in the above accounting of combinations.

This concludes the brief discussion of bosons with their oriented twists. Next are the fermions with the non-oriented twists.

The Fermions

All fermions are left-handed moving along the positive time axis in accordance with the inherent Left-handedness that we mentioned at the moment of Creation, when four complex planes were asymmetrically twisted apart into the p-metric and the s-metric The fermions fall into three classes determined by how many orthogonal ½-twists are in the p-metric.

Name	1	2	3
Neutrino, ν	−½		
Anti-neutrino, ν̶	+½		
Electron, e	−½	−½	
Anti-electron, e̶	+½	+½	
Quark, q	−½	−½	−½
Antiquark, q̶	+½	+½	+½

The fermions having this left-handedness while moving in the positive (complex) time direction along the time axis are called 'matter' fermions and have a spin of −½. The fermions having this left-handedness but moving in negative (complex) time direction along the time axis are called 'anti-matter' fermions and have a spin of +½. While antimatter is moving backwards in complex time, it is the square of the complex value in the p-metric that gives the real-value in external spacetime, and both +1 and −1 when squared are positive. So both matter and anti-matter particles move together along the positive time axis in spacetime.

Three generations

As the chart illustrates, an electron can be thought of as a neutrino with second twist, and a quark as an electron with a third twist.

The neutrino, as discussed so far, has been considered a non-oriented twist that disorients a single spatial dimension, say x. Its full name is the 'electron neutrino,' for it is possible for the twist to disorient two spatial dimensions, say x and y, creating a muon-neutrino, or disorienting all three spatial axes, x, y and z, creating a tau-neutrino.

Adding the second twist to a muon-neutrino creates a muon, adding it to a tau-neutrino creates a tauon. Adding the third twist to a muon creates S-quarks and C-quarks, adding it to the tauon creates B and T quarks. This creates three families of fermions that each come in three generations. Other than the neutrinos whose rest masses are low, all the higher generations are high-energy and unstable, all decaying quite rapidly on the human scale. They were plentiful in the hot big bang, but play little role in everyday life.

	ELECTRON FAMILY	MUON FAMILY	TAUON FAMILY
3	−½	−½	−½
2	−½	−½	−½
1	−½ 1-D	−½ 2-D	−½ 3-D

	ELECTRON FAMILY	MUON FAMILY	TAUON FAMILY
3	U/D quarks	C/S quarks	B/T quarks
2	electron	muon	tauon
1	electron neutrino	muon neutrino	tau neutrino

Adding a second twist to the electron neutrino creates an electron, adding a third creates a D or a U quark. This is the First Generation of fermions.

Adding a second twist to the muon neutrino creates a muon, adding a third creates an S or a C quark. This is the Second Generation of fermions.

Adding a second twist to the tauon neutrino creates an tauon, adding a third creates a B- or a T-quark. This is the Third Generation of fermions.

The generations have all their properties in common except rest mass. The 0.5MeV electron, e.g. can be replaced by a muon (106MeV) or a tauon (1,778MeV) in an atom, and the quarks all have the same color properties and can replace those in a proton. The most massive fermion of all is the T-quark which weighs in at a whopping 173,200MeV.

As matter is composed solely out of 1st-generation fermions, we will only mention the other generations only occasionally.

Chirality

We have mentioned that matter is composed of fermions with complex left-spins of one-half (we will encounter a caveat to this later). The complex spin, however, is only the same as the observed real spin for entities that are moving at the speed of light. This places a limit on observation as you cannot overtake it, turn and observe it coming towards you. We have already noted that clockwise and anti-clockwise rotation are identical, they depend on if you are viewing the circular motion from front or back. As one can never overtake an entity moving at lightspeed, its complex spin and real spin are identical.

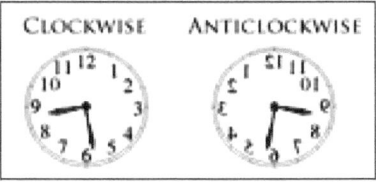

For an entity that customarily moves at sub-sub-lightspeed with a complex left-spin and real left-spin, it is easy to overtake it, turn and see the entity coming towards you with a right spin. A transparent clock goes in the anticlockwise direction if you look at the back. (This is why $+i$ and $-i$ are so similar while $+1$ and -1 are not.) This is why entities with a rest mass inherent energy that move at sub-lightspeed through space seem to come spinning equally right or left. The fundamental left-handedness on the internal level was only uncovered recently by experiment, and its implications have yet to be fully understood.

The Neutrino

The simplest fermion is the (electron) neutrino, a single non-oriented topological defect that disorients a single dimension of space. The left-

Name	1	2	3
Neutrino, ν	−½		
Anti-neutrino, ⩔	+½		

spin neutrino moving backwards in complex time is the right-spin anti-neutrino. The size of the defect in space is on the Planck scale of 10^{-35} meters and, at the resolution of 10^{-18} in current instruments, a neutrino appears as a point.

This is a matter-antimatter pair of a left neutrino and right neutrino with negligible energy as they spin in the favored left sense of the p-metric. A right neutrino has never been observed, and calculations suggest that if it did, it would have an enormous energy, surpassing even that of the W-bosons.

The neutrino is a closed wave so does not have the boundary energy of the open-wave W-bosons. The misalignment energy of a few pixels of space is so minimal that the rest mass energy of the neutrino is only known to have

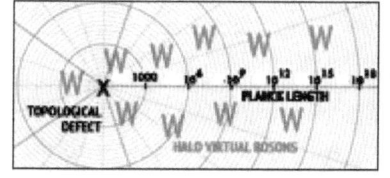

less than 10eV and probably less than 0.01eV of rest mass energy. The other generations are just a tad more energetic, the muon-neutrino at < 0.3 eV and the tau-neutrino at <31 eV. With such low rest mass, it does not take much to get neutrinos moving fast and at essentially lightspeed.

There is not much a neutrino can do, it is as close as anything gets in the physical realm to its abstract origins. Just about all it can do is to flip off virtual

Name	1	2	charge
Neutrino	−½		Weak charge

bosons with a single oriented defect, and this halo of virtual bosons is the 'weak charge' of the neutrino. This size of this halo about the pL-sized topological is 10^{17} pL, which sounds extensive but is only 10^{-18} meters, tiny even on a proton scale.

This is why the neutrino is said to be able to travel through a lightyear of lead without being influenced at all, the chance that its weak halo will intersect that of another fermion in all that lead for any length of time is negligible.

The earth, naturally, is just as transparent to them and they flood upwards through us by at midnight as they do downwards through us at noon. This delight of nature inspired the writing of a poem about the neutrino by a major poet, and odes to fundamental particles are rare indeed. We are quite oblivious to this flux, but neutrino detectors are being constructed (underground ironically) as they provide an unparalleled view of the sun's incandescent core unobscured by the cooler layers it is massively swathed in.

The sun actually coverts ~2% of the energy generated in its million-degree core into a flood of of neutrinos, the remaining 98% is hard gamma photons. The neutrinos rarely have a close encounter as they leave the core at almost lightspeed and flash though the outer layers of the sun in two seconds and reach the earth in about 8 minutes.

While the sun is transparent to neutrinos, it is utterly opaque to all forms of photons. It takes an average gamma ray about a million years to make the journey from the core, losing energy as it interacts along the way, until it is released from the 6,000° surface layer of the sun as a yellow photon which then takes 8 minutes more to reach the earth.

> **COSMIC GALL**
> *John Updike*
>
> Neutrinos they are very small.
> They have no charge and have no mass
> And do not interact at all.
> The earth is just a silly ball
> To them, through which they simply pass,
> Like dust-maids down a drafty hall
> Or photons through a sheet of glass.
> They snub the most exquisite gas,
> Ignore the most substantial wall,
> Cold-shoulder steel and sounding brass,
> Insult the stallion in his stall,
> And, scorning barriers of class,
> Infiltrate you and me! Like tall
> And painless guillotines, they fall
> Down through our heads into the grass.
> At night, they enter at Nepal
> And pierce the lover and his lass
> From underneath the bed – you call
> It wonderful; I call it crass.

These detectors have even measured the tsunami of neutrinos released from an aging massive star when it becomes a supernova at a distance of thousands of light years. Betelgeuse, the pink star in Orion, is soon—in star terms—to go supernova and will become as bright as the full moon in visible light and flood the neutrino 'telescopes' with information about the first few minutes of the star's collapse.

The Electron

An electron is a neutrino with a second non-oriented left-twist added to it at right angles. A positron (anti-electron) is an anti-neutrino with a second non-oriented right-twist added to it at right angles.

Name	1	2	3
Neutrino, ν	$-\frac{1}{2}$		
Anti-neutrino, $\bar{\nu}$	$+\frac{1}{2}$		
electron	$-\frac{1}{2}$	$-\frac{1}{2}$	
positron	$+\frac{1}{2}$	$+\frac{1}{2}$	

While having a single non-oriented twist in the p-metric involves just a little 'rest energy' energy (~1eV), adding another non-oriented twist at right-angles places stresses on the p-metric.

These stresses can be relieved by resonating with another configuration. The misaligned energy of the second –½ spin of an electron can shift to boundary energy in a –1 twist of the neutrino axis. The electron resonates between 'two closed waves' state and the 'one open wave' state, and this resonance stabilizes the energy of the electron with a rest mass of ~500,000 eV. This resonance form of the electron has a spin of –½, an electric

Name	1	2	charge
Electron	$-\frac{1}{2}$ ↑↓ -1	$-\frac{1}{2}$ ↑↓ 0	• Left spin • –1 electric charge ↑↓ • Down magnetic spin

Name	1	2	charge
Positron	$+\frac{1}{2}$ ↑↓ $+1$	$+\frac{1}{2}$ ↑↓ 0	• Right spin • +1 electric charge ↑↓ • Up magnetic spin

charge of –½ (which is confusingly called –1), and a Down magnetic dipole.

The energy of an anti-electron is exactly the same, but the directions of spin are just reversed, a right spinning entity with a positive unit charge and an Up magnetic spin.

The electron is also surrounded by a halo of virtual bosons. The tiny halo of W bosons is the 'weak' charge of the electron and the extensive halo of virtual photons is the electromagnetic' of the electron.

Charge of electron

The two-twist virtual photons that are flipped off a two-twist electron have no rest mass; the wave travels in space but not in time. Unlike the time limit placed on the massive W-bosons, the virtual photons are eternal and have an unlimited range. While the halo of W-bosons is tightly confined, the halo of photons has an unlimited reach with their density falling off with distance.

In the ½ & ½ state, the electron flips off electric and magnetic waves that would wave in-phase with each other, unlike regular photons where they wave out of phase with each other. Instead, this 'spin polarization' sends the electric and magnetic components on separate courses.

The electric components combine as bosons to create the polarized 'lines of electric force' that radiate outwards from the topological defect. This aspect of the virtual halo is called the radial 'negative electric field' of the electron. The polarization of the virtual photons is the opposite of electrons.

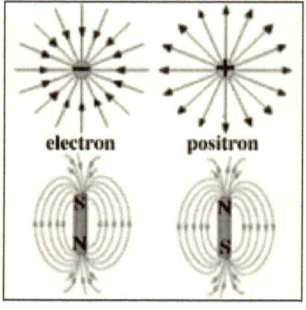

The magnetic components combine as bosons to create the 'lines of magnetic force' that exit the topological defect from the clockwise side and return to the anticlockwise side. This aspect of the virtual halo is the N-S dipole 'magnetic field' of the electron.

Together, the electric and magnetic fields constitute the 'electromagnetic charge' on an electron.

In the –1 state of the resonance, the electron flips off W-bosons, and this aspect of the virtual halo is the 'weak charge' on the electron. This is as ineffectual as the W-halo of a neutrino. Essentially all the interactions of an electron are when its electromagnetic halo intersects with the halo of another fermion.

The topological defect of the electron is on the Planck Length, its halo of W-bosons is almost as tiny, while the electric and magnetic are far reaching before the fields fade out towards zero.

This rest mass of an electron is 0.511 MeV, so it takes a gamma ray photon of over 1 MeV to create an electron/positron pair of real particles.

The Quarks

The quarks are entities with a third non-oriented defect at right-angles to both the second electron non-oriented defect and the first neutrino non-oriented defect. The state of three orthogonal ½-twists is of very high energy, and an isolated quark would theoretically have an infinite amount of energy. As quarks are separated their energy goes up exponentially, and it was this separation of color charges by the inflation of the universe that caused the brake of the initial inflation into a hot big bang with an enormous energy density.

The state of three orthogonal ½-twists is of very high energy and this state decays into in lower energy state by shifting twist from the first onto the second one as the whole framework shifts into the hexagonal relation as discussed in the gluon section.

As discussed, this involves a rotation of the real axis by $\frac{1}{6}$ into a quasireal axis in the hexagonal frame. This rotation can be accomplished in either of two ways with an electron and an anti-electron.

Name	1	2	3
Neutrino, v	$-\frac{1}{2}$		
Anti-neutrino, $\bar{v}$	$+\frac{1}{2}$		
electron	$-\frac{1}{2}$	$-\frac{1}{2}$	
positron	$+\frac{1}{2}$	$+\frac{1}{2}$	
quark	$-\frac{1}{2}$	$-\frac{1}{2}$	$-\frac{1}{2}$
Anti-quark	$+\frac{1}{2}$	$+\frac{1}{2}$	$+\frac{1}{2}$

For the electron, a rotation of the $-\frac{1}{2}$ twist axis of the third defect by $-2/6$ in the favored left direction of the p-metric, shifts $+2/6$ onto the second $-\frac{1}{2}$ electron defect resulting in $-\frac{1}{6}$ defect on the electron level. This final state has an electromagnetic charge of $-\frac{1}{3}$ and is called the D-quark.

For an anti-electron, a rotation of the $+\frac{1}{2}$ twist axis of the third defect by $+\frac{1}{6}$ in the disfavored right direction of the p-metric, shifts $-1/6$ onto the second anti-electron $+\frac{1}{2}$ defect resulting in $+\frac{1}{3}$ defect on the electron level. This final state

Name	1	2	3	charge
	$-\frac{1}{2}$	$-\frac{1}{2}$	$-\frac{1}{2}$	
D-quark	$-\frac{1}{2}$	$-\frac{1}{2}+\frac{1}{3}$ $= -\frac{1}{6}$	$-\frac{1}{2}-\frac{1}{3}$ $-\frac{5}{6}$	• weak charge • $-\frac{1}{3}$ electric charge • Color charge
	$+\frac{1}{2}$	$+\frac{1}{2}$	$+\frac{1}{2}$	
U-quark	$+\frac{1}{2}$	$+\frac{1}{2}-\frac{1}{6}$ $= +\frac{1}{3}$	$+\frac{1}{2}+\frac{1}{6}$ $= +\frac{2}{3}$	• weak charge • $+\frac{2}{3}$ electric charge • Color charge

has an electromagnetic charge of $+\frac{2}{3}$ and is called the U-quark. The anti-matter U- and D-quarks are the same, except that all the signs are reversed.

This is where the earlier caveat about 'matter is made of matter fermions' comes as the U-quarks (which are by far the most numerous in the universe) actually have anti-matter components, but components so twisted that they no longer complement with electrons so they cannot annihilate, they can no longer untwist into bosons.

The 1ˢᵗ twist and the 2ⁿᵈ twist resonate as in the electron. The quark defect is surrounded by a tiny halo of virtual W-bosons, its weak charge, and an extensive but decreasing halo of virtual photons, its electric charge and magnetic spin similar to the electron and positron but in proportionate amounts.

Depending on the orientation of the quasireal axis in the x, y, z frame, the color charge of the quark can be +R, +G or +B, and the energy of this is got rid of by flinging off real gluons (and anti-gluons which are identical) and absorbing them. But gluons have a plus and minus color, and a color can only depart with a negative.

So a +R quark can fling off its color off by emitting a +R–B or +R–G gluon but, as a consequence, it becomes a +B or a –G quark. It can just as easily get rid of the +R by absorbing a –R+B or a –R+G boson and also becoming a +B or +G quark. The quark flings off its color onto gluons so quickly that its overall color falls below the Planck resolution and it becomes essentially colorless. When a +R quark emits or absorbs a +R–R gluon it remains unchanged, and it is incapable of emitting or absorbing the +B–G or –B+G gluons.

Three quarks, each a different color, can emit and absorb each other's colors quickly that all their color energy ends up in the blaze of real gluons, in a surface so pixellated with color and anticolor that it falls under the Planck resolution and appears as a hollow sphere with all its energy in a colorless surface. That lost amongst this haze is an extra pixel of red, green and blue is irrelevant to this glue-ball surface.

As fermions fit nicely in complementary-spinning pairs but not at all as triplets, so there are just two possible sets, the nucleons, of gluon-interacting three color-complementary quarks, the pair +U and–U with a D called a proton which has an overall electric charge of +1, or a pair +D and –D with a U called a neutron which has an overall electric charge of zero.

The total color charge of the proton, the sum of the UUD quarks, that has to be lost among the haze of gluon pixels is $+\frac{2}{3} +\frac{2}{3} -\frac{5}{6} = +\frac{1}{2}$ while the color charge of the neutron with its DDU quarks is $-\frac{5}{6} -\frac{5}{6} +\frac{2}{3} = -1$, and this difference shows up in the energy imparted to the glueball surface, the rest mass of the proton and neutron. The rest mass of the proton is 938 MeV while that of the neutron is ~0.1% at 940 MeV. This is sufficient to make the isolated proton a stable composite entity and for the isolated neutron to be unstable. As neutrons have to have more energy if protons are to be stable and result in an interesting universe, is a sign of the design of the Logos, the laws that govern the internal waves and the external result. If this were not so, the physical universe would be all neutrons.

The neutron decays into a proton when a D quark emits a W⁻-boson and becomes a U quark. This real boson is so ephemeral that its mighty energy does not involve a pixel of action, and it falls apart into an electron and an anti-neutrino.

In composites with the other generations of quarks, the extra color is swamped by other considerations, and the 2ⁿᵈ & 3ʳᵈ generation equivalents of the U-quark, the $+\frac{2}{3}$ C- and B-quarks, are more massive than the 2ⁿᵈ & 3ʳᵈ generation equivalents of the D-quark, the $-\frac{1}{3}$ S- and T-quarks. The masses of the fermions are tabulated with the heavier quarks shaded.

	Family			
Generation	**1**	**2**	**3**	
1	Electron neutrino <2.2 eV	Electron 511 keV	D quark 4.8 MeV	U quark 2.4 MeV
2	Muon neutrino <170 keV	Muon 106 MeV	S quark 104 MeV	C quark 1.27 GeV
3	Tauon neutrino <15.5 MeV	Tauon 1.78 GeV	B quark 4.2 GeV	T quark 171 GeV

Just as it takes a disturbance of some sort for a gamma ray to decay into an electron/positron pair, it takes an overlapping of weak halos before the excess energy can be jettisoned as a real particle, albeit a very ephemeral one.

As far as size goes, the surface tension in the energy of the colorless glueball that surrounds the three quarks in a nucleon makes it a perfect sphere of size ~10⁻¹⁵ meters or 10^{30} pL. This is the size of nucleons.

A sphere that is 10^{30} pL across has a volume of 10^{90} cu. pL which gives the colorless quarks plenty of colorless space for them to roam in. The topological defects of a quark are on the Planck scale, and the accompanying halo of W-bosons only has a volume of 10^{50} cu. Pl. The colorless space available for the quark to move around is 10^{40} the size of the weak halo. This is akin to three peas in a cube with sides the earth-sun distance which is why the quarks hardly ever intersect. It is only when they oh-so-rarely intersect that the excess energy can be dumped into a W-boson and the neutron can beta-decay into a proton. The 11-minute half-life of the neutron, an eon in quark time, is the result of this 'weak decay.'

All the color energy of the quarks is in the glueball surface and the inside the hollow sphere is colorless. The quarks are also colorless, and they only interact with each other by their electromagnetic charges and magnetic spin. They have "asymptotic freedom" from the strong color force that is also called "ultraviolet slavery" for woe betide a quark that ventures to near the colored surface as its energy goes up exponentially.

If a quark were to be knocked out of a nucleon, the energy would suffice to make an antiquark/quark pair and the departing quark would be a quark-antiquark pair in a glueball, a ball of grey-of-grey gluons called a 'pion.' Pions can be uncharged, such as +U–U and +D–D with a rest mass of 135 MeV, or charged pion such as a +U–D with electric charge $+\frac{2}{3} +\frac{1}{3} = +1$, or a +D–U with a charge of $-\frac{1}{3} -\frac{2}{3} = -1$. The charged pions have a 1% greater rest mass of 140 MeV.

The matter-antimatter combination is unstable and a negative pion decays in just 10^{-8} seconds into a muon and an anti-muon neutrino. The neutral pion has an even shorter half-life of 10^{-17} seconds and the quark and antiquark unwind into two photons that zip off in opposite directions.

At resolutions where a proton becomes a point, it behaves exactly like a positron that had put on mass from 0.5 MeV to 938 MeV. But electrically and magnetically, a positron and a proton are identical at an atomic resolution.

This brings us to the topic of the interactions between the fundamental particles. We have already touched on this as quarks cannot be discussed individually but only as colorless combinations. The quarks rid themselves of color by constantly creating a hollow glueball of complementary colored gluons within whose insurmountable walls they lead a colorless, if confined, existence of low energy. In the language we will establish in the next section, the quarks interact by 'coupling' with gluons, the aptly-named Strong fundamental force.

The Fundamental Forces

Modern science recognizes only four fundamental forces in the Physical realm out of which all other forces are derived:

1. **The Force of Gravity**
2. **The Weak Force**
3. **The Electromagnetic Force**
4. **The Strong Force**

These four play very different roles in the functioning of the Universe. Gravity is a global entanglement that is responsible for the large scale structure of stars and galaxies. The weak force is an extremely local effect of W-bosons and is responsible for the stately rate of the thermonuclear reactions that power a star in its long middle age. The electromagnetic force involves photons and is far reaching, but as it tends to cancel out, is responsible for the local structure and chemistry of atoms and everyday matter. The strong force is very local, involving gluons, and is responsible for the structure of the atomic nucleus and the variety of stable and radioactive elements.

Gravitational interaction

Gravity is a result of the internal entanglement present in the p-metric at the beginning that binds everything together in the Physical realm. The energy that is the waves and distortions of external spacetime send out virtual bosons that, like photons, have an unbounded range. This is the internal aspect of what is called a graviton. It is a symmetrical boson, it has an oriented twist, but unlike the other bosons, it has a spin of 2. To return the same state when rotating in this takes only ½ a rotation. If you go the whole distance back to the start, you will have rotated twice.

Gravity is a global effect wherein the entangled bosons of the entire p-metric combine and blend into a single entity called the curvature of spacetime.Where they are abundant, spacetime is more curved; where they are scarce, it is less curved.

The entity emitting virtual gravitons, which like virtual photons carry momentum but no real energy, is also influenced by the gravitons that are flooding by.

Inertia and Mass

The influence on the internal wave of a rest mass energy by the internal aspect of gravitons is called 'inertial mass' and is the 'gravitational potential' of the spacetime being curved by virtual gravitons. The effect of this is to resist any change in momentum, the object has a linear and angular momentum that it takes energy to change. They are distortions on four separate complex planes, so their density and influence falls off with the distance, d.

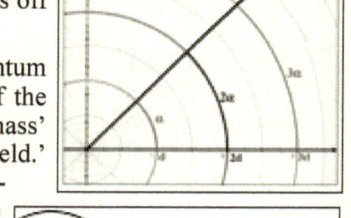

The influence on the external aspect of momentum is in external space and falls off as the square of the distance. This influence is called 'gravitational mass' and the overall influence is the 'gravitational field.' The external transfer of momentum between mass-energy entities causes them to draw together; there is a gravitational force between them.

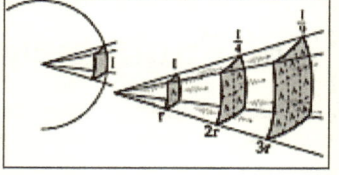

The virtual gravitons all blend together into the structure of spacetime, which is essentially flat except where it is distorted by large concentrations of mass-energy. And they do have to be in very large concentrations since the momentum carried by virtual gravitons is minuscule compared to that carried by virtual photons. The strength of the electromagnetic force is ten trillion-trillion-trillion times greater than the gravitational forces. The magnetic attraction of a weal refrigerator magnet, for instance, holds it firmly attached to the door, and the gravitational force of the entire earth pulling on it is insufficient to dislodge it.

The inertial influence of gravity falls off with distance while the mass aspect falls off as the square of the distance. This makes quite a difference.

To illustrate, consider a universe in which all the mass-energy is spread out at a constant density. The mass in a sphere of radius 1 and volume $4/3\pi^3$ surrounding you has an inertial and a mass gravitational influence which we can both set at 1.

A surrounding sphere of radius 2 adds a greater volume, and the volume of the shell is the volume of the sphere minus the volume of the first shell. A second shell of radius 3 is the volume of the third sphere minus the volume of the second. Another shell of radius 4 has it volume minus that of the 3rd sphere. The inertial influence of each added shell falls off with distance while the mass influence falls off as the square of the distance.

In can be seen from the graph that the inertial influence increases rapidly as the shell gets bigger, while the mass influence falls to a constant. While the universe is most certainly not homogenous on a local scale, at cosmological scales the distribution of matter is essentially uniform, and resolution enters the picture.

The implication is that the distant galaxies play a greater role in the inertial mass than the closer ones do. This is why inertial mass involves the entanglement of all the mass in all the universe, as in Mach's principle of inertia. Momentum is a vector property where direction counts, and the global influence of the graviton entanglement imposes what is called the Conservation of Momentum. In any situation the overall momentum is a constant. The momentum before and after an interaction is the same. This is why a matter/antimatter pair with zero overall momentum annihilates into a pair of photons that speed off in opposite directions so that the overall momentum is still zero. This is another example of 0 turning into +1 and –1. A single real photon has momentum, so annihilation into a single photon would be unbalanced and create momentum, which is forbidden.

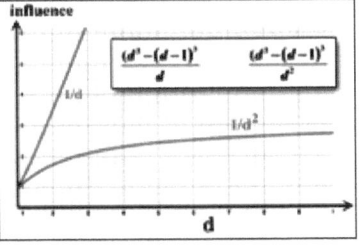

As noted, the velocity of the internal wave is a constant lightspeed. For entities with a rest mass, this velocity is almost entirely along the time axis and zero along the spatial axis. If energy of kinetic motion is added to an everyday object and its velocity approaches a substantial fraction of light speed, its velocity through time diminishes in proportion.

Acceleration through space involves rotating the direction of the constant velocity away from the time axis towards the space axis. For example, when this rotation is 45° the velocity through time and space are equal. The velocity through space is 0.71c —a speedy 473,478,700 mph—while the passage of time is 71% the rate of everyday objects, called 'time dilation.'

The energy added adds to the mass and inertia, such that as the velocity approaches the speed of light the mass increases without bound, and would be infinite at the speed of light. This is why objects with an inherent mass energy cannot travel at the speed of light. A proton moving at essentially the speed of light can have the inertial and gravitational effect of the planet earth, and greater.

In the maelstrom of energy in the first few moments of the hot big bang, all particles were moving at essentially lightspeed, and gravity was as strong as the other forces but quickly weakened as the Universe cooled and expanded.

Conservation

Much of the development of modern fundamental physics is driven by such forbidden transitions, events with a probability of zero. The golden rule is that anything that is not forbidden is compulsory; given enough trials, any probability greater than by zero by even a tiny amount will eventually occur.

It is the conservation of momentum that underlies Newton's 3rd law that states that action and reaction are equal and opposite.

The gravitational influence is local, the influence of the distant galaxies is about the same as the nearby ones, and the pull from one side tends to cancel the pull from the other side. External energy, being the square of the internal momentum, is a scalar property where direction does not count, and the imposition of the entangled gravitons that energy can be neither created or destroyed, the Conservation of Energy, it can only change form and distribution in any interaction. This is called the 1st law of thermodynamics. The gravitational effect of energy is so very small that it takes a very large amount of it to be appreciable, which is reflected in the enormous planet-scale amounts it takes to be noticeable.

On a cosmological scale, gravity opposes the expansion of the universe, and the effect of all the mass in the universe is to slow its rate of expansion. We will return to this when we get to the topic of 'dark energy.' It is now known that all the visible matter in the universe, the stars and galaxies, accounts for just a fraction of the total energy in the universe; the rest is to be found in the confusingly-named 'dark matter.' In just what quanta this energy is to be found is currently unknown but is possibly in the 'relic-neutrinos' from the big bang. Like the 'relic-photons' of the 'cosmic microwave background' radiation, they outnumber the electrons and quarks by a factor of 100 billion, and their tiny masses all add up.

The unit of energy we have been using is the electron-volt, the amount of kinetic energy an electron gains as it moves across a potential difference of one volt. The photons of visible light have energy on this scale.

As there are two trillion trillion trillion eV of energy in a gram, and a gram of mass is considered small, an eV would be an awkward unit for everyday use. Energy in huge, concentrated amounts is called mass; the conversion between the two scales is Einstein's famous equation.

$$E = mc^2$$

The 2nd law of thermodynamics arises from the driving force of interactions called the Principle of Least Action, that entities interact so as to minimize the overall action created by energy-in-time. The inverse of this is called the entropy of a system and the 2nd law is often stated as the entropy of a system always increases to a maximum.

We observe the principle of least action as the tendency of all things to minimize their energy. To get rid of as much energy as possible. To move from an excited state to the ground state.

The principle of least action is at the very foundations of the Logos, of the laws of nature. This is why change in the physical world is said to be driven by energy transfer as all things seek to minimize their 'free' energy, the basis of thermodynamics.

Free energy simply refers to the energy that can be altered, given the circumstances. In chemical changes, for example, the energy that is tied up in the 'mass' of the entities is not able to change, only the energy in electron waves can alter. It is this energy, then, that governs the changes in the waves in chemical interactions.

The implications of the 2nd law are quite different for interactions involving gravity—which always pulls things together—and interactions that involve electromagnetism—which can pull together and push apart. The state of least action and maximum entropy for gravity is when interacting things are all concentrated together, while for electromagnetism the least action is when the interacting things are smoothly spread out.

It is a balance between these opposing tendencies that keeps a massive star in an extended state while it is generating photons during its long maturity, and the gravitational collapse that occurs when the ability to generate photons eventually ceases.

Quantum gravity

It must be noted that the current understanding is still an open question as no way has yet been found to link the local and global aspects of gravity into a unified view.

The pixels of a 'perfect vacuum' have an internal wave aspect that is usually cancelled out by all the others around. Its quiescence is only barely disturbed by the virtual particles it very occasionally turns into. The pixel also has an excited, but not twisted, state. This high-energy state is called the Higgs Boson, and it has no spin, no charge, and no color. A search is currently on to detect a Higgs, and the current lower bound for its rest mass energy is ~160 GeV, about 160 times the energy in the gluon field inside a proton. A virtual Higgs would have an influence even more limited than the W-bosons, and it is thought that it is coupling on the Planck scale with virtual Higgs that causes energetic defects in the p-metric to radiate gravitons.

So just like the other three forces, gravity on the ultra-local level can be described as coupling with virtual Higgs bosons, and as coupling with gravitons on a local level.

This local aspect of gravity is the quantum description of gravity. On a global level, spacetime can be treated as a continuum, and the mathematical tools dealing with the curvature of a smooth continuum are well developed. This perspective was pioneered by Einstein who described the global aspect of gravity as a bending of spacetime in his epochal General Theory of Relativity.

Unfortunately, the local quantum description and that of global General Relativity have yet to be reconciled. Richard Feynman attempted such a unification when he considered the link between local coupling with spin-2 gravitons and the global coupling with spin-0 bosons. This was well before even the possibility of the Higgs had entered the scientific lexicon, and he did not pursue the concept. He did, however, perfect our current view of the electromagnetic force, the topic of the next section.

Electromagnetic interaction

When the virtual halo of polarized electric photons moving out around the electron (its negative electric field) intersects with the virtual halo of oppositely-polarized electric photons moving out around the positron (its positive electric field), they have the same sense of rotation. The spiral leaving the electron has the same sense as the spirals arriving from the positron, and they connect up. The magnetic photons of the electron and positron are in opposite directions and they tend to cancel each other out.

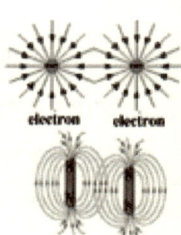

The external result of the momentum exchange via the virtual photons is to move the waves of the two particles together, just as gravitons do.

When two electrons intersect their fields, the leaving and arriving virtual photons have the opposite spins, and they cancel rather than augment each other. It is the magnetic photons that are going in the same direction and the magnetic 'lines of force' are bunched up together. The external result of the momentum exchange via the virtual photons is now to move the waves of the two particles apart, which gravitons with their 2-spin non-polarizable state do not do.

This influence of virtual photons is the cause of the observation that like electric charges and magnet poles repel each other, and that opposite charges and poles attract each other.

This cushion of virtual photons can be directly experienced by attempting to unite the N poles of two strong magnets. The experience of the invisible cushion keeping them apart is about as tangible as virtual photons get.

Atoms

When an electron and positron are near, their probability amplitude waves combine into a single resonant standing wave that is called an 'atomic orbital.' The probability density of them within this standing wave is equal in a sphere of about 10^{-10} meters—small by human standards but vast on the scale of electrons and quarks.

This composite entity is called an 'atom of positronium,' and it is another example of 'system building,' when simple systems interact together to create composite systems. Three quarks interacting together to create a proton is another example.

In the simple terminology we will find useful, positronium is a composite system of two subsystems (the electron and positron) that are coupling with subsystems from their own structure (the virtual, connected photons). The system has an internal standing 'system wave' (the resonance of the internal probability amplitude waves of the electron and positron) and an external substantial form (the composite of the overlapping probability densities of the electron, positron and virtual photons). Even simpler, we will say that the internal system wave confines the history of the subsystems into an external form that reflects the internal form.

If the electron and positron happen to end up too close together, their complementary defects untwist into real photons and the positronium 'atom' decays into gamma photons. We earlier noted that a proton behaves exactly like a massive positron but with twists that do not complement those of an electron.

The combination of an electron and a proton is the stable hydrogen atom, and they both have a probability density in the same system wave. The proton, being 1800 times as massive as the electron, shifts only slightly and jitters around its probability density at the center of the standing wave very much less than a positron would. The light electron has the same probability density as in positronium.

In an atom in which the electron is replaced by a 3rd generation tauon, the tauon which is just as massive as the proton, quivers in much the limited space as the proton, and can be said to be 'orbiting' within the proton. The 'tau-atom' is neutral and it can get very close to other such atoms. The brief lifetime of the tauon is the only barrier to it being a catalyst for nuclear fusion.

The hydrogen system is composed of two subsystems that interact by coupling with their subsystems; it has an internal standing wave called the atomic orbital, and an external form that is a composite of the probability density. The atom is 'substantial' because of the electron cloud, it is massive because of the proton (which itself is massive because of the energy in the blaze of gluons).

The hydrogen atom has an internal system wave and an external system form of confined particles that reflects the form of the internal.

INTERNAL	EXTERNAL	
SYSTEM WAVE	SYSTEM FORM	
Resonance of:	Confined probability density of:	SYSTEM
1 electron wave	electron	Interacting subsystems
1 proton wave	proton	
virtual photon waves	Virtual photons	Coupling sub-systems

We have already seen that the system wave of a proton perfectly confines its quarks and gluons but imperfectly confines the virtual photons that spill out as the electric charge and magnetic dipole of the proton.

The system wave of a hydrogen atom does not perfectly confine the subsystems, and this imperfect confinement is responsible for the chemical behavior of the hydrogen atom. The orbital is capable of holding a pair of oppositely-spinning electrons in a resonant, low-energy state but it only contains a 'singlet' electron.

A example of an almost perfect confinement is to be found in the helium atom with a nucleus of positive charge two and a pair of opposite-spin electrons filling the orbital. The helium system almost confines its subsystems as well as a proton does its quarks. It has no imbalance to give it any chemical properties, and it is the closest that reality comes to the massy-spheres of solid matter considered primal in classical science.

Helium atoms behave almost exactly as classical science expected two tiny billiard balls to behave. This is looking only at the external aspect of interaction, but all the interesting stuff is actually happening on the internal level. The two helium atoms have nothing to gain by resonating together, and the waves bounce and each other and go off in the opposite direction. The well-confined quarks, gluons, electrons and connected virtual photons that are a helium atom preclude any interaction and they are so barely sticky that only a temperature near absolute zero will condense a gas of them into a liquid. This is the source of the everyday experience that matter is solid, even though it is not. Two helium atoms colliding will elastically bounce off each other just like two tiny billiard balls, just as if they are solids. But it is actually the self-contained waves of virtual photons that fill each atom that are doing the bouncing, not anything solid.

The mass aspect of 'matter' is the consequence of a blaze of gluons; the substantial aspect of matter is a blaze of virtual photons. They are emergent properties not fundamental properties. Such are the philosophical implications of modern science that are, as yet, only partially digested.

Quantum mechanics

The area of quantum mechanics that deals with the interaction of photons and electrons is known as Quantum Electro Dynamics (QED).   Richard Feynman's book, *QED, The Strange Theory of Light and Matter* is an excellent overview of its triumphs. In QED, the internal aspect is called the 'probability amplitude' and is described and measured with complex numbers.

Quantum mechanics describes the way that waves of probability amplitudes combine and interfere by the arduous, if accurate, method of adding and multiplying thousands of complex numbers to calculate the final probability amplitude, p@a. The probability density of the particle aspect is then calculated by the elementary step of calculating p^2.

Slit experiment

One of the earliest experiments that revealed that there was more to the physical universe than just the external aspect, and thus was instrumental in the emergence of quantum science that included the internal, was the slit experiment. This experiment was originally designed to determine if a fundamental entity was a particle or wave. The entities are collimated by passing through a narrow slit, and the number reaching a detector on the far side is recorded. A second slit is then opened close to the first, and the detector records that result.

With waves, there is interference of the waves, and an 'interference pattern' is created at the detector. In particular, there are places that detected arriving entities with one slit open but detect nothing at all when two slits are open. This does not happen with particles, opening another slit can never stop particles reaching the detector through the first slit.

The slit experiment confirmed that light was a wave, as is easily confirmed by monochromatic light— all of one frequency, or color, such as the yellow generated by electrons jumping about in the orbital of sodium—when the 'up' amplitude of one wave can be exactly cancelled by the 'down' amplitude of the other wave.

We have seen that a simple wave is described by its amplitude and its phase, just where it is along the sine wave. This will depend on how far the wave has traveled, x, and its phase which will depend on is period of the wave, how far it has traveled, sin(t(x)). When the length of one path to x differs from the length of the other path by ½ a wavelength, the waves will be exactly out of phase and interfere destructively. Conversely, the brightest places will occur at spots that differ in distance by exactly one whole wavelength.

At first, it was thought that it was the external wave that that created the interference pattern, but then came the development of sources and detectors that could deal with single photons at a time. The result was confusing. These single photons behaved like a particle in that they always arrived as a single spot and not spread all over the detector. But, as the pattern of spots built up one by one, an interference pattern emerged. The single particle-like photon was interfering with itself, and the wave was going through slits at the same time.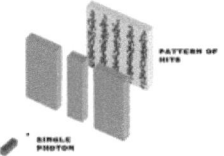

This was not an external wave, it was the internal wave that determined the probability of a photon reaching a spot on the detector. Such a wave did not exist in classical science, which caused endless confusion. The amplitude of the final wave at a spot on the detector determined the probability that the photon would interact there. This is how the phrase 'probability amplitude' entered the scientific vocabulary.

At first, the slit experiment seemed to show that electrons were particles. Then it was realized that the distance between the slits had to be commensurate with the wavelength of the wave, and that separations that were suitable for light would not work for electrons; only atomic size separations would work. The regular planes of atoms in a crystal, such as salt, are just this size, and electrons passing through these 'slits' also had a diffraction pattern, now a 2-D pattern (as illustrated). An electron also had a wave of probability amplitude associated with its particle. The electron also interfered with itself and seemed to pass through both slits at once.

The basic rules used in QED for calculating the final probability amplitude are simple and similar to those of classical probability.

1.   If there are alternative ways of it happening, add the probability amplitudes. (Classical theory says add the probabilities.)
2.   If a series of intermediate steps is involved, multiply the probability amplitudes for each step. (Classical theory says multiply the probabilities.)
3.   The square of the magnitude to the final probability amplitude (the p in $p@a$) is a real number that is the probability, p^2, of it occurring. (Classical theory omits this step—adding and multiplying real numbers always results in real numbers.)

All of quantum mechanics is just repetition of these two operations, just millions and billions of times. Much of the sophisticated math used in QED is just shortcuts to doing this endless iteration of adding and multiplying complex numbers.

This method works just as well for standing waves. If you add and multiply all the probability amplitudes for an electron and proton, the end result is the 1s orbital as calculated from the Schrodinger Equation. The adding and multiplying method is called the 'sum over history' method and it is mathematically equivalent to the differential wave formulation.

What we have described so far is sufficient to understand all nonliving systems. The form of internal wave gets more and more intricate, and the probability density of the interacting subsystems follows along. These intricate waveforms express more and more sophisticated emergent properties that are inherited from the Logos.

But the basic principle is the same: an internal wave determining the external form to the probability density of electrons and atomic nuclei.

Just how intricate a form can result from simply adding and multiplying complex numbers together? Luckily, research in this direction has already started, and the answer to this query would be, "Very." We shall take a brief look at these simple beginnings at exploring the properties of forms generated by adding and multiplying complex numbers.

Sophisticated internal forms

This possibility of exploring this aspect of mathematics only really opened when computers were developed which could add and multiply millions of complex numbers a second. (Some early pioneers of the field actually added and multiplied thousands of numbers by hand!)

As we saw earlier, the endless and fascinating forms of the Mandelbrot Set are likewise created by repeatedly adding and multiplying complex numbers, so we should not be surprised at the sophistication involved in everyday matter. If our computers were powerful enough, they could use the equations of QED to calculate the forms of ordinary objects.

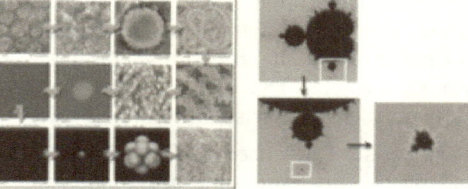

The illustration compares the forms that result when magnifying the Mandelbrot and a powers-of-ten zoom-in on a plant leaf showing how the patterning changes in a similar way. The photomicrograph is recording the result of photons from a source interacting with the electrons in the specimen, and this is just what QED is good at.

Julia sets

The above illustration of the Mandelbrot was generated by what is called 'serial computation,' each complex number in the chosen range is taken one at a time. The central processor then iterates it in the equation and, if the sequence that results is a dust or a bounded sequence, it instructs the graphic processor to color the pixel corresponding to the number appropriately. Then the next number is tested until all of the numbers in the range have all been serially processed in order and the results sent to the screen.

There is another way of generating the Mandelbrot set, and this is by 'massively parallel processing,' where all the numbers are tested at the same time. As mentioned, the numbers in the set create bounded forms on the complex plain, and the form that the number creates is the 'Julia set' for that number. The number 0 has the most boring of Julia sets, when you feed it into the equation, it just stays zero

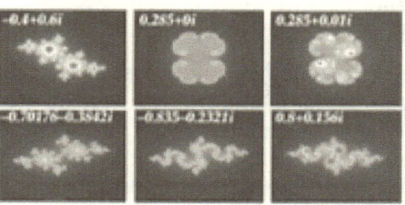

no matter how many times you iterate it.

Other numbers create bounded forms of intricate shapes, for example the six complex numbers in the illustration shapes.

The parallel computer we will consider has a processing unit within each pixel—and this will have to be a thought experiment as current computers can manage only eight central processing units (CPU) at a time. Each computer is numbered by its position in the pixel array. Each pixel receives two pieces of information: the complex number it corresponds to, and the program to iterate the equation.

Each pixel now computes the Julia set for its complex number and writes its result to the display. While this parallel process is more sophisticated than the serial computer, the serial computer can be programmed to model the parallel computer. The result is illustrated for a small array.

The Julia sets are combining with each other into the Mandelbrot set. This is somewhat similar to the Fourier process already described where any shape wave can be broken up into into a set of sine waves, and any shape wave can be created by combining a set of sine waves. The Julia sets are akin to a Fourier analysis of the Mandelbrot set.

QED is likewise couched in terms of the addition and multiplication of complex numbers, and the waveforms of matter are directly related to QED. Furthermore, the properties of matter are a reflection of the waveforms. A simple form, such as a sphere, gives helium its exotic properties, so we should not be surprised if sophisticated forms such as the Julia sets have quite unexpected properties.

Describing Form

The intricate forms of the internal wavefunction are reflected in the composite probability density of subsystems, the external form of composite entities. To understand this external form we need to add an infinitesimal amount of calculus to what we have already discussed. We have already encountered many curves in the form of waves. The calculus provides a way to measure curvature, and how the curvature is changing.

Basic calculus

We start with the concept of the 'slope of a line' such as the two on the graph. If the change along y is Δy when you change x by Δx, then the slope is defined as the ratio of the two. The graph of a car traveling at a constant speed is a straight line when time is plotted against distance travelled, and the speed is the ratio of 'how many miles' to 'how many hours.' The lines on the graphs have different

$$\text{slope} = \frac{\Delta y}{\Delta x} \qquad \text{speed} = \frac{\Delta miles}{\Delta time}$$

slopes. It can be seen by inspection along the x axis, that when the red changes x by 3, the y changes by 4. When the green changes x by 2, the y changes by 4. So the slope, rise or gradient, of the red line is 4/3 while that of the green is 4/2 or just 2. This ratio remains constant if instead of moving the green four whole units along x, we move it just a small amount instead. What if we make the small amount really, really tiny; shrink it as close to zero as we can imagine without ever actually reaching exactly zero. The ratio of these two 'infinitesimals' will always remain constant. This limit is the 'derivative' of the curve, and it can be used for curves as well as straight lines.

$$\frac{\Delta y}{\Delta x} = \frac{4}{2} = \frac{0.000004}{0.000002} \xrightarrow{\Delta x \to 0} \frac{dy}{dx}$$

If the red line is a distance-time graph of a car, and it travels 40 miles in 2 hours, we know that its speed is 20 mph. We observe the rather trivial point that the instantaneous speed (derivative) at 2 hours is also 20 mph, which is to be expected at constant speed.

Next we will look at graphs with curves in them, such as the magenta line which is $y=x^2$. What can we say about its derivative, its instantaneous speed, at the point where x=1. The red line only approximates the slope, and it crosses the magenta in another place. If we rotate the red line counterclockwise a little, however, the two points merge into one point. The green line is the 'tangent' to the curve at that point. The slope of this line is the differential of the quadratic curve at the point where x=1.

It is clear that the tangent is different at each point, for instance it is a horizontal line when x=0 so the differential is also 0, and it has a negative slope when x is negative.

$$\frac{dy}{dx}x^n = nx^{n-1} \qquad \frac{dy}{dx}x^2 = 2x^1$$

Now calculus has been around for a few centuries, and mathematicians do not go around measuring things, they have equations. A particularly useful one gives the derivative of any power of x. It gives the correct answer of 2 without measuring the graph.

The derivative, or slope of the graph is itself a function of x. Whatever x is, the slope will be always be 2x. For example, at 0 where 2(0)=0, we have our horizontal line as the tangent. At x=100 the slope will be 200, while when x is negative, the slope will be negative, a downwards slope. We can graph the derivative against x and get the line y=2x.

We can now ask about the rate of change in the derivative, the 'second derivative' of the magenta curve. This is the slope of the derivative graph, and we simply use the formula on the derivative to get 2, a constant. This is constant acceleration, and it has a derivative of zero, it is a horizontal line.

If the magenta line is a distance/time graph of a race car, it starts off at high speed as it enters the arena, and decelerates at a constant rate until it is zero at the royal box, where it turns around and accelerates at a constant speed until it leaves the arena. The calculus notation for the second derivative, the acceleration, is similar to that of the first.

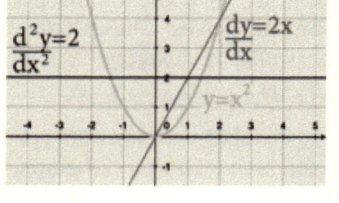

$$y = x^2 \qquad \text{distance at time}$$

$$\frac{dy}{dx} = 2x \qquad \text{speed at time}$$

$$\frac{d(\frac{dy}{dx})}{dx} = \frac{d^2y}{dx^2} = 2 \qquad \text{acceleration at time}$$

Just as addition and multiplication have their inverses in subtraction and division, so differentiation has its inverse called integration. For instance, if you are given the time and speed of a car, how do you calculate the distance traveled? You integrate the speed, v, which we have shown is the derivative of the distance function.

Integrals

Differentials give a measure of how a curve is changing; integrals give a measure of the area encompassed by a curve. We start with the simplest case, a constant curve where the x contribution is always one, x to the zero power, such as the green line on the graph which is y =2. What we want is the area under the graph from x=0 to x=6. For this simple graph, there is an elementary way and a calculus way to do it.

The elementary way is basic geometry, the area, A, is just A=xy=2×6=12. This is the integral of the function y=2 from 0 to 6. The calculus way is to note that the area swept out by the graph when x changes a little bit, Δx, as it does in the red rectangle, the area is just yΔx. And the total area is just the sum of all these little rectangles from 0 to 6. Being somewhat sloppy in the use of symbols by letting Δy stand for what y is at Δx, the sum of these little rectangles is also the total area under the 'curve'.

$$\sum_{0}^{6} y\Delta x \ = \ 12$$

Inverse Calculus

The simple geometry method will not do for finding the area under the curved blue graph, but the calculus method will. We let the small change in x, Δx, tend to zero and an infinitesimal width. As Δ x, the width,

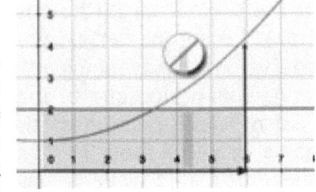

tends to zero, the height of the rectangle is y to a greater and greater accuracy. So the area of the infinitesimal rectangle is ydx. The value of y changes with x at the rate, dy/dx, the first differential, so y is a function of (dy/dx). The total area is simply the sum of these infinitesimal rectangles symbolized by an elongated S, and this is the integral of the curve.

The two processes are interrelated such that if F has the differential D, the integral of D is F. Knowing that the differential of y=x² is 2x, then we know that the integral of the line 2x is x². We need go no further with integration as this is all we need to understand the discussion.

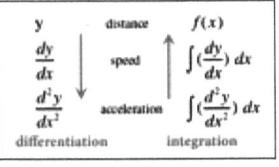

We are interested in the curvature of waves, such as sin(x), and we have already discussed the derivatives of the sine and cosine waves. The first derivative of the sine function is the cosine function.

x	sin (x)	slope	cos (x)
0	0	+1	+1
180°	0	−1	−1
+90°	+1	0	0
−90°	−1	0	0

We can do the same for the cosine function, and the first differential of cos(x) is negative sin(x). This also means that the second derivative of the sine function is the negative sign function.

In plain terms, when the sine curve is at its maximum extent, its gradient is zero, and the change in the gradient is maximally in the opposite direction. When the sine is at its minimum extent, the gradient is maximal and does not change at all. Further derivatives reveal that they come in a cycle of four which repeats no matter how high a derivative you take.

x	$\cos(x)$	slope	$-\sin(x)$
0	+1	0	0
180°	−1	0	0
+90°	0	−1	−1
−90°	0	+1	+1

Furthermore, the minus sine and the plus and minus cosine are identical to the sine wave phase shifted by either a ¼ or ¾ of the period, $\pm\pi/2$ (the cosines), or by a ½-period (the minus sine). Taking the derivative of one of these circular functions is equivalent to advancing its phase by ¼-period (90°, $\pi/2$ radians). This periodicity of four in the sine derivatives is reminiscent of multiplication by the rotation operator, i, which also has a cycle of four: $+i$, -1, $-i$ and $+1$. Everything about the shape of a sine wave is just a variant of the sign wave itself; it is a remarkably self-contained curve.

$$\frac{d\sin(x)}{dx} = \cos(x)$$

$$\frac{d^2\sin(x)}{dx^2} = -\sin(x)$$

$$\frac{d^3\sin(x)}{dx^3} = -\cos(x)$$

$$\frac{d^4\sin(x)}{dx^4} = \sin(x)$$

$$\cos x = \sin(x + 1\pi/2)$$
$$-\sin x = \sin(x + 2\pi/2)$$
$$-\cos x = \sin(x + 3\pi/2)$$
$$\sin x = \sin(x + 4\pi/2)$$

We will shortly be dealing with a very important equation in quantum physics that is of the form shown on the right. Of course, it is not as simple as this, but given the above introduction to calculus, it should be clear that we solve this 'second degree, differential equation' by integration to obtain the first derivative $y = x^2 + k$. The k is a constant to account for the fact that the graphs x^2, $x^2 \pm 1$, $x^2 \pm 2$.... are a family of curves that all have the same derivatives. Going one step further, and knowing from tables that the derivative of $x^3/3$ is x^2, we know that the integral of this will be of the function itself, $y = x^3/3 + k$.

$$\frac{d^2y}{dx^2} = 2x$$

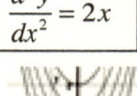

Compared to this differential equation, the only integral equation we are going to encounter is stating the simple fact of confinement within a curve. This simple equation states

$$\int y\,dx = 1 \qquad \int y\,dx = 0$$

that the integral bounded by a curve, S, is 1, and that the integral outside the curve, ~S, is zero. Put simply, all of y is inside the curve S, while none of it is outside. We need to extend this simple calculus just a little to understand how science currently describes the form of the wave.

Calculus with Multiple Variables

The first step is to note that all the curves, or functions as they are more properly called, we have mentioned are functions of just one variable, e.g., $f(x) = x^2$. It is but a small step to visualize functions and derivatives over many variables, such as $f(x,y,z,t)$. The derivatives now involve slopes along more than one dimension, which if we needed to, would take us into the realm of partial differentiation, which fortunately, we don't. Integrals are no longer 2-D areas but 3-D volumes, and 4D and onwards, hyper-volumes.

$\Psi = \Psi_s = \Psi(p_x, a_x, p_y, a_y, p_z, a_z, p_t, a_t)$		Internal, 1s orbital	
$\Psi^2 = \Psi_s^2 = \Psi^2(p_x^2, p_y^2, p_z^2, p_t^2)$		External, probability density	

We mentioned earlier that quantum scientists use the Greek letter ψ (psi) to stand for a quantum wavefunction, such as the 1s orbital. This is a quantum wave over the four complex dimensions, so the wavefunction is a function of eight variables, the wave at a point in internal space and time. As the subscripts get annoying, they are often collapsed into one symbol, or just left out altogether and implicit, as we shall do.

The Schrödinger Equation

In the discussion so far, we have used simple 1-D waves as examples. This can, however, only hint at a more sophisticated analysis as waves in multiple dimensions can be quite complex. The illustration is of a standing wave in two dimensions (such as a drum head) and waves in 3-D can be even more complex. We will meet some 3-D waves when we get to discussing orbitals.

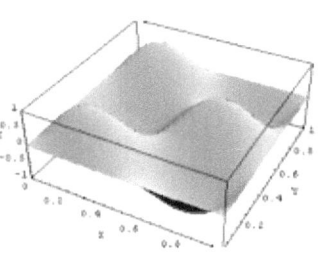

With the little calculus we have discussed, the reader is hopefully not be intimidated by the expressions listed that are central to the quantum description of form and, in a general sense, the description of mind and body in Unified science.

Ψ_s	&	Ψ_s^2	form of mind & body
$\dfrac{d\Psi_s}{ds}$	&	$\dfrac{d\Psi_s^2}{ds}$	curve of mind & body
$\dfrac{d^2\Psi_s}{ds^2}$	&	$\dfrac{d^2\Psi_s}{ds^2}$	change of curve

The psi function, ψ, is the internal wave/ mind, and the external form is ψ^2, the composite density of the body. The 1st derivative gives the curve to the form, whilst the 2nd derivative gives the change in the curve. Higher derivatives are possible, but in practice the second is all that is needed. (This is true for cars as well. Speed is the 1st derivative, acceleration is the 2nd, and a 'jerk' is the name sometimes used for the 3rd if it is ever needed.)

Potential and kinetic

There is just one more thing before we reach the important equation, and that is potential and kinetic energy. This has nothing to do with the enormous energy tied up in the atomic nucleus, just the relationship between position and velocity in an interaction.

We are going to do a 'thought experiment' where you fall into a mine shaft that pierces through the center of the earth. We are ignoring any air resistance (it's a perfect vacuum) and any problem with heat at the center (the earth is stone cold). At the very start, all the energy is in the 'tension' of the relationship, the potential energy, P, and you are pulled very strongly towards the center. At this very start you have no velocity, and your kinetic energy is 0.

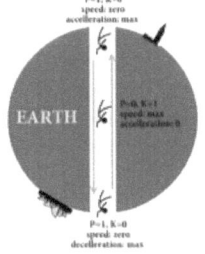

As you speed up and approach the center, the force of gravity diminishes and your acceleration decreases. At the very center, there is no force of gravity but you are moving at thousands of miles an hour. All the energy of the interaction is now in the released speed (kinetic energy) while the tension has disappeared entirely, the potential energy is zero. As you rapidly pass through the center, the force of gravity increases pulling you back to the center. You decelerate as kinetic energy is transformed into potential energy. Your head will briefly appear at the antipodes (speed is 0 and gravity is full strength) where all the energy of the interaction is potential again. The speed is zero at either end and maximum in the middle, while the potential energy is a maximum at either end. Yes, we are dealing with sines and cosines again. The kinetic energy follows a closed sine form, it had a zero node at either end. The potential energy, as we saw with the gluons, is an open cosine form with all the energy at the boundary. (More rigorously, we are dealing with (co)sine squared waves.)

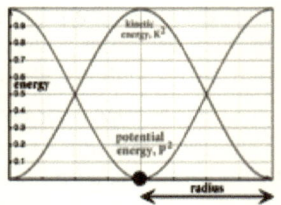

This is a purely classical argument that ignores the internal, but as classical and quantum science do not disagree on such major topics, it suffices to allow the equation to be formulated.

You then repeat the fall in the opposite direction. Without any friction to draw away energy, the conservation of energy dictates that the sequence repeat endlessly.

In a similar way, we can treat the electron as if it is a ball interacting with the nucleus. All the energy of the interaction of a 1s orbital electron is potential energy when it is at the boundary node, and all the energy is kinetic when it is at the nucleus (which is transparent to electrons). The nucleus is also going to move in a complementary way, but being so massive it can be treated as unmoving without sacrificing much accuracy. The inertial mass, m, is just the reluctance of the electron to change its velocity (classical picture) or rotate its direction in spacetime (relativity).

At last we are ready to appreciate two very important equations of modern science, one an integral and the other a differential equation. The first equation is simple and states that, if p^2 is the probability density of the electron in the wave, ψ, the electron is 100% within the wave and not outside it.

$$\int_{\psi} p^2 = 1$$

Integer Solutions

The second equation is more sophisticated, it relates the kinetic/potential energy of the electron's wavefunction to the negative second derivative of the wavefunction (how the form of the wave decelerates, so to speak), along with the mass, the familiar constant π, and the quantum of action, h.

This is important because it is the scientific description in precise math language, of the mind of the simplest entity out of which the material world is constructed. It is the description of the

$$-\frac{d^2\Psi}{ds^2} = \frac{8\pi m^2}{h^2}(K-P)\ \Psi$$

mind of an atom, and is called the Schrödinger Equation. All of the following discussion will be based on this foundation.

Rather like the way that d^2y/dx^2 gave rise to a family of solutions, so too the Schrödinger Equa-

$$\Psi_n = \Psi_1, \Psi_2, \Psi_3 \ldots$$

tion leads to a family of solutions that are numbered by the Principal Quantum Number, integer n. We have already encountered this number, it is the '1' in the name of the 1s orbital.

Without going into detail, we can change all the constants (such as mass, π and h) into units

$$-\frac{d^2\Psi}{dr^2} = (\cos^2 r - \sin^2 r)\sin r$$

where they combine into 1. We can also set the free energy (this is the energy that changes during an interaction, the 'non-free' constant energy being the mass) as 1. We have seen that the share of potential and energy relate as the squares of a sine and cosine wave. Substituting all this into the equation, we get the much simpler expression that has an interesting graph. All the change in the form of the wavefunction occurs near the center, the boundary changes much less so.

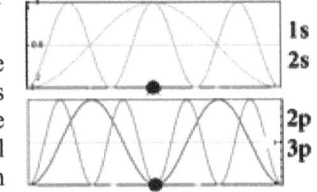

Simple orbitals

This is only a simple introduction, but it should be clear how the equation of the wavefunction can be solved for a symmetrical ½-wavelength that fits with a maximum at the center and just one node, the boundary node.

To rephrase, current techniques are able to solve Schrödinger's Equation for the 1s orbital. This is the good news. The bad news is that we are unable to solve the equation for waves that have internal nodes (although approximations can be useful) such as the 2s (which is like the 1s but has an internal node) or the 2p which has an internal node at the very center. The 's' family of orbitals has no nodes at the nucleus, the 'p' family has 1 node there, the 'd' family has two there, and the 'f' family has three nodes at the center. The primary quantum number equals the total number of nodes in the orbital. These families of orbitals correspond to the blocks in the periodic table of the elements.

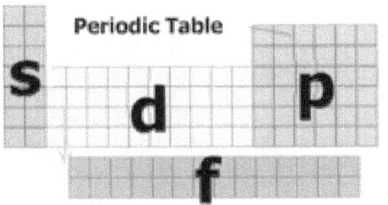

	s				p				d		
	bound node	Internal Nodes			bound node	Internal Nodes			bound node	Internal Nodes	
		off center	on center			off center	on center			off center	on center
1s	1	0	0	2p	1	0	1	3d	1	0	2
2s	1	1	0	3p	1	1	1	4d	1	1	2
3s	1	2	0	4p	1	2	1	5d	1	2	2
4s	1	3	0	5p	1	3	1	6d	1	3	2
...	1	...	0	...	1	...	1	...	1	...	2

Such is the way that the electromagnetic interaction structures the structure and function of everyday objects. The number of electrons in a neutral atom, however, is determined by the atomic nucleus, and this is the realm of the strong force.

The Strong Interaction

As it does not make sense to consider an isolated quark or gluon, we have already discussed discussed the Strong force. The probability that a quark will absorb or emit a gluon is 1, it always does so. In comparison, the probability that an electron will absorb or emit a photon is much less, at ~1/137. This probability is the square of the internal probability amplitude, which is ~1/–11.7, and it is called α, the 'fine structure constant.' It is proportional to the square of the electric charge on an electron.

These probabilities are a relative measure of the strength of the interactions, so the strong force is considered 137-times more powerful than the electromagnetic force. The probability of a weak coupling is severely limited by distance, while gravity is intrinsically weak. The relative strengths are tabulated. Two of the interactions are long-range and two of them are

Coupling Constants			d
Strong	α_s	1	short
Electromagnetic	α	1/137	long
Weak	α_w	10^{-6}	short
Gravity	α_g	10^{-39}	long

short-range. We have seen that it is a balance between gravity and electromagnetism that is responsible for the structure of stars. At the other end of the scale, it is the balance between the electromagnetic and the strong that is responsible for the structure of the atomic nucleus in the stable and radioactive elements.

The strong nuclear force that holds the atomic nucleus together—and the positive charge of the proton is always trying to get it away from any other protons—is a spillover from the strong force that binds the quarks together. A nucleon can exchange virtual pions, a quark and antiquark in a glueball. The virtual pions have a rest mass so they cannot exist for long or travel very far before disappearing before they amount to a pixel of the action. As the lifetime of the charged pions is 10 billionths of a second while the neutral pion's is only a trillionth-trillionth of a second, the charged pions have a much greater reach than the neutral pions.

Nucleons

Even for the charged pions, they cannot reach much more than the diameter of a proton. This is why the strong nuclear force is short range. The gluon shells merge somewhat and the two nucleons settle into a less-energy state by shedding real photons.

The two nucleons both have 'extra energy' problems. We have already seen that the extra pixel of color makes the neutron more massive than the proton. The close confinement of the positive charge in the proton also involves extra energy—it would take energy to force three ⅓ positive charges together, and this adds to the proton's mass, but not quite enough to make up the color energy.

The positive charge makes it impossible for the strong force to hold two protons together, the strong force makes them somewhat sticky, so they can oscillate for a very short time before the electromagnetic force drives the two apart.

The feeble neutral-pions are also unable to hold two neutrons together, the binding energy is too feeble and they fall apart.

If, however, in the brief sticky oscillation of two protons—a low probability—one of the protons does a reverse beta decay into a neutron—a weak process so also of low probability—a proton and neutron can exchange charged pions so that there is a resonance between the two. Both the problematic electrostatic energy of the proton and the problematic color energy of the neutron are reduced by being spread out over twice the surface. The energy saved by this resonance is more than the energy of neutron decay, and such a 'deuteron' is stable with the relatively small binding energy of 2.2 MeV.

This convergence of two events with low probability results in the very low probability of two protons fusing into a deuteron in the sun's core, and the resultant slow and steady conversion by the sun of nuclear energy into electromagnetic radiation. Even though the probability is almost infinitesimally small, the LLN compensates as there are an enormous number of protons in the million-degree core of our sun. There sufficient protons combine into deuterium so that 160,000 tons of gluon-energy is released as electromagnetic energy each second.

The deuterium nucleus picks up another proton in about a millionth of a second, and more energy is released as the problematic electrostatic and color energy spreads out over three nucleons. This diproton-neutron is a helium-3 nucleus. Over a longer stretch of time, and by various pathways, two of these combine into a four-nucleon helium-4 nucleus with two protons and two neutrons. So perfectly do the two pairs of fermions resonate in the nuclear wave, and so perfectly is the problematic energy spread out, that the helium-4 nucleus is so particularly stable that it is called an 'alpha particle' and is often shed by other 'radioactive' nuclei whose protons and neutrons are not well balanced.

The overall energy that powers all stars on the 'main sequence' is the conversion of four protons to one helium-4 nucleus, in which seven-tenths of 1 percent of the original mass is released as energy. We shall the creation of more complex nuclei, such as carbon and oxygen, when stars age and leave the main sequence in a later chapter.

The Weak interaction

The weak interaction, unlike the other three fundamental forces, is not system-building, it does create composites of simpler systems. Its role in the current era is the 'charged current' weak coupling involving the ±W-boson in the reverse-beta decay of protons and the beta-decay of neutrons. The weak 'neutral current' where neutrinos bounce off each other by coupling with uncharged Z-bosons has been experimentally observed.

To summarize, systems interact by coupling with their subsystems. The mutual change in the internal wave (which is immediate) and the external probability density (which follows over time) are the consequences of the interaction.

Thermal Radiation

To conclude this section on interaction, we will look at the simplest of interactions, two spheres of matter bouncing off each other. When two chemically-indifferent atoms (or collections of atoms) such as illustrated by the collision of two helium atoms, we saw the two bounce off each other rather like two billiard balls, the high-action state of the compressed area of the collision sending them both off in opposite directions. The collision is not perfectly elastic, not all the energy goes into the rebound, as the disturbed virtual photons can escape as real photons. The colliding helium convert some of their kinetic energy of movement into the energy of real 'thermal' photons.

The reverse situation can also happen: a helium atom 'surfing' the wave an incoming thermal photon can absorb the energy and the atom accelerates to a higher speed. In a gas of many helium atoms, we end up with a steady state where the loss of kinetic energy to photons in collisions is matched by the gain of energy from surfing the photons.

This bath of photons is called "thermal radiation" and it is always present unless the atoms are perfectly stationary. The amount and energy of these photons depends on the kinetic energy of the atoms, the 'temperature' of the gas.

In the steady state just mentioned, any atom moving faster than average will tend to lose more energy in collisions than it gains from absorbing the photons around. Conversely, slow atoms will tend to gain more energy from the photons than they loose from collisions. The steady state that is reached between the ther-

mal radiation and the kinetic energy is called 'thermal equilibrium' and the thermal radiation depends solely on the temperature,
not composition (only strictly true for a 'black-body' that absorbs
and emits with equal facility).

$$\frac{1}{2}mv^2 = \frac{1}{2}MV^2$$

$$\frac{v^2}{V^2} = \frac{M^2}{m^2}$$

$$\frac{v}{V} = \frac{\sqrt{M}}{\sqrt{m}}$$

In a steady state of a mixed gas of helium atoms and heavier
cousins, the average speed of the two will be different, but the
average energy involved is the same. The kinetic energy, K, is
equal to ½ the mass times the square of the velocity. So a helium
atom accelerating to twice its speed has four time the kinetic en-
ergy than it started with. From this we can calculate the ratio of the speeds in situations of entities with different mass as proportional to the square-roots of their
masses. In a mixed gas of chemically-indifferent helium atoms (He = 4 daltons)
and oxygen molecules (O_2=32 daltons) the helium atoms will be moving almost
three times faster than the oxygen molecules. (As the speed of sound is governed
by this velocity, speaking in a helium atmosphere creates a squeaky, high-pitched
voice.)

The thermal velocity can also apply to single neutrons during their 11-minute
lifetime. A high-energy neutron, such as those released from a fissioning uranium
nucleus, is moving at high speed and the probability amplitude wave of the nucleus is very localized. The probability density of the neutron is very small and
concentrated. The neutron passing through ordinary water encounters many hydrogen nuclei (protons) and in the collisions the energy is redistributed with the
neutron probably losing energy and the protons
gaining it. The neutron is slowed down by the
'moderator' to thermal speeds, and the neutron
wave spreads out.

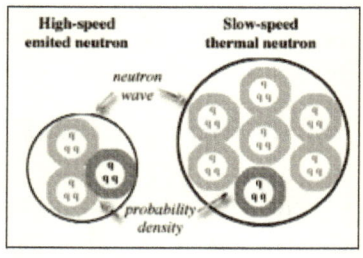

The mass of a neutron is about that of a hydrogen atom, and the best moderators are small
atoms. An illustration of this is provided by the
classical behavior of billiard balls. A fast ball
colliding with a ping-pong ball will send it
shooting away while its own progression is
barely altered. A fast ball colliding with a stationary cannonball will barely shift it
and just bounce of with hardly any alteration in its speed. But if a fast ball hits a
stationary billiard ball, its kinetic energy is probably going to be shared about
equally between the two.

After a few bounces in a moderator the neutron's speed is much reduced and
the probability density of the thermal neutron becomes larger in size than an entire
uranium nucleus of 238 nucleus. The 'cross-section' of the thermal neutron to be
absorbed by the uranium nucleus is far greater than it was for the high-speed neutron. The absorption of the neutron creates such an imbalance that the uranium
nucleus falls apart into two smaller nuclei and this fission releases a few neutrons
that, in turn, are moderated and cause a chain-reaction of uranium fission. This is
the basis for the controlled fusion in nuclear power plants.

At thermal equilibrium, the kinetic energy of the particles and the energy of the
thermal photons is equal. There will be a spread of velocity and kinetic energy,
naturally, but in any sizable quantity of gas there are so many collisions that the

Law of Large Numbers dictates that this spread of is that of the calculated classical probability, the normal, or Bell curve.

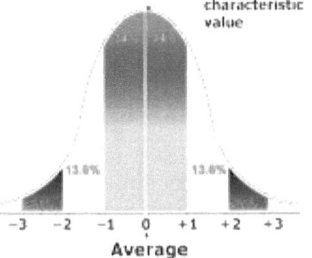

The precise measure of deviation is called the 'standard deviation,' SD, and the Bell curve states that 99% of the speeds will be within ±3SD from the average speed, 95% will be within ±2SD, and almost 70% within just ±1SD of the average. The measure called 'temperature' is proportional to the average kinetic energy. The 'absolute scale' of temperature starts at absolute zero where the entities are motionless, and for each degree Kelvin, K (which is ~C°+273) the temperature rises, the kinetic energy goes up by 'Boltzmann constant,' which is ~10^{-4} eV per degree K.

This is a classical calculation about the external aspects, and it works very well. So it was a great puzzle to scientists when it was found that the 'blackbody spectrum' of photon energies did not have a 'normal curve' spread. It was skewed, as is shown in the diagram for four different temperatures, one of which is that of our sun. The high energy end of the curve was 'pushed in,' so to speak, and hardly any photon had an energy greater than 1SD of the average. It was this disparity that inspired the very start of the quantum revolution in science. This occurred when Max Planck hit upon the idea that photons were 'quantized,' that they each had just one unit of action. When he included this in his calculations he arrived at the skewed-spread of the thermal radiation. This epochal start of the 2nd scientific revolution is commemorated by Planck's Constant being the name given to one quantum of existence, and his name attached to all sorts of pixels.

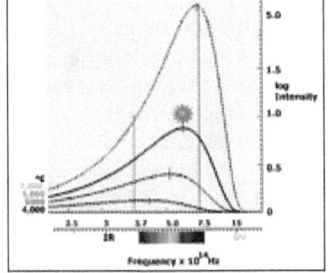

This skewing by quantization is much more likely to end up in a low energy rather than a high energy quanta. The average energy of the photons is the same as in the collisions, but the characteristic energy is not the average energy of the photons.

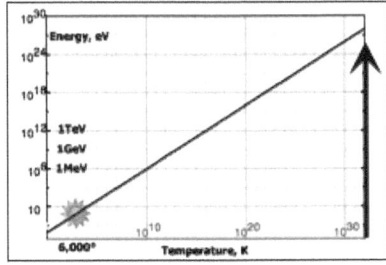

As the temperature rises, two things happen to the thermal radiation. The density of thermal photons increases (the intensity rises), and the characteristic frequency gets higher. This is the difference between the dull red and the bright white poker at low and high temperatures.

Using Boltzmann's constant, it is possible to graph the characteristic energy of photons all the way up to the extremes of temperature during the Big Bang, the theoretical limit being the Planck Temperature of 10^{32}K and photons with an energy of 10^{28}eV and a period of 10^{-44} sec, a period of one Planck time. From the graph it can be seen that at a temperature of 10^{14}, the photons have the same energy, 1 GeV, as that locked up in the rest mass of a proton. At this temperature, there is more than enough energy to mangle spacetime into a proton-antiproton pair, and above this temperature there are as many protons and antiprotons as there are photons, which at this temperature, are so numerous that their density exceeds that of water.

At even higher temperatures with densities approaching that of the atomic nucleus, the quarks, antiquarks and gluons are flying free and are also as numerous as the photons. As we are interested in falling temperatures, the point at which the temperature falls to where the energy is below the rest mass, and matter and antimatter annihilate faster than it is being created, is called the freezing out of, say, protons and neutrons. At energies much above their rest mass, all entities are moving at essentially the speed of light.

Entropy

There is another aspect of heat and temperature that is more abstract than kinetic energy, and that is entropy. The second law of thermodynamics states that the entropy of an isolated system of interacting entities can either be constant (at equilibrium) or must be increasing. This is just a technical way of saying that, absent an outside interference, the direction of change is always from a less probable state to a more probable state (the 'probability' being invoked is the regular, average kind, not the boson/fermion kind of fundamental entities). As there are a lot more disordered states than there are ordered states, entropy can also be described as the tendency to move from ordered to disordered states.

In the diagram, A is a system where all the kinetic energy is aligned in one direction—the whole system is moving in that direction. When it hits the ground, this kinetic energy is transferred to the random movement of the atoms, and the block heats up. There is only 1 ordered state A, but there are many disordered B states, only 2 of which are shown. Getting them all aligned again is going to take work, an input of energy. As there are often a huge number of disordered states, the measure of entropy is related to the logarithm of the numbers involved. If there are, say, 1,096 random B states, then the entropy change, ΔS, in going from the single A state to the random B states, will be proportional to 7 (as the already-encountered transcendental number e gives 1096 when multiplied together seven times, the numbers being chosen for convenience).

$$\Delta S \propto \log(1,096) = 7$$
$$e^7 = 1,096$$

We have already seen that probability is as real as anything else is in modern physics. We have already seen that it is real enough to prevent the collapse of our sun when it ends its life as a black dwarf the size of the earth. So, although it is based on the rather abstract concept of probability, entropy is as real as energy and has to be taken into account. In chemical change driven by the principle of least action, for instance, both the energy, E, and the entropy, S, need to be taken into account in understanding chemical transformation at a temperature of T Kelvins. This combination is called the free energy, H, and it is this quantity that is important in chemical changes. While kinetic energy can explain why a gas expands into a vacuum, it is the entropy that insists on it staying as a homogenous gas and not ever reassembling in one corner, leaving behind a vacuum.

$$H = E - ST$$

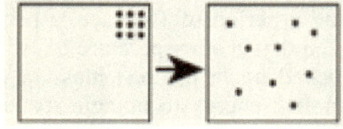

Gravity will collapse a sufficiently large assemblage of atoms (as in the formation of stars) but then the probability is reversed, and the collapsed state is more probable than the extended states. In this case, the calculations involving the balance of energy and entropy are somewhat different.

Goldilocks Wave

The interactions of countless entities results in the waves that rule the current state of the environment. Sometimes these waves combine as bosons into a wave of unusual size. This phenomenon of extreme constructive interference has only recently become the topic of intense study. Unfortunately, there are only two familiar examples, and both are destructive as far as humans are concerned, one in 2-D and one in 3-D.

Rogue wave

A 2-D example is when the waves on the ocean combine into a 'rogue wave' that is much larger, and much more destructive, than its component waves.

The wiki description is that rogue waves (also known as freak waves, monster waves, killer waves, extreme waves, and abnormal waves) are relatively large and spontaneous ocean surface waves that are a threat even to large ships and ocean liners. The rogue waves are not necessarily the biggest waves found at sea; they are, rather, surprisingly large waves for a given sea state.

Perfect Storm

A storm is a 3-D example, when the pressure waves driving weather, with a timecycle on the order of days, combine into a mighty and devastating storm. The experience of being at sea in such a storm is captured in the movie *The Perfect Storm*.

A "perfect storm" is an expression that describes an event where a rare combination of circumstances will aggravate a situation drastically. Since the release of the movie, the phrase has grown to mean any event where a situation is aggravated drastically by an exceptionally rare combination of circumstances. Perfect storm is nearly synonymous with "worst-case scenario" although the latter term carries more of a hypothetical connotation.

Goldilocks Wave

There are times in history when the environment is in the Goldilocks zone for systems to interact and become the subsystems of a more sophisticated system. The overall wave creates an environment which does not have too much or too little of what the interaction requires. The wave creates an environment that is 'just right' for system building to occur—hence its name—the environment is an 'eden' in which the more sophisticated system can emerge. A 'just right' goldilocks wave is synonymous with "best-case scenario."

An example is the creation of 'primordial' helium in the helium-eden a few minutes after the hot big bang.

Before this time, the universe was a dense soup of free protons, neutrons and electrons in a bath of gamma rays with energy greater than that of the binding energy of a deuteron. Any proton and neutron that did cling together would be quickly smashed apart by an incoming photon.

After this time, the thermal radiation could no longer prevent a deuteron from forming but the density had now fallen so low that protons and neutrons hardly ever encountered each other.

During the short period of the goldilocks wave, however, the conditions were 'just right' for the helium-eden when all the neutrons ended up in helium-4 nuclei and

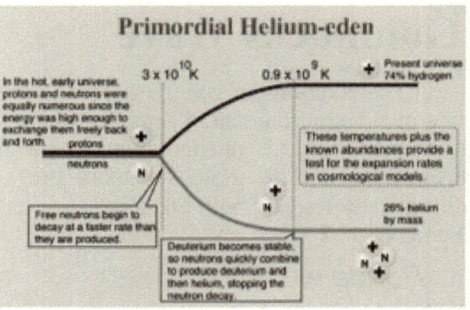

those that did not get bound up decayed into more protons and electrons. A similar goldilocks wave created an eden for atoms some million years later when the thermal radiation had fallen into the infrared and the electromagnetic force could work to unite the protons and helium nuclei into atoms. The result was the 'dark age' when the Universe was 74% hydrogen, 26% helium in a bath of infrared photons. The eden for complex molecules took quite a few more billion years, a tale we will take up again when we return to discussing history and the Big Bang.

EDEN OF THE ELEMENTS

With the preceding overview of the fundamental entities out of which all things in the current physical universe are composed, we can proceed to an overview of how it all came to be. The history of the universe as it is currently understood.

This brings up the topic of the passage of real, positive time to make a history which is subtly different from the complex time of the p-metric (as witnessed by the difference between matter and antimatter which move in opposite directions in complex time while moving in the same direction in real time).

Time and history

One problem that arises in 'dating' the history of the first three minutes of the universe is that the temperature was so extremely hot and the average kinetic energy so extremely high that everything was moving at essentially the speed of light. The rule being that if the thermal energy is a few multiples of the rest mass of a particle, then the particles will be as abundant as the photons. We early noted that the boundary energy of open cosine wave of the real W and ±W pair gave them a rest mass of 90 and 160 billion eV, so if the thermal energy is one trillion eV, the real W-bosons will be as abundant as the photons, and all traveling at essentially lightspeed. the weak force is on a par with the electromagnetic. The thermal energy during the very first moments was much greater than this, and the two forces are said to be a unified electro-weak force.

Even today, 'extreme cosmic rays' with an energy of a billion-trillion eV have been observed to strike the atmosphere, creating a shower of every type of boson and fermion-pair imaginable.

Absolute dating

Energies in the first few moments were much greater, implying that during the first few moments the internal wave of everything was moving essentially along the spatial axes and not at all along the time axis. It was almost a timeless moment for matter. Luckily, this state is normal for photons, so we can 'date' the first three minutes by the thermal spectrum of the photons, the blackbody radiation, and a history as it changes as the universe cools. The blackbody history also tells us what particles were in thermal equilibrium with the photons, and when they 'froze out' and could no longer be created by photons.

Most of the gamma ray photons generated by the hot Big Bang have yet to interact with anything; they have been flying unimpeded since they were released. These photons are all around in a vast abundance to this very day, 13 or so billion years later. By measuring their thermal spectrum we get an absolute measure of time since the moment of creation.

During all those billions of years, however, the energy of the photon has a gravitational effect, it has been attracting all the other quanta of energy (in bosons and fermions) in the universe. The universe, of course, has been tugging back during all that time. The photon has lost energy in opposing the expansion of the universe which has been slowed down. The gamma wave has lost energy and has been 'stretched' by the expansion of the universe into a microwave photon.

These 'relic' photons of the Big Bang—which outnumber the atoms in the universe by a hundred-billion to one—are the 'cosmic microwave background' radiation (CMB) that provide an absolute measure of time. For masses that have moved at substantial fractions of the speed of light in space since the 'decoupling' of radiation and matter, this blackbody radiation with a temperature just a few degrees above absolute zero will be skewed in predictable ways. Luckily, most things have only moved at tiny fractions of lightspeed since that epoch, and the time frame is barely shifted. So matter in the current universe is all on universal time. The CMB is the same no matter which direction you view the universe from earth, it takes sophisticated instruments to detect the slight shift caused by our rotation about the galaxy and its rush towards the Great Wall of galactic superclusters.

Primordial Atoms

At last we can return to the moment of creation. We saw that this involved the creation of an abstract Logos to control the internal wave. This operated on Nothing to create four orthogonal complex planes that were asymmetrically twisted apart into two sets of four axes, the p-metric with a left(−) twist and the s-metric with a right(+) twist.

The Logos operated on the p-metric, and the internal waves driven by the Principle of Least Action combined, changed and developed to generate the external history of the physical universe, the topic of which we will now discuss. The action of the Logos on the s-metric and the history of the substantial Spiritual realm will be taken up towards the end of this work.

Planck period

In the beginning, it was all Planck scale and, to avoid a plethora of Planck, we will just abbreviate it as P. The p-metric was one P-length along each of the three imaginary spatial axes, and one P-time along the real time axis with an inherent chirality of left. It was at the P-temperature of about a trillion trillion trillion degrees, and had a P-energy of about ten thousand trillion trillion eV, about a microgram in mass units.

This 'inflaton' pixel of the 'false vacuum' as it is called had a negative pressure. Driven by this negative pressure, this pixel entered a period of inflation during which pixels doubled themselves in every tick of time and the energy was in all sorts of particles of every kind of charge. When the inflation started to wrench the color charges apart, the inflation was slowed as its outward rush was converted into the real energy of every kind of matter/antimatter pair and bosons. The universe was reheated to almost the P-temperature again by this energy dump into the universe which had expanded to solar-system size before the breaking kicked in.

TOE period

The kinetic energy was such that gravity was as powerful as the other forces, and this is called the TOE period when the forces were indistinguishable. Supermassive boson-fermions, called the X- and Y-bosons, with all axes a-jangle and an electrical charge of ±4/3 and a color charge, were in thermal equilibrium with the photons.

The now-slowed expansion caused the density and temperature of the universe to drop below the freezing-out point of these supermassive boson-fermions, and the matter and antimatter Xs and Ys decayed into electrons and quarks. The left-preference of the p-metric made this decay slightly asymmetric, and a tiny fraction of the −4/3 particle turned into an electron and a D-quark while the +4/3 particle turned into two U-quarks.

The temperature and density were so high that the thermal gluons and quarks were not confined and the three forces were on an equal footing.

Freezing-out forces

As the temperature and density continued to fall, the spatial limitation of the strong forces emerged and all the quarks became confined as protons, neutrons and others. This is the end of GUT period when the strong force froze out leaving the final two forces unified as the electroweak force.

Eventually the thermal energy fell below the mass of the W-bosons, and the weak force froze out leaving only the electromagnetic, and the four separate forces we recognize to this day. The neutrinos and antineutrinos that had been in thermal equilibrium, and in numbers equal to the photons, 'decoupled' from the other particles and went their separate ways.

The entities of composite quarks continued to be in thermal equilibrium, and matter and anti-matter were essentially equally present (there was that tiny imbalance from the X-boson decay).

Freezing out matter

Eventually, the thermal energy fell below that of protons and neutrons, and all the matter and antimatter nucleons annihilated leaving behind just the slight excess of matter. During the helium-eden period, some ended in the composite nucleus but most of them ended as free protons. The still-energetic photons were a hundred billion times as abundant.

Much later, the thermal energy fell below the electron/positron threshold and they annihilated leaving behind the electrons that exactly balanced the charges on the hydrogen and helium.

Much, much later when the gamma had stretched into the near infrared, the electrons combined with the protons and helium nuclei to create neutral atoms. The plasma froze into matter that only interacted feebly with the photons, and matter and radiation decoupled.

The atoms gravitationally aggregated while the photons went on to become the CMB; the neutrinos that had earlier decoupled from matter, and were as abundant as the photons, went on to be the best candidate yet for the 90% of the mass-energy in the universe, the quanta of the 'dark matter' that rules the large scale structure of the universe.

The initial pixel of spacetime in the inflationary phase doubled each Planck tick of time for a time period that is estimated to have lasted for 10^{-35} seconds which, while brief by human standards, is 10^9 quantum ticks. And the number of pixels doubled each tick. The final number of ticks is roughly $2^{10^{10}} \sim 10^{33,333,333}$. As the current volume of the visible universe has $\sim 10^{30,000}$ cubic Planck units in it, so we are only able to see a fraction of the entire physical universe. The rest of the universe, though invisible, is at the same universal time.

Stars

The decoupling of matter and radiation happened about a million years after the big bang. The radiation had previously kept the plasma homogenous; once it lost its influence the atoms started to clump together by gravitational attraction.

The much more abundant dark matter, which had decoupled from radiation a million years earlier, had already clumped together into the scaffolding that attracted the atoms into the large-scale structure of visible matter that are the voids and strands of superclusters of galaxies that we see today. In the following illustration of the large scale structure of the universe,[140] each pixel of white is a cluster of galaxies similar in size to our home galaxy, the Milky Way.

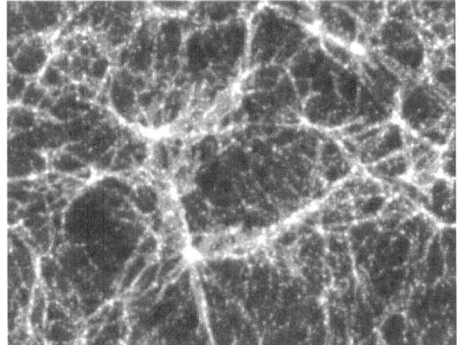

The relic neutrinos, just like the photons, had lost energy in opposing the universe's expansion, but unlike the photons they have a rest mass energy, which is constant and cannot be drained away. As we have seen, the rest mass is very small, only a few eVs, but there are a 100 billion of them for each atom, so their influence is considerable. If their rest mass is as small as 0.1 eV, the mass energy of the relic neutrinos would be ten times that of the atoms, which is why these relic neutrino's are the best candidate for the Dark Matter that is ~10 times as great as that of the atoms in the universe. The nature of the dark matter must be considered to be still an open question as there are problems to be resolved.

1st Generation Stars

It was only after this decoupling from the overwhelming number of photons that clouds of ~75% hydrogen-1 (protons) and 25% helium-4 atoms could start to condense around the dark matter.

The gravitational instabilities in these clouds fragmented them into galaxies and then into stars. The gravitational potential energy of the in-falling and collid-ing atoms was converted into kinetic energy, and the clouds started to warm up and emit thermal radiation. This radiation pressure opposed the gravitational collapse. Hydrogen and helium are not good radiation emitters, so the first generation of stars involved clouds that were ~100 the sun's mass before gravity could impose as contraction over the poorly-radiated away electromagnetic heat that kept it ex-tended.

As this collapse continued, the atoms re-ionized back into a plasma, and when the core temperature and pressure reached high enough values, the protons started to fuse together as a deuterium (hydrogen-2) nucleus (with the emission of a posi-tron which combined with an electron into gamma rays) and so on to helium-4. The star ignited.

The radiation pressure outwards was now as great as the inward pull of gravity, and the star reached an equilibrium where the two were balanced. It took up a po-sition as O class stars on the Main Sequence determined solely by its mass. As massive stars have a greater inward pull, they have to be at a higher temperature to reach equilibrium, and blaze with an intense violet-white light radiating 10,000 times as much energy as our sun does.

Before the stars ignited, the universe had been dark because all the gamma photons were now in the infra red, and this period is called the Dark Age. This period of darkness ended with the ignition of the first stars, and visible light reap-peared in the universe. (It should be noted that all the billions of stars in the bil-lions of galaxies, over 13 billion years of shining, have added but a tiny fraction to the number of photons in the CMB.)

Radiating so much energy, the 1st generation of stars ran out of hydrogen fuel in less than a million years as compared to the tens-of-billions of years for our G-class sun which is much more frugal with its initially-less fuel than such spend-thrifts. When the hydrogen of the first generation of stars was gone the stars left the main sequence and started burning their helium.

The Main Sequence

The equilibrium between gravitational collapse and radiation inflation is reached at the same temperature for stars of the same mass powered by hydrogen-to-helium thermonuclear burning at the core. This relation of temperature and mass is called the Main Sequence. Our sun is about midway on this sequence, bracketed by massive stars radiating intensely in the UV-violet above, and low-mass stars dimly radiating in the red below.

A blue-white O-type star, with a mass 100 times that of our sun, has a surface temperature of 30,000K and emits 100,000 times as much energy as the sun does each second. It runs through its hydrogen in only ten thousand years before leaving the main sequence. A red M-type star with a mass just one-tenth of our sun, dribbles out just ten-thousandth of the sun's energy and can last over ten trillion years.

When the core of a star is so depleted of hydrogen that it can no longer create photons to oppose the inward gravitational pull, the core contracts and heats up as gravitational potential energy is converted into kinetic energy. The core heats up from the tens of millions of degrees in main sequence stars to hundreds of millions of degrees. The outer layers respond to the ferocious heat of the core by expanding and cooling. The aging start leaves the main sequence and the ultra-hot and dense core expands enormously the outer layers which cool off and the star becomes a red giant. The release of photons by helium burning now keeps the star expanded against gravity. Because the increased amount of energy is spread out over a larger area, each square centimeter will be cooler. The surface will have a red color because it is so cool and the surface will be much further from the center than during the main sequence.

Betelgeuse, the pink star in Orion, is an example of a red giant—actually a supergiant, as as it is much more massive than the sun and in its maturity it would have been an O-class star. Betelgeuse now has a size so great that its surface would almost engulf Jupiter's orbit. It is 150,000 as luminous as the sun, but much of this is in the infrared. This prodigious expenditure of energy is fueled by helium burning in the core, but exactly how far along it has progressed is currently unknown.

The energy released by helium burning liberates only a fraction of that released by proton fusion, and further stages release even less, so the lifetime is proportionally much less. Betelgeuse is expected to run out of all its fuel in the next million years, "any day now" by astronomical standards, and meet its spectacular end, to be shortly discussed.

Helium burning

Helium-4 will not undergo any change at the temperatures attained during hydrogen burning because it is so utterly stable. The quarks and gluons are fully-confined and in such a state of low energy that it takes what is called the "triple-coincidence" to get it to where system-building to carbon nuclei can occur. The development of this carbon-eden with a Goldilocks wave at the core of a star that is 'just right' for this to occur is directed by the Logos, the natural law that governs the development of the wavefunction.

A proton that tries to combine with the helium-4 to make what would be lithium-5 is quickly ejected after a brief stickiness, and the isotope has a half-life measured in fractions of a zeptosecond (10^{-23} s). Two helium atoms that attempted to merge into a beryllium-8 nucleus would vibrate in a strong stickiness but briefly and then fall apart into two helium-4s again. The half-life for the 'double-alpha' decay of ^{8}Be is only 6×10^{-17} seconds.

When the core of a star contracts and becomes hotter and denser, helium nuclei are fusing together at a rate high enough to rival the rate at which their product, beryllium-8, decays back into two helium nuclei. This means that there are always a few beryllium-8 nuclei in the core, which can fuse with yet another helium nucleus to form carbon-12, which is stable

Ordinarily, the probability of this 'triple alpha process' would be extremely small. However, the beryllium-8 ground state has almost exactly the energy of two alpha particles and the kinetic energy with which the helium's collide at 100 million degrees is just right to put the beryllium-8 into an excited resonance that when combined with that of a third helium, $^8Be + {}^4He$, has almost exactly the energy of an excited state of ^{12}C. These resonances greatly increase the probability that an incoming alpha particle will combine with beryllium-8 to form carbon. The existence of this resonance was predicted by Fred Hoyle before its actual observation, based on the physical necessity for it to exist, in order for carbon to be formed in stars. Experimental verification of these energy resonances gave very significant support to Hoyle's hypothesis of stellar nucleosynthesis, which posited that all chemical elements were formed from primordial hydrogen.

The final "coincidence" in the carbon eden of a star is that the Logos does not provide such a resonance for the carbon-12 to absorb another of the abundant helium and create an oxygen-16 nucleus leaving no carbon behind. Only a fraction turns into oxygen.

This is the final stage for a medium-sized star such as our sun. The core continues to shrink and heat up and the outer layers of hydrogen blow away as a planetary nebulae. But the core never reaches densities or temperatures sufficient for carbon and oxygen to start fusing together.

The star has a layered onion-like structure where the outer layers are still undergoing fusion. The star becomes a white dwarf as the hot innards are exposed, and once all the fuel is used up the star slowly cools to a black dwarf star. It is estimated that this takes a long time, and the universe is not yet old enough for such a white dwarf to cool off into the infrared.

As the star shrinks under the lash of gravity, another factor comes into play that prevents a total collapse, the fermion nature of the electrons in the star's plasma.

As mentioned earlier, the quantum probability of two fermions being in the same state is zero; it is impossible. As the volume of the white dwarf shrinks, the electrons reach a "degenerate" state in which they are on the verge of being forced to enter into the same state. The impossibility of this happening prevents any further collapse. As the white dwarf continues to cool, the volume remains a constant, held up by the 'degeneracy pressure' of the electrons, as it slowly cools through yellow heat to red heat and ends as a black dwarf. The mass is still that of the sun, but the volume is about the size of the earth's. The universe is too young for even the 1st generation of G type stars to have turned into black dwarfs.

Supernovae enrichment

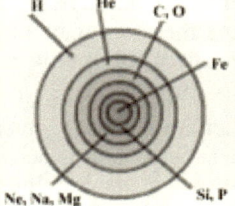

When the core runs out of helium, it heats up to almost a billion degrees and the nuclei fuse into oxygen, then neon, sodium and magnesium, then silicon and phosphorus, and finally into iron and nickel. For O-class stars such as the 1st generation of stars and our neighbor Betelgeuse, this is the end of the line, as iron has a minimum binding energy, and no more energy can be released by further fusion. There are no more photons to hold up the star against gravity and the core starts to collapse. The rest of the star has an onion layer structure, with the lighter elements around the core.

As the core fills up with iron, everything is happening so quickly that this final stage only lasts for minutes. The collapse continues and the core reaches a temperature such that the thermal energy is sufficient to create electron/positron pairs and reverse beta decay becomes possible and neutrinos are produced in abundance. These leave the core removing energy, and the core collapses accelerates.

Bereft of support, the outer layers collapse and this sudden release of gravitational energy heats the entire star, and it explodes as a supernova. In this maelstrom, a flood of neutrons are released and the thermal energy so great that some of the energy is absorbed in constructing atomic nuclei more massive than iron. The star shines out with the brilliance of 100 billion regular stars, and the elements heavier than helium (all called 'metals' by cosmologists) are scattered, enriching the primordial hydrogen and helium with all the other elements.

The force of this explosion so compresses the remaining core and smashes through the electron degeneracy pressure. Rather than be forced to do the impossible, the electrons take the only possible route and merge with the protons to create neutrons.

If the mass of the core remaining after the ejection is not too great (~2 solar masses), the neutrons, being fermions, also have a degeneracy pressure, and refuse to do the impossible and share a state. This can stop any further collapse, and the result is a neutron star that retains its size as it slowly cools off from the millions of degrees at formation as a gamma star through X-ray, UV, white, blue, yellow, red stages towards blackness.

The neutron star has the mass of the sun in a volume the size of a comet. The density of this 'neutronium' is that of the atomic nucleus. The neutrons are stable since it is the proton/electron pair, in such a situation, not the neutron that has the extra energy.

If the remaining mass is too great for the neutron degeneracy to resist the intense gravity, the collapse continues and a 'black hole' is the end result. To understand a black hole, we need the concept of escape velocity.

Black Holes

We have already discussed the interconversion of kinetic and potential energy in the simple pendulum in a gravitational field. A similar thing happens when a bullet is shot straight up in the air. As it rises in the gravitational field of the earth, it slows down until it comes to a momentary halt. All the kinetic energy of the bullet leaving the gun has been sapped away and stored as potential energy. It then falls towards the earth with increasing speed until (ignoring the frictional heat lost to the air) it hits the ground with its original speed. The energy is now all kinetic, and when it hits the ground this is randomized as heat.

The consequence of the gravitational disturbance of physical energy is attractive, and this applies to all forms of physical energy, including that of a photon. Just like a rising bullet, the gravitational field drains energy from a photon (a speeded-up version of how the big bang gammas became the microwaves of today). As a photon is always moving at lightspeed, it does not slow down, the energy is drained from the photon. It is redshifted as its wavelength is stretched, its period increases and its frequency decreases.

Gravity is, however, an extremely tiny effect. Two positive protons will, for instance, repel each other with a force that is a billion trillion, trillion, trillion times the attractive force of their mutual gravitation as a consequence of their physical energy.

For example, the earth and a human body both have equal numbers of positive and negative charges, and the virtual photons are all well-confined. The electromagnetic interaction between the two is minimal, and even a minuscule imbalance is quickly restored by a flow of current, such as a static spark or lighting strike.

Unlike electromagnetism, however, which has an equally powerful repulsion between like charges, physical energy is only attractive, and does not have an expansive effect. (We will encounter an expansive form of gravity when we get to the section that explains why we are using the term 'physical energy' rather than just plain energy.)

Gravitation is cumulative, and even an infinitesimal value can amount to a large value if it is multiplied by a big enough number. Most of the mass-energy of an atom is in the blaze of gluons that is the nucleons, and it is this energy en masse which gives rise to the gravitational mass of a composite body. The human body is a unified wave confining and giving form to ~30,000,000,000,000,000,000,000,000,000,000 quarks, electrons and an even greater number of gluons and photons. The earth has 2×10^{22} as many of them confined. Each one of these quanta of energy is gravitationally attracted to all the others, and the consequence is the force of gravity that keeps us on the earth's surface.

Sir Isaac Newton, while entertaining no hypothesis about the internal workings of gravity, derived these very useful approximations about the external nature of gravity that are still most useful to this day:

1.   The mass of an extended rigid body can be treated as localized at a point within the rigid body, its 'center of mass' which, for an isotropic sphere is coincident with its geometric center.
2.   The attractive force between two extended bodies is proportional to the product of their masses.
3.   The attractive force between two extended bodies is inversely proportional to the square of the distance between their centers of mass.

The surface of the earth, and thus the position of the center of mass of the human body, is ~6,400 km from the center of the earth. The tiny force of attraction between all those zillions of quanta sums up to our 'weight' that holds us to the earth's surface, a force of 9.8 newtons for each kilo of mass.

$$9.8 \times \frac{(6{,}400)^2}{(6{,}410)^2} = 9.77$$

In an airplane flying at 10 km (33,000 ft), we are at a greater distance from the earth's center, and the force is proportionately reduced. As the the proportional change is very small, however, and a person of 170 lbs, only weigh 8 ounces less than on the ground.

$$9.8 \times \frac{(6{,}400)^2}{(6{,}770)^2} = 8.76$$

Even on the International Space Station, at an altitude of 370 km, the force is still 8.76 and I would weigh 152 lbs. It is by no means a gravity-free environment. The weightless experience is a product of 'free-fall' and is

$$9.8 \times \frac{(6{,}400)^2}{(376{,}400)^2} = 0.003$$

akin to the weightless feeling momentarily experienced on a roller coaster. At the distance of the Moon, the force of attraction is reduced to just 0.003 newtons/kilo.

MATHEMATICS, PHYSICS AND CHEMISTRY

The Moon is 1/80th as massive as the earth, but the surface is just ¼ of the distance to the center, so the surface gravity of the moon is 1.62 newtons, about 1/6th that of earth's, not 1/80th.

The sun is 332,950 times the mass of the earth, but its visible surface is 109 times as far from the center, so the surface gravity of the sun is just 28 times that of the earth's. Just as a bullet fired from the earth loses kinetic energy rising against the pull of the earth, so a photon leaving the sun's surface loses energy and is red-shifted, but only by about one millionth, and the red shift on leaving the earth is proportionately even smaller.

A bullet fired from the earth's surface is slowed until it stops, and then starts to fall back. The faster the bullet rises, the higher it will climb. But as the bullet rises, the force of gravity falls off. At a high enough initial velocity, the bullet will rise so far that the gravity is too weak to stop it so it never comes to a full stop but continues to climb. This initial velocity is called the 'escape velocity' (EV) and is a useful measure of the gravity gradient of a body.

body	EV (km/s)
Earth	11.2
Moon	2.4
Sun	618
White dwarf	5,200
Neutron star	100,000
Black hole	299,792

The earth's EV is 11 km/sec, while a bullet from an M16 rifle travels at 10 km/sec and would be almost fast enough to never fall back again if air resistance didn't sap much of its energy.

It would, however, easily escape the Moon as its EV is only 2.4 km/sec. At the sun's visible surface, the EV is 618 km/sec, but the concept can be applied to any distance from the center. At the 'surface' of a sphere as large as the earth's orbit, the escape velocity is down to 42 km/sec, so even though the bullet can escape from the moon, it cannot shake free of the sun.

The surface of a white dwarf has an EV of ~5,000 km/sec, and a photon rising against this gravitational field experiences a significant redshift. The surface of a neutron star has such an intense gravitational field that the EV is >100,000 km/sec, or ⅓ the speed of light. When a neutron star implodes, all the mass collapses through a surface, the event horizon, at which the EV is the speed of light. A photon attempting to escape from the event horizon has an infinite red shift that is indistinguishable from no photon at all. Nothing, not even photons, can escape from the event horizon. It emits no light, hence its name.

Structure of a Black Hole

No one has ever seen inside an event horizon, even in a thought experiment, so we can only theorize. Some think that the mass collapses into the infinite density of a point, a singularity. Quantum physics, however, suggests something that does not involve an infinite quantity as it connects distance with energy. The closer two entities get, the higher is their kinetic energy and thus the temperature.

As the collapsing star breaks through the neutron degeneracy pressure, rather than do the impossible, they first dissolve into a quark-gluon plasma. As they are compressed ever closer, the temperature reaches the stage at which average energy is capable of creating electron/positron pairs and then nucleon/anti-nucleon pairs. At even higher temperatures, the electromagnetic and weak interactions unite as the kinetic energy soars above the rest mass of the weak bosons and so on back to the earliest stages.

These are the last stages of the Hot Big Bang, and as the temperature rises in the collapsing core, the stages of the Big Bang reappear in reverse order. The outward pressure of this recreated primordial plasma prevents further collapse. Just how many stages are necessary to stop the collapse depends on the in-falling mass. The very

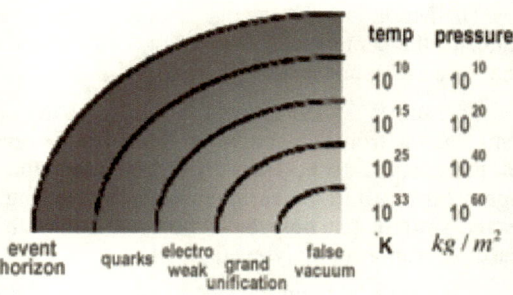

	temp	pressure
	10^{10}	10^{10}
	10^{15}	10^{20}
	10^{25}	10^{40}
	10^{33}	10^{60}
	K	kg/m^2

event horizon · quarks · electro weak · grand unification · false vacuum

largest Black Holes >billion suns probably have to go all the way and recreate a few pixels of false vacuum at the very center whose enormous expansive pressure holds up the layers above.

2nd and 3rd Generation stars

The first generation of massive stars within a few million years ran through their lifecycle ending up as neutron stars or black holes.

The black holes at the very centers of collapsing galactic-sized gas clouds merged into supermassive black holes with an enormous release of energy that it is thought to power the quasars that were commonplace in the first billion years of history and can be observed at the current limits of technology The formation of galaxies and their supermassive black hole centers is still an open question. As our galaxy has a billion-sun black hole at its center, we can suspect that the Milky Way home galaxy was a quasar in its early youth.

The now-enriched clouds of hydrogen/helium collapsed rather more quickly into the somewhat smaller 2nd generation of stars. Many of these are still around in the galactic halo of type 2 stars. Those that were large enough went supernova and further enriched the interstellar medium.

This metal-enriched H/He in turn collapsed into a 3rd generation of stars that are the type 1 stars (the somewhat inverse naming reflecting their order of discovery) of which our sun is a member.

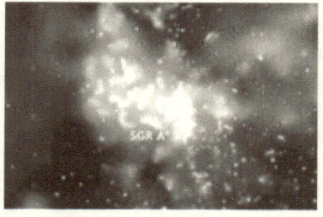

Our galaxy went through its early stages until everything settled down ~10 billion years ago. The black hole that lives in the center of our galaxy is named *Sagittarius A** (pronounced "Sagittarius A-star"). It is ~26,000 light-years from Earth and its event horizon is measured to be about 14 million miles across. This black hole would fit inside the orbit of Mercury and is estimated to have the mass of ~4 billion Suns.

OUR HOME GALAXY
"THE MILKY WAY"

The rest of the 100 billion stars in our Home Galaxy, the Milky Way, rotate roughly in a plane about this central point with a period of ~250 million years. If we call this a galactic year, our earth which is ~4 billion earth-rotation years old, our home is just 16 years old on the appropriate timescale.

All of the processes just described—from galaxy formation to 'metal' production in stars—occur under the direction of the Logos, and the end result was that the clouds out of which the 2nd generation of stars formed had a smattering of the 'metals' in them. The 'metals' are much more efficient at coupling with photons, and the 2nd generation stars tended to have smaller masses than the 1st generation. They were still large, however, and by about 5 billion years ago, they had also gone supernova and further enriched the interstellar medium.

While only about 1%, there were enough 'metals' around that when the 3rd generation of stars ignited and blew away the outer hydrogen and helium, there was enough of them around to form dust, which amalgamated into planetesimals and then into planets. The 3rd generation of stars to emerge were enriched with metals. Our sun is a 3rd generation star and has a significant enrichment in the 'metals' at 71% hydrogen, 27.1% helium, 1.3% carbon and oxygen, and less than 0.7% all the other 'metals'.

In the terms we have been using, the external development of the physical universe under the internal rule of the Logos created a Goldilocks wave that was a set of edens in which all the elements of our world were created. The stars were a set of 'wombs' in which the elements took their characteristic forms that were reflections of the abstract forms in the Logos; their forms and functions were 'inherited' from the Logos.

Element	Abundance (percentage of total number of atoms)	Abundance (percentage of total mass)
Hydrogen	91.2	71.0
Helium	8.7	27.1
Oxygen	0.078	0.97
Carbon	0.043	0.40
Nitrogen	0.0088	0.096
Silicon	0.0045	0.099
Magnesium	0.0038	0.076
Neon	0.0035	0.058
Iron	0.0030	0.14
Sulfur	0.0015	0.040

Here we will pause the progress of history of the universe and discuss some of the elements out of which everyday matter is constructed. Once freed from their stellar wombs, the atomic nuclei interacted with electrons to create the neutral atoms which predominate outside of stars.

It should be noted that while our local environment in which the sun and earth emerged is enriched with the astronomer's *metals*, on a cosmological scale the amount of primordial hydrogen and helium that has been processed into them is as yet a small fraction of the total. I have seen somewhere an estimate that it will take the universe at least another 100 billion years before the primordial storehouse is seriously depleted.

As mentioned, the entire universe is the same age as our bit a space, so the shortage will be a global, not a local challenge. As what has been done once can be done twice, the technological challenge of that far distant age will be to repeat the Big Bang—there is an inexhaustible amount of Nothing to convert into something, after all—and create fresh universes to expand into.

Atomic Nuclei

The multi-nucleon elements that are created during the excesses of supernova are extremely diverse, but only those that were stable or with very long half-lives made it into our everyday world. The resonance of proton and neutron is essential if the neutron is not to beta decay, and the composition of all small nuclei is ~50/50.   As the nuclei get larger, the long range electromagnetic repulsion starts to overwhelm the sort-range strong force and extra neutrons have to be added to shore up the cohesion of the strong force, but this means the neutrons do not have an equal proton-partner and their instability becomes important.

Eventually, no balance is possible, and all elements larger than Bismuth and Lead are unstable, they are radioactive.

The atomic number, N, is the number of protons—which determines the chemical character of the element—and the atomic weight, Z, is the sum of the protons and neutrons together, with the carbon nucleus with 6 protons and 6 neutrons being given a mass of 12 daltons, where 1dalton is ~900 MeV, the rest mass of hydrogen (the proton and the tiny contribution of 0.5 MeV by the electron). $^Z_N X$ $^{12}_6 C$

Two nuclei with a different number of neutrons but the same number of protons (and thus the same number of electrons and chemical properties) are 'isotopes' of each other. Some elements have many stable isotopes—tin being the champion with ten in all—while others have only one, gold being an example. The stable elements can have radioactive isotopes such as a carbon where adding two extra neutrons to the usual six protons, six neutrons results in the carbon-14 isotope, and the neutron instability gives it a half-life of 5,730 years. At some point, a weak interaction between the quarks occurs, the neutron does a beta decay into a proton and the nucleus becomes a stable nitrogen-14 isotope. $^{12}_6 C$ $^{14}_6 C$

The binding energy of helium is very large, and large nuclei sometimes behave as if some of the constituents are a helium nucleus interacting with the remainder. This 'alpha particle' is what leaves a nucleus in the alpha decay of large nuclei such as Uranium.

Radioactive Decay

The binding energy reaches its maximum at iron-56, to make larger nuclei than this, energy has to be added. Although all larger nuclei, such as gold, are theoretically unstable, their decay rate is so small as to be undetectable even over time periods greater than the age of the Universe.

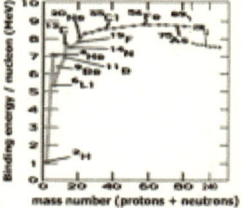

The most commonly observed decays of massive nuclei are:

1.   Beta decay: When there is an excess of neutrons, their inherent instability comes into play. This excess energy is expelled by one of the D quarks as a virtual W⁻ which decays into an electron and antineutrino which both leave the nucleon, leaving behind a proton. The fact that the electron does not carry away all the excess energy was the clue that led to the discovery of the otherwise unobtrusive neutrinos.

2.   Alpha decay: A helium 'alpha particle' can 'tunnel through the surface tension—a node in the wave—as the wave has a non-zero value outside. The probabilities involved are so very small, such as 10^{-44} for an alpha particle reflecting off the surface barrier 10^{33} times a second as in a uranium atom with a half-life of a billion years.

3.   Gamma decay: The nucleons are often left in an excited state by the departure or arrival of a neutron, an alpha or beta particle: this energy is shaken off as gamma rays. This leaves the nucleus in the ground state.

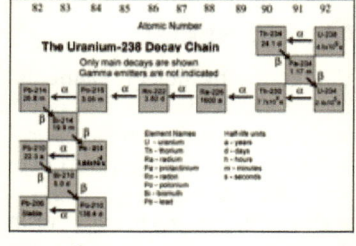

Loss of a helium 'alpha particle' leaves behind a nucleus with an atomic number minus two, and an atomic weight minus four. Loss of an electron 'beta particle' leaves behind a nucleus of atomic number plus one and unchanged atomic weight. Loss of a gamma results in an unchanged nucleus at the ground state.

Large nuclei, such as uranium, can split roughly into two when hit by a neutron—nuclear fission is a process successfully modeled by the 'liquid drop' model where the surface tension plays a major role.

In all cases of nuclear decay, the products have a greater balance between the strong, electromagnetic and weak decay. The process, such as in the decay of uranium-238, can have many steps until a stable balance between these forces is reached in a Lead or Bismuth nucleus.

The stable nuclei that are greater than just trace subsystems in living systems are just 11 in number, and the atomic weights and atomic numbers of their most common isotopes are here tabulated, along with their relative number in the earth's crust, EC, and a human being, HB.

All the atomic nuclei that are involved in the system-building interactions we will be discussing are stable. There are a few unstable nuclei around, such as carbon-14, and the energy released in the nuclear decay is almost always sufficient to disrupt whatever higher system it is a part of.

To summarize, the atomic nucleus is a system of interacting nucleons coupling with pions. The system wave firmly confines all but the virtual photons which escape to tangle with those of any nearby electron.

Element	# p	# n	Earth crust	Human being
			Per 100 atoms	
H-1	1	0	0.22	63
O-16	8	8	47	25.4
C-12	6	6	0.19	9.5
N-14	7	7	<0.1	1.4
Ca-40	20	20	3.5	0.31
P-32	15	17	<0.1	0.22
Cl-35	17	18	<0.1	0.03
K-39	19	20	2.5	0.06
S-32	16	16	<0.1	0.05
Na-23	11	12	2.5	0.03
Mg-24	12	12	2.2	0.01
All others			41.9	<0.01

Atoms and Chemistry

The atomic nuclei created in the star eden and released to enrich the primordial H/He attracts electrons and become neutral atoms. The electrons and nuclei interact by coupling with virtual photons, and they are all confined by a standing wave created by the resonance of all of their individual internal waves. The electron density and the density of coupling photons together have an external form that reflects the form of the internal wave, the system wave, which is directly determined by the Logos.

It becomes tedious to repeat phrases such as "the system wave is a composite standing wave resonance of the internal complex waveforms of the interacting subsystems and their coupling subsystems which is altered by subsystems entering or leaving the system" and "the overall external form of the system is the composite probability density of the interacting subsystems and their coupling subsystems confined by the standing system wave." The behavior and properties of a system are determined by its external form which is determined by the internal system wave, and its tendency to gain and lose subsystems in interaction is also determined by the internal system wave. We shall lump all these aspects together and refer to the complex wave as the 'internal character' of the system, and the composite real density of the interacting and coupling subsystems as the 'external form' of the system.

Helium

When the confinement of the subsystems is almost perfect, as it is in the neutral helium atom, the character of the system is utter inertness, and the external form is essentially that of the classical bit of massy matter, a tiny solid sphere that behaves like a classical billiard ball.

It is only at temperatures close to absolute zero that the character of helium shrugs off its classical mask and reveals its true 'odd-to-classical-eyes' nature. Each proton and neutron has three quarks with a spin of ½, and the sum of these is always a spin of ½, so they behave as fermions, not as bosons. The neutral helium-4 atom has six fermions in total—two protons, two neutrons and two electrons—and these can sum to an integer so the helium-4 atom has a boson character. When the thermal energy is small enough not to mask this tendency, all the helium atoms settle into the same state; they all are in the same internal wave and the liquid behaves as a single, unified system.

"Liquid helium behaves like a fluid without viscosity and with extremely high thermal conductivity. It appears to be a normal liquid, but will flow without friction past any surface, which allows it to continue to circulate over obstructions and through pores in containers which hold it, subject only to its own inertia. Since even gases have viscosity, superfluids have less resistance to shear than a gas does. Despite its lack of viscosity, the liquid still has surface tension, which allows it to rise up the sides of its containers without any normal frictional restrictions to flow. This allows the liquid to flow up the sides of containers, over the top, and down to the same level as the surface of the liquid inside the container, in a siphon effect."[141]

At high temperatures, such as those found in the Big Bang and in stars, the average kinetic energy is so high that electrons cannot unite with atomic nuclei; if they do so, they are quickly knocked off again. This is the state of matter called a 'plasma.' When things are sufficiently cool, enough electrons can form standing waves around the nucleus to cancel out its positive charge and create a neutral atom.

Having so much rest mass, the atomic nucleus only jitters over a fraction of the center of the wave. The nucleus and the light electrons both resonate in the same wave, but the spatial extent of the massive nucleus is 10^{-12} the volume of the electron cloud. For this reason, when discussing the structure of the atom it is customary to ignore the quivering of the central nucleus and treat it as an unmoving point. The focus is on the shape of the surrounding probability density of electrons. When an atomic wave bounces off another, however, the nucleus follows along, just like the electrons do.

If both the positive and negative charges are massive, both quiver in the atomic wave at the very center. If, for example, a tauon replaces an electron in an atom such as in deuterium, the mass of the tauon is equivalent to the mass of the deuterium nucleus, and it is also confined to the very center of the wave. It spends all its time inside the deuterium nucleus which is now electrically neutral. Two such tauonic deuterium atoms can approach, the nuclei can touch and fuse into the stable He-4 nucleus with the liberation of the binding energy. This would open the way to fusion power if the 'technical problem' of the very short lifespan of the tauon can be overcome.

Orbitals

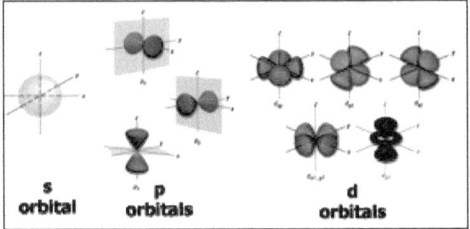

s orbital   p orbitals   d orbitals

As fermions find it impossible to share the same state, each standing wave can hold just two electrons, one going clockwise, one going anti-clockwise. As mentioned, the standing waves of the electrons are classified by the number of nodes the wave has at the center where the nucleus is. There are the 1s, 2s, 3s... orbitals with no center nodes, the 2p, 3p... orbitals with one center node, the 3d, 4d... orbitals with two central nodes and the 4f... with three. The nodes are spatially arranged, giving at every level one s-orbital, three p-orbitals, and five d orbitals. The external shape of the probability density of these is illustrated—as promised, these 3-D shapes can be quite more complex than any 1-D standing waves might suggest (and are generated by simply adding and multiplying complex numbers together to get the internal form of which the external form is but a reflection).

As we traverse the periodic table of elements in their neutral state, adding one proton at a time, each additional electron is added to either an orbital to make a pair, or to an empty orbital next highest in energy. Two electrons, while retaining their left-spin on the internal level, can share the 1s orbital by externally going clockwise and anticlockwise. The internal waves, usually symbolized by the Greek letter ψ (psi), combine into a resonance with a probability that is four times greater, not twice, than that of the single wave, so the paired electrons are very stable. Such paired electrons are in a low-action, high-probability state. Most of chemistry is driven by electrons seeking to enter into this paired state.

$$\psi \rightarrow \psi^2$$
$$2\psi \rightarrow 4\psi^2$$

The single s orbital can hold 2 electrons, the triple p orbitals can hold 6, while the quintuple d orbitals can hold 10. It is this sequential filling of orbitals, and its repetition at each primary quantum number, that gives the familiar arrangement of the periodic table.

Chemical character

The character of the other 'noble gases', such as neon, argon, krypton, xenon and radon is almost as inert as helium with their subsystems almost as well confined by the system. These elements have their s-orbitals and three p-orbitals filled with pairs of electrons and are in a low-action state of stability.

In the other elements, such balance is not attained in their atoms, the wave does not perfectly confine the subsystems, and the Principle of Least Action drives the search for stability in the atom's interactions with other atoms. This is the 'chemical character' of the atom and this is inherited from the Logos via the form of the stable waves that result.

In fact, almost all of simple chemistry can be explained by a drive to attain a stable 'noble gas' configuration for the electron waves where everybody gets to have a complete set of electron pairs in the innermost, least energetic s- and p-orbitals.

Depending on the character of the atom, the internal confining wave, there are just three basic characters to atoms, the way they usually get their electrons all into the noble, stable, inert configuration.

The various elements have these abilities in various degrees. We will discuss the internal character and external form of a representative selection of the elements and their chemical character as we proceed.

1.	Abandon electrons
2.	Adopt electrons
3.	Share electrons

Hydrogen

We have already discussed how the internal traveling waves of an electron and a positron resonate together to form a standing wave that is called the 1s orbital. The equal-sized external probability density of the two creates the composite external form of a positronium atom.

To conserve momentum, the two particles coupling with connected virtual photons tend to occupy different halves of the sphere oscillating about the center at a frequency, that if a photon, would be in the ultraviolet.

Positronium is an unstable entity, but it can be made stable by replacing the positron with a proton, which behaves exactly like an overweight positron. The wave still confines the two, but while the electron density remains extended, the overweight proton density is now so localized that it is often considered a point on the atomic scale. The momentum is conserved as the electron and proton oscillate in the system wave of hydrogen.

The single electron of the hydrogen has the probability density that has the form of a 'singlet' electron in the ground state 1s orbital.

The wave has just one node, the boundary of the sphere, and the electron is confined within this boundary. Most of the virtual photons of the electron are connected to those of the proton, except for those that escape as the hydrogen atom, although overall neutral, has an electrical dipole from the charges tending to opposite sides. The external aspect of this sphere is a blend of the probability densities of the electron, proton and their coupling virtual photons. This is the state of lowest action, the ground state of the hydrogen atom.

A real photon with real energy can enter into this mix and kick the wave into an 'excited state' with two nodes, a boundary node and an internal node, and a greater spatial extent. This is called the 2s orbital. The external now has the form of two concentric spheres, the single wave has shells with a forbidden zone between them.

Excited states

Addition of more photons can create a wave with three nodes (one boundary and two internal) called the 3s, the 4s with four nodes and so on. The diagram illustrates the orbitals up to the 15s, although they can be much larger. It can be seen that they get closer together as they get bigger.

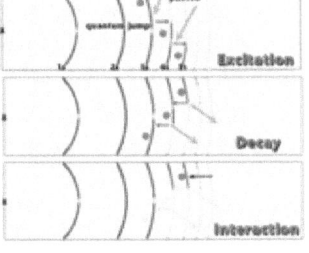

An electron in an orbital, say the 3s, can absorb a photon and add its twist, and the wave becomes a 4s wave. Another photon can add on making it a 5s wave. Precise experiments with lasers have put electrons into the ~100 level wave, a so-called excited 'Rydberg atom' that can be millimeters across.

Such an excited state has a high action and a low probability state, and the electron wave quickly spits out a photon and 'decays' to a lower energy state.

From the 5s orbital, it can emit a photon, and jump to the 4s orbital. With a series of jumps, the electron reaches the 1s orbital. The photons emitted are the characteristic 'emission spectrum' for that element. Hydrogen can also absorb these same photons 'that fit' and jump to an excited state, the 'absorption spectrum' of hydrogen, as illustrated.

The hydrogen atom is very lopsided, as can be seen from the crude diagram. Such a 'singlet' atom is highly chemically active as the hydrogen seeks a state of lower action.

Hydrogen chemistry

There are three ways in which the hydrogen can rid itself of the unbalanced, high-action single electron.

1. Accept an electron so the singlet becomes a pair, a negative hydride ion (H⁻). Only powerful electron donors, such as sodium, can force a hydrogen atom into this state with two electrons managed by one proton.

2. Share the singlet electron with another atom so that both end in a low-action state. The simplest case is two hydrogen atoms, sharing their electrons as a matched pair. This is almost like a helium atom with two separated positive centers and paired electrons filling the 1s orbital. The 1s atomic orbitals of each atom blend together into a 1s molecular orbital. The composite electron density is somewhat dumbbell-shaped. The pair of electrons the two protons are sharing is a 'single chemical bond.' It is not as stable as a helium atom, and the heat of a match will break it up into 'free radicals.'

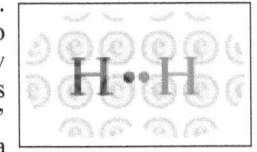

3. Get rid of the electron and use some other atom's pair of electrons. Two hydrogen atoms can share their two electrons with oxygen's six to make a super-stable state of four pairs of electrons for oxygen and a pair of electrons for each hydrogen.

The hydrogen in a water molecule can abandon its electron and hook up with a pair of nonbonding electrons of an oxygen atom in another water molecule. The abandoned water molecule becomes a negative hydroxyl ion and the three hydrogens now on the receiver resonate so they are identical, and the positive charge is spread out over all three hydrogens.

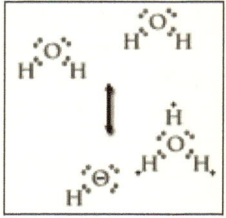

Most of the hydrogen outside of stars in the physical universe is in the form of neutral hydrogen molecules. In everyday life, however, the chemical character of the hydrogen atom in liquid water is of paramount importance, and it gives rise to the simple chemical concepts of 'acidic' and 'basic.'

Acids and Bases

The tendency of the hydrogen atom is to reach a balanced state by abandoning its electron and attaching to a molecule with a 'lone pair' of electrons to share. Such lone pairs are to be found in oxygen and nitrogen atoms, where some of the outer p-orbitals are filled with electron pairs. A water molecule can ionize; a hydrogen abandons its electron and fills its orbital with a lone pair of an oxygen of another water molecule.

This is usually written, ignoring the role of the second molecule, as: $H_2O \leftrightarrow H^+ + OH^-$

In pure water, the concentration of ionized water molecules is 10^{-7}—it has a pH of 7— and the product of the hydrogen and hydroxyl ions is 10^{-14}. A chemist measures probability using the "Law of Mass Action" that, in the case of water ionization, states that the product of the concentration of hydrogen and hydroxyl ions will always be this 10^{-14}.

$$[H^+][OH^-] = 10^{-7} \times 10^{-7}$$
$$= 10^{-14}$$

The tendency for the hydrogen atom to abandon its electron and associate with a water molecule is much greater in the hydrogen chloride molecule, and a strong solution has a hydrogen ion concentration of 0.1 and a pH of 1. A solution of hydrogen chloride in water is a strong acid, $H_2O + HCl \rightarrow H_3O^+ + Cl^-$. The concentration of the hydroxyl ion is proportionately mopped up by this excess of hydrogen ions, but their product is always 10^{-7}.

$$[H^+][OH^-] = 10^{-1} \times 10^{-13}$$
$$= 10^{-14}$$

The ammonia molecule, on the other hand, has a nitrogen lone pair that is very available. The hydrogen abandons its electron and attaches to the nitrogen atom. In strong ammonia, the hydroxyl concentration is high and the hydrogen ion concentration is a low 10^{-13}, a pH of 13. Their product is always a constant. Ammonia solution is a strong base.

$$NH_3 + H_2O \rightarrow NH_4^+ + OH^-$$

$$[H^+][OH^-] = 10^{-13} \times 10^{-1}$$
$$= 10^{-14}$$

The ability to make electron pair bonds with other atoms is called 'valence' in chemistry, so hydrogen, which can make only one bond, is said to have a valence of 1. Chemists use a simple dash to represent a pair bond, so a hydrogen molecule can be written as H–H.

Character of subsystem

To summarize, an atom is a system composed of a set of interacting subsystems—the electrons and atomic nuclei—that are coupling with their subsystems—the virtual photons—all of which are confined and given external form by an internal atomic wave. In situations in which the confinement is not perfect, the atom can interact with other atoms by coupling with some of its subsystems, the electrons and virtual photons. If these interactions involve sharing a pair of electrons in a covalent bond, the atoms become the interacting subsystems of a higher system called a molecule.

A molecule is very similar to an atom (which can be considered a molecule of one atom), it is an array of subsystems, the atoms, interacting by coupling with their subsystems, the electrons and virtual photons. The interacting subsystems and their coupling subsystems are confined, and given form, by an internal molecular wave. If the confinement of the subsystems is not perfect, then some of the subsystems can be used as couplers in the interactions with other molecules.

The character of the molecule is inherited from the Logos, and a more sophisticated set of emergent properties manifest, one obvious example being to use entire atoms as couplers of interaction, such as in the hydrogen bond.

Systematic hierarchy

We see here the emergence of two classes of subsystems, each with a different character and role in the overall system. To illustrate this, we recall what was discussed earlier about music. There we also had two roles: the instrument that generated the wave, and the air as a resonator that responded to the wave that was generated. The instruments were massive and barely moved, while the air molecules were light and fast moving.

We shall call the subsystems in the generator role the G-subsystems and those in the resonator role, the R-subsystems.

System	G subsystems R	
	Wave generator	Wave resonator
	Few, massive, unmoving	*Many, lightweight, mobile, coupling*
Atom	Nuclei	Electrons

As we have seen, both nuclei and and electrons contribute to the unified system wave of the atom. The nuclei are the G-subsystems—they are located at the center of the wave (as if they were generating all by themselves), and they are heavy and slow moving. The electrons are the R-subsystems—they are light and respond to the wave, giving it an external form, and it is the electrons that couple the interactions of atoms.

The wave generators in a symphony are digitally programmed as to what wave to create (the score). In this sense, the atomic nuclei are also digitally programmed, although permanently by their atomic number, as to what wave they generate.

The musical analogy breaks down when we come to the next way of classifying subsystems by their confinement by the wave. In other words, while the G-subsystems are almost always firmly confined, some of the R-subsystems are only loosely confined and can be used as coupling subsystems to interact with other systems.

While a system interacts with other systems by coupling with its subsystems, not all the subsystems participate. Atoms interact by coupling with electrons (and photons) but never couple with atomic nuclei (or pions, quarks and gluons). Atoms only couple with their outer electrons, the "valence" electrons, but they firmly confine the electrons in the filled inner orbitals.

For instance, a neutral sodium atom will interact with other atoms by coupling with its lone, barely confined electron in the 3s orbital, but even fluorine cannot get at the electrons in the filled, inner 1 and 2 'shells'. These are core subsystems, and remain constant, while the valence electrons are variable and available for coupling. The G subsystems are always core subsystems but, while only some of the R subsystems are core, some are potential couplers.

System	G subsystems R	
	Wave generator	Wave resonator
	Few, massive, unmoving	*Many, lightweight, mobile*
	Core	Coupler
Atom	Nucleus	Electrons Virtual photons
Molecule	Nuclei	Atoms Electrons Virtual photons

The progressive filling of higher-energy orbitals is the cause of the periodicity of chemical character in the atoms as we progress up the Periodic Table of the elements. An atom with a lone electron in the 2s orbital has an internal character that is similar to one with a single electron in the 3s orbital, while an atom with 5 electrons in the three 2p-orbitals, and sorely lacking an electron to make a balanced three pairs, has a similar character to an atom with 5 electrons in its 3p orbitals.

This is how the Logos directs system building at the level of atoms. As each new system emerges, a more sophisticated set of emergent properties are inherited from the Logos. The opposite process of system dissolution is also under the direction of the Logos.

Dissolution

The opposite of systems coming together as subsystems of a higher system is the dissolution of a system into its subsystems as a set of lower level systems. Whatever emergent properties the internal character of the higher system had inherited from the Logos disappear in the dissolution. For instance, the very basic emergent character of being substan-

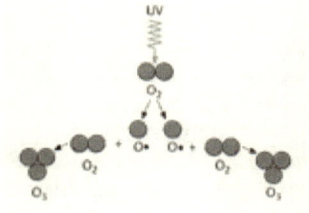

tial, essential to the experience of life, that is inherited from the Logos, and is considered a fundamental character in classical physics, completely disappears when an object is dropped into the sun and becomes a plasma of free electrons and nuclei.

A photon of energy with just the right timecycle can split the system wave of an oxygen molecule into two atoms of oxygen, each with a singlet electron. In an excess of oxygen molecules, these free radicals convert an oxygen molecule into a triple-molecule with the wave and properties of ozone.

Another example of dissolution is electrolysis. If a current is passed through water, the ions go in opposite directions. The H^+ is attracted to the negative electrode where it picks up an electron, pairs up with another H, and bubbles off as hydrogen gas. The OH^- is attracted to the positive side where it gives up an electron, pairs up with another to create water and an O atom which pairs up and bubbles off as oxygen. Passed through molten salt, sodium metal appears at the negative while chlorine gas bubbles off the positive.

$$4H_2O \quad \rightarrow \quad 4H^+ + 4e^- \rightarrow 4H \rightarrow 2H_2 \qquad\qquad 2Na^+ + 2e^- \rightarrow 2Na$$
$$+ \quad 4OH^- - 4e^- \rightarrow 4OH \rightarrow 2H_2O + 2O \rightarrow O_2 \qquad 2Cl^- - 2e^- \rightarrow Cl_2$$

The process of breaking a polymer into its monomers occurs when a water molecule is added to a bond; it is hydrolyzed by enzyme activity. Proteins are hydrolyzed to amino acids, nucleic acids to nucleotides and polysaccharides to simple sugars. This is the process of digestion. The simple monomers are then reassembled into polymers or broken down for their free energy. A living system rarely repairs old components; they are usually broken down while a new component is made afresh.

The constant cycling of material means that almost all of a human body is replaced by new material every six months. The flow of subsystems in and out, however, makes very few changes to the system wave. Our bodies are constantly being broken down and built up; the mind and personality are not.

In the discussion we will focus more on system building and tend to neglect dissolution, but both are aspects of the workings of the Logos in the Physical realm. Much of the internal chemical character of the various atoms, and their capacity for system building, can be understood as a yearning, driven by the Principle of Least Action, to be a resonant wave that emulates that of the Noble Gases, so we start the discussion of the atoms of everyday life with these elements.

The Noble Gases

The s- and p-orbitals are similar in energy level and in their space-filling shape about the nucleus. As mentioned, the filled 1s-orbital of helium firmly confines all its subsystems, and the helium atom does not couple with any of them—it has a valence of zero and is electrically neutral. The filled s-orbital is not so stable when there are empty p-orbitals close by in energy. But a very similar stability to helium is reached when both the s and p-orbitals are filled. These are the 'noble gases' that have little if any tendency to couple with their subsystems.

So stable is this arrangement that much of chemistry can be understood as being driven by the attempt of unbalanced electrons to reach a noble gas configuration, becoming a part of the unchanging core subsystems that are not used as couplers. A convenient shorthand in describing the electron in atoms is to show the 'noble gas' core with the outer electrons around it.

He	: $1s^2$
Ne	: [He] $2s^2 2p^6$
Ar	: [Ne] $3s^2 3p^6$
Kr	: [Ar] $4s^2 4p^6$
Xe	: [Kr] $5s^2 5p^6$
Rn	: [Xe] $6s^2 6p^6$

Alkali Metals

The elements lithium, sodium, potassium, rubidium and cesium are all similar in having a singlet electron in an s-orbital. This is the valence R-subsystem electron about a noble-gas core. This electron is easily lost, leaving behind a sodium ion, a noble configuration but with a positive charge.

Li	= [He] 2s^1	Li$^+$	= [He]$^+$
Na	= [Ne] 3s^1	Na$^+$	= [Ne]$^+$
K	= [Ar] 4s^1	K$^+$	= [Ar]$^+$
Rb	= [Kr] 5s^1	Rb$^+$	= [Kr]$^+$
Cs	= [Xe] 6s^1	Cs$^+$	= [Xe]$^+$

The alkali metals all have an electropositive valence of 1, they all have a tendency to give up the singlet electron becoming a positively-charged ion, this tendency increasing with atomic number. This is exemplified by the reaction of the alkali metals with water: Lithium generates bubbles of hydrogen; sodium reacts violently; potassium spontaneously ignites; while cesium explodes. This is the formation of sodium hydroxide (a strong base) from sodium and water. Na + H$_2$O → Na$^+$ + H$_2$ + OH$^-$

SYSTEM	G SUBSYSTEMS R	
	Wave generator	Wave resonator
Alkali metal	Core	Coupler
	Atomic nucleus & inner electrons	s-electron

As far as living systems are concerned, only sodium and potassium ions from the alkali metals are essential subsystems (while lithium does have medicinal uses). These similar elements are all metals, a broad class of elements which are characterized by their loosely held outer electrons, the metals.

Metals

In cosmology, all the elements other than primordial hydrogen and helium are metals. In everyday life, and all other scientific disciplines, the metals are the many elements that, in macroscopic form, reflect light and conduct electricity.

When a number of atoms are together in a solid, the outer ½-filled orbitals merge together into a continuum of energy levels. The lower energy orbitals are being all filled with electron pairs—the valence band of orbitals—and the upper ones are empty—the conduction band. The atoms have a positive charge and are strongly cemented together by the sea of electrons. This gives many metals a high melting

		MP	BP			MP	BP
1	Hydrogen	-259	-253				
2	Helium	-272	-269	36	Krypton	-156	-152
3	Lithium	180	1317	47	Silver	962	2212
10	Neon	-248	-246	54	Xenon	-112	-107
11	Sodium	98	892	55	Cesium	28	690
18	Argon	-189	-186	74	Tungsten	3407	5927
19	Potassium	64	774	78	Platinum	1772	3827
26	Iron	1535	2750	79	Gold	1064	2940

point—although mercury is an exception—and it takes a very high temperatures to turn them into a gas of free-flying atoms. The chart lists melting points (MPs) when the solid turns into a liquid, and boiling points (BPs) in centigrade when the liquid turns into a gas for the noble gas elements, the alkali metals, and some of the familiar metals.

It is the sea of electrons that gives metals their shine and ability to reflect light. The electrons are so mobile that the electromagnetic wave of an incoming photon sets them in motion, and like an antenna, sends up the exact same wave which radiates the same photon back. A photon of light striking a metal will cause the mobile conduction electrons to oscillate. This oscillation generates a photon that leaves the metal, i.e., the photon is reflected by the metal. This is why untarnished, polished metals reflect light rays, the principle behind the workings of a mirror.

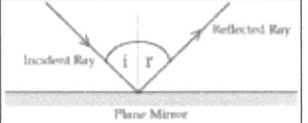

The phenomena is summed up in a law of optics: the angle of incidence, i, equals the angle of reflection, r.

The gap between the valence and conduction bands determines the mobility of electrons in the solid, and its ability to conduct an electron current. An 'electric current' is the bulk flow of electrons that have been kicked by thermal energy into the conduction band.

• If the gap between the two bands of orbitals is large, the solid is an insulator.

• If the bands overlap, the valence electrons can move freely in the conduction band through the solid—it is a conductor.

• If the bands are close, a few electrons are in the conduction band and the solid is a semiconductor.

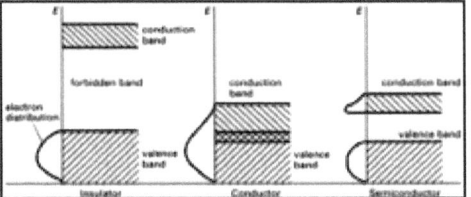

States of matter

This brings up the state of atoms and molecules when there are a great many of them together, all having the same average kinetic energy and the same temperature. Some of the systems we will discuss have a strong interaction with each other, they tend to stick together. Others hardly interact and fly free. Unless otherwise stated, we will always be discussing systems at the standard temperature and pressure, STP, which is roughly that experienced in everyday life.

Systems that are only barely sticky are gasses. An example is the hydrogen molecule which is in a helium-like self-satisfied state. Their kinetic energy is more than sufficient to overcome any tendency to stick together, and the molecules fly about freely, bashing into each other without harm, and bouncing off the walls of any container (the gas pressure). The average distance between the molecules is hundreds of times larger than the size of the molecules, and gases are easily compressed. The kinetic theory of gases uses statistical methods and classical probability to derive relationships between temperature, T (the average kinetic energy), pressure P, (average effect of bouncing off the walls), and container

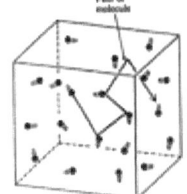

volume,V. The ideal gas law (for molecules of zero size and zero stickiness) is that the ratio, PV/T, is a constant. At absolute zero, 0 K, -273°C) the kinetic energy is zero, as is P and V (which is clearly impossible for actual molecules).

Helium atoms are so indifferent to interaction that a temperature close to zero is needed before the not-quite-perfect cancellation of charge and the resultant tiny dipole is sticky enough to overcome the very small kinetic energy. At this point, the gas turns into a liquid, the boiling point of the molecule. This inherent stickiness of all molecules is called the van de Walls force, and it is at an absolute minimum in helium atoms.

Helium-4 is a boson, it has an even number of ½-twist entities in its structure. At a low enough kinetic energy, the atoms all enter the same state—a giant molecule held together by one quantum wave. This state is quite unusual, with properties such as superfluidity, and quantized rotation. Helium-3, with an odd number, is a fermion, and its properties near absolute zero are even more bizarre.

A hydrogen molecule is not quite as balanced as the helium it is emulating, and they are somewhat more sticky. They have a higher boiling point (which is the same as the condensation point, just coming from the other direction). All the noble gasses have low boiling points, the stickiness increasing with the atomic number as the confinement by the wave becomes less and less perfect. Nitrogen and fluorine diatomic molecules also have tight confinement, and they have even lower boing points than monatomic argon.

He	-269 °C	Ar	-186 °C
H	-253 °C	O	-183 °C
Ne	-246 °C	Kr	-153 °C
N	-196 °C	Xe	-108 °C
F	-188 °C	Cl	-35 °C

At the other extreme, the element osmium, a dense cousin of platinum, is so sticky that it does not fly free as a gas until the temperature reaches 5027 °C. At such temperatures, the average kinetic energy of the atoms and thermal photons is sufficient to knock electrons out of atoms, and above 10,000 °C, all atoms are ionized and we have a plasma of electrons and atomic nuclei, the so-called fourth state of matter. (The strange boson/fermion behavior of bulk liquid He-3 and He-4 probably merits a few more recognized states.)

Liquid and solids

When a gas condenses into a liquid, the kinetic energy is not sufficient to overcome the stickiness. The atoms are in contact, but otherwise are free to shuffle around. The are only two elements whose bulk state is a liquid at STP—bromine and mercury.

The atoms are pulled in all directions by the surrounding molecules, except at the boundary with something else, say air. Any imbalance at this boundary causes the phenomenon known as surface tension. For liquid mercury, the imbalance of forces at a glass-air boundary, the result called a meniscus, is convex shape, as is also seen with liquid water at a wax-air interface. In contrast, for liquid water at a glass-air interface, the imbalance of forces causes a concave meniscus.

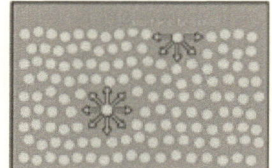
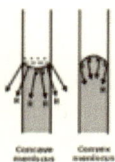

As the kinetic energy drops, the intermolecular forces become strong enough to hold the atoms and molecules firmly in place, the liquid freezes into a solid. All the elements become solids at a low enough temperature, except helium-3 which needs pressure to overcome its fermionic tendencies.

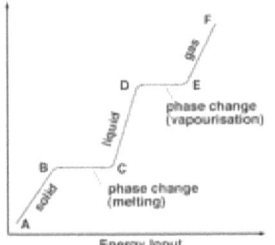

The melting points, where the transition from liquid to solid occurs, varies widely with the element, and some elements, like iodine, do not enter the intermediary liquid state but go straight from a solid to a gas, or vice versa. This is called 'sublimation,' and it also occurs in 'dry ice,' or frozen carbon dioxide.

There is a binding energy when molecules change from the gaseous state to the liquid state, and from the liquid to the solid state. Added energy that does not go into kinetic energy, or raise the temperature but into the phase change is latent energy, and it has to be supplied from outside when melting or evaporating. Latent energy is a potential, not kinetic, form of energy. This is why the 'phase change' from one state to another occurs at a fixed temperature, the melting/freezing point, and the vaporization/condensation point. This potential energy is called the latent heat of melting and vaporization, and it is

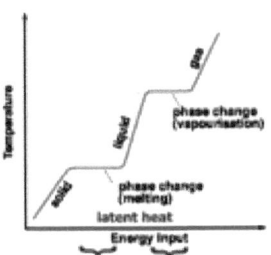

returned in full in the opposite direction as the latent heat of freezing and condensing.

At normal temperatures and pressures, the eighty or so stable elements in their pure form are mainly solids, most of which are metals. Just ten of them are gases and only two of them are liquids.

The solid state of bulk elements can come in different forms, called allotropes, and these can have a quite different internal character and external form.

Carbon, for example, has (at least) four common allotropes: diamond (where the carbon atoms are bonded together in a tetrahedral lattice arrangement), graphite (where the carbon atoms are bonded together in sheets of a hexagonal lattice), graphene (single sheets of graphite), and fullerenes (where the carbon atoms are bonded together in spherical, tubular, or ellipsoidal formations). They have a wide variety of properties, from the black, slippery solid of graphite to the crystalline durability of diamond.

For some elements, allotropes have different molecular formulae which can persist in different phase—for example, two allotropes of oxygen (oxygen, O_2 and ozone, O_3), can both exist in the solid, liquid and gaseous states. Conversely, some elements do not maintain distinct allotropes in different phases—for example, phosphorus has numerous solid allotropes, all of which revert to the same P_4 form when melted to the liquid state.

As we continue to discuss the elements, we will be dealing with standard temperature and pressure unless otherwise indicated.

The Halogens

Complementing the alkali metals are the halogen gases: they have one electron too few to com-

F	= [He] $2s^2 2p^5$	F⁻	= [Ne]⁻
Cl	= [Ne] $3s^2 3p^5$	Cl⁻	= [Ar]⁻
Br	= [Ar] $4s^2 4p^5$	Br⁻	= [Kr]⁻
I	= [Kr] $5s^2 5p^5$	I⁻	= [Xe]⁻

plete a noble core of electron pairs, rather than the one too many of the alkali metals. The halogens are fluorine, chlorine, bromine and iodine.

These are all avid acceptors of an electron to complete the noble shell, a configuration with a negative charge. The halogen atoms all have an electronegative valence of 1.

The tendency to gain an electron decreases with atomic number, a fluorine atom being so avid it can wrest an electron from almost any element, including xenon, and destroy any living thing. Iodine, on the other hand, is a mild and benign antiseptic.

Sodium chloride is the result when an electron slips from the sodium to the chlorine with the release of much free energy. The sodium ion is in the low energy neon configuration, and the chlorine is now in the argon state, which is much larger. Their opposite charges make them stick together by electronic attraction. They couple with virtual photons, and this 'ionic bond' results in high-melting solids with a very regular, crystalline shape. The cubic shape of common salt crystals is a reflection of this atomic arrangement.

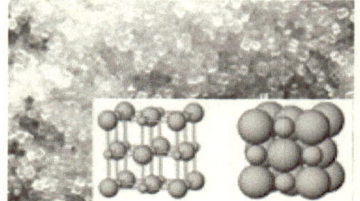

$$Na\ +\ Cl\ =\ Na^+Cl^-$$
$$[Ne]\ 3s^1\ +\ [Ne]\ 3s^2\ 3p^7\ =\ [Ne]^+\ [Ar]^-$$

As far as living systems are concerned, only chlorine and a trace of iodine ions from the halogens are essential subsystems (while fluoride does have dental uses).

Alkali Earths

Be	= [He] $2s^2$	Be^{++}	=	[He]$^{++}$
Mg	= [Ne] $3s^2$	Mg^{++}	=	[Ne]$^{++}$
Ca	= [Ar] $4s^2$	Ca^{++}	=	[Ar]$^{++}$
Sr	= [Kr] $5s^2$	Sr^{++}	=	[Kr]$^{++}$
Ba	= [Xe] $6s^2$	Ba^{++}	=	[Xe]$^{++}$

The properties of metals with two extra electrons in the outer s orbitals—such as calcium and magnesium—behave in a similar way to the alkali metals, just that they have a valence of 2. They are called the alkali earths. They are not so anxious to lose their electrons, however, so they are milder in their properties than the alkali metals. (An object made of sodium would not survive the first rainstorm, while magnesium is metal in wide use.) They form very similar ionic solids and crystals to those of the alkali metals.

$$Ca\ +\ 2Cl\ =\ Ca^{2+} + 2Cl^-$$

The sold ionic 'salts' are held together by electrostatic attraction which gives them a high melting and boiling points. As far as living systems are concerned, only magnesium and calcium ions are essential subsystems.

	MP	BP
Sodium chloride	801	1413
Calcium chloride	772	1935

Oxygen family

O	= [He] $2s^2 2p^4$	O^{2-}	=	[Ne]$^{2-}$
S	= [Ne] $3s^2 3p^4$	S^{2-}	=	[Ar]$^{2-}$
Te	= [Ar] $4s^2 4p^4$	Te^{2-}	=	[Kr]$^{2-}$
Se	= [Kr] $5s^2 5p^4$	Se^{2-}	=	[Xe]$^{2-}$

The family of atoms that lack two electrons are not so alike as the halogens are. In water, the oxygen ion is unstable and quickly picks up a hydrogen to form a hydroxide ion. The sulphur atom is much less grasping and can take up a variety of oxidized and reduced states depending on the circumstances. Only oxygen and sulphur are essential subsystems in living systems.

Oxygen needs two electrons to complete its noble-gas core, it has electronegative valence of two (as does sulphur). It can form ionic bonds with electropositive elements, such as calcium. $\text{Ca} + \text{O} = \text{Ca}^{2+} + \text{O}^{2-}$

Water

Oxygen can also form 'covalent' pair bonds in which each donate an electron pair to a molecular orbital that embraces many atoms, such as the three atoms in water, $2\text{H} + \text{O} = \text{H}_2\text{O}$.

Water can couple not only with electrons and photons, it can also couple with hydrogen atoms as well, and it does so readily with other molecules, particularly other water molecules. The hydrogens switch allegiance from one O to the other, they oscillate back and forth. This H-bond makes the water molecule tend to stick together at preferred angles—the cause of the shapes of snowflakes. A molecular wave has a ground state, that gives form to the molecule, and excited states with extra nodes and waves that give form to the excited state of the molecule. Liquid water forms an intricate pattern of such 'hydrogen bonds' and this extra stickiness gives water anomalously high freezing and boiling points for its atomic weight.

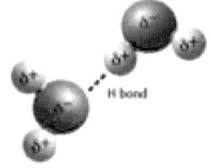

Heavy molecules move slower than light ones so their phase transitions would be expected to occur at higher temperatures. This is true for molecules that only interact by the van der Waals force. Such molecules are incapable of hydrogen bonding; their hydrogens have no dipole and are firmly held. This inability to H-bond is a characteristic of the hydrocarbons, and the observation that, "Oil and water do not mix."

Light hydrocarbons have low temperature transitions, heavy ones have higher temperature transitions as illustrated by the boiling points of methane and ethane.

The ability to H-bond makes molecules very sticky, so they have unusually high transition temperatures for their molecular weight. This is seen in the boiling points of water and ammonia, with the oxygen having a greater H-bonding ability than nitrogen. It takes the hefty molecule heptane to match the boiling point of water. This increase of phase transition temperature is not so great as that caused by ionic bonds, metallic bonds or covalent bonds.

molecule	weight	BP, °C	bond
He	4	−269°	Waals
CH_4	16	−160°	Waals
C_2H_6	30	−90°	Waals
NH_3	17	−33°	H-bond
C_7H_{16}	100	+98°	Waals
H_2O	18	+100°	H-bond
LiF	26	+1676°	ionic
Al	27	+2467°	metallic
Diamond	12	+4827°	covalent

The properties of water are of great significance for living systems, such as ourselves, in which the majority of the subsystems are water molecules. The internal character of the water molecule inherits qualities from the Logos that makes it the perfect resonator-subsystem that takes up the form of the composite wave from an array of generator-subsystems.

We have already encountered two types of systems with a small number of generator-subsystems creating a composite wave whose form is expressed in a large number of resonator-subsystems:

1.   A symphony where an array of musical instruments create a composite wave of air pressure that expresses the form of the music.
2.   A molecule where an array of atomic nuclei create a composite wave of electron standing waves that expresses the form of the molecule.
3.   To these examples, we will add a third example. Living systems where an array of proteins create a composite wave of water structure that expresses the form of the living system.

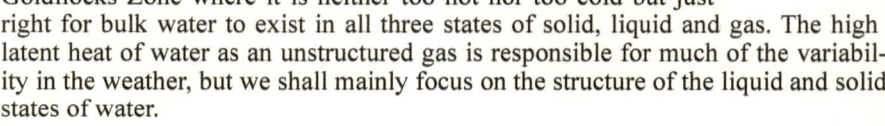

As water-structure is to living systems as air-pressure is to a symphony, it behooves us to understand the character of the water molecule in more detail. For a start, the planet earth to which we are currently confined is said by astronomers to inhabit the Goldilocks Zone where it is neither too hot nor too cold but just right for bulk water to exist in all three states of solid, liquid and gas. The high latent heat of water as an unstructured gas is responsible for much of the variability in the weather, but we shall mainly focus on the structure of the liquid and solid states of water.

Structured water

We have described the interaction of hydrogen and oxygen as sharing a pair of electrons in a single covalent bond. It is not, however, an equal sharing as the electron wave is not symmetrical along the H-O axis, it is pulled asymmetrically towards the oxygen. The positive charge of the H-nucleus is only partially shielded, and the molecule is polar. There is a separation of charge. There is a positive side of the molecule where the two hydrogens stick out at an angle of 105°, and there is a negative side where the oxygen is.

It takes energy to overcome the electrostatic attraction between molecules. When the thermal energy is low enough, the angle between the two hydrogens dictates that the state of least energy is a hexagonal 'chair' that stacks up to create the open structure of ice crystals.

The covalent bond between hydrogen and carbon, unlike that with oxygen, has the electrons evenly shared and the 'hydrocarbon' molecules are not polarized. The thermal energy at −10°C is sufficient to disrupt the solid form of sluggish dodecane, a molecule of 12 carbons, 26 hydrogens and a mol. wt. of 170, while it takes another 10° rise in temperature to disrupt the ice-structure of water molecules with just a tenth the mass.

The ice melts into liquid water in which the molecules are mobile and the thermal energy is sufficient to break up any macro ice structure that attempts to reform on the microlevel. The liquid is more dense than the solid, and ice floats in water. This is almost a unique property of water since most molecules form solids that are denser than the liquid form, and this is particularly important for life as it keeps the oceans from being almost all ice with only a thin layer of liquid water on top.

The water molecules in the liquid are strongly attracted to one another, and this gives water a strong surface tension at an interface with air. Air has no capacity to form hydrogen bonds, so the molecules at the surface are exposed to an unbalanced force being pulled into the liquid but not out by the air molecules.

The same thing happens with an oil and water interface—the surface tension of water squeezes out the oil and they separate. The oil is hydrophobic and the water will move such molecules so as to minimize the surface between them.

On the microlevel, a proportion of the water molecules in the liquid state will be in the preferred hexagonal form, and this proportion falls as the temperature and kinetic energy rises. This holds for pure liquid water, but the proportion in the preferred form can be altered by other molecules. The simplest situation is when a molecule is soluble in water.

Solubility

While an ionic solid such as sodium chloride has a high melting point due to the attraction between the ions, they can reduce this by attracting water molecules and spreading their electric charge over a volume of water. The ions, now coated with a 'hydration shell' of water, fall apart and the salt is soluble in water.

Many salts are soluble in water—sodium chloride will form a clear solution even at 33% concentration—but some ionic salts have such an intense attraction that water does not stabilize the ions. Some salts, such as calcium sulphate, are insoluble and only 0.01% will go into solution.

Molecules that can H-bond are also soluble in water. Examples are molecules with hydroxyl groups, such as alcohols and sugars, which are good at giving a hydrogen bond, and the amines that are good at receiving hydrogens.

In some situations, no water is involved and the most stable state is a hydrogen bond directly between the nitrogen and oxygen, linking their attached molecules together in another level of system-building interactions directed by the Logos. Many of the emergent properties in the subsystems of life depend on such a balance of hydrogen bonding with water or with some other entity, even another part of a large covalent molecule.

Sulfur

Unlike oxygen, sulfur is a solid; it is not so avid as oxygen and tends more towards more equal bonds with itself and hydrogen. In the elemental form, the sulfur shares electrons with four others and they form a ring of eight bent into a crown. The molecular wave firmly confines the electrons, and solid sulfur is an excellent insulator that will not allow any current to flow even under a powerful voltage drop.

The counterpart of water is hydrogen sulphide, H_2S, which is a (noxious) gas as there are only weak hydrogen bonds to hold the molecules together. In bulk, it stays a gas down to $-60°C$ when it turns into a liquid that solidifies into crystals at

−80°C. The liquid form has only a 20° range compared with the 100° range for water. While the sulfur equivalent of the hydroxyl radical, the sulfhydryl radical, is not as commonplace in living systems, the more covalent character of sulfur plays an essential role in many of life's subsystems.

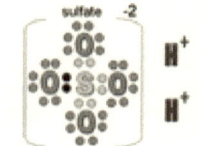

Sulfur will unite covalently with four oxygens to generate the powerfully-acid sulphate ion. While only sparingly used in living systems, the emergent properties of this molecule make it of preeminent importance to industry.

Nitrogen family

The family resemblance between atoms lacking three electrons in the outer p-orbital is even less than in the oxygen family.

N	= [He] $2s^2 2p^3$	N^{3-}	= $[Ne]^{3-}$
P	= [Ne] $3s^2 3p^3$	P^{3-}	= $[Ar]^{3-}$
As	= [Ar] $4s^2 4p^3$	As^{3-}	= $[Kr]^{3-}$
Sb	= [Kr] $5s^2 5p^3$	Sb^{3-}	= $[Xe]^{3-}$

Nitrogen and phosphorus tend to form covalent bonds, and with oxygen they unite to form the acidic nitrate and phosphate ions.

The atom of nitrogen is three electrons short of the balanced, low energy state of filled electron pairs. It can merge its sp^3 valence orbitals with the 1s orbital of three hydrogen atoms to create the molecular wave of ammonia. It has a lone pair which readily shares with the H^+ ion making ammonia a base and nitrogen a good complement to oxygen in hydrogen bonding capacity. Ammonia is almost as good as water in forming H-bonds, making it exceedingly soluble in water.

The ammonia wave has a resonance mode in which the N atomic nucleus 'tunnels' across the plane of the 3 hydrogens. This is an excellent example of how the wave determines the particle's history, the massive nitrogen teleporting back and forth in the oscillating wave. The smeared-out density of the single N nucleus is, just like the 2p orbital, in two lobes of 50% with a node of 0 in the center. This is an example of entanglement over atomic distances of a relatively massive entity.

We will see, on a larger scale, this ignoring of the spatial distance between particles when atoms jump about in a wave in the structure of macromolecules essential

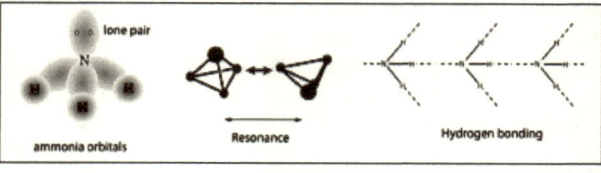

to life, the proteins and the nucleic acids. Both nitrogen and phosphorus are essential subsystems for life; the nitrogen with hydrogen in the ammonia-like state, and phosphorus with oxygen in the form of the phosphate ion. Of particular significance are the high-energy polyphosphate bonds whose making and breaking is the energy currency of life. The polyphosphate bond is high-energy compared to most of the other chemical bonds and breaking the phosphate bond can drive the formation of most bonds.

Carbon family

C	= [He] $2s^2 2p^2$	N^{3-}	= [Ne]$^{3-}$
Si	= [Ne] $3s^2 3p^2$	P^{3-}	= [Ar]$^{3-}$
Ge	= [Ar] $4s^2 4p^2$	As^{3-}	= [Kr]$^{3-}$
Sn	= [Kr] $5s^2 5p^2$	Sb^{3-}	= [Xe]$^{3-}$

The family of elements with just two electrons in the p-orbitals are quite dissimilar in their internal character, as exemplified by the great difference between carbon and tin. An isolated carbon atom is in a high-energy, asymmetrical state. It has an atomic number of 6 balanced by 6 electrons. If the orbital filling was simple, its electronic state should be: $1s^2\ 2s^2\ 2p^2$

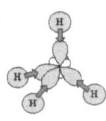

$2s \qquad 2p_x + 2p_y + 2p_z \qquad 4\ sp^3\ orbitals$

This, however, is highly-asymmetrical and the mismatched 2p electrons would have a very high energy. To balance things out, the s and p waves combine together to form four identical waves, called sp^3 hybrid orbitals, with the form of lopsided lobes with one central node that point to the vertex of a tetrahedron. Each orbital has a singlet electron jittering about in it.

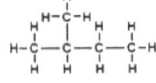

A hydrogen atom can merge its singlet 1s wave with a singlet sp3 to create a molecular wave filled with a balanced pair of electrons. Four such chemical bonds gives the molecular wave called methane. This system wave governs the density of five atomic nuclei, 10 electrons and innumerable virtual photons. The containment is almost perfect and methane is not a very reactive molecule. The carbon atoms can link up in chains with a molecular wave that can span less than a dozen atoms, as in isopentane. The wave can be much larger, embracing billions of atoms as it does in the polymer called DNA.

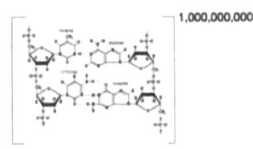

1,000,000,000

Handedness

A carbon atom that is bonded to four different groups comes in a left-handed and a right-handed form. The molecular wave is different in the two forms, and they can have quite different properties.

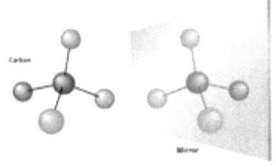

All living systems, for instance, are built of left-handed amino acids and right-handed DNA bases. Right-handed amino acids and left-handed bases are non-nutritious and often poisonous.

Just Carbon

The four coupling waves of carbon can link up in chains, and rings of covalent bonds are preeminently suited for system building. This is observed in the two common allotropes of pure carbon, two quite different system waves embracing a huge number of carbon atoms. These are diamond and graphite.

In graphite, the carbon nuclei lie at the corners of 2-D hexagons; in diamond, they lie at the corners of 3-D tetrahedrons.

In diamond crystal, the molecular wave confines the vast number of carbon nuclei into one vast tetrahedral molecule. This is what gives carbon its strength. The electrons are rigidly constrained, so diamond is an insulator. The electrons are

in such a stable state that they retard, but do not absorb, photons. Diamond is transparent and slows light down to 43% of lightspeed.

The structure of diamond is a direct reflection of the tetrahedral sp^3 orbitals. The graphite structure does not reflect this tetrahedral structure because it involves carbon's penchant for blending waves into new waves.

Delocalized waves

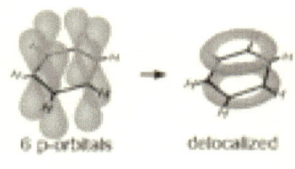

6 p-orbitals   delocalized

Carbon atoms can use 3 of their 4 valence waves to create 3 'regular' covalent bonds with other atoms. The fourth wave can blend to form a 'delocalized' orbital in which the probability density is spread out all around the molecule.

The 'aromatic' refers to the alternation of double bonds which allows a single electron wave to span the whole molecule. The simplest example of this is the benzene molecule of six carbons and six hydrogens. Each carbon contributes one of its waves to the delocalized waves, the electrons being delocalized around the whole ring. Compounds in which this delocalization occurs are called 'aromatic' (as they often have strong odors).

Graphite is a vast flat plane of hexagonal carbons—benzenes without the hydrogens—the whole being one great 2-D molecular wave. The planes stack up but hardly interact, making graphite an excellent lubricant and conductor of electricity. Electrons flow easily along the planes and readily absorb any incident photons, making graphite jet black.

It is the different forms of the wave governing the valence electron density that give diamond and graphite their very different properties, not the subsystems (which are identical in both systems).

Diamond	Graphite
It is the hardest natural substance known	It is soft and greasy to touch
It has high relative density (about 3.5)	Its relative density is 2.3
It is transparent and has high refractive index (2.45)	It is black in colour and opaque
It is non-conductor of heat and electricity	Graphite is a good conductor of heat and electricity

Silicon

Below carbon in the periodic table is silicon. It also lacks four electrons to enter a noble-gas stability, but it is less reactive than carbon and prefers to bond with other atoms rather than with another silicon. Unlike carbon, it is not a subsystem of life but its internal character, particularly its affinity for creating diamond-like structures with oxygen, is essential in the formation of the planets that are the eden and home for life's flourishing. The tetrahedral form in crystal silica is identical to that of diamond except that there is an oxygen bond between each silicon atom. Glass is similar except that the bonds are random and not regular.

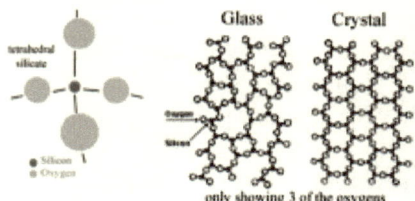

Glass   Crystal

only showing 3 of the oxygens

"Silicon is the eighth most common element in the universe by mass, but very rarely occurs as the pure free element in nature. It is most widely distributed in dusts, sands, planetoids, and planets as various forms of silicon dioxide (silica) or silicates. Over 90% of the Earth's crust is composed of silicate minerals, making silicon the second most abundant element in the earth's crust (about 28% by mass) after oxygen."[142]

Iron

In most elements, the magnetic virtual photons play a small role in the macroscopic properties. It is all too simple for two electrons to rotate so that their magnetic dipoles are aligned, causing attraction, confining all the magnetic virtual photons.

This cancellation is not always possible. In a cold bar of iron, each atom has 1 electron whose magnetic dipole is not aligned and cancelled, and tiny domains in which all these dipoles are aligned. This is state of least energy in which the magnetic virtual photons all add together as boson and create a small magnetic field.

In an external magnetic field, the domains align with each other. The magnetic virtual photon bosons pile up and reach far into space. The bar of iron has become a magnet.

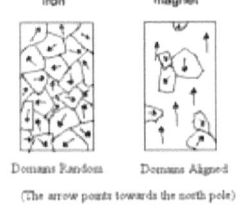

The probability wave that determines the density of these virtual photons is a "chord" composed of millions of tiny generators working together.

In musical terms, we have an array of wave generators creating a 'wave in full.' This wave can be visualized by sprinkling iron filings on a paper over the magnets, each speck of iron becoming a magnet and aligning with all the others.

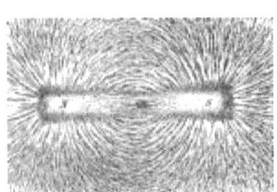

The constructive and destructive flow of 'magnetic lines of force' causes:

- Like poles to repel
- Opposite poles to attract

The quantum probability wave and the consequent density of the virtual photons— the magnetic fields— can be experienced directly with two strong magnets by attempting to force two like poles together or separating opposite poles.

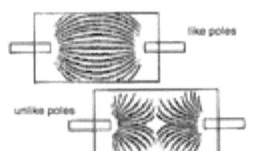

Electricity

Electricity is the flow of conduction electrons in a metal. The simplest, and first to be noticed historically, is static electricity and its everyday manifestation as static cling and shocks. This is caused by charge separation on a macro scale. When electrons get separated from the nuclei they normally neutralize—such as by being mechanically rubbed off a surface—there is a separation of positive and negative charges. This separation takes energy, and this potential energy creates what is called a 'potential difference' between the two that is measured in volts. If allowed to, the electrons will flow to bring this situation back to overall neutrality, and this flow is called an 'electric current,' measured in amperes (amps), that flows from the negative to the positive.

It is almost impossible to observe static electricity with metals, as the conduction electrons rapidly adjust things back to equilibrium. It is only with materials that do not conduct electricity that this static separation of charge can be maintained.

A van de Graaf generator scrapes off electrons from a source and carries them onto a belt into a sphere, where they are scraped off. The electrons build up until the voltage is so great that the air molecules are ionized and conduct the current as brief spark. The ionized nature of a plasma makes it an excellent conductor, as do the ions in a melted salt or in solution.

Dynamic electricity

For historical reasons, the flow of 'electric current' is designated as flowing from positive to negative, exactly opposite the flow of electrons. This makes little difference, in practice, as a flow of positive charge in one direction is equivalent to a flow of negative charge in the opposite direction.

In static electricity, the flow of current quickly equilibrates the charge difference, the voltage drops to zero, and the static electricity is 'discharged,' as in the spark from a van de Graaf generator.

If there is a way of maintaing the voltage, then the current flows continuously, as in the case of everyday electricity. There are two basic ways of generating such a constant voltage: a moving magnet can be used to sweep the electrons along, as an electric generator; or chemical energy can be expended to maintain the potential difference, as in a battery.

When a metal moves past a magnet, the changing magnetic field drags at the electron, and an electric current is generated. The magnetic field, the movement of the wire, and the generated electric current are all at right angles to each other. This is the principle of generators where mechanical motion is converted into electric current.

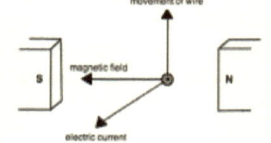

This generates an alternating current, AC, with the voltage and current in phase and changing as a traveling sine wave.

Complementing this, a flow of electrons through a coil of wire generates a magnetic field. This is an electromagnet. A flow of charge in the earth's outer fluid core generates the earth's magnetic field. The relation of current to field is the 'right-hand rule.'

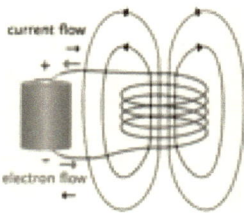
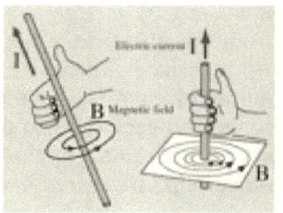
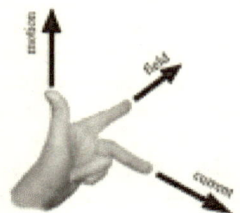

Finally, a wire carrying a current in a magnetic field experiences a force and will move if free to do. This is the principle behind the electric motor. The current, field and motion of the wire are connected by the left-hand rule.

Chemical potential energy can also be converted into electrical potential energy in a battery. Two metals are involved, one with a greater tendency to lose its electrons (more electro-positive) than the other (less electro-positive), such as zinc and copper.

When immersed in a solution of zinc sulphate, the zinc has a tendency to dissolve as its atoms give up their valence electrons and enters solution as zinc ions. Copper, in copper sulphate, has the opposite tendency, it gains electrons and comes out of solution. This disparity shows up as a voltage difference between the two electrodes.

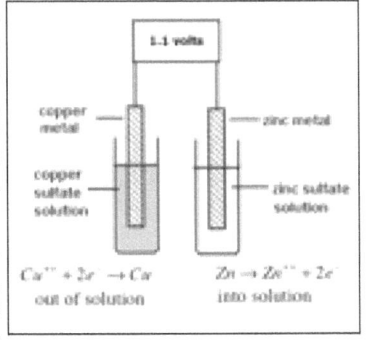

If the two solutions are connected (by a saline bridge or a porous container, for instance), this potential difference can drive electrons around an external circuit. This is a direct current, DC, where the potential difference and the resultant current are constant and in one direction, and can continue until all the zinc has dissolved or all the copper sulphate is converted into zinc sulphate, and the battery goes dead.

Resistance to current

DC electricity is relatively simple, and only one factor, other than voltage and current, needs to be taken into account.

A wire of a pure element has a resistance to the flow of electrons. They bump into atoms and are slowed down, their lost energy of motion adding to the random thermal motion of the atoms. The wire gets hot and its resistance increases.

For a given DC electric potential difference, or voltage, V—set up by a battery or a generator—the amount of current, I, is inversely proportional to the resistance, R, in the external circuit. This relation is known as Ohm's Law:

$$V = IR$$

The energy used to overcome the resistance is turned into heat and, in watts, this is:

$$W = VI = I^2R$$

So doubling the current quadruples the heat generated. The chart gives examples of the wide range of bulk resistance to electron flow by pure elements.

Element	Symbol	Resistivity
Silver	Ag	2×10^{-8}
Iron	Fe	1×10^{-7}
Silicon	Si	6×10^{2}
Sulfur	S	1×10^{15}

Silver and iron have a very small resistance; they are examples of conductors. Sulfur has an enormous resistance and is an example of an insulator. Silicon, with an intermediate value, is an example of a semi-conductor.

To force 1 ampere of current (a very large number of electrons/second) through a standard block of each element takes different voltages and generates very different amounts of heat.

Silver: 10^{-8} V & 10^{-8} W   **Silicon**: 600 V & 600 W   **Sulfur**: 10^{15} V & 10^{15} W

It goes without saying that, with DC electricity, a break in the wire—a gap with a very high resistance—in the external circuit will stop the flow of electrons. This is not the case in AC electricity where the frequency of the sine wave that is the voltage and current is important. The sine wave is most simply described by complex numbers (where, to avoid confusion with the 'I' that is used for current, they use 'j' instead of 'i' to denote the rotation operator).

Along with resistance, there are two other factors in AC circuits that determine the relation between voltage and current—capacitance and induction.

Consider a DC circuit with a break in it. When a voltage is applied, no current will flow. If two metal plates are placed on opposite sides of the gap, however, a current will briefly flow as the electrons build up on one plate until there are so many of them that they repel any more from arriving. Current flows briefly in the circuit (in the opposite direction to the electrons) but nothing crosses the gap.

In an AC circuit, the electrons flow back and forth onto the plates with their capacity to hold electrons, and the faster the voltage changes direction, the more electrons can get on before the limit is reached. This is a 'capacitor,' and the larger the plates, and the closer they are together, the greater is the capacity, measured in farads.

The effect of a capacitor in an AC circuit puts the voltage and current out of phase with each other.

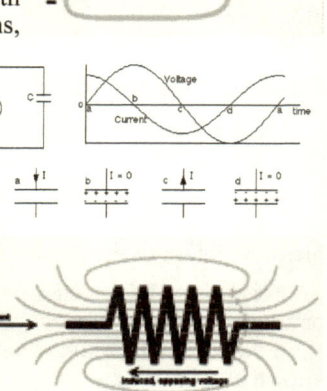

The opposite effect is created by an 'inductor' in the AC circuit, such as a coil of wire. The current through the coil generates a magnetic field, and this change induces a voltage that opposes the flow of current. The inductance, measured in Henries,   increases with the frequency and the size of the coil.

An inductor in an AC circuit, like a capacitor, puts the current and voltage out of phase, but does so in the opposite direction.

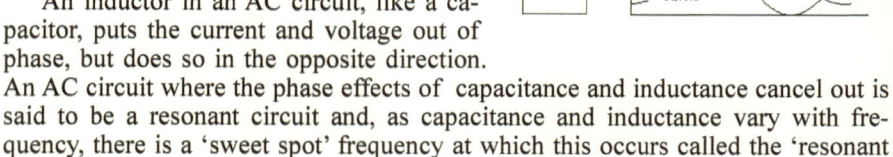

An AC circuit where the phase effects of capacitance and inductance cancel out is said to be a resonant circuit and, as capacitance and inductance vary with frequency, there is a 'sweet spot' frequency at which this occurs called the 'resonant frequency' of the circuit.

Excited atoms

That concludes our brief overview of the elements and some of their interactions by coupling with electrons and their virtual photons. Atoms are also able to absorb and emit real photons. We have already discussed how thermal photons of low energy move atoms around bodily and alter their kinetic energy. At the other extreme are high frequency UV and X-ray photons that strip electrons from atoms they encounter—they are ionizing radiation.

At intermediate values, a photon wave can combine with an atomic wave and put the atom in an excited state, as distinct from the ground state. This can only happen if the photon fits, and is just the right size, for instance, to convert a 1s wave into a 2s wave with an extra internal node. The electrons jitters about in the whole wave, seemingly teleporting from one lobe to another without ever being at the node.

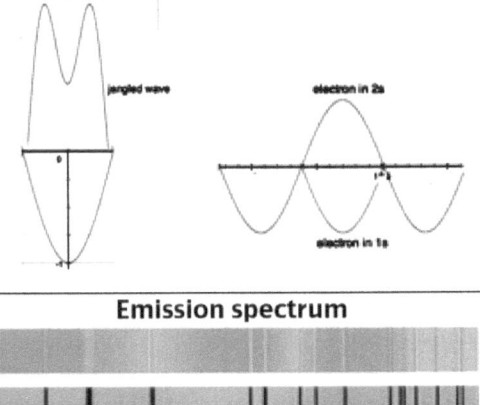

Being in an excited state, the wave quickly reverts to the ground state and spontaneously emits the photon at exactly the same frequency as before. As the smashing together of thermal motion can also kick the wave into such a state, the result is that atoms emit a characteristic set of frequencies at a hot temperature (emission spectrum), and absorb the exact same set of frequencies at a cooler temperature (absorption spectrum).

Each element has its own characteristic 'spectrum,' and this is the basis of spectroscopy that can identify the elemental composition of a lab sample or the composition of the distant stars and galaxies, by the frequencies of light they absorb and emit.

Red Shift

The period of a red photon is approximately twice as long as that of a blue photon. They both have just one quantum action. The blue photon is just more tightly wound. The blue photon has a higher frequency, and a shorter wavelength.

The differences between the photons, however, can be abolished by the relative motion of the source, Sc, and the observer, Ob. If Sc sends a red photon towards Ob, and the distance between them is constant, Ob can count the arrival of each crest, and calculate the frequency of the photon. It is observed to be red. If Ob is moving towards the source, however, he will count more crests arriving each second, and see a yellow photon. If the distance is decreasing at an even faster rate, the rate of arrival will be even greater and Ob will now call it a blue photon. This change in observed frequency with relative motion is called the Doppler shift and; as the frequency shifts towards the higher, blue end, this is the blue shift of photons observed when the distance is decreasing; they have a relative velocity towards each other.

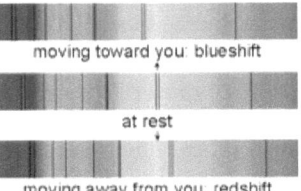

The opposite occurs if the distance between is increasing. Less crests are counted per second and the photon is red-shifted. This terminology is used for radio waves and gamma rays, even though the colors are now in the wrong direction. A red-shifted infrared photon actually moves away from the red spot of the spectrum and becomes a radio wave, while a blue-shifted X-ray moves away from the blue spot and becomes a gamma ray.

Combined with the absorption spectrum of the elements which shift against a continuous spectrum, this effect allows a calculation of the relative velocities of stars and galaxies relative to the earth.

This shifting of colors is only significant at high speeds; a substantial fraction of lightspeed is required to make a blue sweater look red.

Expanding universe

Exactly the same effect occurs if the space between source and detector is expanding. The inflation and expansion of the universe is not in space, but in the *creation* of space. The universe of 10 minutes after the Big Bang was densely filled with gamma rays just after the final annihilation of antimatter —electrons and positrons. There with 100 billion photons for each electron that escaped destruction. These photons are still filling the universe around us, but the expansion of space makes them now look like short radio waves (microwaves), the cosmic microwave background that fills all of space evenly. Nothing has touched these gamma rays since they were emitted, but to us they look like low frequency microwaves.

No energy was lost, of course, they still have their one quantum of action, but it is now spread out over a greater amount of space. A gamma photon has a short wavelength, and it passes by quickly. A microwave photon with a long wavelength takes a 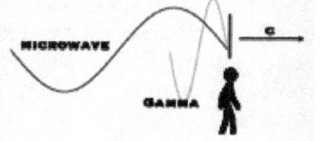 lot longer to pass by, as it lazily waves along. In either case, the wavefront is always observed to be moving at exactly lightspeed irrespective of the speed of the observer.

It was an epochal discovery that all the galaxies in the visible universe (except for a few, gravitationally-bound in the Local Group) were all fleeing away from our galaxy, and the further away they were, the faster they were fleeing. They all had a red shift. Speculation that our galaxy was diseased in some way was squelched when astronomers realized that the universe was not, as had heretofore been assumed, eternal and static. They realized that the universe was expanding, it had a beginning and it had a dynamic history.

Rydberg atom

The ground state of an atom is of a very small size. This is relative, of course, for if an electron particle is a grain of sand, and the atomic nucleus Manhattan Island size, the atom would be the size of the earth. But on our scale, they are small enough at between 60 and 600 trillionths of a meter in diameter. The radius of an atom is more than 10,000 times its nucleus, at 2–20 billionth billionths of a meter, but less than 1/1000 of the wavelength of visible light at 400–700 billionths of a meter.

The radius of the orbit can be huge on the atomic scale, the $n = 137$ state of hydrogen has an diameter ~1 millionth of a meter, a billion times larger, and a billion billion billion times the volume of the ground state atom.

Because the binding energy of a Rydberg electron is proportional to $1/r$ and hence falls off like $1/n^2$, the energy level spacing falls off like $1/n^3$ leading to ever more closely spaced levels converging on the ionization energy. These closely spaced Rydberg states form what is commonly referred to as the *Rydberg series* as shown in the diagram.

This vast distance over which the system wave of a Rydberg atom governs the subsystem—about the size of a bacterium—is an example of the importance of excited states in making it easier to interact with loosely held subsystems.

Light and matter

The wave-confined electron density in atoms can be influenced by photons. The twists of an electron can combine with the twists of a real photon. What happens depends on the frequency of the photon.

An X-ray or gamma ray photon, with a large energy and small time cycle, will rip the electron away and send it flying away at high speed. The now-less energetic X-ray photon continues on its deflected way. The atom is ionized, becoming an ion.

$$H \rightarrow H^+ + e^-$$

An infrared photon, on the other hand, with low energy and a long time cycle, will just jiggle the whole atom back and forth, adding to its energy of motion as an atom thus adding to its "thermal motion" and increasing its temperate. This is why infrared electromagnetic radiation is also called 'heat rays.

Photons of light, with energy and time cycles in the range of the electron waves in atoms, readily combine with the electron wave. The photon waves and the electron waves momentarily jangle together.

We have already discussed one possibility—the wave combination fits as a standing wave of higher energy. This leads to the specific absorption and emission spectrum of an atom (or molecule)

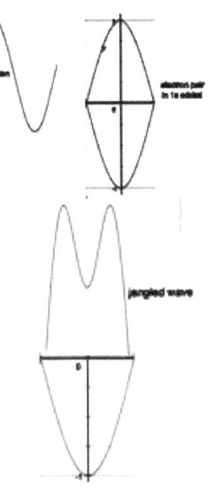

The other possibility is that the wave does not fit, and the photon wave untangles itself and continues on its way. The photon is retarded in its progress, and this shows up in the everyday world as the refractive index of transparent materials.

Refractive index

The electron-photon wave is not a good fit so the photon continues on its way with exactly the same energy and time cycle as it started out with. This holds for molecular as well as atomic orbitals. For a brief moment, the photon wave was detained in its passage through space.

In passing through the trillions of atoms in a gas, liquid or transparent solid, this retardation builds up and light appears to move more slowly than it would through a vacuum. The ratio of the speed of light in a vacuum to the speed of light in a transparent medium is called its 'refractive index,' RI.

In a gas, this effect is so slight that the index can be taken as 1. The effect can be considerable, however, and mount up to a significant slowing, as in these examples:

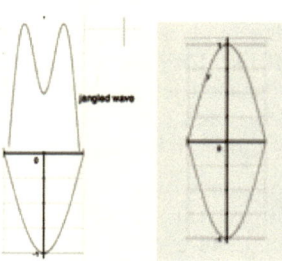

Light photons move through water at 75% of light speed, through glass at 56% of lightspeed while the photons are so retarded by the molecular orbitals of diamond that they travel at only 41% of lightspeed.

Medium	RI	Speed of wave
Vacuum	1	c
Air	1.000277	0.999723 c
Water	1.333	0.75 c
Glass	1.8	0.56 c
Diamond	2.42	0.41 c

A charged particle, such an an electron, can be accelerated to almost lightspeed and then sent through a diamond. The electron is now traveling faster than the photons which have, so to speak, no chance to get out of its way. A shock wave of photons builds up and radiates from the path of the electron. This 'Cherenkov radiation' is often how the detectors used in high-energy physics function. A similar shock wave is created when an airplane or the tip of a bull whip travel faster that the speed of sound, ~760 mph, which is called Mach 1.

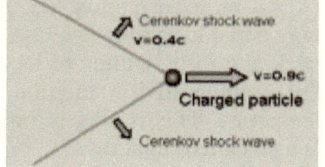

A host of photons traveling through space together is called a light ray. The study of these is called 'optics.'

This is one of the earliest 'hard' disciplines to be developed in science. A scientific discipline is 'hard' when it has a well-developed mathematical description; it is a 'soft' science when its description is in a natural language, such as English, with its fuzziness that promotes much hand-waving, or worse.

English has a plethora of words to describe the complexity of things, digital objects. It has very few words to describe the complexity of waves. As wave concepts predominate in a dualistic science, there is a mismatch of vocabularies. Most of the more sophisticated wave words are to be found in music, which is why we introduced the topic so early in the discussion.

The wavefront of a ray of light passing from air into glass is bent in its course by the speed reduction. This phenomena is summed up in the Law of Refraction, $\sin i = n \sin r$, *(where n is the refractive index of the glass).*

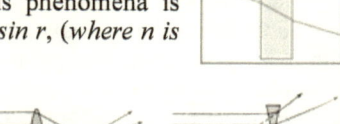

If the surface by which the ray exits the glass is parallel to the entry surface, exactly the reverse happens, so looking through a pane of glass does not change anything. If the exit surface is not parallel to the entrance, the bending does not cancel out, and we have a lens.

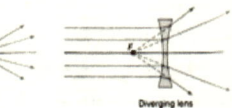

Going from a less dense to a more dense medium, the light is bent towards the normal. Going in the opposite direction, the ray is bent away from the normal. A limit is reached at the incident angle that would bend the refracted ray at 90°, and we get 'total internal reflection' where, instead of leaving the denser medium, it reflects off the surface. Diamond, having such a high refractive index and a 'critical angle' at which this occurs, bounces light around inside the crystal before it is released as the sparkle so admired in diamond jewelry.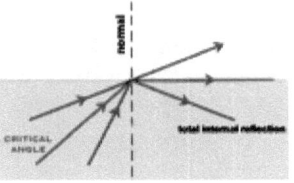

A hot body such as the sun (6,000°) emits a white light, a mix of photons with a spread of energies. As far as visible light is concerned, these photons can be separated into a spectrum of colors by a prism, where the sides are not parallel and the refraction does not cancel out as it does when passing through a pane of glass,

The slowing down of the passage of photons by a transparent substance depends on the energy of the photons, the high-energy (blue) photons being retarded to a greater extent than the low-energy (red) photons. The critical angle for each color light is also different, which explains why a diamond can sparkle in many colors.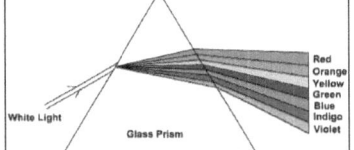

Eden for Molecules

We have discussed how the Logos directs interaction towards a series of edens in which systems can interact and become subsystems of a higher system of a higher 'sophistication.' The higher system has a set of emergent properties inherited from the Logos that were not to be found in the lower systems.

The eden for the creation of the atomic nuclei of hydrogen and helium was the first three minutes of the universe's history. The eden for the creation of atoms was the 'recombination' of electrons and nuclei about one million years on. The eden for the creation of the other elements was the aging of two generations of stars over the next eight billion years, and the enrichment of the interstellar gas.

Molecules do form in these enriched clouds, but the atoms are so few and far between that the probability of them getting close enough to each other to interact is on a par with the weak interaction. Many simple molecules, such as water, have been detected by their emission and absorption spectra. Until recently, the rates of reactions in interstellar clouds were expected to be very slow, with minimal products being produced due to the low temperature and density of the clouds. However, organic molecules were observed in the spectra that we would not have expected to find under these conditions, such as formaldehyde, methanol, and vinyl alcohol.

The eden for molecules more complex than these, however, is to be found on planets that concentrate the elements and allow for chemical interactions in abundance. Before we leave the realm of the universal and focus on our home, we will address a few philosophical questions.

The Anthropic multiverse

The philosophical problem is that the natural laws at work in the universe are finely tuned to allow for living systems to emerge and flourish. There are innumerable examples, so we will just mention a few of the 'fine-tuning' situations we have already encountered.

• A tiny change in the relative strengths of the four fundamental forces does not allow for the sequence of edens ending in the familiar elements, nor would it allow stars such as our sun to form and burn sedately.

• The 'triple-coincidence' that allows for helium to burn into carbon but not all of it into oxygen is plainly exemplary. That massive stars end in a supernova to spread their abundance depends on the precise balance between all four of the fundamental interactions—even neutrinos have a role to play.

• The internal characters of hydrogen and oxygen allow for the formation of liquid water which is to life as air is to music. Liquid water in itself has dozens of characteristics that are essential for life.

• The equal and opposite H-bonding tendency of nitrogen and oxygen allows for the digital processing of analog information by nucleic acids, and allows for long-term storage and transmission of digital memory. It is the different reduction states, determined by their system wave, that give the closely-related nitrogen and phosphorous their essential, and quite different, roles in living systems.

• The facility with which the internal character of carbon allows it to form low-energy covalent bonds with itself and a host of other elements. Most of chemistry (and all of biology) is 'organic,' dealing with carbon-containing molecules; all the rest is 'inorganic chemistry.'

• The particular characteristics of left-handed amino acids and right-handed sugars that in combination have emergent properties that are exactly suited to be the basic subsystems of life. No scientist has yet to construct even the simplest metabolism out of D-amino acids or L-sugars, and probably never will as the combination does not have the right character.

The problem only gets worse the more we understand about living systems in detail. One current example is the quite unexpected sophistication in the emergent properties of RNA. They can do just about anything because they are chemically versatile. The macromolecule has a finely-tuned set of internal characters that make it the active core of living systems. DNA, in comparison, does few things well as the passive memory core of living systems.

Every year it seems a new type of RNA is discovered, and there are now dozens of known types, each with a different role to play as a unique subsystem of life. They routinely read, write and manipulate digital information about analog forms stored on DNA, are master manipulators of metabolism, and are even considered to be the first macromolecules that interacted as subsystems of a sophisticated system that had the emergent properties we associate with life, such as digital manipulation of memory of analog wave-generators that resonate and structure water.

So, the philosophical question that science has to deal with is: How come the universe is so finely tuned to allow for our everyday existence?

The Abstract realm

We have taken a commonsense approach based on the understanding of the physical world we have already outlined (in reverse):

- A system is a set of interacting subsystems coupling with their subsystems.
- The external form of a system is the composite probability density of the subsystems.
- The subsystems are confined by an internal character, the composite resonance of all the internal waves contributed by the subsystems. This external form of the system, by the Law of Large Numbers over time, is a reflection of the form to the internal character.
- The form of the internal character is directly and precisely determined by natural law. As this is more akin to the 'idea of a symphony' than to a simple linear relationship, we place the usual natural laws at the foundation of a very sophisticated abstract entity, we call the Logos.

The commonsense view is that the emergent properties are expressed because they were put there in the elaboration of the Logos. We often use music as an illustration since it has a sophisticated approach to complex waves. When we appreciate qualities such as pathos, excitement and sorrow in a piece of music, we are aware that these were placed there by the craft of the composer. The performance has inherited these qualities from the composer.

Using this as a guide, we expect that the qualities we appreciate in everyday objects—such as being substantial, having finely-tuned valence and chemical behavior—are inherited from the Logos because they were put there in its elaboration. I certainly ascribe the wonderful qualities exhibited by the MacBook on which I am writing—it being a few pounds of aluminum, silicon and plastics—to the expertise of those who drew up the blueprints for their assembly into a MacBook.

Simply put, this again led to a scenario of an Abstract Creator, creating an abstract Logos that acted upon Nothing to create the p-metric history we are discussing and the s-metric we have yet to include. The qualities are expressed by systems because they were put into the Logos with the express purpose of leading to life.

Many intellectuals, however, are more uncomfortable with the concept of an Abstract Creator with a Purpose for creating than they aret with the classically-weird description of the substantial world offered up by modern science. They have offered alternative explanations for the increasingly apparent fine tuning of our universe for life.

Anthropic Principle

Simply put, the 'anthropic principle' answer to the question is another question: "How could it be any other way?" After all, the sun is shining and we are here discussing the question, so how could the universe be any other way. If it were not as fine-tuned as is, we would not be here and the question is moot. This is a little like looking at the *Mona Lisa* and thinking, "That's delightful; but as it is there before me, how could things be otherwise?"

Such thinking does have its uses, however, as it was just this perspective that enabled Hoyle to think the sequence, "There is carbon, so helium must convert into carbon, there has to be a triple-coincidence of nuclei strong resonances for the triple-alpha process to happen." And so there was. Looking at the painting, we could anthropically deduce, "Someone put paint on canvas to express such personality" and "people have cherished the painting for centuries or it would not have survived in such condition," but not much more than that.

Multiverse

The multiverse explanation for the fine-tuning is that our universe is just one of many universes. The emergent properties are not the result of a universal law, but are randomly assigned and it just so happens that, by chance, all the emergent characters are just right for life.

There are, however, a lot of things that could be quite different, so it takes a lot of them that are not right to account for one that is just right. The discomfort with a purposeful Abstract Creator is apparently less than the proposed number of universes in the multi-universe which is about 10^{500}, a number whose ridiculousness can only be appreciated by reading it aloud: "our universe is one of a trillion, trillion [repeat 'trillion' another forty times] trillion, trillion universes in which the laws are just right for living systems like us."

This perspective is akin to having a profound experience of Beethoven's 9th Symphony and then assuming that it is just one selected from a trillion trillion random compositions.

I personally prefer the commonsense explanation, i.e., the qualities are there because they were put there by great expertise. That is the end of the philosophical detour, and we now return to what is known about the history that occurred around the sun, our third generation star.

The Earth

The gas that condensed to form the earth and sun 5 billion years ago had been enriched by a recent supernova that created an abundance of stable nuclei as well as a host of radioactive elements that have all decayed by now except for a few almost-stable with particularly long half-lives, such as uranium and thorium.

A small fraction of the collapsing gas did not fall into the sun. The metals remained in orbit and combined into molecules, grains and higher aggregates, while the hydrogen and helium was blown away by the solar wind. These aggregates condensed to form the earth about 5 billion years ago.

A collision with another such condensation created the Moon (at ½ the distance it is now). The original crust was like that of Venus—all of one piece—but this impact cracked the shell into tectonic plates that are shifting around to this day. The Moon also stabilized the axis of the Earth's rotation at ~23° inclination, instead of it being all over the place as happened to the planet Mars.

While the water molecule is small, it is chemically attracted to minerals and aggregated with them. Along with the water contributed by the comets during the heavy bombardment phase of the earth's formation, the water condensed on the cooling Earth into oceans that were saturated with reduced, soluble iron in the ferrous state. The Moon raised enormous tides which eroded the land and created great beds of porous clay in the cracks in the ocean floor. These faults between the tectonic plates circulated seawater through the crust as black and white smokers perfusing the clay beds. The intense UV light and lightning drove many organic reactions in the dense atmosphere of carbon dioxide, ammonia and nitrogen.

The history of the universe up to this point has evolved under the direction of the Logos, particularly that aspect we earlier described as the Perfect Wave, i.e., many waves combining over time into a wave with unique properties. As the science describing the formation of 'rogue waves' and 'perfect storms' is in its infancy, this period is as yet only imperfectly described.

But one thing is known for certain. The end result of this 10 billion-year history was an eden perfect for the origin of life to occur with relative rapidity over a few hundred million years. As we shall see, the influence of the Moon was crucial to the rapidity of the origin events that led to life. As very few planets can be expected to have such a large moon, we can expect that life was kick-started on the Earth, and has only slowly developed on the moonless majority. It is to be hoped that most of these moonless planets will have at least developed photosynthetic bacteria and an oxygen atmosphere by the time we get to them.

The emergent properties that were inherited from the Logos made the earth a perfect eden for the emergence of life, entailing the transition from systems characterized by chemistry to a new set of systems that were predestined to be the subsystems of Life.

Biological foundations

In the Book Two we will look at the biological sciences, and the study of living systems. Unfortunately, the 2nd scientific revolution has yet to reach them. The biological sciences have yet to include consideration of the causal internal aspect of things as uncovered by the physicists—the biological sciences are all classical and deal solely with external appearances. So, while the biology of our era is proud of its firm foundations in the 'hard' sciences (those amenable to mathematical rigor), the physics in which current biology is rooted is the classical physics of Darwin's day.

"It is most ironic that today's perceived conjunction between physics and biology, so fervidly embraced by biology in the name of unification, so deeply entrenched in a philosophy of naive reductionism, should have come long past the time when the physical hypotheses on which it rests have been abandoned by the physicists."[143]

If the truth be told, much of this slow progress is because the physicists are uncomfortable with their irrefutable discovery of the internal aspect and constantly attempt to explain the basics with classical concepts plus a dash of 'weirdness' that is to be ignored as quickly as possible.

For it was with the greatest reluctance that classically trained scientists faced up to the implication that their description of objective reality was horribly inadequate. The establishment of the current worldview of physics was not based on theoretical speculation. The current view vanquished the old not for theoretical reasons but because that ultimate arbiter of science, experiment, insisted on it.

"The quantum era had arrived but it did not bring an end to controversy. The interpretation of the new quantum kinematics was, and still is, a source of both conceptual discussion and experimental exploration of its consequences in places where it contradicts deep-rooted intuitions of physicists and others, especially for questions of physical reality and causality. So far, all the experimental tests have decided in favor of the quantum kinematics. More than that cannot be said."[144]

HIERARCHICAL LOGOS

All that we have discussed so far can be summarized in the general terms of systematics.

- An entity that holds together long enough to be dignified with a name is a system. (Short-lived entities are considered resonances or excited states of a 'ground state' system. As witnessed by their role in the 'triple coincidence' that allows for carbon atoms, they can play a role in system-building in an appropriate eden. We shall later encounter another such 'triple coincidence' when three types of procaryotes—bacteria-like living systems—came together to form the precursor of the eucaryotes—all other types of living systems, including ourselves.)

- A system has a quantum wave that confines and structures a set of interacting subsystems into the characteristic form of the system. This form gives the system its characteristic properties (which is how we recognize the system). This composite internal hierarchy of subsystem waves is the character of the system, and is what we will be calling the 'mind' of living systems.

- The form of this internal character is absolutely determined by natural law and is expressed as the external behavior of the system.

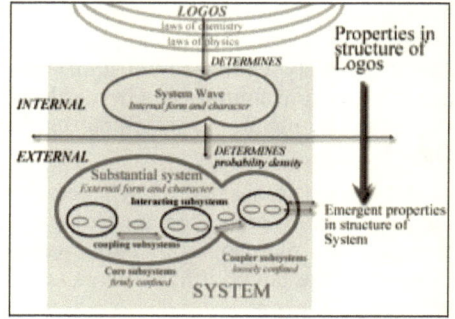

- Incomplete confinement of some of the subsystems allows the system to couple with these subsystems, and thus interact with other systems. This is the coupling wave of the system, and it is this that merges with the coupling wave of another to create the interaction wave.

- Subsystems, being systems, have a wave that either totally confines its sub-subsystems, the core, or partially confines them, the coupling subsystems.

- The waves of the interacting subsystems resonate and harmonize as a wave in full, the mind of the system, and this is given external form as the probability density of the subsystems.

This systematic description of inanimate things creates a hierarchy of systems, with less sophisticated systems being the subsystems of more sophisticated systems. At the base of the hierarchy are the wave and particles of the fundamental entities. At each level a new set of emergent properties are expressed as inherited from the structure of the Logos, and these qualities have been specifically engineered to allow for the higher levels of living systems to emerge, in their turn, as each eden appears in history.

The course of this development involves the emergent properties of molecules in water, so it will be the chemistry of solutions that we will be focussing on. In Book Two we will review the earth's history and the hierarchical sequence of edens in which each step of the system-building occurred.

Emergent inherited properties

We have discussed the origin of the universe, and the origin of the fundamental entities that are the basic subsystems of the physical universe. It took a period of expansion and cooling until the average kinetic energy was sufficiently low enough to allow nucleons to stick together as atomic nuclei (<3 minutes), and for electrons and anti-electrons to complete their almost complete annihilation into gamma rays (<10 minutes). This briefly heated the universe, but it continued to expand and cool until the temperature had fallen sufficiently for electrons to form stable standing waves about the nuclei (~500,000 years). This is the period of re-combination, the Origin of Atoms, when the next level of system building beyond fundamental entities began.

Essentially all the nuclei (~90% single protons, ~10% helium-4 nuclei) hooked up with all the electrons and the entire universe underwent a phase transition from a plasma to a regular gas of neutral hydrogen. While a plasma of free charges interacts strongly with all kinds of photons, a gas of neutral atoms is almost transparent photons. The photons 'decoupled' from matter, and have gone their separate ways ever since. This period is also called the dark age as the universe was as transparent as our atmosphere, but there was nothing to see.

Emergent Properties

Something else also happened during the Origin of Atoms. We see a set of emergent properties appearing in the universe, properties that were entirely absent up to that point of history. They emerged on a scene that was designed for their emergence.

We shall discuss a few of the emergent properties (or, more generally, 'emergent qualities') that appeared for the first time in the universe with the Origin of Simple Atoms. (A similar emergence occurred with the Origin of Nuclei, but atoms are more familiar and easier to discuss).

1. Solidity: Before there were atoms, there was no such thing as a substantial entity in the universe. Hydrogen molecules and helium atoms have the property of claiming a relatively large space for themselves and of excluding other atoms and molecules from sharing that space. The nuclei generator subsystems generate a wave that is a million trillion times larger than themselves, and this is 'fleshed out' as the body of the atom by the probability density of the electrons, the resonator subsystems. Solidity is a quality not observed in electrons and only on a tiny scale by nucleons.

2. Chemical properties: Electrons and nucleons do not have chemical properties, they do not have the capacity to interact by coupling electrons. (Although much of chemistry and biochemistry talk of hydrogen ion concentration and transport, this is not ever found as a single proton but always as a subsystem of a hydroxonium ion, H_3O^+, or another H-bonding molecule.) All the properties we discussed in our brief survey of the elements emerged during these origin events. (This is a process that continues to this day, so we shall make a distinction between an Origin event (the first of its kind to emerge) and origin events that follow after, a distinction that is irrelevant for non-living systems, but of great significance in living systems.)

3. Molecular properties: The bulk properties of atoms, such as liquid and solid emerged when the universe was cool enough for aggregations of atoms to stick together. The eden for molecules we are interested in is planets, such as earth, that are in the Goldilocks zone about a star that allows for liquid water.

As we have described, all these emergent properties are a reflection of the external electron density and its confinement by the internal wave; it is the form of the wave that gives these qualities to the atom. The form of all these internal waves—be it system, subsystem or coupling wave—is completely determined by natural law, the lower levels of the abstract Logos.

Hierarchical edens

We have defined the eden for a specific level of system building to be a time and place in which there is an internal coming together as a perfect wave dictated by natural law in which the subsystems are plentiful and the environment is suitable for their interaction as subsystems of a higher system.

This is the cycle, with its levels of increasing sophistication, that unfolds as the Logos directs the course of history and is expressed over time.

1. Given a population of systems, S1, with a set of interaction capabilities that defines a sophistication of Level One. This population is in an environment, E2, that is just right in terms of conditions such as temperature and concentration for them to have a non-zero probability of interacting and becoming subsystems of a system, S2, with a larger set of more sophisticated interaction capabilities as dictated by natural law.

2. The S1 systems explore their interactions as their waves overlap and combine in the E2 eden as dictated by natural law.

3. By the Law of Large Numbers, the non-zero probability of system S2 is expressed, and the first of its kind with its form and properties determined by the Logos emerges in the environment. This is the Origin event for the S2 system.

4. The S2 system becomes commonplace in the E2 eden and contributes its wave to the development of the perfect wave that is to be the eden for the S3 type of system. For non-living systems, this process of becoming abundant S2 systems is just a repetition of the Origin event of the first, and is just as directly determined by the Logos. Living systems have a more sophisticated way of becoming abundant. They just copy the first system to make two of them, copy these two to make four of them, and so on in exponential duplication that multiplies S2—one of its emergent properties is the ability to directly assemble system S2 out of the S1 systems without the direct involvement of the Logos. This

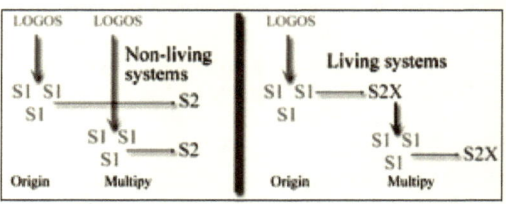

ability to duplicate is the preeminent emergent property inherited from the Logos that determines the boundary between living and non-living systems. The duplicated systems created by this multiplication are directly determined by the duplication mechanism and only indirectly determined by resonance with the Logos and inheritance of its qualities.

5. The process then repeats itself. The abundant S2 systems interact and contribute their waves to the perfect wave development of the E3 eden for the next level of sophistication to emerge. Under the direction of the Logos, there is eventually an Origin event of an S3 system, followed by either path of multiplication, followed by Logos-directed interaction and emergence of a E4 eden...

Mutability and stability

There is one more thing to be said about a systematic hierarchy that applies to both the abstract and physical realms: Systems depend on the stability of their subsystems.

In math, this is called consistency. The entire structure of math depends on the lower levels of the hierarchy not entertaining any inconsistency. If, for instance, the lowest levels allow for an integer with an odd and an even count of 2s in its prime factorization, the all of math collapses into a shambles, like Sauron's fortress bereft of its founding ring, and it cannot exist. The existence of the unified systematic hierarchy at the highest levels depends on the immutability of the lower levels.

The same applies to the systematic hierarchy of physical things. A simple example is the dependence on the stability of the carbon atom. Most of the carbon dioxide molecules in the atmosphere have a carbon-12 nucleus as a core subsystem. A tiny fraction, however, have a carbon-14 nucleus (created earlier by the impact of a high-energy cosmic ray proton liberating a neutron which was absorbed by a nitrogen-14 nucleus). This isotope has a half-life of ~5,700 years and will eventually revert back to nitrogen-14. The energy liberated is more than sufficient to disrupt the CO_2.

This hardly makes a difference to the atmosphere, but if the carbon is a subsystem of a large biological molecule, say the DNA that digitally stores a crucial bit of information, the consequences for a complex system such as the human body can be as disruptive and serious as cancer. Our bodies depend absolutely on the stability of the atomic nuclei that are its subsystems, just as math depends on unique factorization.

2ⁿᵈ Scientific Revolution

The First Scientific Revolution, started by Newton and ending with Einstein, dealt with the external realm, the only realm in the classical view of the world. This is a world of material particles. Its characteristic is that it is described by real numbers.

The Second Scientific Revolution will complete the picture by adding a description of the internal realm. This complements a world of external particles with a world of internal waves. Its characteristic is that it is described by complex numbers.

Internal wave and external particle—these are the dual characteristics possessed by all things according to modern physics. This duality cannot be described by the familiar real numbers in which all of classical science is expressed. It can only be expressed fully in the complex numbers which have the duality of linear size and circular rotation (or real and imaginary) in their structure and properties.

This new view of the universe is more sophisticated than the old and it adds a whole new level of mathematics that has to be mastered. All the calculation in the new, unified science is done with complex numbers. The result is always a real number, the external aspect that the old science dealt with. To ask in unified science, "Is an electron a particle or is it a wave?" is akin to asking in mathematics, "Is a complex number real or imaginary?" In either case, the only correct answer is, "It depends."

Since mastery of this new level of mathematical sophistication is not easily obtained (at least for those with a classical education, hence the phrase "quantum weirdness"), it is arbitrarily declared "not necessary" is the less advanced sciences. As scientific thought is hierarchical, with one level of sophistication expressed in terms established by lower levels of sophistication, there is a transition zone from a science that uses complex numbers (which include the real numbers) to a science that uses only real numbers.

This transition zone is currently found, a century after the inception of the modern unified view, in biochemistry which will sometimes be couched in external terms such as 'lock and key' and 'thermal motion' and sometimes in the unified terms of orbital waves and excited quantum states.

The external foundations of science were laid in the 18th century, while the internal foundations were laid in the 20th. It is important not to confuse the everyday, and fuzzy, use of the words 'internal' and 'external' with the precise definition of these terms used in the sciences. The aspects of nature that are described by complex numbers are 'internal,' while those that are described by real numbers are 'external.'

classical science external only	current science divided	unified science internal & external
Mind-brain	Mind-brain	Mind-brain
Evolution	Evolution	Evolution
Genetics	Genetics	Genetics
Biology	Biology	Biology
Biochemistry	Biochemistry	Biochemistry
Chemistry	Chemistry	Chemistry
Physics	Physics	Physics

Logos and Natural law

All the sciences accept the fact that there are abstract laws involved in the functioning of the universe. It is the goal of all the sciences to uncover these natural laws and to derive a mathematical description of them.

The external foundations of classical science required the unspoken assumption that natural laws act directly on the external aspect.

The unified view is more sophisticated. Natural law acts directly only on the internal aspect. For instance, the wavefunction of the electron is exactly determined by natural law and has a precise mathematical description involving complex numbers. This internal wave aspect governs the external probability density of the particle aspect. Within the confines of this external projection of the internal aspect, the particle is found at random.

It is this random aspect that gave quantum mechanics the reputation of uncertainty. This plays only a small role in the systems of everyday life. It is the confinement by the wave that plays by far the main role.

Unified science adds an extra level of causality that is absent in the classical view of the connection between the control by natural law and its expression in the external aspects. As already mentioned, the projection of the internal aspect onto the external is always the absolute square of the internal, so all unified science interest is focused on the internal wave and its developments.

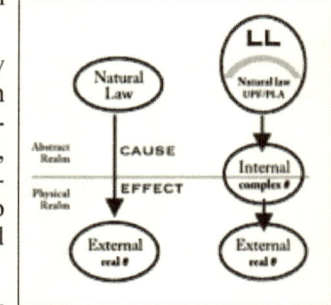

As we are going to show that the laws that govern the internal aspect are hierarchical and not at all like the simple linear image conjured up by the term 'natural law,' we will call the abstract structure of hierarchical laws that emerges in the discussion the Logos.

Basic Principles of Unified Science

To summarize, the external interaction of two systems changes the internal waves of both systems. Adding this aspect to the relationship between internal and external, we have now covered the following basic principles of Unified Science:

1.   *A system has an internal and an external aspect.*

2.   *The internal wave is determined by Natural Law.*

3.   *The external aspect is a set of subsystems.*

4.   *The internal determines the form of the external over time.*

5.   *Systems interact externally by coupling with their external subsystems.*

6.   *Interaction modifies the internal wave, and this change has consequences.*

7.   *When systems interact in an eden, they can become subsystems of a higher system with a set of novel properties that are inherited from the Logos*

This illustration is an outline of the hierarchy of system building by Logos, and a few of the properties inherited from the Logos that will be important in the later discussion. Note that the only difference between atoms and molecules is the number of core wave generators: an atom has only one, a molecule has more than one.

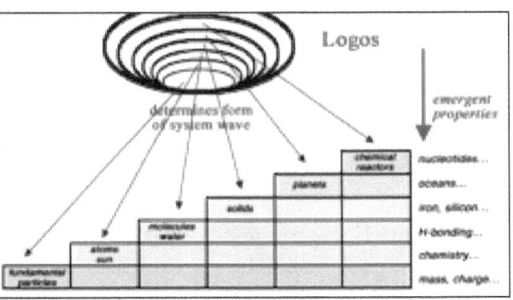

This is a hierarchy of waves with their wave-within-waves structure. It is an intricate structuring of the quantum waves of fermions. The electrons and nuclei are just following along and taking up the density predetermined by the waves. All the important developments take place on the wave level. It is the quality of the wave that gives properties to arrays of nuclei and electrons.

Unified View

The Logos, like mathematics, inhabits the Abstract Realm. Like the Logos, mathematics is a unified, hierarchical structure, and some of the mathematical structure is intimately tied in with the lower workings of the Logos (and the "unreasonable effectiveness of mathematics in the sciences" is not unreasonable at all).

The view here presented of reality, which includes an abstract Creator and a sophisticated Logos running things, is likely to offend both materialists—who do not admit to an Abstract origin for emergent qualities and only simple laws—as well as Christians and others who consider the Creator to be directly running the Universe. The view is likely to offend both as it implies that both views will have to adjust before we have a unified view in which both are satisfied. To encourage both sides to embrace a larger view we offer each a suitable exemplar, Lord Kelvin for the science-minded and St. Paul, for the religion-minded.

Lord Kelvin was a great scientist of the late 1800s. Just one of his achievements was to establish the Absolute scale of temperature as commemorated by 'degrees Kelvin' used in all the sciences. His expertise in what was known, however, led him to a famous case of hubris when, in 1899 conference (held just 20 miles from my hometown), he declared that all of science was now known, and all that remained was more and more precise measurements. This was less than a decade before quantum mechanics, relativity and the Big Bang universe entered into the scientific mainstream. A current example of such hubris is Richard Dawkins, who similarly declares that all the basic principles of evolution are understood and all that remains is a more precise understanding of how they worked. We can predict that he will be another famed example of misplaced confidence as the foundation of his view is the Central Dogma of Genetics—that information is read from, but not written to, digital memory. This foundational concept, however, is currently being steadily eroded by the emerging science of epigenetics that does include the writing of information by biological systems to their digital memory store.

The religious exemplar is St. Paul who is so central to Christian theology that excerpts from his letters are regularly read out loud around the world on Sundays. His insistence was not hubris, but a humble admission that "now we see through a glass, darkly" and that there is lot deeper understanding to come in the future.

The ratchet of Interaction

We have already discussed how the wave changes in an interaction. But the wave only determines the density of the valence subsystems, not where they are exactly now. If a hydrogen atom flies past an oxygen, the wave will change but the electron particle might be on the other side. The hydrogen is passed before the electron is on the other side and ripe for coupling.

The wave only determines the probability of a coupling occurring with a valence subsystem. It is the subsystem's situation that determines if the coupling actually does occur. Coupling is a very black-and-white digital affair: a subsystem was coupled OR a subsystem was not coupled. This is the ratchet of time. It goes as follows.

1. *A system has a structure and coupling wave, A. Time is reversible.*
    All the known laws, and thus the Logos itself, are time reversible.
2. *A system couples with a valence subsystem.*
3. *The system wave changes to B. This is the irreversible ratchet of time*
4. *The system has structure & coupling wave B, time is again reversible.*

This is systematic time and it is different from physical time, although the two are often used interchangeably. Physical time, for instance, can go in both plus and minus directions, as it does in matter and antimatter twists. Systematic time is one-way. Like the integers under addition, it is one way and can only get bigger.

Physical time only began when the particles with rest energy created in the Big Bang had cooled sufficiently so that their kinetic energy fell into the range of their rest energy. Before that, all had been moving at essentially the speed of light in space and there was no extension of physical time.

A cosmic ray proton in the current era is an example. It can have a kinetic energy of over 10^{20} eV, far greater than its rest energy of 10^9 eV. The rest mass of the proton is inconsequential and it moves at lightspeed through space and so is motionless in the internal aspect of physical time. When it interacts with an atmospheric nucleus, however, and ejects a neutron before continuing on with a little less energy, it moves forward in external systematic time.

SAMPLING THE WAVE

As we have established, all things are composed of external fundamental particles that are confined by an internal wave. As the fundamental particles are, in all cases, identical, all the qualities of the many disparate things are to be found in the internal wave.

The divide between the inanimate systems described so far, and the living systems that will be dealt with in Book Two is to be found in the internal wave, not the external particles. This necessitates a further extension of the science of sound waves to set the foundation for the discussion to follow.

The great American inventor, Edison, was the first to put this aspect of science to practical use when he invented the first method of sound recording. The method involved a thin membrane that moved back a forth with the change in air pressure, and attached to this was a needle that made a scratch on a wax-coated cylinder.

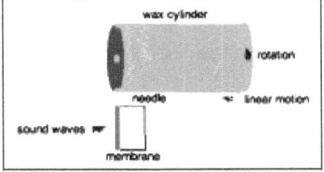

This cylinder was put into two motions. It rotated at a moderate, constant speed, and it also moved slowly from left to right. In quiet air, the needle inscribes a smooth helix about the cylinder. Placed in the room with the tuning forks resonating to the A-major chord, the needle scratches out a wavy line just like our machine did.

Note what Edison has done with his phonograph that combines constant linear motion with constant circular motion: he has taken a sample of the resonance filling the room. This is a huge and complex structure and the phonograph sensor, the membrane, takes only a very small sample of this three-dimensional vastness. It outputs a simple compound sine wave, a linear wave in just one dimension.

We have sampled the huge complex wave-in-full and output a much simpler, smaller wave. This is called 'down-sampling' a wave. We can call this down-sampled, simpler wave the wave-in-image of the wave-in-full. The wave has been recorded as a scratch in the wax, so the process can be called 'recording' a wave-in-image to linear memory.

Edison did not stop there. He first turned the soft wax impression of the wave-in-image into a hard wax form and then—brilliant when you think about it first—he ran the apparatus backwards.

The back and forth movement of the needle in the groove was transferred to the membrane that vibrated along with the needle. This membrane transferred its vibrations to the air and, voila, a sound could be heard filling the room. The sound in our example would be the A-major chord.

We have turned a small, linear, one-dimensional wave into a huge three-dimensional resonance of the air in the room. The simple wave memory has been transformed into a huge and complex wave that is a replica of the original wave-in-full. The wave-in-image has been up-sampled to the wave-in-full. This is "re-playing" a simple, linear wave-in-image into the large complex resonating replica of the original wave-in-full.

The simple mechanism of Edison's phonograph is an example of a reversible 'transducer' that can down-sample a wave-in-full into a wave-in-image record and when run in reverse, replay the recorded wave-in-image and recreate a replica of the wave-in-full.

The transducer connects the wave-in-full form of the extended, multiple resonators to a wave-in-image memory of the form. This memory can be replayed to recreate the wave-in-full form of the resonators.

We will call the mechanism through which the complex wave is down-sampled and, when reversed, up-samples the complex wave again, a simple 'transducer' that can run in either the read or write direction. Edison's

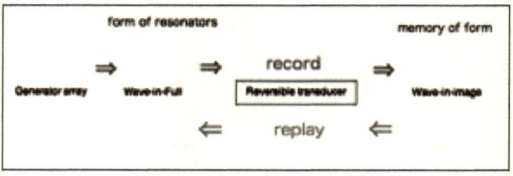

simple method was quite capable, for the first time in history, of recording talk, music and song, and playing it back again as many times as required.

More sophisticated versions of the transducer soon emerged where the writing and reading of the linear wave became separated, specialized and sensitive. The culmination of this was the black vinyl record that dominated the 1950s through 80s, with a 'single' disc that rotated at 45 rpm and the much larger album, or LP, that rotated at 33 rpm. Here, the wave movement of the needle was electrically amplified before reaching the membrane or loudspeaker.

A symphony

Let us go back to the array of sound generators. We discussed a very simple example: three tuning forks together generating an A-major chord. Now we can discuss the array of generators that allow us to create the most sophisticated waves-in-full: the symphony orchestra. For our example we will use the climactic finale of Beethoven's 9th Symphony, the sublime *Ode to Joy,* the anthem of the European Union. For specificity, we will locate the performance in Carnegie Hall, a vast hall with wonderful acoustics. The array of sound generators includes:

400 human singers in banks of sopranos, altos, tenors and basses. 400 string instruments in banks of 1st and 2nd violins, violas, cellos and double-basses. 300 woodwinds in banks of bassoons,   oboes and clarinets, 300 brasses in banks of trumpets, trombones, French and English horns, flutes and piccolos. Drums, pianos, a couple of harps.

Each one of these sound generators, even when sounding the same note, resonates with a different mixture of standing waves: the 1st harmonic, the 2nd, 3rd, 4th, etc. It is this different mix of harmonics that gives an instrument its 'timbre', its unique sound that makes an A on the piano sound so different from an A on a violin. It is what makes one voice sound different from all others.

At the climax of the symphony, pretty much every generator is going full out, pouring out its unique blend of harmonic sound waves. It is quite overwhelming.

The mass of air in the hall is resonating with a constantly changing, extremely complicated, wave within wave within wave. This is the wave-in-full.

There are also an array of samplers, the audience. Each tiny eardrum is picking up a down-sample of this wave-in-full. It transmits a wave-in-image through a series of tiny bones to the inner ear. This does a simple Fourier Analysis of the wave-in-image and sends the results to the brain where the wave can be recreated as we 'hear' the sound. The ear is an example of a sophisticated one-way transducer that can only record but not replay the wave.

Edison's example of a simple reversible transducer is quite capable of recording the wave-in-image of the symphony, and of playing it back, although at a much lower volume in its simplest form.

A perfect wave-in-full can be recorded as a wave-in-image. If the process is reversible, and if there is memory, the perfect wave can be recreated. We shall continue this discussion when we introduce the difference between nonliving systems which get their wavefunction from the Logos, and living systems. Living systems get their wavefunction from the Logos in the origin event, and then store a wave-in-image in memory which can recreate the perfect wave and also be duplicated.

Analog and digital

So far, all the waves we have discussed have been continuous, and they smoothly change their form and shape. They are examples of what we will call 'analog form.' The resonance of the generators, the wave-in-full structuring the form of the resonators, and the memory of it as a wave-in-image are all examples of analog form.

The final thing we need to know about music does not involve analog form. Rather, it involves 'digital information' that is not continuous, i.e., it comes in distinct bits called 'notes' in music and 'quanta' in physics.

Linear digital information always involves a reading convention that is followed by a set of wave generators, which together structure a mass of resonators into a characteristic wave-in-full form.

For sound waves, this convention is called 'music notation.' It specifies the frequency of the 1st harmonic to be sounded (along with any characteristic higher harmonics) and its duration. Each named note has its own place on a 5-lined stave, either in a space or on a line. A simple oval instructs the generators to sound an A, a C# and an E together for a whole beat, the basic rhythm of the piece.

For a complex piece, like *Ode to Joy,* each bank of musical generators gets its own line of digital instructions, one page covering a few bars of music. Each bank reads the score at the same rate, and when instructed to sound, does so. This is parallel processing of digital information—there are many parallel lines each being read by an array of different wave generators.

The constructive interference of all these waves is the singular wave-in-full we call a symphony performance.

For all its magnificent complexity, the Edison phonograph is quite capable of down-sampling the performance into a simple, 1-D wave-in-image.

This is parallel digital information being read in time by an array of banks of generators. The recording and replaying of a down-sampled wave-in-image by a transducer is as before.

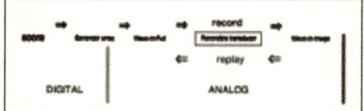

This was the situation of sound recording until the 1990s with the LP record (except for the specialization of the transducers into separate, irreversible recording and replaying mechanisms). The final step was taken with the advent of the CD and the storage of information about the wave-in-image in a digital form.

The wave-in-image from the transducer is fed into an electronic device that measures the height of the wave many thousands of times a second. The analog-to-digital device outputs this as a stream of digital numbers that are written to the CD. An example covering just a fraction of a second would be:

99 98 99 96 90 80 70 71 72 74 75 78 60 50 40 23 etc.

This string of digital numbers is eventually written to a compact disk, a CD, in a binary code, utilizing only the digits 1 and 0.

The CD player reverses this process, it is a DC-AC transformer. Each number tells it how much electricity to send to the amplifier and loudspeaker. As the thousands of numbers per second pour into the player off the CD, the current to the loudspeakers is a recreation of the wave-in-image. The loudspeaker turns this into a wave-in-full that recreates the symphony.

It is easy to see that the process can be made into a circle if the same reading convention is used for the digital information at either end—the instruction of the generators and the recreation of the wave-in-image. An example of this is the MIDI system for music recording and playback that uses the same coding convention for both.

Waves and computers

Using the term 'computer' in its broadest sense, we have four different types of computers involved in the creation of music, and the recording and replaying of music. Using DC to stand for digital information, and AC to stand for analog wave forms, we have:

1.   DC/AC computers. The array of primary sound wave generators (dozens of musical instruments, singers, etc) are of this type. Using a universal reading convention, they take a multitrack digital input (a score, a MIDI track) and convert it into an analog wave that gives form to a mass of resonators. The CD player is also a DC/AC computer: it turns the stream of digital numbers into a voltage wave, the wave-in-image, which drives the secondary sound generator, the loudspeaker. It has been found that a better recreation of the wave-

in-full is obtained by taking two samples of the wave offset from each other. These two wave-in-image recordings are used to energize two loudspeakers separately, resulting in stereo sound. Dolby sound recording involves at least four channels and four loudspeakers. In general, however, there are many more primary generators than there are secondary generators.

2.   AC computers. The simple, reversible phonograph transducer is one of these. In the record direction, it takes a down-sample of the wave-in-full form of a mass of resonators, and outputs a simple wave-in-image. In the replay direction, it inputs the wave-in-image and outputs the wave-in-full that, relatively faintly, recreates the symphony. The phonograph is a ±AC computer. In later, more sophisticated developments, the transducers became irreversible. The writing direction now involves a membrane in a microphone jiggling magnets past each other. The tiny currents this sets up are electronically amplified into an electric current whose alternations are the wave-in-image. It is this current that drives the writing needle in the soft wax or serves as input to the CD writer. The microphone/amplifier combo is a + − AC computer. The reading step involves a needle vibrating in the groove attached to magnets whose current changes, yielding the wave-in-image, are amplified and used to drive a loudspeaker that recreates the wave-in-full. The output wave-in-image from the CD player is treated similarly. The amplifier/loudspeaker combo that up-samples the wave is a − + AC computer.

3.   AC/DC computers. The CD writer is an example of these. It takes a wave-in-image and samples the size of the wave repeatedly. It outputs these almost instantaneous sizes of the wave as a linear string of measurements written in a standard code. A MIDI recorder does something more complex, but its final output is also a sting of binary bits, just using a more sophisticated code.

4.   DC computers. These have digital input and digital output. A computer that can take a MIDI-encoded file and print out a multi-part score for an orchestra is an example of this. This is a feat well with the capabilities of the DC computer I am writing this on, a MacBook Pro, if it were provided with the appropriate software program. This would be a digital program that is fluent in two coding conventions, both MIDI and musical notation, taking the MIDI input and translating into musical notation. All current computers are of the DC variety.

It might appear that my Mac is also capable of outputting AC in the form of sound, photos and movies. This is but an illusion: the output is purely digital and discontinuous. It is the resolution that gives the impression of AC, not DC. It is the thousands of discontinuous pixels on the screen—each either red, green or blue—that "fools" my eyes into seeing smooth and continuous colors. It is easy to see such pixels with a magnifying glass.

Music and Matter

This concludes our discussion of the basics of music and its recording. We have discussed constant rotary and linear motion, and the sine and cosine waves, both traveling and standing. We have discussed how sound

is described as combinations of sine waves, and how it is generated by resonating mechanisms. We have discussed the connection between a digital score, an array of sound generators and the wave-in-full that structures the air in the concert hall.

Further, we have discussed the recording and replaying of the wave-in-image as well as four types of computers that variously manipulate AC forms and DC information. This is summarized in the diagram.

The ratchet of systematic time is sometimes called the 'collapse of the wavefunction.' This is not a very good name because it implies that the wavefunction, the term quantum scientists use, collapses and disappears. This does not happen. Rather, the wave changes from one wave to a another, usually from an extended wave to a localized wave. The term arose from the earliest experiment that detected the wave aspect of matter, the slit experiment. This involved a simple set-up. A particle (such as an electron) is shot out of a gun in a vacuum, passes through a barrier with two slits in it, and hits one of the detectors arrayed on the other side.

The electron behaved as a wave would, not as was expected from a particle. This was hard to understand for those who considered an electron a solid particle, and that 'solid' was not fundamental quality, but one inherited from the Logos by composite systems. This wave-like behavior remained when only a single particle traversed the apparatus at a time. Each electron ended up in a single detector, say a crystal of silver iodide in a photographic plate. The wave-pattern emerged when thousands of electrons were sent one after another. The electron wave was interfering with itself.

Each electron wave, after being split in two by the slit, continued on its interfering way until it reached the detector array. At this point the wave is spread out over the centimeters of the entire array. Hitting one of the detectors, this extended wave immediately disappeared and the electron appeared in the local detector. This is where the term 'collapse of the wavefunction' originated, to describe the disappearance of the extended wave.

It is not really a disappearance, however, so the name is unsatisfactory. What actually happened is that the electron wave began to merge with all the waves of the silver atoms, coupling with virtual photons occurred with an atom where the particle happened to be in its jittering, and the electron wave snapped to being one with the atomic wave of the silver atom in the detector crystal.

The single electron wave passed through the slit into a double-lobed electron wave with a sizable node. The two lobes interfered with each other into a wave pattern of nodes and antinodes covering the detector array. Whilst at a location in the wave, a coupling of virtual photons with the ion of silver at that location occurred. This alteration of subsystems snapped the extended, rippled electron wave into a subsystem wave of the silver atom, a black speck in the silver iodide crystal. At all times, a single electron particle is jittering about in a single electron wave that determines its density over time. The change in the wave-that-fits from plate size to atomic size is instantaneous, but there is always a wave at all times.

This type of demonstration of the wave aspect has recently been reported for buckminsterfullerene, a spherical molecule of 60 carbons. The size of this decidedly a bit of 'matter' makes it difficult to digest when it 'travels through both slits at the same time.' This behavior is difficult to explain by those who think the property of being solid is a given, and not a sophisticated quality inherited by sophisticated system waves from the Logos. The demonstration has yet to be performed on something as ponderous as a tRNA, but a tRNA is more wave than particles.

VOLUME 3.
LIFE,
MIND
AND SPIRIT

SYSTEMATIC HIERARCHIES

In *The Unity of the Sciences, Volume One,* we discussed how modern physics has replaced concepts at the foundations of all sciences, and speculated on how this change in their foundations would alter the biological sciences. The second and third volumes in the trilogy develop these ideas in detail.

In *The Unity of the Sciences, Volume Two,* we dealt with the three realms that make up the cosmos, the entirety of everything. We called these the Abstract Realm, the Physical Realm and the Spiritual Realm. In *Volume Two,* we dealt mainly with aspects of the cosmos that are considered to be firmly established and well understood; mainly mathematics in the Abstract Realm, and physics, chemistry and cosmological history in the Physical Realm.

In this *The Unity of the Sciences, Volume Three,* we will apply these concepts to the structure, function and origins of living systems; the nature of the mind and the realm of the spirit.

A Systematic Hierarchy

Both the physical and abstract realms have the structure of a systematic hierarchy whose levels are distinguished by the sophistication of their emergent properties. The systems are composed of interacting subsystems coupling with sub-subsystems. We will review 'interaction' in a later section.

At the lowest level of the hierarchy are the simplest of systems, s1. The s1 systems interact to create more sophisticated systems, s2, with a set of emergent properties that are not possessed by the isolated s1 systems. We will discuss the source of these emergent properties when we discuss the *Logos*, a sophisticated, systematically structured type of Natural Law.

A Systematic Hierarchy

The s2 systems interact to create even the more sophisticated systems, s3, with a set of emergent properties which, in turn, interact yielding even more sophisticated s4 systems and on up, generating a hierarchy of emergent properties.

We will refer to this ubiquitous type of structure as a 'systematic hierarchy' of entities—the term *entity* simply meaning anything deemed worthy of being assigned its own name. A large part of *Volume Two, Mathematics, Physics and Chemistry*, was taken up with examining the entities at the lowest levels of the systematic hierarchy that comprise both the Abstract Realm and the Physical Realm.

Abstract realm

There are only two entities at the very foundation of the abstract realm, linear extension and circular rotation. In *Volume Two* we discussed both in detail and how they are seamlessly combined in the entity called a *complex number*. Unlike almost all types of numbers, complex numbers are technically *complete* in that whatever you do to a complex number, you always end up with another complex number. The set of complex numbers are a closed and complete system. The integers, on the other hand, are an example of a set of *incomplete* numbers since doing things to integers, such as making fractions out of them or taking the square-root of minus-one, does not result in another integer.

The study of the lowest levels of the abstract realm is mathematics. We discussed how the concept of *Absolutely Nothing* is inconceivable inasmuch as it implies, by Set Theory, the concept of One, and this in turn implies the counting integers, the simplest of all numbers. These are at the foundations of math and have emergent properties such as addition and multiplication. We discussed the systematic hierarchy of the integers, the rational, the irrational and the transcendental numbers—all of which are purely linear extension—and their combination with angular rotation to create the complex numbers. The highly sophisticated properties of these complex numbers were illustrated by the forms created by simply adding and multiplying them together to generated the intricate forms expressed in the Mandelbrot Set. The connection of 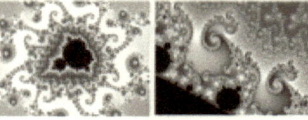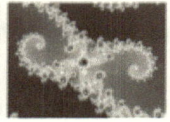 this abstract level with modern physics was noted in that spacetime, in special relativity, is described by complex numbers, and the behavior of fundamental particles is accurately predicted, in quantum mechanics, by repetitively adding and multiplying complex numbers.

Cause and Effect

The lower and the higher levels of the abstract realms are inextricably inter-twined so that it is impossible to say which level is the cause and which level is the effect. Put another way, it is impossible to say which came first.

An example is the simple Euclidean concept of singular parallel lines on an infinite flat plane. Euclid started his systematic hierarchy with a set of axioms, all of which were simple and obvious except for this "parallel axiom." Unlike the other axioms, his statement about this one was quite convoluted and, when simpli-fied, amounted to the assertion that, given a line and a point *not* on that line, there is one, and *only* one, parallel line passing through that point.

Later, a sophisticated manipulation of theses Euclidean axioms proved the sophisticated Py-thagorean relation between the areas of squares sitting on the sides of a finite triangle.

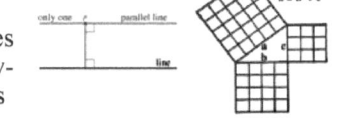

It is impossible to say which is cause (came first) and which is effect (came after). Either can be taken as fundamental, i.e., as an axiom. It is impossible to prove the parallel postulate from any simple concepts—many have tried; all have failed—and there are no flat infinite planes available on which to test the assump-tion. So you have to take it as an axiom, a given about the way the cosmos works. But it is equally possible to take the Pythagorean theorem as the given axiom and easily prove the parallel postulate using it. Cause and effect are not aspects of the Abstract Realm. In the systematic hierarchy of the Abstract Realm, the simple levels are cause-and-effect and the sophisti-cated levels are also cause-and-effect.

s4	Transcendental numbers	$\pi\ e\ e^{\pi}$
s3	Irrational numbers	$\sqrt{2}\ \sqrt{3}\ \sqrt{5}$
s2	Rational numbers	½ ⅔ ¾
s1	Positive integers	1 2 3

Note that a systematic hierarchy cannot be changed at any level or the whole structure collapses. As it was assumed that space was flat, it was thought that Euclid's Parallel Postulate was an example of an unshakeable and absolute truth about the universe; once the concept of *curved space* was included, the axiom of a single parallel held true only in a flat space but was untrue in a curved space. A convex space has an infinite number of parallels, while a concave space has zero parallels. Euclid's problematic axiom has a limited *domain* of flat space, it does not have a universal domain.

Universal Truths

$$d^2 = 1^2 + 1^2 = 2$$
$$d = \sqrt{2}$$

Some truths, however, do have a universal domain, they apply everywhere. An example is the *existence proof* of num-bers that are not the ratio of two integer. Pythagorus was a great believer in ratios, but he was confounded in attempting to find the ratio that measured the diagonal of a perfect unit (flat) square. His own Pythagorean Theorem stated that the square of the length of this diagonal equalled the sum of the squares of the other two unit sides, equalled 2. So the diagonal was the number, that when squared, resulted in exactly 2. So the

Pythagoreans then searched for the ratio of integers that, when squared, resulted in exactly 2.

$$\left(\frac{n}{m}\right)^2 = 2$$

They failed in their search. As it is an instructive example of the abstract systematic hierarchy at work, we will pause to see one of the ways of proving that the square-root of two, $\sqrt{2}$, cannot conceivably be the ratio of two integers:

First, a few simple foundations:

1. An *even* integer leaves a remainder of 0 when divided by 2. Examples are 0, 2, 4, 6....

2. An *odd* integer leaves a remainder of 1 when divided by 2. Examples are 1, 3, 5, 7....

3. Both even and odd integers when multiplied by 2 become even integers.

4. Any integer has one, and only one unique factorization, the set of prime number factors $\neq 1$ that, when multiplied together, result in that integer, $n=\{p^n, q^n, r^n...\}$ where n is the count of each prime in the factorization. As this is not obvious, this is simple subproof of prime-factors uniqueness.

    a. One of Euclid's proofs is that if an integer, N, is the product of two or more integers, {a, b, c...}, and N is divisible by a prime number, p, then p must also divide at least one of the integers {a, b, c...}

    b. Assume that there is an integer, N, that has two different prime factorizations,

$$N = \{p_1 p_2 p_3 \ldots p_n\} = \{q_1 q_2 q_3 \ldots q_n\}$$

    c. Prime p_1 divides N which is obtained by multiply all the qs together, so p_1 must divide one of the qs. But all the qs are primes, and only have the factors $\{1, q,\}$ so p_1 must be identical to one of the qs. Renumbering the qs so that $p_1 = q_1$, and then dividing N by $p_1 = q_1$, and repeating it for all the ps. If all the ps and qs are identical, and all cancel out, then the result is 1. If all the ps and qs did *not* cancel we get the absurd situation where a prime number not equal to 1 is equal to 1. The ps and qs must be identical. The factorization of every integer, note the domain, is unique.

$$\frac{N}{p_1 p_2 \ldots p_n} = 1 = q_?$$

5. This unique prime factorization, P.f, of any integer has a count, n, for each prime in the factorization that can be:

    a. An all-odd count of prime factors. Examples are all the prime numbers, with just 1 factor, e.g., {2}, {3}, {5}..., or 27 with a P.f of three 3s, {3x3x3}, or 54 with one 2 and three 3s {2,3,3,3}

    b. An all-even count of factors present, e.g., 4={2,2}, along with the even number, 0, of prime factors that are all absent, $4=\{2^2, 3^0, 5^0...\}$

    c. A mixture of even count factors and odd count factors, e.g., 18={2,3,3} with a count of 1 and a count of 2.

5. When any integer is squared to create another integer, the number of its prime factors is doubled. An integer that is a perfect square has an all-even

count of prime factors.

$$N = \{p^n, q^n, r^n\}$$

6. When a single prime is added to the prime factors of an all-even square integer, the count of that prime factor in the result becomes odd.

$$N^2 = \{p^n, q^n, r^n, p^n, q^n, r^n\}$$
$$= \{p^{2n}, q^{2n}, r^{2n}\}$$

$$N^2 \times p = \{p^{2n+1}, q^{2n}, r^{2n}\}$$

$$\frac{n}{m} = \sqrt{2}$$

$$\frac{n^2}{m^2} = 2$$

$$n^2 = 2 \times m^2$$

$$N = N$$

$$even = odd$$

On these simple foundations, it proves inconceivable that the number that measures the length of the diagonal in a unit square, the square-root of 2, is the ratio of two integers is quite elementary:

If √2 equals the ratio of two integers, squaring both sides of the equation gives 2 as the ratio of two square integers, both with an all-even prime factorization. In particular, both square integers will have an even number of 2s in the count of their prime factors.

Multiplying both sides of the equation by the integer m^2 results in an integer, N, that has two prime factorizations, one with an even count of 2s, the other with an odd count of 2s.

As it is impossible that an integer not have a unique prime factorization—the *Fundamental Theorem of Arithmetic*—it is also impossible for √2 to be the ratio of two integers, a *rational* number. Rather, √2 is an example of an *irrational* number, and the diagonal of the unit square cannot be measured by the ratio of two integers to the chagrin of the Pythagoreans.

Measuring rotation

Moving on from linear extension in *Volume Two*, we then discussed circular motion and angular rotation, in terms of i, the rotation operator and the *imaginary* square-root of minus-1, $i^2 = -1$. Then we dealt with the sine and, cosine entities as open or closed, bound or unbounded waves.

$$z = (x + yi)$$
$$= me^{i\alpha}$$
$$= m(\cos\alpha + i\sin\alpha)$$
$$= m @ \alpha$$
$$= \nearrow$$

Next, we combined linear extension and angular rotation into a discussion of the entity known as a *complex number*. In complex numbers, both size and rotation are seamlessly combined into one number. As complex numbers might seem to be exotic to some readers, we went into the many sophisticated emergent properties of the complex numbers, starting with an explanation of the various ways of expressing the same complex number, z, each with its particular usefulness.

Finally, we looked at more examples of inextricably intertwined simple-sophisticated levels such as: the counting numbers and the irrational transcendental numbers; the distribution of the prime integers and the infinite sum of the zeros of the Zeta Function.

s4	Bound waveform	ψ^2
s3	Sine wave	ωt
s2	Sine function	$\sin(a)$
s1	Rotation operator	i

We then discussed how the Abstract Realm could give rise to the other two realms by creating two complementary structures at the initiation of the Big Bang.

A complex number is usually illustrated as an arrow on a plane of two dimensions. While this is extremely useful—the popular book by Richard Feynman on quantum mechanics is full of little such arrows—it is misleading because it suggests an external rotation in two dimensions, x and y. A more realistic illustration of a complex is an arrow with a length and a twist along a single axis.

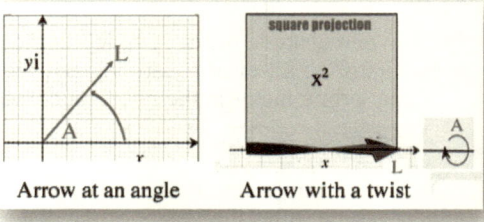

Arrow at an angle   Arrow with a twist

This twisted arrow represents a single, complex dimension. Much of the common perception of quantum science as being "strange" and "weird" is based on the common assumption that reality can be explained with simple dimensions, when in reality the universe can only be described by complex dimensions, and components of complex dimensions. This will become clear as this discussion progresses.

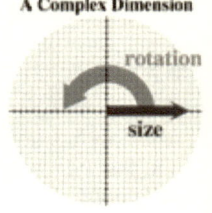

A Complex Dimension

A very important property is the square projection (technically, the absolute square) of a complex number which, as the name suggests, is not a length but an area. In this case, the twisted arrow is a better illustration as the square projection is simply the square sitting on the arrow. If along the x-axis, the square projection is simply, x^2.

The relation between internal and external is as the relation of a complex number to its square projection, which is why so many important equations in physics involve the square of a parameter. The equation that gives the external probability of an event in quantum mechanics, for example, gives the square projection of an internal probability amplitude that is the result of adding and multiplying complex numbers. Squaring a line alters dimensions and it becomes an area. The absolute squaring a complex number alters dimensions, and internal becomes external.

THE PHYSICAL REALM

The Physical Realm, like the Abstract Realm, is also systematically hierarchical in structure with simple systems at the lower levels (e.g., electrons, quarks, photons, etc.) which are interacting subsystems of sophisticated systems at a higher level (e.g., water, aminoacids, nucleotides).

Unlike the two-way Abstract Realm, the relation of "which came first" is strictly one-way in the physical realm where the systematic hierarchy is built over time from simple levels to sophisticated levels.

In the physical realm, simple systems come together to create more sophisticated systems with emergent properties, and these systems, in turn, come together to create even more sophisticated systems. (It is this decidedly one-way nature of our physical experience that probably accounts for our tendency to assume that the counting integers came first and that the Mandelbrot Set came after when, in fact, they are inextricably both cause and effect.)

Complex dimensions

At the very foundations of the physical hierarchy are time and space. In classical physics, this pair was taken at face value, and their external differences made them as different as chalk-and-cheese. They were obviously very different aspects of the experience of the physical realm.

Physics found things to be more sophisticated when it started exploring regions beyond everyday experience. The real numbers of classical science had to be replaced by the sophisticated concepts of Special Relativity, which unite time and space dimensions into a subset of a construct in four complex dimensions, four orthogonal complex planes. A complex dimension embraces both linear extension and angular rotation in a unity that can only be described by complex numbers.

$$(x \circ y) \circ z \neq x \circ (y \circ z) \quad \text{nonassociative}$$
$$x \circ y \neq y \circ x \quad \text{noncommunative}$$

While the mathematical entities on the complex plane—a single complex dimension—have been explored, witness the Mandelbrot Set and the Zeta Function, the mathematical structures in four complex dimensions have hardly been examined. A start has been made in the exploration of the octonions, one of the few algebras that allow for division. The octonions involve sets of eight numbers and have the unusual property of being *nonassociative* along with exhibiting the the more common *noncommutative* behavior. So, when performing the basic operations on octonions, shifting brackets or altering the order can produce quite a different result:

It is the mathematical exploration of such spaces, as well as the two subsets we will be assigning to the physical and spiritual metrics, that will broaden our understanding of spacetime.

Twisted spacetime

While not yet a mainstream concept, much about the nature of the fundamental particles that make up the physical realm can be explained by considering them as twists in spacetime that mix one dimension with another. Such twists have been suggested relatively recently:

> In the 1970s and 1980s, [Dr. John Moffat] further explored modifications to general relativity, including a 'non-symmetric gravitational theory,' in which extra terms are added to Einstein's equations (think of it as giving space a 'twist' in addition to being curved).[145]

General Relativity established that this complex spacetime could be curved on a large scale. The bosons and fermions uncovered by high-energy physics, out of which matter is constructed, all behave as topological twists in this complex spacetime on a the truly-minuscule scale of ten -trillion-trillion-trillionths of a meter. The tiny atom is a trillion-trillion times larger than this fundamental pixelation of spacetime.

These tiny twists can be 'oriented' twists (*or-twists*—in which no dimensions get mixed up—or 'nonoriented' twists (*nor-twists*) that do mix up dimensions. A transparent Moebius strip is a simple space with a nor-twist in which the two sides get mixed up. The nor-twist changes the two-sided space into a space with only one side. Traveling once around a Moebius strip turns you upside down, and you have to make another full circuit to get right-side up again. This defining characteristic of an nor-twist is called a spin of ½ as it takes two cir-cuits to get back to the starting state. (The spin

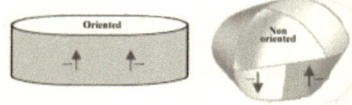

of a no-twist is 0, while a single 360°-twist has a spin of 1). Transfer through, or by, a nor-twist always flips a spin into its opposite spin.

The fundamental systems at the foundation of the physical systematic hierarchy are twists in the physical metric. The oriented twists in spacetime are called the *bosons*, and the non-oriented twists in spacetime are called the *fermions*.

The spatial dimensions are twisted along the complex time axis and, as we shall see, can twist in either direction. As there are only three orthogonal space dimensions, the tiny twists that are the boson and fermion fundamental particles both come in sets of three.

Bosons

The bosons are simple or-twists in external spacetime. They are symmetrical open 'cosine' waves with a spin of 1 that does not mix dimensions. Bosons can have a single twist, or two orthogonal twists, or three orthogonal twists.

0 or-twist: The hypothetical Higgs Boson is expected to have zero twists.

1 or-twist: The simplest boson has a single or-twist, and is called a Z-boson. Being an open wave, all its energy is at the boundary which then abruptly falls to zero. This abrupt energy change at the boundary stresses the spacetime and this gives the Z-boson an enormous mass-energy of ~90 GeV. This energy can twist a second dimension, and there are two resonances, the W^+ and the W^-, that also have electric charge and a lesser mass-energy of 85 GeV.

2 or-twists: A boson with two or-twists at right angles is called a photon. It escapes the obeseness of the Z because the two twists resonate together out of phase, as a sine wave driving a cosine wave, and as a cosine wave driving a sine wave. The energy of one open boundary is constantly being transformed into the other open boundary. The energy in either boundary does not last long enough to amount to a Planck's Constant of *the action*. So the double-wave energy is virtual, not real, and there is no open real wave and no real boundary energy. The two aspects are called the electric and the magnetic, respectively, and a photon is an electromagnetic wave.   The only energy is inherent in how tightly wound up is the double-helix in spacetime. Each photon has just one quantum of the action—a measure of energy-in-time—distributed between energy and time period. Existence, like space, time, and energy is also pixelated. For example, a radio photon is loosely wound; it is low in energy and long in time. A gamma photon is tightly wound; it is high in energy and short in time. But both have exactly one quantum of action.

3 or-twists: A boson with three oriented twists, all at right angles, in space along time is called a gluon. The three directions are differentiated by the three quantum colors—Red, Blue and Green—and the three quantum anticolors—Cyan, Yellow and Magenta—or antired, antiblue and antigreen. A gluon has one twist along positive complex time, a quantum color, and two pointing along negative complex time, as quantum anticolors. Using color terminology, the pairs are as in the chart. As discussed in *Volume Two,* the energy in these open waves twists the local spacetime out of its usual relectilinear configuration into a hexagonal configuration where only one direction remains as before—the positive color—while the other two shift by 30° or 60° as negative anticolors. In the illustration is a red quark as it is unchanged and orthogonal to the time axis, the other two axes are at only 60° or 30° in a hexagonal relation to the time axis.

All three axis time-orthoganal

One axis time-orthoganal, Two axes not time-orthoganal

So gluons come in different varieties, such as red-antired (cyan), red-antigreen (magenta), red-antiblue (yellow), red-antired (which is indistinguishable from blue-antiblue or green-antigreen). Having an odd number of twists, a resonance is not possible and a single gluon has a tremendous boundary energy. This is reduced when a large number of resonating gluons in a sphere—called a glueball—creates a surface in which the colored ends are crowded together with ends of other colors and anticolors. Each pixel of surface is constantly changing in colors and anticolors, it never remains the same for more than an instant. The color at a location never stays around long enough to amount to a pixel of the action, so the energy of the bit of colored surface is virtual. The surface of the glueball sphere is still where all the energy resides but, by this rapid flickering of color and anticolor, is now *virtually* colorless and the interior of the sphere is colorless and has zero energy.

	COLOR	ANTICOLOR
x	Red	Cyan
y	Green	Magenta
z	Blue	Yellow

It is not considered unusual for the mathematics developed in one field is successfully applied to another, quite different field. This is an example. In *Volume Two* we explored the mathematics of color technology—the positive RGB colors of TVs and computers, and contrasted them with the negative CMY colors of printing [146] —and applied its insights to the open cosine waves of gluons.

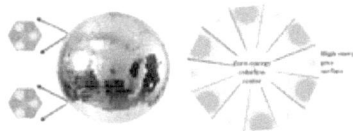

Gravity's Boson

These are the three simple bosons with, with oriented twists of spin 1, that mediate the three fundamental interactions, the Weak, Electromagnetic and Strong, that are the foundations of quantum physics. In *Volume Two,* we also discussed the boson with integer spin-2, called the graviton, that is the *local* description of that faint universal tendency of all energy to clump together. This is best described as a *global* entanglement on the complex level, and the curvature of spacetime fabric quantified by the equations of General Relativity. Gravitons are so low energy as to be undetectable, although the death throes of massive stars are expected to generate gravity waves of synchronized gravitons that should be detectable.

There is also the putative Higgs Boson with integer spin-0 that would be involved in giving the fermions their (naked) mass-energy. If found, this boson will be an excited, but untwisted, pixel of spacetime.

Fermions

Fermions are composite, not singular entities like the bosons. At the core is a non-oriented twist, and this jitters about, shedding a halo of *virtual* bosons. It is as if the nor-twist is constantly trying to flip off its nor-twist but only sheds or-twists as a nor-twist is permanent (unless it meets an ant-fermion with exactly the opposite twist). The shed virtual bosons do not have pixel of energy-in-time action, the Planck's Constant necessary for real existence in spacetime, so they do not have a *real* existence—hence the name—and exist solely in complex spacetime. This halo of not-real, virtual or-twist bosons in which the nor-twist fermion is enshrouded is called the *charge* of the fermion.

This halo can be directly experienced by attempting to force together the N-poles of two strong magnet. The cushion of virtual bosons is quite tangible, if invisible and insubstantial.

1 nor-twist: The simplest fermion is a single nor-twist in spacetime and is called a neutrino. It is surrounded by a halo of virtual 1-bosons which gives the neutrino a *weak* charge. If the twist is to the *left* along the time axis, it is a *matter neutrino,* if the twist is to the *right*, it is an antineutrino, an example of antimatter.

2 nor-twists: A fermion with two orthogonal nor-twists is called an electron. The electron is surrounded by a halo of virtual 1-bosons, which gives it a weak charge, and a halo of polarized 2-bosons, which gives it an *electromagnetic* charge. The polarization gives the electron a *negative electric* charge, a *down magnetic* charge, and a left spin. If it has a right spin, it is an antimatter *positron* with a positive electric charge and an up magnetic charge

3 nor-twists: A fermion with three nor-twists puts so much stress on spacetime that it collapses from its regular square-symmetry into the hexagonal symmetry that characterizes quantum color.

The three hexagonal nor-twists are called quarks that, depending on the pattern of collapse of 30° or 60°, can be either a D-quark or a U-quark alters the electro-magnetic nor-twist. The shift gives the U-quark ⅔ the positive electric charge of the positron, and the D-quark ⅓ the negative electric charge of the electron. The U-quark is a variant built on a positron, so it could actually be considered to be ⅔rd antimatter.

The quarks are surrounded by a halo of virtual 1-bosons, which gives a quark a weak charge, plus a halo of polarized virtual 2-bosons—negative charge and down magnetism for the D, positive charge and up polarization for the U—and a halo of 3-bosons which, while individually flickering in and out of existence too rapidly to ever amount to a pixel of action, collectively have a real, and high energy, the energy that gives matter almost all its mass.

Depending on its hexagonal structure and which of the three spatial dimensions remains orthogonal to the time axis, a quark can have a red, blue or green *color*. Inasmuch as the energy of an isolated quark is unboundedly enormous, the existence of an isolated quark is impossible since it would have an infinite energy. This accounts for what physics calls the *confinement* of quarks; a free quark is impossible. Quarks can only exist in two types of virtually colorless confinements where all the energy is in a glueball surface pixelated with virtual colors and anticolors:

(1)   **Mesons**. These are composed of a color quark and an anticolor antiquark, and are surrounded by a halo of gluons. They are bosons as they have an integer spin. The simplest are the pions.

(2)   **Baryons**. These are composed of three quarks of each of the three colors and surrounded by a halo of gluons. They are fermions as they have a half-integer spin. The simplest are the nucleons, the proton of two U-quarks and one D-quark, and the neutron of two D-quarks and one U-quark. Combinations of three U-quarks or D-quarks have been observed in high-energy physics.

Surface-neutral glueballs

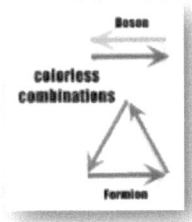

In either case, all types of quarks throw off the now-finite color energy into gluons so rapidly that they are essentially colorless and behave as if a colorless 2-fermion. In this color-free state they are confined to the colorless center of the glueball of real gluons they generate about them. Over 99% of the mass-energy of pions or protons is in the pixelated surface of the glueball, a hollow center in which the now-colorless quarks jitter about.

As an aside, note that if the U-quarks have an anticolor, a colorless black in a neutron is attained with two colors and a third anticolor, while in the proton, a colorless white is attained with one color and two anticolors. The neutron has ~1% more mass-energy in its gluon halo than the proton, and it has a small probability of decaying into a proton, an electron and an antineutrino, with a half-life of about eleven minutes.

The bosons are symmetrical 'cosine' wave-twists in the space dimensions along the complex time dimension, and going backwards or forwards in along the complex time axis, they look the same.

This is not true of the fermions which are asymmetrical 'sine' wave-twists in space along the complex time dimension. A fermion going in the positive direction in complex time is called a 'matter' particle; a fermion going in the negative direction in complex time is called an 'antimatter' particle. So there is an anti-neutrino, an anti-electron (positron) with an opposite polarization, and antiquarks with anti-color that congregate into anti-protons and anti-neutrons. The pions are a quark and an anti-quark in temporary alliance so look the same both ways.

Three Generations

There is yet another layer of complexity to the composite fermions: they come in three generations. The first generation is based on the neutrino that has a nor-twist that mixes up a single complex dimension which, unlike a simple dimension, has a sidedness to it. The line of real numbers has a plus-imaginary side and a negative-imaginary side to it. It is these two sides that get mixed up in the nor-twist of a neutrino. As might be expected, this is not a high energy state and, for a time, the neutrino was considered as massless as a photon.

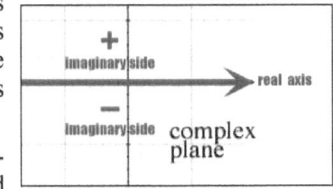

Adding a second nor-twist to this 'electron-neutrino' creates an electron, and adding a third creates a D or U quark. The foundation neutrino of this first generation , the *electron neutrino,* has a nor-twist in a single complex spatial dimension.

If the nor-twist of the foundation neutrino rotates two complex spatial dimensions as a unit, it is a *muon neutrino.* Adding a second nor-twist generates a *muon,* and adding the third generates a C or S quark.

If the nor-twist of the neutrino rotates all three complex spatial dimensions as a unit, it is a *tauon neutrino.* Adding a second nor-twist generates a *tauon,* and adding the third generates an B or T quark.

The nor-twist of the foundation neutrino of each generation mixes the sides, by half-twisting one, two or all three complex spatial dimensions *left* along the time axis.

Generation # nor-twists	1	2	3
Three	U quark D quark	C quark S quark	B quark T quark
Two	Electron	Muon	Tauon
One	**x** e-neutrino	**xy** μ-neutrino	**xyz** τ-neutrino

The rest energy mass of the neutrinos is small and difficult to measure. The muon is just like an electron except that it has ~200 times the mass-energy of an electron—105 MeV vs. 0.5 MeV. The second generation of quarks are like the first generation quarks but have a greater rest mass-energy. The C-quark is a very overweight D quark, while the S is an overweight U. While the collapse into the hexagonal form makes the U less massive than the D—of great importance for proton and nuclear structure—the collapse makes the C more massive than the S.

The tauon is an obese electron, while adding the third nor-twist results in the third generation of quarks, the B-quark that is an obese D-quark and the T-quark which is a morbidly obese U-quark.

The quarks are not symmetric. In the first generation, it is the positive baryon, the proton, that has less mass-energy than the neutral baryon. This is usually said to result because the D-quark is more massive than the U-quark (even though examination of neutral first-generation mesons, the pion,

	QUARKS		
	1st	2nd	3rd
Charge + ⅔ Spin ½	**U** 0.002 GeV	**C** 1.270 GeV	**T** 171.2 GeV
Charge – ⅓ Spin ½	**D** 0.005 GeV	**S** 0.104 GeV	**B** 4.2
Mass ratio	*0.4*	*12.2*	*40.8*

finds them to be 50% a +U–U pair and 50% a +D–D pair). The other possibility is the difference between the black and a white colorless canceling in the neutron and proton. Whatever the reason, it is swamped by other concerns, since in the second and third generations it is the neutral baryon that has less, and the positive baryons the greater, mass-energy, and usually exponentially so.

All the fermions in the three generations have their anti-matter counterparts. The second and third generations played a role in the Big Bang, but everyday life involves only the first generation.

Room temperature fusion

The tauon has interesting possibilities for technology as it is quite capable of taking the place of the electron in the hydrogen atom. The rest mass of the tauon at 1,776 MeV is almost twice that of the proton at 938 MeV, so the 1s-orbital of the tauon is confined deep inside the proton, far below the massive surface, jittering with the UDU color-free quarks almost as an equal. The isolated tauon is unstable, and falls apart in 30 trillionths of a second into a tau-neutrino and a W$^-$ boson. The W then decays into leptons 35% of the time, and 65% into a neutron, an electron and a neutrino. Inside a proton, it is probable that these pathways are inhibited and the confined tauon stabilized by its environment.

This is a novel nucleon with zero electric charge as all the virtual photons of the electromagnetic interaction are as confined as the gluons are. But the system still has the chemical imbalance of a hydrogen atom and has the usual valence and reactivity of a hydrogen atom. Two tau-hydrogen atoms will chemically unite into a tau-hydrogen molecule.

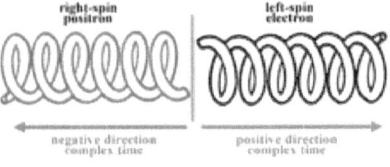

In this tau-hydrogen molecule the two nuclei are touching. This two-proton combination is unstable as it has excess nuclear energy over the more stable one-proton, one-neutron *deuteron* combination. A D quark beta-decays, liberating a couple of 2 MeV energetic gamma photons, while the two protons have united to become a single deuteron. This concludes our brief foray into the possibilities of room temperature, hydrogen fusion.

Antimatter

As mentioned, the twists in spacetime are in the three spatial dimensions along the time axis. An oriented boson has an identical projection from the internal to the external going in either direction along the time axis. A photon, for instance, is externally identical going forwards or backwards in complex time. $+t^2 = (+t^2) = (-t^2)$

This does not hold for the nonoriented fermions. Twists along the positive time axis are matter fermions, while twists along the negative time axis are antimatter fermions. An electron's two nor-twists are both along the positive complex time axis, while a positron has the same twists but along the negative complex time axis.

This is not really time travel because the external, observable movement through time is the square projection of complex time, and this is always positive.

Even though the electron and positron are traveling in opposite directions through complex time they both travel in the same direction in observable time.

If a left-spinning electron were reflected in a (hypothetical) time mirror, it would be observed as a right-spinning positron.

A Chiral universe

There was an assumption shared by both classical science and early quantum mechanics, an assumption that was so commonsensical that no one even realized it was an assumption. It was one of the "unknown unknowns" and not one of the "known unknowns" in our theories of how the universe functioned. I am sure the reader will agree that it is a most reasonable assumption. The universe has no preference for rotating in one direction as opposed to rotating in the opposite direction. In more technical terms: the universe shows no *chiral* preference for rotating clockwise or counterclockwise, for spinning up or spinning down, or for twisting to the right or twisting to the left. All else being equal, the universe is not inherently chiral.

Angular rotation, unlike linear size, can appear to be different depending on how you look at it. From the front, for example, the hands of a clock appear to be going clockwise, but if you walk past the clock and look back at it, you will see that the hands are moving in an counterclockwise direction. Its chirality is dependent on your point of view. The only situation in which the clock has an absolute chirality is if it is moving away from you at the speed of light so there is no possibility of overtaking it and viewing it from the other side. This is true for subatomic particles: only those traveling at light speed have an *absolute* chirality. Fundamental particles moving at sub-light speeds have a *relative* chirality as you can always move and view them from the back side.

The non-oriented twists of the sub-light fermions do come in two relatively opposite chiral forms, and they like to pair up as chiral pairs they have no overall chirality. For electrons and protons with an appreciable rest-mass, it is only with difficulty that they ever get close to light speed, so their chirality is always relative.

The only exception to this is the neutrino, which has, at most, only 1eV of rest mass-energy. If it created with a lot of kinetic energy, say 1 MeV, it will shoot away at light speed to an accuracy of a dozen decimal places. In such a situation, the neutrino will have an absolute chirality. Neutrinos that are generated in the decay of radioactive nuclei have kinetic energies in this range, and it was expected that such neutrinos would be found to have an absolute chirality that was either right-handed or left-handed in equal numbers.

Some odd aspects of the weak force that was responsible for the decay could be explained by a chiral preference, but it took the genius of Madame Wu, a Chinese-American *grande dame*, to come up with an experiment, and then perform the difficult experiment (involving a magnetic radioactive isomer of cobalt in an ultra-cold high-vacuum), to put the assumption of chiral equality to the test.

It would be an understatement to say that the results of this investigation into absolute chirality shook the scientific edifice—all the neutrinos were spinning to the left; the physical universe was fundamentally left-handed. Furthermore, as theory digested this fact about absolute chirality, it turned out that while a left-neutrino had a minuscule rest mass-energy, a right-neutrino would have a truly enormous mass-energy far beyond that reached in any experimental apparatus. To this day, no one has yet detected a right-neutrino.

The implication was that the electrons and quarks also had an absolute left chirality, a chirality that was lost in their normal state of sub-light speed.

The tiny fraction of quarks and electrons that emerged unscathed, after all the electrons and anti-electrons (positrons), quarks and anti-quarks created in the maelstrom of the Hot Big Bang had annihilated with each other into gamma rays, were of high-energy, and all the fermions in the universe had an absolute left-spin. As the universe cooled and expanded, the fermions interacted with each other, shedding their kinetic energy as extra photons, and the sub-light speed world of atoms appeared on the stage of history. The absolute chirality was lost in the familiar universe that has no preference for left over right angular rotation.

The absolute 'left-handedness' of the early universe that distinguishes ever-present matter from extremely rare antimatter (that spins in a right-handed direction along negative, complex time) must either have been established at the singularity of the Big Bang or it was established in the first few ticks of quantum time. It could not have been later since all that followed was symmetrical and could not alter a chirality in any way.

While current theory looks for a 'just after' scenario to explain the asymmetry, it will be simpler in the approach we are developing if we consider the 'right from the very start' scenario. The structure of the complex physical metric is left-handed, although this is lost in the square projection of observable space and time.

Chiral relations are not possible in one dimension or two dimensions, but they are possible in three dimensions. Consider the three orthogonal axes, colored red, green and blue in the 2-D illustration, each axis extending out in either direction If the construct is separated at the zero point into two separate spaces as shown, then the two spaces are chiral complements, that are equal and opposite, as +1 and −1, or R and L.

No matter how you rotate the spaces, it is impossible to superimpose all three colors—when two of them are aligned, the third is always pointing in the wrong direction.

Modern science is quite comfortable with zero decaying into equal and opposite parameters. The virtual foam that is the vacuum is an example of the decay, where the parameters

$$0 \rightarrow +1 \ \& \ -1$$
$$\rightarrow +i \ \& \ -1$$

can be a unit of ±spin, ±electric charge ±magnetic moment, or ±strong-force color, ±momentum, etc. It is really a generalization of Newton's 'action and reaction are equal and opposite' to the realm of quantum physics and the vacuum.

Chiral World

Because the Logos involves complete, not component, complex dimensions, it is not chiral. Its effect in the left-handed complex spacetime of the physical realm is decidedly chiral. The sequential establishment of the physical systematic hierarchy involves many situations in which the chiral foundations reveal themselves.

For instance, the complex structures on the midlevel of the Logos, which resulted in the subsystems uniting in the form we call life, differentiate between left and right.

1.   Carbohydrate is a fundamental subsystem of all living systems and it plays a vital role in life's functioning. Every single one of these mole-

cules—from the ribose-sugars in virus DNA, to the glucose-cellulose in trees, to the glucose-sugar for tea and coffee—is a specific chiral form; they are all D-sugars—the reference compound rotates the polarization of light to the right.

2.   Protein is also a fundamental subsystem of all living systems and it plays a vital role in life's functioning. Every single one of these molecules—from the aminoacids in the yeast enzymes that turn sugar into alcohol, to the muscle fibers in a tender salmon steak, to the proteins that are pumping ions around in your eye and brain as you read—is a specific chiral form; they are all L-aminoacids (the reference compound rotates the polarization of light to the left).

The natural law, the abstract structure in complex space, that makes life probable is decidedly chiral. Only the D/L carbohydrate/protein combination is present at this level in the Logos; it does not have a D/D, an L/L or an L/D combination.

This suggests that all the living systems we eventually come across as we explore the universe will have the D/L structure; all things considered, they will be fundamentally fine to eat (the L sugars and the D aminoacids are poisonous.)

Quantum probability

To summarize, observable reality involves a subset of four complex dimensions and its twists that are the fundamental bosons and fermions. The square projection of these four complex spacetime dimensions is the observable location of a fundamental particle in real space and real time. As it is the square that is observed, even though they are moving through complex time in opposite directions, the square is the same for both and they travel together through observed time.

Quantum physics adds more complex dimensions to the picture. The square projection of these 'higher' complex dimensions is the observable probability of a fundamental particle being at that location. These higher complex dimensions are what scientists call natural law, a topic we will later discuss in more detail.

The probability is determined precisely by natural law, but quantum probability is not at all like the classical probability that is exhibited by the tossing of coins. At the foundations of classical probability is the concept of 'independent assortment.' For coins, this means that the result of the first coin toss has no influence on the result of the second coin toss, i.e., the probability of the second is independent of the result of the first.

A coin is composed of zillions of bosons and fermions, so it is somewhat surprising that neither bosons nor fermions obey the rules of classical probability and independent assortment. They follow two opposite types of 'probability statistics,' and classical probability is just the average of the two.

The three probability statistics—classical, boson and fermion—can be illustrated with a thought experiment with three types of coins that can come up heads, H, or tails, T, when tossed. The classical, boson and fermion coins all behave the same when dealt with singly, they each come up 50% H and 50% T. The dramatic difference between the three types of coins only shows up when two or more coins are thrown together.

1.   The classical-coin pair has four equally-possible combinations that can result—HH, TT, HT, or TH—when tossed together. Two of these are

even combinations with both the same (HH or TT) and two are *odd* combinations, the mixed (HT or TH). The result of independent assortment when tossing a classical coin pair is 50% even and 50% odd combinations.

2.　　　The boson-coin pair when tossed together behave quite differently. A toss always results in *even* combinations—HH or TT—100% of the time. There is no independent assortment; this is an example of *dependent* assortment.

3.　　　The fermion-coin pair when tossed together also behave differently but in an opposite way to the bosons. The toss always results in *odd* combinations—HT or TH—100% of the time. Again there is no independent assortment, but a dependent assortment known as the *Pauli Exclusion Rule*.

The departure from commonsense classical probability is even more marked when a third coin is thrown. The boson coins continue with even combinations—HHH or TTT—a behavior that continues even when zillions are tossed, always an even combination. This is the quantum probability that results in the zillions of identical photons that compose laser light and radio broadcasts.

The third fermion coin is even less commonsensical in its behavior. It comes up 50% H or T but it refuses to land on the table, instead floating above the other two coins, and it is impossible to force it down to mingle with the other two coins.

Coins	1	2	3	4
classical	50% H, 50% T	50% even, 50% odd	25% even	12.5% even
boson	50% H, 50% T	100% even	100% even	100% even
fermion	50% H, 50% T	100% odd	$50\%\frac{H}{HT}, 50\%\frac{T}{HT}$	$50\%\frac{HT}{HT}$

Electrons behave like this, and Pauli's Exclusion Rule of a zero probability of two electrons ever being in the same state that results in the sequence of ever-larger quantum orbitals that, when electrons inhabit them, give rise to the systematic hierarchy that is the periodic table of the elements and much of chemical behavior. The various elements all have a different, if periodically varying, set of emergent properties because of Pauli's rule.

In classical science, probability was a human construct that had no real existence. In the new physics, probability has the iron rule of absolute law and has a very real existence. It is quantum probability, for example, that will maintain the extension of our sun at planetary-size when it runs out of fuel and cools down from a white dwarf star to a black dwarf star 20-odd billion years from now. It will be quantum probability alone that will prevent any further gravitational collapse as the star remnant cools towards zero, Quantum probability is mighty indeed, and very real.

All this is enough to cause classically-trained scientists to throw up their hands and mutter about the weirdness of the quantum realm, the 'strange theory' that precisely explains so much about the world. But there is more the science has discovered about this Physical realm that is even stranger to the classical mindset.

While spacetime and particles involve just a real and imaginary subset of four complex dimensions—we will discuss the complementary subset later in the discussion—quantum probability involves all of a complex dimension, the first level of natural law and the foundation of the systematic hierarchy that is the Logos.

We shall refer to the spacetime aspects of a system that are described by simple numbers as being 'external,' while the natural law aspects of a system described by complex numbers as being 'internal.' Quantum physics states succinctly that the fundamental entities of the physical realm are quanta of energy-twists in spacetime with an external particle aspect and an internal wave aspect, the projection of the internal determining the history of the external in spacetime.

Quantum waves

$$z = \frac{\begin{array}{ll} x + yi & \text{rectangular} \\ me^{\alpha i} & \text{polar} \\ m(\cos\alpha + i\sin\alpha) & \text{trigonometric} \end{array}}{}$$

Both unbounded traveling waves and bounded standing waves are combinations of sine and cosine waves. The internal aspect of a physical system is a wave that is described by complex numbers. The trigonometric form of complex numbers explains why this internal aspect of fundamental physical systems is called its wavefunction, ψ (psi).

The wavefunction, of complex numbers $p@a$, is also called the *probability amplitude* because is absolute square, p^2, is the external probability of finding the electron density.

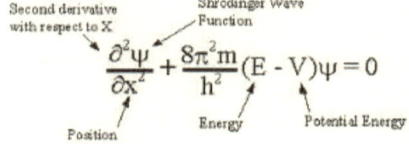

$$\frac{\partial^2\psi}{\partial x^2} + \frac{8\pi^2 m}{h^2}(E - V)\psi = 0$$

Over time, the electron has an overall average density around the nucleus that reflects the complex form of the wavefunction. When the wavefunction has the form a closed sine wave, the electron density around the nucleus—and the electron is rotating trillions of times a second—is exactly that of a closed, sine-squared wave.

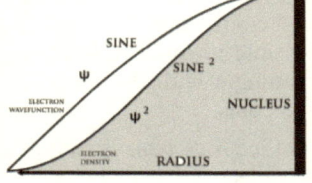

It would be quite possible by adding up all the little arrows to come up with the shape of a helium atom. A shortcut called the Schrödinger Equation combines all the adding and multiplying of complex numbers, and condenses it all into a single, elegant equation. We explored in detail this equation in *Volume One*. While it looks formidable to the non-mathematician, the equation is actually stating something that is simple enough to be put into words for a helium atom.

First, because a helium atom is perfectly spherical, we can replace the position x with r, the distance from the nucleus. Second, it states that the sum of adding the two main terms (involving the second derivative of the wavefunction, and the wavefunction) is always zero. These two terms are always equal and opposite.

The first term is the second derivative of the wavefunction, which is very simple calculus. We are perfectly familiar with speed being the rate of distance travelled over time, and with acceleration/deceleration as the rate of change in speed with time. Acceleration is the rate of change in velocity, the change in the rate of distance travelled with time. A mathematician would say that velocity is the *first* derivative of distance with time, acceleration is the *second* derivative of *distance* with time (and the first derivative of *velocity* with time).

So, given that the wavefunction is the overall shape, the first term is a measure of the rate of change in the shape, how the curvature changes with distance from the nucleus. This term can be thought of as the acceleration of change in the overall form.

The first third of the second term is a constant that includes a multiple of π-squared, the rest mass/energy of the electron, and the square of Planck's Constant, the pixel of action and existence.

The terms within the brackets are E, the energy of interaction, and V, the potential energy at point r from the nucleus, the center of the interaction. Together they are a measure of how the interaction energy is distributed with distance along the radius from the center. This is akin to swinging pendulum bob with energy that is conserved but oscillates between two extremes, all as kinetic energy or all as potential energy in the gravitational field. At the center of the arc, the energy is all in the speed of the bob as it rapidly sweeps through the center. At the end of the arc, the bob is momentarily stationary before it swings back, and all the energy is in the gravitational field.

The electron in a helium atom is similar. When it is at the boundary, all the interaction energy, E, is in the electromagnetic field of the coupling virtual bosons, V. The velocity at the boundary is zero and the all the energy is potential, V=E.

When the electron is streaking through the center, all the energy E is kinetic, and none of it is potential energy, V=0. The electron oscillates up and down the axis which is also rotating in all three spatial dimensions.

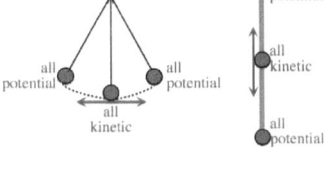

The two electrons move in phase to minimize their mutual repulsion: either both are at opposite ends and as far away as possible, or streaking past each other at the center so briefly that they hardly interact. The rotation rate is a trillion, trillion times a second.

The Schrödinger equation simplifies for the helium atom, and can be solved. The deceleration in the curvature of the form at distance r from the nu-

$$-\frac{d^2\Psi_r}{dr^2} = M(E - V_r)\Psi_r$$

cleus equals the enhanced mass times the form at distance r times the kinetic energy at distance r. As the total energy always equals the sum of the kinetic and potential, the bracketed term is simply the energy in kinetic motion at a distance from the center.

The final term is simply the complex number that is the wavefunction at that distance. So Schrödinger's Equation can be simply expressed in words—At every distance from the center, the rate of change in the form of the wavefunction, when added to the wavefunction mass times the kinetic energy, will always sum to zero.

Like any equation, there are only a small set of solutions to this equation, and these solutions are precisely the orbitals of the elements. Each of the odd-looking 4f orbitals—uranium and plutonium have electrons in them—is a ψ that is a solution to the equation.

For a traveling wave, the only difference between a sine and a cosine wave is phase; otherwise they look the same. There is a significant difference between

closed standing waves—exemplified by asymmetric sine waves with zero energy at the boundaries—and *open* standing waves—exemplified by symmetric cosine waves with all their energy at the boundaries. We saw a consequence of this in the massiveness of the Z boson and the hollow colorless centers of nucleons.

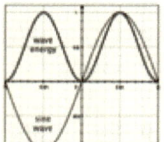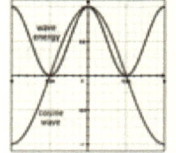

The wavefunction, the internal aspect of fundamental particles, is a waveform that can be either an unbounded traveling wave or a bounded standing wave. Most of the standing waves we will be dealing with in sophisticated systems will be closed, asymmetrical sine waves.

As we saw with the boson coins, bosons get together in even combinations where they are all in the same state. Their individual waves unite together with the amplitudes adding together arithmetically as 2, 3, 4, 5 ... N, while the intensity of

the wave—being the square of the amplitude—exponentially increases as 4, 9, 16, 25 ... N^2. The intensity of the final wave can be enormous, as is exemplified by laser beams that slice through steel. The photons are all in the same state.

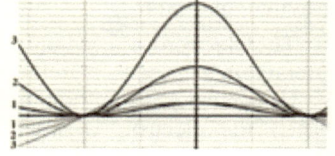

Both classical probability and both kinds of quantum probability are alike in one respect—they all obey the Law of Large Numbers. A classical coin toss illustrates the basic principle, i.e., the greater the number of tosses, the more accurately the actual result will precisely reflect the abstract probability.

Just throwing one coin results in either 1H or 1T, which does not reflect the 50% probability at all. Throwing ten coins you sometimes get 5H+5T, but often the result is 6H+4T or 4H+6T, or even 7H+3T or 3H+7T. Occasionally the even 10H or 10T will appear. But, if you throw a trillion fair coins at a time, however you will find that the result is 50±0.000,001%.

The Law of Large Numbers (LoLN) states that the probable deviation of the observed result from the true probability will be proportional to the square root of the number of trials. So a fair coin thrown a million times can be expected to result in 500,000±1,000 H and T. So a result of 450,000H and 550,000T would be highly unlikely if the coin was fair.

For classical science, this is somewhat a circular argument since the only way to test the fairness of a coin is to toss it repeatedly, the more tosses, the more accurate the measure. In quantum science, the probabilities can be calculated directly by squaring the complex probability. At present, this can only be done precisely for simple systems, such as the electrons of hydrogen and helium in their standing waves, the orbitals, about the nucleus.

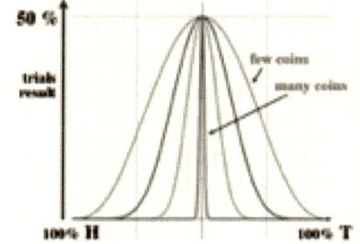

The electron wave sweeps around about a trillion trillion times a second, each second there are 10^{24} trials. The electron appears over even brief periods of time to be smeared out around the nucleus in an electron-density that has a boundary where it falls to zero—an orbital is a closed sine wave. The calculated density is exactly that of the observed density.

Confinement and form

We have already encountered confinement in the quarks and gluons of protons, neutrons and composites as nuclei. The internal wavefunction is a standing wave such that the probability of a quark or gluon leaving the nucleus is exactly zero. As described in the mention of white dwarf stars, the power of quantum zero is not to be trifled with.

A better-understood example is the confinement of electrons by the internal wavefunction of atoms. For example, the probability of an electron, proton or neutron leaving a helium atom is essentially zero. The internal aspect of the helium system, a set of resonating standing waves, confines all the subsystems as the external aspect of the system. The form of the internal aspect, by the LoLN, is expressed as its square in the external probability density of the electrons.

The electrons are subsystems of the helium system, and it is the internal aspect of the helium atom that determines the external density of its subsystems. This principle holds for all systems: The form of the internal aspect determines the external form of the interacting subsystems over time.

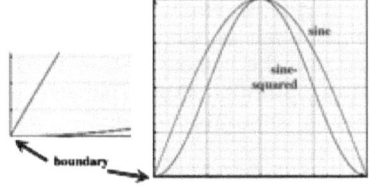

In bounded waves, the LoLN has a direct relation with time. In a wave, the same state occurs with each cycle, so the number of trials in a second will depend on the frequency. We can define the characteristic period of any system to be the time in which sufficient trials occur for the LoLN to express the internal form. The internal wave confines the subsystems of the helium atom into a perfect sphere, just as the wave of the atomic nucleus confines the quarks and gluons.

While the wave of the helium is a three-dimensional sine wave, it will suffice for our purposes to illustrate it with a simple sine wave. While the internal sine wave is rapidly changing at the boundary, the external sine-squared wave of the electron density has a zero rate of change there (more technically, while the sine wave has a derivative of 1 at the boundary, the sine-squared wave has a zero derivative there). While the internal form of the wavefunction is that of a sine, the external form of the electron density is that of a sine-squared wave.

This perfectly spherical density of electrons is greatest at the center and zero at the boundary—it is as close as modern physics gets to the "massy spheres" of solid matter at the foundations of classical physics. The helium atoms in thermal motion behave just like billiard balls; they bounce off each other in 3-D just as solid balls do on a pool table. This is why helium is as close as reality gets to being a perfect gas.

The helium's electrons are in a standing spherical wave called the s-orbital with a zero electron density at the boundary and the maximum density around the center. A simple sphere, however, is not the only form that the standing waves can take up. While the standing wave of the 1s-orbital involves only 1 ½-sine wave, the 2s-orbital involves 3 ½-sine waves. The density of the electron is zero at the boundary, but there is now a second node of zero density between the boundary and the center. The electron density is now in both the center and a shell about the center.

2s orbital

The 3s-orbital has five ½-sines, the 4s has seven, etc. It is these s-orbitals, with only a singlet electron in the neutral atom, that give hydrogen, lithium, sodium, potassium their chemical properties that make them essential in living systems. The illustration is that of a single electron in the 2s orbital state with a lobe at the center and one on either side.

The s-orbitals involve a standing wave with an odd number of ½-wavelengths. There is also a family of orbitals involving an even number of ½-wavelengths called the p-orbitals. When the simple s- and p-orbitals are filled with electrons, the result is a very low-energy, and chemically inert. The simplest p-orbital is the 2p which, along with a boundary node, has a single inner node at the very center where the probability density of a 2p-orbital electron is exactly zero.

The probability density is in two lobes on either side of the center, and a 2p electron is 50% of the time in one lobe and 50% in the other. There are three p-orbitals that can fit at right angles about the central nucleus,

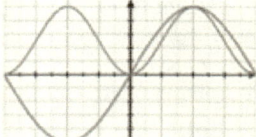 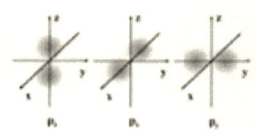

when they are full of electrons the density is almost as perfectly spherical as that in helium. The neon atom has its s- and p-orbitals filled, and the electron density is almost as perfect a sphere as that in helium. Helium is the first, and neon is the second, of the 'noble' gases that form one column in the periodic table of the elements. The p-orbitals play a significant in the chemistry of the elements that are the main subsystems of living systems, such as carbon, nitrogen, oxygen, sulphur, phosphorus and chlorine.

In 3-D, standing waves can take on some unusual forms, such as seen in the d-orbitals that have two nodes at the center. Four of the forms are variations on the p-orbital theme but one of them has a most unusual donut form.

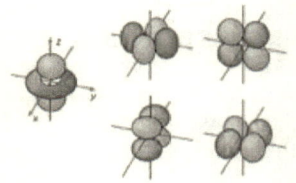

Entanglement

We have already seen that when scientists started to probe into a detailed understanding of nature they found that the 'nuts and bolts' running beneath everyday reality were a lot more sophisticated than any simple classical concepts could handle. The orbitals introduce what is probably the least commonsensible, most classically impossible way the very simplest, fundamental systems in the physical hierarchy behave.

An single electron in a p-orbital has a 50% probability of being in either lobe and a zero probability of being at the center. A single electron in a d-orbital, as illustrated, is 50% of the time in the red lobe and 50% in the green lobe, and 25% of the time in each of four-lobed forms. It does not travel between the lobes; it is sometimes in one lobe and sometimes in another, but it is never in the space between them. Some would call this jittering back and forth while ignoring the space in between—a simple form of teleportation.

While classically strange, this is the way things are, and a simple experiment suffices to illustrate such nonclassical behavior. When a wave passes through two slits that are close together, the wave splits into two waves that move apart in space. The external separation, however, does not change the internal wave and its associated probability density. A single electron goes 50% through one slit and 50% through the other one.

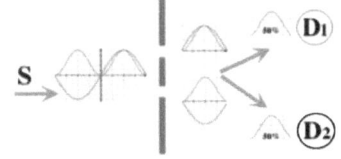

If an electron is sent one-by-one from a source, S, through the slits to detectors D1 and D2, the wave splits in two and the electron density is now in two lobes that are substantially separated. The electron is 50% in one lobe and 50% in the other lobe with a zero probability of being in-between the two lobes. It does not travel between lobes, it is either in one or the other.

It is just a matter of chance which lobe it is in when the wavefront intersects the detectors and the electron interacts with it. The interaction instantaneously alters the wavefunction of the electron. The distant lobe disappears—called *the collapse of the wavefunction*—and the wave changes to form localized in the firing detector. D1 will fire 50% of the time, and D2 will also fire 50% of the time. Like a coin, it takes the LoLN to express this probability, so many electrons need to be sent one by one through the apparatus to show this "interference" effect.

Such *slit experiments* in which the classically-impossible feat of passing through both slits at the same time are not confined to simple electrons and photons, decidedly 'bits of matter' molecules of 90 atoms have been successfully passed through two slits at the same time. All 90 atoms being 50% of the time in one lobe, and 50% of the time in the other lobe that is centimeters distant. Teleportation indeed.

It should be clear by now why scientists speak of reality as being permeated with "quantum weirdness" and abandon it for the comfortable, if only approximate, concepts of classical physics. By any common application of the word "weird," the implications are "not normal', "not natural," "not what one would expect." But the fabulously-successful concepts and equations of quantum physics apply throughout the known universe, and as such are the very paragon of what is normal, natural and what one *should* expect. Hopefully, the next generation of physicists will expect the universe as it is and not think anything weird is going on at all.

This phenomenon of being in two places at the same time because of the internal wave is called 'entanglement,' and is particularly fascinating and non-classical when pairs of particles are involved. If two entangled particles, one spin-

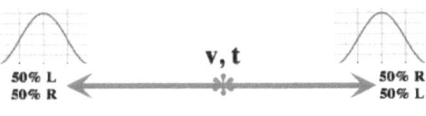

ning left and the other spinning right, move apart at velocity v, for a time period t, they will both be in the same wave but constantly switch locations. The L will be in one lobe 50% of the time while the R is in the other, and vice versa.

The most astonishing thing about such entanglement is that it is independent of both v and t. Theory suggests that even if the velocity of separation is close to light speed and the time since separation is billions of years, the same phenomena holds—L or R will be found in one lobe and R or L will be in the other lobe. That the node externally separating the two lobes of the wave stretches across giga-light-years to the edge of the visible universe is irrelevant to their internal wave-function. The particles are quite oblivious to their external separation in space.

As many natural processes in outer space eject such entangled particles, and many reach us here on earth as cosmic rays moving at essentially light speed. If the separation of the entangled particles occurred ten million years ago then its entangled partner is 20 million light-years away.

The technological implications for interstellar travel of being able to exploit this natural and abundant web of internal connections across the vastness of space are clearly considerable.

Physical Interaction

Helium is an example of a system whose internal wave firmly confines all of its interacting subsystems; in most systems, the confinement of at least some of the subsystems is only partial.

We have already seen such an example in the electron whose internal wave does not locally confine the virtual photons and they fade off to infinity as the electric field created by the electric charge, although it does keep the weak bosons firmly confined.

In a systematic hierarchy, the stability of a system is dependent on the stability of its subsystems. While the internal wave of a neutron confines its subsystems as firmly as a proton does, if one of the quarks decays into another quark, the neutron is also altered. The quarks in the colorless center have a halo of virtual photons and weak *virtual* bosons and, if one of its D-quarks ejects a weak *real* boson—which, before it amounts to a quantum of action, decays into an electron and antineutrino. The result is the the D-quark becomes a less-massive U-quark, the *beta decay* converts the neutron into a proton. The half-time for this 'beta decay' is when the LoLN states that 50% of a large number of neutrons will have become protons. This is ~11 minutes, an age in nuclear time, and the reason why the weak boson was given such a moniker.

An atomic orbital is filled when it contains two electrons spinning in opposite directions. If two fermions with the same spin attempt to get together, their internal waves merge and cancel out by destructive interference; the probability of them getting together is zero. If the fermions have opposite spins, however, the nor-oriented ½-spin flips the sine wave so that the two waves combine, just as the boson waves do. The internal probability becomes a 2sine wave while the external probability becomes a 4sine-squared wave. This high probability gives such a matched electron-pair a low energy and a high stability.

These electron-pairs are so stable that much of chemistry is contained in the admonition, "Thou shalt not break up an electron pair." Much of chemical change is driven by an electron moving to create a stable pair with another.

A hydrogen atom has a single electron in the s-orbital. This is an unbalanced, high-energy and unstable state that gives the atom its a 'chemical valence,' determining how it interacts with other atoms to make a stable electron-pair. The hydrogen with this singlet electron is called a free radical, and it highly reactive chemically.

In the systematic view, the subsystems of any system can be divided into two classes, the 'core' subsystems that are firmly and stably confined by the internal wave, and the 'peripheral' subsystems that are not. For the hydrogen atom, the quarks and gluons are core subsystems, while the single unbalanced electron and halo of virtual photons are peripheral subsystems. The electron is called the 'valence' electron of the hydrogen atom, and the unbalanced virtual photons are its electric dipole.

The hydrogen atom can make a pair by sharing, gaining or losing its electron. A H-atom can share its electron, for example, with another H-atom. Their two atomic waves merge into a single football-shaped molecular wave that firmly confines both paired electrons, all the virtual photons and, of course, the quarks and gluons. This is a stable system at ordinary temperatures and a gas of hydrogen molecules will last indefinitely.

An H-atom can also abandon the single electron and share an electron-pair with another atom. A oxygen atom has two p-orbitals containing electron pairs that the H-atom can share. This is what happens in acids and, when a hydrogen atom leaves a molecule it is a part of, abandoning its electron, and shares an electron pair in the p-orbital of an oxygen atom in a water molecule. The acidity of acids is not free protons, as H^+ would suggest, but the H_3O^+ ion.

Occasionally, the hydrogen atom can be induced to take on an electron (by pushy atoms such as sodium) and become a hydride ion. This is not a particularly stable situation, and NaH reacts with water to release hydrogen gas.

Covering all these ways, hydrogen is said to interact by 'coupling' with its electron. This illustrates a general principle—systems interact by coupling with their peripheral subsystems. An electron and positron interact with each other by coupling with their peripheral virtual photons, forming a 1s 'atom' of positronium momentarily before they meet and annihilate into photons.

A system is capable of coupling in all the ways that its subsystems are capable of, and it is capable of coupling with the subsystems themselves. A hydrogen atom is capable of coupling with electrons and photons; a water molecule can also do this as well as couple with hydrogen atoms. Larger molecules are capable of coupling with fragments of molecules, called radicals, such as OH–, and so on.

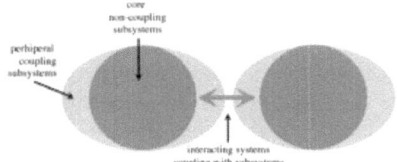

Unlike the symmetrical nonpolar bond

in a hydrogen molecule, the bond between H and O atoms is asymmetrically polarized, and the electrons spend more time about the O nucleus than they do about the hydrogen. This allows water molecules to couple with each other by coupling with hydrogen atoms, the hydrogen bond that is responsible for all the anomalously-useful properties of bulk water and the hexagonal symmetry of ice crystals.

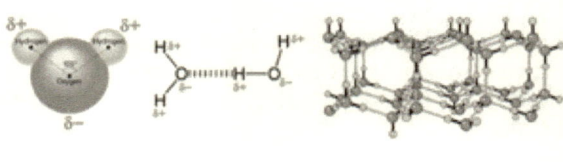
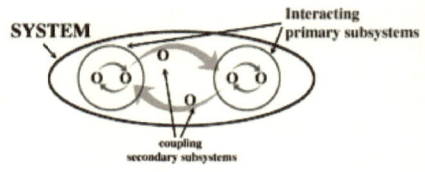

Interacting subsystems

This coupling with subsystems applies to the interacting subsystems that comprise a system in a systematic hierarchy. A system is a set of interacting primary subsystems that couple with sub-subsystems, the secondary subsystems of the primary subsystems.

A systematic hierarchy of systems composed of interacting subsystems coupling with sub-subsystems has a triple-level structure, illustrated by molecules and cells.

This coupling by a system using its subsystems is the external aspect of interaction, but there is also an internal aspect to take into account. For subsystems do what the wave tells them to do over time, and the external

System	molecule	cell
Subsystems Interacting	atoms	organelles
Sub-subsystems coupling	electrons	molecules

is a result of the internal resonance of waves. The Logos is the source of the set of emergent properties that appears at each level during the interaction of the subsystems to create the system.

Confinement

In system building, when a system emerges from its disparate subsystems, the internal waves of the subsystems resonate together, as prescribed by the Logos, as the internal wave of the system. The form of the internal system wave is prescribed by the Logos, and it confines the subsystems into the externally observable form of the system.

This confinement of the subsystems by the system wave can range from totally confined to barely confined.

Total confinement

Simple examples of total confinement are provided by the helium nucleus and the helium atom. The subsystems of the nucleus are two protons and two neutrons coupling with quarks and gluons. At reasonable temperatures, there is zero probability of any of these subsystems leaving the system. The subsystems of the atom are the nucleus and the two electrons coupling with photons. There is also an almost zero probability of any of these subsystems leaving the system. Helium is physically and chemically inert.

Partial confinement

In most systems, the system wave only totally confines a subset of its subsystems, the remainder are only partially confined. The system wave of the helium nucleus, and of all nuclei, is like this, i.e., it confines all its subsystems except for virtual photons of a positive polarization. The helium atom, on the other hand, confines even its virtual photons.

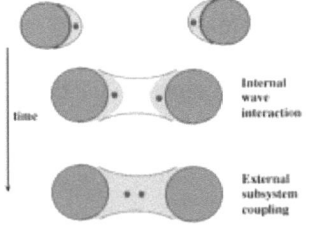

Many more examples are provided by the atoms of the various elements. Each system wave totally confines a 'core' subset of subsystems while the 'valence' electrons are not in a stable state. The periodic table of the elements results from the sequential filling of the orbitals, resulting in atoms where attaining stability ranges from dumping a valence electron, to sharing a valence electron, to adopting a valence electron. These are the emergent chemical properties inherited from the Logos.

Any system that is not totally inert will have this set of partially-confined valence subsystems as well as its core subsystems.

Ordered interaction

There is a sequence to any interaction, first the internal and then the external. First, the internal waves overlap and interfere. The internal wavefunction, z, changes to z*. The external probability alters from $|z|^2$ to $|z^*|^2$. This is the internal aspect of interaction.

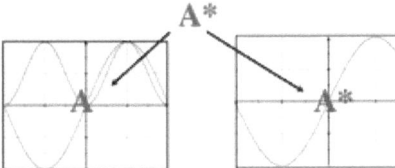

Second, over the characteristic time period the LoLN will alter the external density of the subsystem from $|z|^2$ to $|z^*|^2$, and the external coupling is completed. We can illustrate this with two hydrogen atoms, each with their single peripheral electron and constant core.

Some interactions—in the sense of influencing—between systems are wholly internal, and do not have an external aspect. We have already encountered this in the behavior of fermions and bosons in the way they do not obey independent assortment, on an internal level they are responding to each other. Nothing is exchanged externally, there is no energy involved.

Another example is the excited atoms we have already encountered. A helium atom, for instance, can absorb a specific size of photon and one of its electrons leaves the 1s orbital and enters into the 2s orbital. In a few billionths of a second, this excited atom spits out that specific photon and the helium returns back to its ground state.

In order for a photon to exist between two mirrors, the separation has to be a whole number of ½-wavelengths, otherwise the photon wave experiences destructive interference and cannot exist there. If an excited atom is placed in a box into which its specific photon does not fit, the excited atom is quite stable, the photon is not emitted. Its internal wave is zero in the box and it stays with the atom. There is nothing external happening, but internally the photon is suppressed.

Logos and Natural Law

It is the forms to this multidimensional abstract Logos that progressively guide the systematic confinement of the external particles into the forms we call atoms, the forms we call molecules, the forms we call cells, the forms we call human, etc.

So both matter and natural law are constructs in complex dimensions. Space and time, which is twisted in matter, take up the first four levels. Natural law starts at level five, with the constructs of the quantum wavefunction, the atomic nucleus and orbitals of the Periodic Table of the elements. Level six rules the chemical interactions as molecules, on so on up a hierarchy until we reach the level at which the human form resides, a finite number of levels from 5 to N.

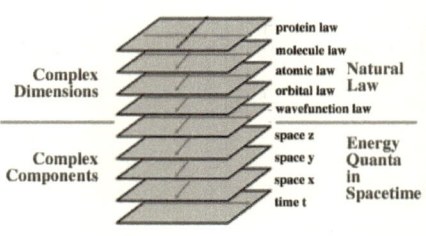

It is this structure of natural law that makes each discipline in science somewhat self-contained, e.g., a chemist needs only a smattering of physics or biochemistry to be a brilliant chemist. That natural law is a complex entity related to time and space, and that its structure rules the realm of quantum probability, explains one of the facts of life often mentioned by Christian de Duve:

1. We know from our study of history that there are millions of discrete steps that had to have occurred in the 4 billion-year progression from simple chemicals on the abiotic earth to the emergence of human beings.

2. We know from probability theory that the probability of each step is unity or less, and that the overall probability of a sequence of steps occurring is equal to the product of the probability of each step occurring.

3. We know from basic arithmetic that multiplying a set of numbers all less than 1 together results in a number that is smaller than any of the members in the set. If one or more of the set is infinitesimal, the result will also be an infinitesimal.

$$10^{-1} \times 10^{-3} \times 10^{-2} = 10^{-6}$$
$$10^{-1} \times 10^{-3} \times 10^{-900} = 10^{-904}$$
$$10^{-1} \times 10^{-900} \times 10^{-900} = 10^{-1801}$$

The implication of these points is that not a single one of the steps from atom to Man could have had an infinitesimal probability, let alone two of them; every single step from atom to Man must have had a significant probability.

While this is impossible to explain in classical science, it is to be expected if the internal space of quantum probability has a structure which, at its very highest levels has the structure of a human being. The Logos has a structure in multidimensional complex space that is expressed in the external form-over-time, the internal wave confining the host of fundamental particles that comprise the human body. Plants and animals do not reach into these upper levels and reflect the forms of abstract structures lower down.

The lowest levels of the Logos deal with the quantum waveforms that, when inhabited by electrons, are the orbitals that give the elements their emergent chemical properties. These forms in the lower levels of the Logos already have a precise and sophisticated form, as illustrated in the bizarre shapes of the higher orbitals. The seven 4f orbitals, with many entangled lobes are an example of structure at the lowest levels of the complex dimensions of the Logos. It is the filling of these 4f orbitals that give the increasingly valuable *rare earth* elements—the *lanthanides* such as cerium, europium and dysprosium—the properties that make them so essential in our technological world

4f orbitals

When the external form reflects the internal form, we call this a state health and happiness, but when it does not, there is unnatural discomfort and disease. In this sense, the concept of the Logos is similar to that of the commonly used concept of "Mother Nature."

Physical systematic hierarchy

The systematic hierarchy of the physical realm is founded on complex spacetime. On this foundation, a systematic physical hierarchy emerged over time—reflecting the abstract hierarchy in the Logos—of atomic nuclei, atoms, molecules and macromolecules, each with a set of emergent properties. In *Volume Two,* we examined the levels up to molecules, and the stepwise expression of the Logos over time from the Big Bang origin event.

While we have numbered the levels in the Logos, the math of multiple complex dimensions is in its infancy, so it is quite possible that, like the spacetime level, each of the indicated levels involves multiple complex dimensions. Just as in the Abstract Realm, it is impossible to change one level without altering all the other levels. The Logos of the Big Bang is the Logos of the present day; the laws of nature have not altered at all.

We next explore the levels above the molecular level—the key macromolecules of life, the proteins and the nuclei acids.

The proteins have a set of emergent properties that make them preeminent at manipulating analog forms, such as the ice structure of water or the chemical transformation of molecules.

The nucleic acids have this analog ability to a degree, but they

Logos level	Systems	Interacting subsystems	Emergent properties
11	Nucleic acids	Monomers & phosphate	Manipulate and transmit digital information
10	Proteins	Monomers	Manipulate Analog form of water & other molecules
9	Molecules	Atoms	Plethora of molecular properties
8	Atoms	Electrons & nuclei	Periodic chemical properties
7	Nuclei	Protons & neutrons	Stable isotopes
6	Fermions	½ twist & halo of 1 twists	Weak, electric & strong charge
5	Bosons	Twisted spacetime	Coupling of weak, electric & strong interaction
1–4	Spacetime		

have a higher set of emergent properties making them preeminent at the manipulation of digital information about the analog form of systems. Like the analog structure of living systems, this digital information about analog form is stored in a
multilevel systematic hierarchy of sophistication.

Resonance

The 'two opposite sides' of spacetime involve four of the five basic parameters
of energy quanta. The fifth parameter, that of probability, is the same for both
physical and spiritual realms in that it involves both components. If the spiritual
and physical were identical, if the wave aspect of the basic spiritual atom was
identical to the wave aspect of the basic physical atom, we would expect some sort
of crossover. Modern experiments with atoms would surely have picked up any
such crosstalk between the two realms.

As the complex components of the physical metric have a fundamental *left*
configuration, we can expect that the complementary spiritual metric has a fundamental *right* configuration. This is imposed by the Logos at the moment of Creation.

We have already seen that the simplest physical quanta of positive energy, the
left neutrino, has a very small mass-energy and moves at essentially light speed,
the upper limit. The *left* neutrino is in resonance with the Logos-imposed chirality.
On the other hand, a putative *right* neutrino is not in accord with the Logos, and it
would have an enormous positive mass-energy, and lumber slowly along close to
the zero-speed, the lower limit.

This suggests that the simplest spiritual quanta of negative energy spin *right,*
and have a low negative energy and move close to the unbounded, upper speed
limit. On the other hand, a spiritual quanta spinning *left* would have an enormous
negative mass-energy, and lumber slowly along close to the light speed lower
limit.

Thermodynamics tells us that the speed of quanta is reflected in the temperature of the surroundings. If actions that resonate with the Logos are *good,* and result in generating *right* quanta, while actions that do <u>not</u> resonate with the Logos
are *evil and* result in generating *left* quanta, the result would be a blazingly hot
heaven of good people and a bitterly cold hell of evil people. These are just suggestions to stimulate further thought about the harmony of religion and science.

The negative energy quanta of the complementary spiritual metric is a barely-
understood topic, so we can only make suggestions. As we shall see in the following section, the difference in the physics of the two realms is often that of an inverse relationship: a parameter that is x in one realm is $1/x$ in the other.

In the systematic hierarchy of the physical realm the wave aspect has the effect
that simple systems are condensed (atomic nuclei) while sophisticated systems
(human body) are diffuse. If the wave aspect of the spiritual realm has the opposite
effect—the simple atoms are diffuse while sophisticated humans are condensed—simple systems would be so different in the two realms that they would be far
apart and not influence each other.

As systems become more sophisticated in the same way, another possibility of
crossover between the two realms emerges, the phenomenon of resonance. Since
we discussed this for waves in general in *Volume Two,* we will just mention an
example here—the diamond allotrope of carbon. The adamantine quality of dia-

mond is not because it is difficult to disturb the bond between just two carbon atoms on the surface. The carbon atoms in a diamond are almost identical, and they resonate together as a single entity. An attempt to disturb the surface is akin to disturbing all the bonds in the crystal. Natural carbon is almost entirely the isomer carbon-12, but there is a small fraction of it that is carbon-13. This 'impurity' does not resonate exactly with the others which reduces the overall unity and stability of the giant molecule. It is theorized that when the technology to create pure carbon-12 diamonds emerges, they will be harder and even able to scratch natural diamond.

Sequential expression of Logos

We discussed the Logos—the laws of physics, chemistry, biochemistry, etc.—a construct in the Abstract Realm. The natural law determines the internal aspect of systems—an aspect that can only be described by complex numbers (e.g., the wavefunction of a photon or the orbital of an atomic electron). This internal aspect is in the Abstract Realm, but it is amphibious in that it has a projection in the Physical Realm known as quantum probability. This quantum probability aspect of a system determines over time the probability density of the fundamental entities that are at the lowest level in the Physical Realm hierarchy—mainly electrons, quarks and photons. This is the external aspect of a system, its external form in the Physical Realm.

While the Logos, as a structure in the Abstract Realm, does not operate with a cause-and-effect relation between the lower and the higher levels in the hierarchy, the structure of the Logos is expressed first at the lowest levels and only later at the higher levels.

In *Volume Two*, we discussed the cosmological history of the Physical Realm and the stepwise expression of the Logos, starting with the nature and origin of the fundamental particles in

LOGOS	Subatomics	Atoms	Molecules	Polymers
Systematic expression as history of Physical Realm				polymers
			molecules	molecules
		atoms	atoms	atoms
	subatomics	subatomics	subatomics	subatomics

the Big Bang, then moving on to the nature and origin of the elements in the first generation of stars, and concluding with a discussion of a third-generation star, our sun, and the earth, and the nature and origin of the pre-life molecules on the early, and abiotic, earth of four billion years ago.

THE SPIRITUAL REALM

Up to this point, we have discussed the systematic hierarchy of the abstract realm from mathematics to the Logos, and the systematic hierarchy of the physical realm that unfolds from simple to sophisticated under the guidance of the Logos. We turn now to the third realm about which very little is known to science, the realm of spirit.

First, we will discuss the theoretical possibility, given modern physics, that another realm coexists with the physical realm. Is this consistent, given what modern science understands about the nature of reality.

Second, we will discuss the experimental evidence already uncovered in modern cosmology that a second realm actually shares the cosmos alongside the physical universe of galaxies, stars, planets and humans.

Theory

Modern physics has gone far beyond the classical concepts of time, space, energy and mass. While an invisible, intangible spiritual realm coexisting alongside the physical is inconceivable using classical concepts, as we shall see, the concepts established in modern science are remarkably amenable to its existence. We will start with the contemporary understanding of the spacetime metric.

Spacetime

It took the genius of Einstein to realize that space and time were so similar that in his Special Relativity they could be transformed into each other. In this experimentally-verified view, velocity through space and velocity through time are very similar when expressed in "natural units."

Velocity through space and through time, in natural units, ranges from zero to one, from absolute zero to the speed of light. In spacetime, the velocity is always 1, but this can be unequally distributed between space and time.

In our quotidian lives, our velocity through space is essentially zero in natural units, we rarely move at significant fractions of light speed in the absolute frame of reference proved by the cosmic microwave background radiation. We are moving through time at essentially light speed. While 1 centimeter and 1 second might both be considered equally small units of time and space in human units, they are vastly different in natural units, 1 second = 30,000,000,000 centimeters!

In high-energy physics, unstable subatomic particles can be accelerated to almost light speed—both absolute zero and light speed are asymptotes—and their prolonged lifespan observed. A particle that would normally decay in seconds at thermal speeds can have its half-life stretched into hours when traveling at essentially light speed. Traveling at high-speed through space involves traveling at slow speed through time, the relation being a simple Pythagorean one in natural units (and condensing the three space dimensions into one for simplicity).

$$1^2 = t^2 + x^2$$
$$1^2 = 1^2 + 0^2 \quad \text{zero in space}$$
$$1^2 = 0^2 + 1^2 \quad \text{lightspeed in space}$$

It can be seen that, at a velocity $\sqrt{\frac{1}{2}}$ light speed (~0.7c), the velocities through time and space are equal and the velocity through spacetime has been rotated by 45°. Increasing a linear velocity in space rotates the direction of the constant velocity through spacetime. Such a combination of linear and angular is the prime characteristic of complex numbers.

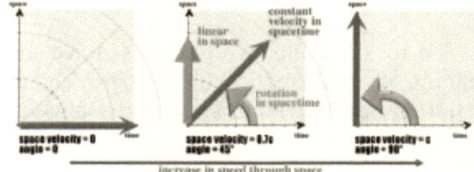

In this unified spacetime, the obvious difference between time and space showed up in the Pythagorean relation that measured the separation between events in spacetime. Two formulations of this worked equally well with a minus sign. This implied that either space or time had to be an *imaginary* number, i.e., a complex number with a zero *real* component. The other, time or space, being a *real* number, i.e., a complex number with a zero *imaginary* component.

In the early days of the science, it seemed obvious to assign space the real number and time the imaginary number. Later consideration made it clear that time behaved as the real numbers while space behaved as the imaginary numbers. For instance, the nonoriented twists of an elec-

$d^2 = t^2 - x^2$	$\{t, ix\}$
$d^2 = x^2 - t^2$	$\{it, x\}$

tron can be along either direction of the complex time axis—it is matter going in one direction and antimatter going in the other. An electron going the opposite way in complex time is a positron. Very different. This is like the different behavior of the real numbers, plus-one and minus-one.

Electrons, on the other hand, traveling east or west in space are different, but not by much. This is like the behavior of the imaginary numbers where the only difference between +i and −i is that of clockwise and counterclockwise, so similar that a different point of view will convert one into the other.

n	n^2	n^3	n^4
+1	+1	+1	+1
−1	+1	−1	+1
+i	−1	−i	+1
−i	−1	+i	+1

So, the currently accepted expression for the 'metric' in human units for measuring separation in physical spacetime is with time as the real complex component, and space as the imaginary complex component.

$$d^2 = ct^2 + (xi)^2 + (yi)^2 + (zi)^2$$
$$= ct^2 - x^2 - y^2 - z^2$$

We earlier mentioned that spacetime embraced four complex dimensions. The caveat is that physical spacetime involves only an asymmetrical subset of the components of four complex dimensions. The physical metric of spacetime is an asymmetric set of the components. Applying the simple principle behind $0 = +1 \,\&\, -1$, the remaining subset of components can be postulated as the metric of the spiritual realm.

We have already seen that the physical metric is left-handed on the internal level, so we can expect that the spiritual metric is right-handed on the internal level.

It is the physical metric that is twisted up in the positive energy quanta of tardyon bosons and fermions. We can

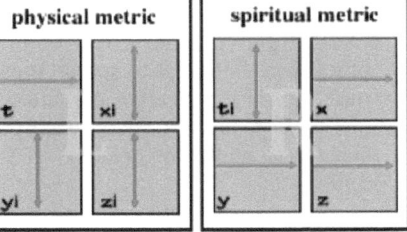

assume that it is the spiritual metric that is twisted up in the negative energy quanta of tachyon bosons and fermions.

Two solutions

The quantum theory that dealt with the internal wave aspect of matter was originally worked out for systems moving at everyday velocities, speeds that were a tiny fraction of the speed of light. It was Dirac who combined the wave aspect of quantum physics with the theory of special relativity for velocities approaching light speed and completed the description of relativistic quantum wave mechanics.

His equation for energy allowed for two solutions, in the same way that both +2 and –2 are both the square-roots of +4. The positive solution to the Dirac equation corresponds to the physical world, which is all about the behavior of quanta of positive energy. When the quantity p (related to momentum) is zero, this positive solution to the Dirac equation collapses into Einstein's well-known identity. The negative solution to Dirac's equation was an enigma. Investigation of the consequences of quanta

$$E^2 = c^2 p^2 + m^2 c^4$$
$$E^2 = m^2 c^4$$
$$E = mc^2$$
$$-E = -mc^2$$

of 'negative energy' revealed that such quanta, called tachyons, would have a very strange set of properties:

1.   Tachyons would have an energy measured by imaginary numbers [but energy, surely, was always positive and real] whose square would be negative.

2.   The velocity of tachyons would also have an asymptotic range from a *lower* limit of light speed to an unbounded upper limit. This was hard to comprehend inasmuch as the familiar limits to the velocity of quanta were the inverse, from zero to light speed. Tachyons with negative energy would have a negative mass and travel at light speed or faster. The familiar tardyons with real energy that compose the physical realm travel at light speed or slower.

3.   As the kinetic energy of a tachyon increases, the slower it moves; as its kinetic energy decreases, it speeds. It made an Alice-in-Wonderland sense that adding negative energy would slow things down.

4.   As a tachyon adds negative energy, it slows down asymptotically to light speed as its negative inertial and gravitational mass increases without limit. This is in contrast to tardyons that speed up asymptotically to light speed by adding positive energy as their positive inertial and gravitational mass increases without limit.

5.   The gravitational interaction between quanta of negative energy is to expand and increase their spatial separation. The gravitational interaction between quanta of positive energy is to contract and reduce their spatial separation, an antigravity.

6.   The gravitational effect of multiple tachyons of negative mass on each other would attempt to spread them out relatively uniformly in an open spacetime metric of negative curvature. The gravitational effect of multiple tardyons with positive mass on each other is to clump them together into a closed spacetime metric of positive curvature. This is an example of a 2-D 'space' with:

a) a closed, positive curvature is the surface of a ball
b) an open, negative curvature is the surface of a saddle.

Allow me to post an historical note here. For classical, commonsense reasons, the negative solution to Dirac's equation obviously could not possibly be taken at face value. Negative energy made no sense, so it was a relief when antimatter was discovered. This, it was decided, must correspond to the negative solution to the Dirac equation.   Negative energy must be positive energy but with all its properties reversed. With the discovery of the antielectron and the antiproton, this seemed a reasonable solution.

This perspective still holds even though later work in quantum mechanics revealed that anti-matter actually involved the same quanta just going in the opposite direction through complex time, made this an unreasonable conclusion. Antimatter is just as much a quanta of positive energy as is matter: both an electron and a positron have a positive rest-mass energy of ½MeV and when they annihilate, they mutually untwist into gamma ray photons with 1MeV total energy.

$$+1-1 \neq 2$$

The problem with this is that it is akin to plus and minus one summing to two, instead of the expected zero.

$$+1-1 = 0$$

Tardyons and tachyons

We shall take the two solutions to Dirac's equation at face value, and add it to our picture of the spiritual realm—tachyons interacting on the other side of spacetime, and the physical realm tardyons interacting on this side of spacetime. One interacts with gravity, the other with anti-gravity. So far, modern physics has been quite at home with the concept of a second, co-existing realm we are calling the spiritual.

In the physical realm, quanta of energy move at fractions of light speed, while in the spiritual realm, quanta move at multiples of light speed. The relationship between velocity in the two realms is a reciprocal one. What is v in one realm is 1/v in the other.

realm	physical	spiritual
logos	natural law	
systems	bottom-up	top-down
metric	t, xi L	ti, x *R*
energy	+E	−E
quanta	tardyons	tachyons
Adding energy	Speed up 1/v → 1	Slow down v → 1
asymptotic speed limits	0 → c	c → ∞
gravity	Collapse ⇓ ⇑	Expand ⇑ ⇓

The reciprocal function in math is a fascinating one in that it can take the infinity of integers and slot every single one of them into the space between zero and one. It does this remarkable feat not by packing them close together, as one might think—the integer 2 takes up fully half the available space—but by decreasing the distance at exactly the same rate as the integers are increasing.

A google at 10^{100} is an enormous integer, greater than the number of atoms in the visible universe, but the reciprocal function finds it a place close to zero, but still leaving room for the truly enormous integers, google^{google} and 1+ google^{google} to have a place.

We discussed earlier the systematic hierarchy of complex dimensions that comprise matter and natural law. We have just discussed the lowest four of these levels in terms of their components. The laws at each level on up, however, are structures in the full hierarchy of complex dimensions, so we can expect the same natural laws to apply in the spiritual realm and the systematic hierarchy founded on the tachyon quanta. The math involves switching plus and minus signs, and multiply and divide signs, but the observed realm ends up looking the same.

Cause and effect

The physical and spiritual realms appear to an asymmetric distribution of properties, somewhat akin to 0 = +1 & −1. The abstract realm is bidirectional cause and effect, while the physical realm is one-way, bottom-up cause and effect. So, we

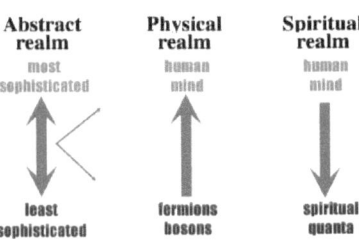

Abstract realm	Physical realm	Spiritual realm
most sophisticated	human mind	human mind
least sophisticated	fermions bosons	spiritual quanta

will suppose that the spiritual realm is also one-way but in the top-down direction, from most sophisticated to least sophisticated. The most sophisticated entity in the physical realm is the human mind, which can impose its form on the less sophisticated levels. This would account for the religious teaching that our character is reflected in our eternal spiritual body. We will return to this much later on.

Missing Cosmological Constant

In *Volume Two*, we discussed *virtual particles*, ephemeral quanta that lack a pixel of existence. They can be experienced directly by attempting to squeeze the N-poles of two strong magnets together. The distinct cushion experienced that resists such efforts is composed of a host of virtual particles.

Just like its twisted cousin, undisturbed spacetime has an internal probability aspect. The probability that the ground state, empty vacuum, will do nothing is very, very high. But it not *exactly* 100%. A speck of utterly empty vacuum of undisturbed spacetime has a infinitesimal 'probability amplitude' to decay into an electron-positron pair, and an even infinitesimally smaller probability to decay into a proton-antiproton or neutron-antineutron pair, etc. The duration these matter-antimatter pairs exist for is also infinitesimal, so they never ever amount to a pixel of existence (more technically, a Planck's Constant of action or energy-in-time).

So the probability that a volume of space will contain a virtual particle pair such as electron-positron with a transitory positive energy of 1MeV is very, very small indeed.

The problem arises because the Law of Large Numbers insists that even a very, very small probability becomes a certainty if extraordinarily immense enough numbers are involved. As an example, we can use some simple numbers. Let the the complex probability of a pixel of space spontaneously twisting into a virtual electron-positron pair be one billionth, with the observed external probability, as usual, being the square of this at one billion-billionth. The decay lasts only a billionth of a second before the particles recombine back into untwisted space without ever amounting to a pixel of existence. So, in a volume of space of a billion, billion, billion pixels—10^{18} pixels—the observed density of particle-pairs is, on average, one pair and the energy in that volume will be 1 MeV. This should show up as an observed energy density in empty space. This energy density is called the *cosmological constant.*

As discussed in *Volume Two*, space only seems continuous because it has a very high resolution. This is rather like the white screen of the computer I am writing on which is an illusion of the resolution of the tiny pixels. My eye is not able to resolve the tiny pixels present in equal amounts of bright red, green and blue, so I perceive a continuous white. While it is known from theory that space is not continuous but pixelated, the resolution is so high that our best instruments are incapable of resolving the separate pixels. While the resolution of my computer screen is ~10,000 pixels (10^4) / sq. inch, the resolution of space is ~a google (10^{100}) / cubic meter. So, even infinitesimal probabilities can become significant.

The actual numbers have been measured quite accurately and the expected energy density of virtual particles in empty space calculated. This is the expected cosmological constant, and it is ~10^{100} times greater that the observed energy density of empty space (which is essentially zero).

A solution to this glaring disparity between theory and reality was soon offered by an area of mathematics called 'group theory' that was very successful when organizing and predicting the host of 'fundamental particles' uncovered by modern experiment.

Group theory

A 'group' is a set of objects, e.g., numbers, that have the following properties:

1) They interact with each other—called an *operation*

2) The interaction always results in another member of the set—the set is *closed* under multiplication

3) There is an identity interaction in which there is no change—the *identity* operation.

Based on these three simple foundations, the astonishingly-sophisticated edifice of group theory is constructed. A very simple example of a finite group is the set of four complex numbers, $\{+1, -1, +i, -i\}$, that interact by multiplication. The results of multiplying any combination of them always results in a member of the set, which is closed as illustrated in the multiplication table. The identity operation is multiplying by +1, as seen in the first row and column. This set of four numbers is a group, and everything that has been proved in group theory will hold for this group.

×	+1	−1	+i	−i
+1	+1	−1	+i	−i
−1	−1	+1	−i	+i
+i	+i	−i	−1	+1
−i	−i	+i	+1	−1

Another representation of the same abstract group is a switch with four settings, and the operation is turning the switch. Its behavior is identical to the set of four numbers.

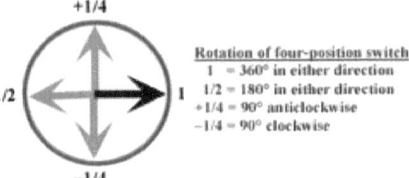

Rotation of four-position switch
1 = 360° in either direction
1/2 = 180° in either direction
+1/4 = 90° anticlockwise
−1/4 = 90° clockwise

Many of the subtle properties of groups are tied in with their subgroups, subsets of the set that are themselves groups under the same operation. Two subsets of any group are the 'trivial' subgroups: the identity element by itself—multiplying +1 an unlimited times is always +1 so it is closed and there is an identity element—and the subset that is the entire set which also obeys the rules. It is the nontrivial 'proper' subsets that have interesting properties.

In the above example, the two trivial subgroups are $\{+1\}$ and $\{+1, -1, +i, -i\}$. Like the group itself, they have a period-four—multiply any element by itself four times and you get back to the starting number.

The only nontrivial subgroup is $\{+1, -1\}$—any other combination, say $\{+1, +i\}$—is not closed under multiplication, so cannot be a group. Even this simple example of subgroups illustrates the emergence of new properties—in this case, the emergence of period-two subgroups that was not a property of the period-four original group.

The rules connecting the properties of groups and the properties of subgroups are precise and well-understood. One of the basic rules governing groups and subgroups is obvious with a little reflection: The period of a proper subgroup is a factor of the period of the group. Perhaps less obvious is that this also implies that a group with a prime-number period cannot have any proper subgroups at all.

Supersymmetry

Without going into any further detail, we will just state that sophisticated group theory was applied to the plethora of things emerging from the smash-up of very high-energy particles colliding head-on.

All the menagerie of subatomic particles could be placed in a closed set because interactions between them always resulted in particles that were members of the set. The barrier to this set being a group was that a single nonoriented particle never turned into a single oriented particle, or vice versa. The two sets had all the appearance of being subgroups of a larger group in which they could interconvert. Adding to the known set of particles another set of theoretical particles, known as 'supersymmetric' particles, would complete this all-encompassing group. Each regular particle has its supersymmetric twin in this larger group. The table is a partial listing of the names they were given.

Particle	SS twin
neutrino	sneutrino
W boson	wino
electron	selectron
quark	squark
photon	photino
gluon	gluino

This set of particles, it seemed, could resolve the glaring factor of 10^{100} in the disparity between the cosmological constant as predicted from the results of many precise experiments and the observed cosmological constant. All that was needed was that the contribution of the supersymmetric particles exactly cancel out the contribution expected from the regular particles.

The first, rather obvious, requirement for such a cancellation is that a pixel of spacetime has an equally small probability of decaying, for an equally brief moment, into a virtual supersymmetric pair as it does for decaying into a regular virtual pair.

The second requirement is also simple if we accept the imaginary complex energy of Dirac's equation at face value, i.e., that the positive rest-mass energy of a regular matter/antimatter pair is equal to the negative rest-mass energy of a supersymmetric matter/antimatter pair.

In this case, the massive, positive cosmological constant of the regular virtual particles in the vacuum is cancelled by the equally massive, negative cosmological constant of the virtual supersymmetric particles in the vacuum.

The resultant cosmological constant would be zero by the simple math of $\left(+10^{100}\right)+\left(-10^{100}\right)=0$. A similar situation applies to each of the protons in our bodies, they are repelled by all the protons in the earth with enormous force, and attracted with an equally enormous force by the equal number of electrons in the earth. When this balance of enormous forces is disturbed, as it is in a thunderstorm, the huge energies involved in this balance is revealed.

Unfortunately, the concept of tachyons with negative energy had fallen out of fashion and a more complex picture was adopted: The supersymmetric particles had real, positive energy and their cancellation somehow occurred because the supersymmetric twin of an oriented boson was a nonoriented fermion, and vice versa. The twin of the doubly nonoriented electron was the doubly oriented photino, and so on.

This theoretical consideration has given rise to the hopeful expectation that such supersymmetric particles will come flying out of the ultra-energetic collisions created in the CERN synchrotron, the LHC, on the border between France and Switzerland. There has been no sign of them yet at energies of tens of billions eV.

Energy quanta	Physical	Spiritual
Nonoriented: *spin ½* *+ EM charge ½ ½* *+ color ½ ½ ½*	neutrino electron quark	sneutrino selectron squark
Oriented: *spin 1* *+ EM charge 1 1* *+ color 1 1 1*	woson photon gluon	swoson sphoton sgluon

This is to be expected if the supersymmetric particles are negative energy, rather than the positive energy being pumped in by CERN, and are the basic building blocks of system-building in the spirit realm in its top-down fashion. In the physical realm, the quanta of positive energy are twists is the physical metric; in the spiritual realm we can expect that the quanta of negative energy are twists in the spiritual metric.

We again find that modern physics can readily accommodate a second realm without effort. Summarizing these theoretical discussions in the table, we now move on from theory to experiment.

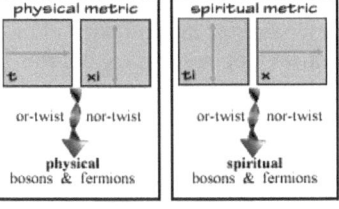

We will assume that the quanta of negative energy are twists in the three spatial dimensions, and that the twists can be either oriented (the bosons) or nonoriented (the fermions.)

That these spiritual quanta do not interact with the physical quanta is not so unusual: the neutrinos, even through they are a part of the physical realm, hardly interact with with anything and sail just as easily through you, the earth, or the sun without interacting. It is, therefore, a problem of philosophy as to whether neutrinos are insubstantial or substantial physical matter.

As the same laws are acting in both realms, albeit in a complementary fashion, we can expect that there is a partial resonance between the two realms on an internal level, that is a resonance of 100% in a human who is in tune with the Logos.

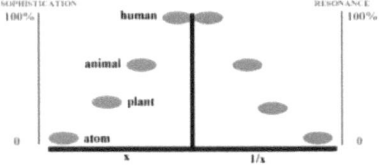

The spiritual quanta composing such a spirit-self resonating with the Logos, would be tachyons of low negative energy spinning *right,* and moving at an unbounded speed, close to the infinite limit, on the other side of the spacetime metric. The spiritual quanta, on the other hand, composing a spirit-self that did not resonate with the Logos, would be tachyons of high negative energy spinning *left,* and moving at a slow speed, close to the light speed limit, also on the other side of the spacetime metric.

The systems in heaven are fast, hot and extended on the other side of spacetime; the systems in hell are slow, cold and condensed on the other side of spacetime. Almost all the great religions teach that. corresponding to the healthy physical diet, there is a spiritually healthy *diet* of right and good actions that involve simple things like:

1. True parental love in the family, true conjugal love between spouses, and education of children in true filial love and sibling love.

2. Live for the sake of others, give and forget you gave.

3. Forgive and forget the mistakes of others. Teat strangers as you would members of your own family.

4. Do not be a foolish immortal. Try to take the big, long viewpoint and don't sweat the small stuff.

The major religions also teach that, just as a rotten physical diet will have unhealthy consequences for the physical self, a rotten spiritual *diet* of self-centered, unloving actions will have unhealthy consequences for the physical self.

It will only be towards the end that we will be in a position to discuss why the vast majority of humans are not in tune with the Logos and do not resonate with the spiritual world except partially in dreams and the like.

The chart summarizes how current theory in physics readily allows for a second, complementary realm.

realm	physical	spiritual
logos	natural law	
systems	bottom-up	top-down
metric	t, xi L	ti, x **R**
energy	+E	–E
quanta	familiar tardyons	supersymmetric tachyons
Adding energy	Speed up $1/v \to 1$	Slow down $v \to 1$
asymptotic speed limits	$0 \to c$	$c \to \infty$
gravity	Collapse	Expand
Cosmological contribution	$+10^{100}$	-10^{100}

This concludes an overview of the theoretical possibility in modern physics of a second substantial realm in the cosmos. With a minimal manipulation of what is already known about substance in the physical reality, we see that modern science is quite comfortable with a second realm coexisting alongside the physical in the cosmos.

Experiment

It is one thing, of course, to state that a spiritually-substantial realm is *theoretically* possible, while it is quite another for scientific experiment to demonstrate that this is actually the way the universe is. Our attempts to understand the structure and function of the natural world began with the stuff that was easily accessible through our senses. The start of what is now modern science studied the three familiar phases of matter—the solids, liquids and gases to be found in the substantial material of the everyday world.

The study of the night sky suggested that there were things beyond the earth that obey a different set of principles—the Moon and planets seemed to move without friction and the sun poured out endless amounts of energy—but Newton was the first to show that the same principle that governed the falling of an apple could also explain the movement of the Moon, and so much else.

It was the exploration of the heavens that revealed that the familiar solid, liquid and gas was actually a minor component of the universe: that most of the visible matter in the universe, contained in the myriad stars, was actually in the form of a *plasma*—a form of matter in which are no atoms, just high-energy free electrons and atomic nuclei flying about freely. A roughly equal amount of almost invisible matter was later found to be distributed throughout the vast spaces between the stars and galaxies in an ultra-low density of gas that was a better vacuum than those created in the most high-tech laboratories.

Both the stars and interstellar gas, however, had a similar makeup: both were composed of 75% hydrogen-1, 24% helium-4 and 1% all the other elements combined. All were made of the same basic stuff: the 'baryons' of the atomic nucleus (protons and neutrons) and the electrons. Since the electrons contributed less than $1/2000^{th}$ to the mass-energy, these components are together called 'baryonic matter.' The amount of baryonic matter in the visible universe is measured in units where our sun has a mass-energy of 1 SU: the visible stars in our home galaxy have a mass of $\sim 10^{10}$ SU, as does the interstellar gas and dust. This is all positive energy, which gives the galaxy a powerful force of gravitational attraction.

A further component to the universe was added with the discovery of the Cosmic Microwave Background, the discovery that all the vastness of interstellar space is filled with of microwave photons. Their numbers are huge: for every baryon in the universe, there were 10^{11} microwave photons. Each photon had a positive energy and thereby also contributed to the gravitational field. But, since each photon had a very small energy, and since the CMB photons were evenly distributed throughout intergalactic space, their contribution could be ignored.

Dark Matter

This 'baryonic' did not stand up to scrutiny when the rotation motions of galaxies were studied in detail. There are two extremes of rotational motion about a center: a solidlike rotation in which the rotation of the parts is constrained, or a gaslike rotation where the is no constraint.

An example of a solidlike rotation is two markers at different distances from the center of a phonograph turntable. The outer marker moves faster than the inner, and they both make one rotation in the same time period. The constraint is the electromagnetic interaction which makes the turntable rotate as a solid unit. In such solidlike rotation, the outer marker goes at a faster speed than the inner one. As the

$$v \propto d$$
$$v = kd$$
$$1 = 1$$
$$2 = 2$$
$$3 = 3$$

distance that each travels with one rotation is $2\pi d$, a marker that is twice the distance from the center will travel at twice the speed. The velocity is proportional to the distance.

A example of a gaslike rotation is the orbits of the satellites rotating about the earth: The international space station orbiting hundreds of miles up and just above the atmosphere moves very fast and orbits the earth 19 times each day. The geosynchronous TV satellites that orbit thousands of miles up move much slower and take 24

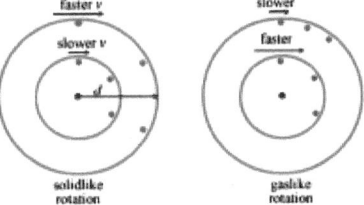

hours to make a complete circuit. The Moon which is hundreds of thousands of miles up moves even slower, and takes a whole month to make a single circuit about the earth. There is nothing constraining them except the inward gravitational tug of the earth keeping them in free-fall. This is the situation in a gas where there are no constraints on the movement of molecules.

This gas-type of rotation also has a defined relation between speed and distance of the markers: the cube of the velocity is proportional to the square of the distance—as the distance increases, the velocity rapidly decreases. To a high accuracy, the satellites orbiting the earth, and the planets orbiting the sun all have a gaslike rotation.

$$d^2 \propto \frac{1}{v^3}$$

$$d = k\sqrt{\frac{1}{v^3}}$$

1 = 1
2 = 0.35
3 = 0.19

When the techniques were developed that could measure the velocities of the stars in the galaxy as they all rotated about the billion SU black hole at the very center, it was expected that the stars at the very periphery would be moving with a speed about the center that has a gaslike relation to the speed of stars much closer in.

This expectation was found to be incorrect; the speeds were found to be almost solidlike, and the galaxy rotated almost as a solid disk. The only explanation was that the galaxy was embedded in the center of a rotating invisible mass that was so much vaster in extent than the inner and outer stars that they were all roughly the same distance from the center. If the radius of this rotating invisible mass is 100, and the radius of the galaxy is 1, then the difference in speed between the peripheral and inner stars would not be that different.

A second surprise emerged when the behavior of clusters of galaxies was explored. Our home galaxy, for instance, belongs to the Local Group, with Andromeda as a similar-sized galaxy and about a dozen smaller-sized galaxies such as the Magellanic Clouds. Clusters and superclusters can have thousands of galaxies in them.

The dynamics of the gravitational attraction between galaxies, however, could only be explained if each one of them had a mass that was 10 times bigger than the baryonic matter in them. For instance, the mass of all the stars in the home galaxy is 10^{11} SU, as is the mass of the tenuous interstellar gas and dust, giving a total mass of 2×10^{11} SU. The same applies to Andromeda. The dynamics of the Local Group, however, can only be explained if the Milky Way and Andromeda both have a mass-energy of 20×10^{11} SU that extends 10 times as far as do the galaxies' visible components. This invisible-to-our-instruments mass is called 'dark matter. The home galaxy is at the rotation axis of an enormous halo of dark matter.

Every galaxy is similarly endowed, and galaxies are often strung out in fuzzy lines along great webs of dark matter, with superclusters where the lanes cross. Between the lanes of dark matter, the holes in the web so to speak, are the Great Voids that span hundreds of millions of light-years with hardly any visible matter within them.

Yet again, a readjustment had to be cosmology to the recognized constituents of the physical realm: Baryonic matter is a ~10% minority in a physical universe whose large scale structure is governed by the 90% majority of dark matter.

Identity of dark matter

At the current time of writing this, the identity of dark matter is a known unknown. A few things, however, are expected to hold for dark matter.

1.   Dark matter is positive mass-energy and has the usual attractive gravity.

2.   All energy is quantized. It comes in integer amounts, so dark matter is quanta of positive energy that has a complex wavefunction that determines its history-over-time.

3.   It is expected that the quanta of dark matter are either already known or at least compatible with the current Standard Model of subatomic physics.

4.   The quanta must not interact with any type of photons; it must be 'dark' at all radiations from radio waves to visible light to gamma rays.

5.   It is expected that the density of dark matter at the center of the halo is greater than at the periphery, so the Milky Way, and hence our laboratories, is embedded in a higher-than-average number of dark matter quanta. Yet no experiment as yet performed has been sufficiently sensitive to notice this plethora of particles. The quanta of positive energy can interact with regular matter by gravity, but not in detectable amounts by the strong, the electromagnetic, or the weak fundamental forces (coupling with bosons).

6.   There are two basic possibilities for the dark matter quanta: Either there are relatively few of them, each with a large rest mass-energy, or there are a numerous host of them, each with a tiny rest mass-energy.

7.   The current theory of the Big Bang origin of baryonic matter must also explain the prevalence of dark matter.

Neutrinos

The neutrino, in all its forms, satisfies all of these requirements, except for one caveat we will discuss at the conclusion.

The neutrino is a single nonoriented twist in spacetime, a quanta of positive energy that is so small it is difficult to measure. The rest mass-energy of a twist-along-time in one spatial dimension, the electron-neutrino, is <1 eV; the rest mass-energy of a twist in two dimensions, the muon-neutrino, is probably >5 eV; and the rest mass-energy of a twist in all three dimensions, the muon-neutrino, is probably >15 eV.

In comparison, the rest mass-energy of an electron is ~500,000 eV and that of a nucleon is ~900,000,000 eV. The baryonic matter in the universe is ¾ hydrogen (with 1 nucleon) and ¼ helium (with 4 nucleons). If the mass of dark-matter neutrinos is to be 10 times that of the baryonic mass, there has to be a host of them that outnumber each nucleon a billion fold. For every hydrogen atom in the Milky Way, there would have to be billions of neutrino's for them to comprise the dark matter. Is it possible that, for every molecule of air, there is a haze of a billion neutrinos, but no one has noticed it? This is quite possible.

The neutrino can only interact with other quanta by coupling with virtual weak bosons. These have such an enormous mass-energy of 70,000,000,000eV that the quantum probability of a neutrino coupling with one while passing near an electron or quark is infinitesimal. A neutrino could sail through 10 light-years of solid lead with a high probability of not interacting with anything.

The neutrino is utterly indifferent to photons except when they ever-so-briefly decay into a virtual electron-positron pair. It is also utterly indifferent to gluons, except when they ever-so-briefly decay into a virtual quark-antiquark pair.

So, as far as neutrinos are concerned, there is no difference between sailing through intergalactic space and sailing through the earth, and a billion fold haze of neutrinos would be currently quite undetectable. A neutrino approaching the earth will accelerate in the gravity well, and pass through the center of the earth at high speed. As it climbs up out of the gravity well, it will decelerate and leave the earth's opposite surface at the speed it entered. When it is far from the earth, it will have the speed it originally had before the earth exerted any influence.

To complete the neutrino's case for being dark matter, a billion fold excess of neutrinos in the Big Bang is almost required by the Standard Model of subatomic particle physics.

Negative Energy

In fact, for many years the great debate in cosmology was about the overall curvature of the universe and its eventual fate. There were three possibilities:

1. An Open universe has a negative curvature and the expansion continues while getting incrementally slower. There is not enough positive energy—bosons, photons and dark matter—to completely slow down the expansion.

2. A Flat universe has a zero curvature and the expansion eventually stops, but it takes an eternity to actually get there. The expansion slows asymptotically to zero as there is just enough positive energy to stop the expansion given an eternity. The fate of the Open or Flat universe is essentially the same. This is not so for the third possibility.

3. A Closed universe has a positive curvature and the expansion eventually stops. There is more than enough positive energy to completely stop the expansion. Like a ball thrown upwards, the universe does not stop there; it starts to contract. The CMB starts gaining, not losing, energy and the temperature starts to rise. Cosmic history would repeat itself as the universe headed towards the Big Crunch. From there it could rebound with another Big Bang and so the cycle could continue again and again without any loss during the cycle—there wasn't anywhere to lose it to (and who is to say whether this was the first time around or not). It was this possibility that motivated the measurement of the historical rates of expansion.

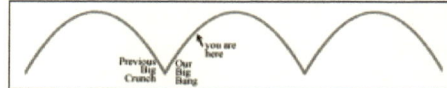

Measuring the expansion

All of these fascinating speculations had to be abandoned when sensitive techniques were developed that could peer into past eras and actually measure the expansion rate as the universe matured.

It should be remembered at this point that, while all the universe is the same age, the cosmologically-slow speed of light means that the farther away things are, the younger is the stage that we observe them at. For instance, a couple of billion years after the Big Bang (~10 billion years ago) was the epoch of galaxies forming, each with a massive black hole at the center. So massive is the energy output during the formation of the black hole that we can observe them as quasars at a distance of 10 billion light-years, while the protogalaxies are hardly visible. (They are quasi-stellar objects because, at that distance, they appear as points, as do most stars even when viewed through the most sensitive instruments.)

If we could instantaneously teleport an observer to one of the quasars, however, he would find himself in a quiet, mature galaxy much like our own. Moreover, looking back across the void at the Milky Way galaxy, he would observe it in early stages when the central galactic black hole was forming and it would appear to be a quasar.

The result of such peering back into the expansion history of the universe was quite unexpected. This is what was found:

1. For the first third of the universe's history, the first 'trimester' of 5 billion years, it behaved as expected, the rate of expansion fell off with time.

2. In the second trimester, the expansion was constant and unchanging.

3. In the third trimester, our current era, the universe started to expand at an ever increasing rate.

A curve of this shape can be obtained by adding two curves together, the (natural) log function and the exponential function. The log function starts off fast but has an exponentially decreasing rate of growth similar to that of a universe of positive energy with a closed, positive curvature. The exponential function, on the other hand, starts off slow but has an ever-increasing rate of growth. The sum of the two creates a curve similar to that of the universe's rate of expansion. Because it is not possible for positive energy with gravity to increase the rate of expansion, cosmologists realized that there was another component to the universe made of negative energy with an antigravity action.

This negative energy is currently called Dark Energy, and since it is currently overwhelming regular matter that is trying to slow down the expansion, there is more negative energy than positive energy in the universe. To my mind, the term *Dark Energy* is so easily confused with *dark matter* that it is better dropped. While dark matter and dark energy are both dark because neither interacts with photons, the great difference is in the gravitational effect. For this reason, we will call it negative energy to make this distinction clear.

Current techniques of detecting dark matter are so crude that we have little idea as to how this dark energy is structured, but it seems to be relatively uniformly distributed throughout all of space. The best contemporary estimate of the current composition of the universe is that the physical realm of positive energy amounts to 30%, and that the realm of negative energy accounts for the remaining 70%.

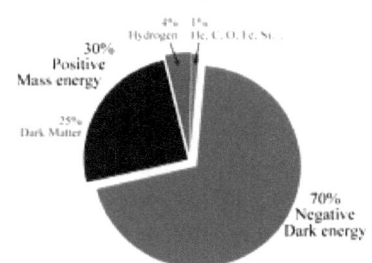

Quantized energy

Some thinkers have proposed that this negative energy component of the universe is a continuum. As noted earlier, this goes against the very core of quantum theory that all energy is quantized. Unless shown otherwise, we can assume that any negative energy in the universe will be quantized, and that such energy will appear in one of the ways in which the 'other' metric can be twisted. As noted, quanta of negative energy will be either tachyon particles or speed-of-entanglement couplers moving, as we have suggested, in the spiritual metric that complements the physical metric.

We suggest that the quanta of dark energy are the fundamental components of the substantial spiritual realm, and that its structure fills all of space. This behavior is the exact opposite to that of physical energy that clumps into suns which, while large from our parochial point of view, are small on the scale of interstellar and intergalactic space.

As mentioned earlier, our physical substantial bodies are composed of entities that are moving at high-speed. Electrons move at thousands of miles a second while photons move at light speed. Yet, when they are confined by the internal quantum wave, they result in bodies that have small relative motions.

In a similar way, to suggest that the substantial spirit body is composed of elements that move at greater than light speed is not to suggest that spirit bodies are zooming around at high speed. When confined by the spirit mind, the tachyon fundamental components of the spirit body can also make up a body at rest.

So, while spirit 'atoms of solid' have constituents moving at multiples of light speed, the wave confines them into structures that can be at rest, albeit ones that are much more extended

One Universe, Two substantial realms							
small condensed structures emergent substantial physical realm				vast extended structures emergent substantial spiritual realm			
30% Dark & regular matter regular tardyon particles quanta of positive energy				70% Dark Energy SUSY tachyon particles quanta of negative energy			
xi	yi	zi	t	x	y	z	ti
physical metric				spiritual metric			

than those in the physical world. If there is a second substantial realm, we will, as the simplest of assumptions, expect it to also have a similar structure to that of the physical realm:

1.   Spiritual systems are a hierarchy of interacting subsystems.

2.   The foundation of the hierarchy is a set of fundamental spiritual entities.

3.   The fundamental spiritual entities interact with each other, but they do not interact with the fundamental physical entities

4.   The fundamental spiritual entities are quanta of an energy that is not identical to physical energy. The quanta are twists in a spacetime that cannot be identical to physical spacetime.

5.   The spiritual quanta are confined by a progressively more-structured and intricate internal wave.

6.   The spiritually-substantial form of any system over time is a reflection of the form of its internal wave.

7. We shall tentatively suggest that the internal wave aspects of the physical and spiritual are identical.

Substantial experience

From all accounts, the subjective experience in the substantial spirit world is remarkably similar to that experienced here in the physical realm. If our thesis is correct, then the substantial spirit body is composed of tachyons interacting by speed of entanglement couplers all confined by a quantum wavefunction. We shall now examine whether this could possibly be the basis for a subjective experience of a substantial world.

To see that this is possible, we start with an overview of our subjective experience of the physical world as being substantial and solid. Even the most erudite of scientists has difficulty experiencing everyday life as a being of quintillions of quarks and electrons all coupling with photons and gluons, confined by a quantum wavefunction. While this is the reality, we do not experience it directly; we do so through our senses. Our physical senses are triggered by a multitude of electromagnetic interactions with the objects around us. The quality of these interactions is converted into digital information that is sent along nerves to the brain where it is further digitally processed. The results of this digital processing are converted back into analog form which then merges with our mind.

A very similar sequence can be proposed for the subjective sensing of the spirit world as a substantial realm just by replacing physical particles with spiritual particles; the rest is the same. In either case, there is nothing really 'solid' there but, in our mind. our experience is of substantial solids.

Two beings composed of sub-luminal particles in the physical metric experience a handshake just like two beings composed of super-luminal particles in the spiritual metric. In either case, the experienced solidity of the handshake is all in the mind.

Creation of Spirit Realm

Unification Thought is not specific about the order of creation. Some think that the spirit world was created first and the physical world created later on that foundation. The process of origin-division-union, however, suggests something more elegant; that creation involved Nothing being separated into two complementary realms at the same moment. In this thesis, we suggest that four of the eight components of complex space-time were twisted into a left-handed configuration—the physical metric—while the remaining four were twisted into a right-handed configuration constituting the spiritual metric. This would account for the distinct left-handedness of particles and, it is thought, the excess of matter over antimatter in the physical realm—a fact otherwise to explain in current theories.

> It was a revolution whose implications are not completely resolved today. Wolfgang Pauli captured the idea best in a phrase calling God 'a weak left-hander!' He initially used those words to dismiss the results, and did not accept the fall of parity until he read papers produced by physicists who confirmed [the] result in various other experiments.[147]

We can speculate that the left-handed metric at the moment of Creation was the tiny bit of 'false vacuum' popular in current cosmology. This Planck-sized speck inflated rapidly creating the space of the physical universe. This exponential inflation was abruptly slowed by the strong force which converted the inflation into positive energy, the hot Big Bang. A more sedate expansion continued, matter and antimatter annihilated, and the remnant matter eventually cooled and condensed into galaxies, stars and planets.

> False-vacuum-dominated inflation is dramatically different from the usual true vacuum case, both in its cosmology and in its relation to particle physics.[148]

We can speculate that the right-handed entity followed a quite different history. It doubled in size at a constant sedate rate and has been doing so ever since. It seems that the (moderate) inflation phase of spirit-world history never ended and never will.

The inflation phase of the physical world was very rapid and stretched the false vacuum at a constant energy density: at the strong braking, this accumulated energy was dumped into the energy of the hot Big Bang and has been a constant ever since.

For the first few billion years, there was a preponderance of positive physical energy in the universe and this slowed down the residual expansion. The spirit world, with its much slower but never-ending inflation, added negative energy to the universe at a constant rate. Since the positive energy in the universe is a constant, at some point in time the amount of negative energy equaled and then surpassed the constant positive energy. This apparently happened ~5 billion years ago, ~8 billion years after the Big Bang. At this point, the decelerating influence of the positive energy on cosmic expansion was replaced by the accelerating influence of the excess negative energy.

Theology suggests that the modest inflation phase of the spirit world is eternal, that it will never stop as did the violent inflation phase of the physical realm.

So, we conclude that current theory is quite comfortable with two substantial realms—one an emergent property of wave-confinement of positive energy quanta in the physical metric—the regular particles—and the other an emergent property of wave-confinement of negative energy quanta in the spiritual metric—the SUSY particles.

The physical realm of our everyday lives includes qualities of real *substance*—e.g., weight, color, sound, touch, etc.—and qualities of abstract *mind*—e.g., love, generosity, hate, jealousy, etc. Since we have postulated that the physical and spiritual realms are complementary, it suggests that, in the spiritual realm, the reverse is the case. Qualities of substance are love, generosity, hate, jealousy, etc., while qualities of mind are weight, color, sound, touch, etc.

One Universe, Two Realms

When the physical world alone is taken into account, the universe is almost empty. The 100 billion stars in our home galaxy, for instance, take up a negligible amount of space even when their solar systems are added into the calculations. Over 99.999% of the galaxy has no physical structures in it except for wisps of gas. Why all that 'wasted' space in God's creation.

This is an even more pressing question when we look beyond the galaxy, inasmuch as there is even more seemingly wasted space out there. This is not immediately obvious when looking into the night sky, which seems quite populated. Almost everything we see there with our naked eye, however, is local and belongs to the home galaxy, the Milky Way.

If all the other galaxies in the universe disappeared, we would notice no difference in the night sky to the naked eye other than the absence of a small hazy patch in Andromeda. This is the Andromeda Galaxy with 100 billion stars and is as big as the Milky Way. But it is almost invisible to the naked eye for, while it is huge and packed with stars, it is 4 million light years away—a vast, empty distance. All the other billions of galaxies God has scattered across space are far too distant and far too faint to be seen without a telescope. Even with the most sophisticated telescope, their domain is limited to the galaxies in the *visible* universe, since the universe is not yet old enough for light from the rest of the physical realm to reach us here on earth.

Intergalactic space is vast and physically empty. A simple calculation reveals that the regular matter in the physical universe (mainly in the form of stars) takes up only 0.000,000,000,000,000,000,000,000,0002% of the volume of the visible universe[149].

Now a fundamental axiom of Unification Thought is that the universe is designed, and in fact, is specifically designed for human beings. So there must be a reason for all that vast space, especially as it seems to otherwise conflict with the 'home for Man' concept. Did God create all this empty vastness just to make travel between the stars and galaxies through the hostile-to-life vacuum difficult for Man? This does not make sense in light of the following reasoning based on the Divine Principle:

God's plan was for Adam & Eve not to Fall. If they had not fallen, they would have multiplied a family of True Love—presumably quite a large family. Their descendants would also multiply such families of True Love. Assuming large families as the norm, it is not unreasonable to expect that the human race would double in size every 25 years or so, since war would be unthinkable and disease would be controlled far better than it is today.

The population by year N00 would be 2^{4N} and the population would reach the current levels of 10 billion in just 800 years after the first humans! Given that 100,000 years have passed since Adam, the current human population would be $2^{4,000}$ or, roughly, a trillion trillion trillion trillion trillion trillion billions. Clearly the earth would NOT be room enough for such a vast population, and either stringent birth control would ensue or, God having planned ahead, a way to expand to all the billions of earth-like planets out there would be easily found. One way, of course, would have been to have all the other planets close at hand: but they are not.

As birth-control is not DP-compatible, we can assume that there is a easy way to reach other earth-like planets. The best possibility for such easy intergalactic travel that I can envisage based on current knowledge involves the phenomenon of entanglement. An entangled pair of particles have an instantaneous connection that can span the universe. Many phenomena over the past 13 billion years have created such pairs of particles that have since been separately cruising through the universe at light or sub-light speed.

The Earth (or better, the Moon without an atmosphere to destroy the entanglement) is bombarded each second by billions of entangled particles whose partners are spread across the visible universe.

A technology based on this gift from God could open the way for travel between the stars and galaxies. Admittedly, science has just started to explore this phenomenon so it is a ways off yet. At this point, however, it is instructive to remember a legend from the birth of the electric age. In the mid 1800s, William Pitt, the Prime Minister of Great Britain, was touring the laboratory of the Royal Society when he encountered Faraday playing around with wires and magnets. "Of what possible use is all this?" he politely enquired. "I have no idea," replied Faraday, "All I know is that you will be taxing it in 50 years."

As Father Moon has often pointed out, the entire physical universe was created as the home for True Love mankind. So, such speculation is not unreasonable based on both science and religion.

Just an infinitesimal fraction of all the vastness of space is used for life in the physical world involving the stars and planets. In this paper we suggest that all this vastness is not wasted space; it is where the spirit world resides. It is certainly parsimonious to suggest that, rather than wasting all this vast space, God had a purpose for such vastness to contain the spirit realm. We have one universe containing two complementary substantial realms.

> "Do you know how infinite and unchanging the spirit world is? We are now aware of the vastness of the universe. The universe is over 22 billion light years across. How big is that? Light travels 300 million meters in a second. In one second, light can go around the Earth seven and a half times. The distance light covers at that speed in one year is called a light year. Light takes 22 billion years to cross the universe, not 200 days. So how vast is the universe? The entire universe is the stage for our activities…." [150]

CREATION OF THE PHYSICAL REALM

In *Volume Two*, we discussed the Big Bang in detail. Following is a summary of the sequence of events that took place in the Big Bang.

Time Zero

In the beginning, there was an abstract hierarchical structure of a finite number of complex dimensions. The four dimensions at the foundation of the hierarchy were asymmetrically separated into two orthogonal chiral constructs, the L-physical metric and the R-spiritual metric. The structure in the remaining upper dimensions is the Logos, the natural law. The two realms followed an external history determined by the Logos.

Exponential Inflation

In *Volume Two*, we discussed the pixelation of space—the *Planck Length*, pL, measuring ten trillion, trillion, trillionths of a meter—and the pixelation of time—the *Planck Time*, pT, measuring ten billion, trillion, trillion, trillionths of a second. We also discussed the physical limits on temperature that can only be approached asymptotically by regular matter. The lower limit of temperature is called *Absolute Zero*, 0°, while the upper limit of temperature is called the *Planck Temperature*, pK, one hundred million trillion trillion degrees Kelvin or 1.4×10^{32} K which corresponds to thermal photons with the Planck energy, pE, of 1.2×10^{28} eV = 1.2×10^{19} GeV.

The p-metric construct that emerged from Time Zero is called the 'false vacuum.' It had an extension in the spatial dimensions of 1 pL and an extension in time of 1 pT. This pixel of metric was at temperature pK, and the energy amounted to the Planck Mass of ~1 microgram. The inherent Leftness gave the false vacuum a negative pressure, which drove an exponential inflation of the spatial dimensions. A separation of 1 pL became 2 pL each pT.

The era of inflation lasted until ~10^{-36} seconds after time zero, a period of 100 million pT, during which the doubling occurred 100 million times. A separation of 1 pL, in this brief instant, became $2^{100,000,000}$ ~$10^{30,000,000}$ pL. The tiny speck of physical metric became enormous, on the order of $10^{29,999,965}$ meters = $10^{29,999,949}$ light-years

In natural units, light speed is 1 pL/1 pT, so during this time light speed influences could only travel 100,000,000 pL = 10^8 pL, which is infinitesimal compared to the exponential distance. The contents of the false vacuum were separated at hyper-luminal velocities.

Creation of energy

This hyperinflation did not have any consequences due to the weak and electromagnetic interactions between the quanta in the false vacuum because both decrease in energy as separation increases. The gravity of the initial μgm. of energy is negligible and also falls off with separation.

The color interaction between the quarks and gluons, however, has the rather counterintuitive property of increasing with separation. While the energy in the electromagnetic interaction is *inversely* proportional to the square of the separation, the energy of the color interaction is directly proportional to the sixth power of the separation.

At the Planck Temperature, the two forces have the same energy, and we can set the energy of both interactions at 1. At the end of inflation, the energy of the electromagnetic interaction will be infinitesimal, while the energy between the separated quarks and gluons will be cosmically enor-

	$t = 0$	$t = 10^8 P\sec$
E_{EM}	1	$\dfrac{1}{\left(10^{60,000,000}\right)^2} = 10^{-120,000,000}$
E_{QC}	1	$\left(10^{60,000,000}\right)^6 = 10^{360,000,000}$

mous. The impetus of inflation flashed over into positive energy at almost the Planck Temperature. This enormous positive energy had an equally-great gravitation that opposed the inflation, and the exponential increase was braked to a more moderate expansion rate.

The inflation slowed from an exponential rate to an ever-decreasing rate of expansion. The now-considerable-sized physical universe was an ultra dense plasma of every and all kinds of particles. This would also hold for the ultra massive X-bosons with all three fundamental charges, predicted by group theory, that interconvert quarks to and from neutrinos and electrons. Theorists who do not accept an asymmetric origin pin the prevalence of matter on an asymmetric decay of these hypothetical X-bosons. The X-boson, however, predicts that the proton should be unstable but, while avidly searched for, the decay of a proton has never been observed.

All the bosons and fermion/antifermion pairs were equally represented, the original Leftness now being expressed in the slight inequality between matter and antimatter fermions.

$$\frac{matter}{antimatter} = \frac{100,000,000,001}{100,000,000,000}$$

Freezing Out

So, just 10^{-36} seconds after the Origin Event, the physical universe was an expanding -dense plasma of every kind of subatomic particle whose history was determined over time by the wavefunction reflecting the very lowest levels of the Logos.

The era that followed the braking of inflation generation of the hot Big Bang by quantum color separation involved a series of 'freezing out' of particles. This is a technical term that, while based on the freezing of ice, has nothing to do with cold. It is based on the phase-change that occurs with falling temperature as the water turns to ice. In this general sense, steam freezes at a specific temperature to water, and liquid iron freezes at a specific temperature to a magnetic state (even though this Curie Point is almost 1,000°F).

As the universe cools from the trillion-degree temperatures of the start (where the characteristic energy is in the trillions of terra-electron-volts) over a period of ~10 minutes to where the characteristic energy is less than 1 MeV, the universe goes through a series of phase-changes as particles 'freeze-out.'

The temperature-energy at which a particle freezes out depends on the rest mass-energy of bosons (twice this for fermions that only come and go in particle-pairs). If the temperature-energy is above this rest mass-energy, then the particle (or particle-pair) is in equilibrium with the photons and all the others that are currently present, and it travels at light speed.

The rule is, at temperature-energies well above the rest mass-energy, the particle is abundant and travels at essentially light speed. The Planck Temperature-energy of 10^{28} eV is far, far above the rest mass-energy of every particle discovered so far in any experiment.

The hot Big Bang was a menagerie of every known particle and antiparticle in an ultra dense state which rapidly (but not exponentially) continued to expand, cool and become less dense. All three quantum interactions, even the weak, were on an equal footing. This is the state above the Grand Unification Temperature (GUT). This equivalence did not last long.

As the universe grew less dense, the quarks, antiquarks and gluons avoided being separated by collecting into colorless hadrons—such as a proton or pion. The color force froze out of the GUT state. As the temperature-energy was still vastly greater than the rest mass-energy of the hadrons, they continued in equilibrium with all the others.

The weak bosons (Z, W^+, W^-) have a mass-energy of ~80 GeV, so when the temperature-energy fell below this, these previously abundant bosons disappeared from the plasma. Only the tiny halos of virtual weak bosons remained, and the weak force froze out leaving only electromagnetism. Ignoring gravity for the moment, since it is not a simple quantum force, the three forces progressively froze out of the GUT state.

The rest mass-energy of the nucleon pairs (proton and neutron) is ~2 GeV and, when the temperature-energy fell below this, the matter/antimatter pairs froze out of equilibrium as they annihi-

$$\frac{nucleon}{antinucleon} = \frac{100,000,000,001}{100,000,000,000}$$

$$\xrightarrow{freeze} \frac{1 \ proton}{200,000,000,000 \ photons}$$

lated into photons. The resultant plasma had 1 proton and 1 neutron to half a trillion photons (which equilibrated with the others that had yet to freeze out).

When the temperature-energy fell below 10 MeV, the neutrons (which, being heavier, were less numerous and unstable) could stick to the protons by a derivative of the strong color force forming deuterons (heavy hydrogen-2), almost all of which rapidly united together as very stable helium-4 nuclei. It is the tiny fraction of neutrons that did not end up in helium-4 that has allowed science to accurately understand this phase in the physical universe's history. The end result was ~70% hydrogen-1 and 30% helium-4.

About 10 minutes after the Origin Event, the temperature-energy had fallen below 1 MeV, the rest mass-energy of the electron-positron pair. They disappeared from the universe, as they fell out of equilibrium and annihilated into photons, leaving behind the small excess of electrons that electrically balanced the small excess of protons.

Proto-universe

All that emerged from the Big Bang were the tiny number of hydrogen and helium nuclei plus the electrons and a huge excess of photons, neutrinos and antineutrinos (of all three generations).

The key point is that the neutrinos had fallen out of equilibrium once the weak force had frozen out. A neutrino is a single nonoriented twist and it can combine with an antineutrino to form a single oriented twist, a weak boson. Because a neutrino/antineutrino pair do not have sufficient mass-energy for this, this is not possible. Also, they cannot unite as a photon, since this involves two oriented twists.

The only remnant of the once abundant antimatter was in the anti-neutrinos. Ignoring the antimatter difference, for every nucleon that emerged unscathed from the Big Bang, there were 100 billion gamma ray photons and 100 billion neutrinos/antineutrinos.

If the mass-energy of the neutrino is 1/10 eV, and there are one hundred billion of them for every nucleon, the ratio of boson mass to neutrino mass is within the range of the experimentally-calculated ratio of regular matter to dark matter in the universe.

$$\frac{mass \ bosons}{mass \ neutrinos} = \frac{1}{10^{11}} \times \frac{10^9 \ eV}{0.1 \ eV}$$

$$= \frac{1}{10}$$

$$= \frac{regular}{dark} \ matter$$

Resisting Expansion

The energy in the Big Bang photons and neutrinos is all positive, and it has an attractive gravitational effect which opposes, and slows, the expansion of the universe. Energy is lost in opposing this outward thrust. As the photons lose energy, their period increases and they fade from gamma ray through X-ray, UV, visible light, IR to the microwave photons that flood the universe as the Cosmic Microwave Background, CMB, of our current era.

The neutrinos lost energy by slowing down becoming the slow, relic neutrinos that also pervade the universe. It is a matter of debate as to whether or not this slowing down was sufficiently rapid for their mutual gravitation to amass them into the condensations and filaments of dark matter that are currently observed.

What is not up for debate is that matter and relic photons/neutrinos are all positive energy that has been opposing, and slowing down, the expansion of the universe—somewhat like a ball being thrown up in the air, slowing down under the influence of gravity. It seemed hardly worthwhile to test the implication of then-cosmology that the universe had been slowing down in its rate of expansion for all the 13 billion years since the Big Bang.

The diagram is what the early expansion of the universe was expected to look like if and when it was actually measured.

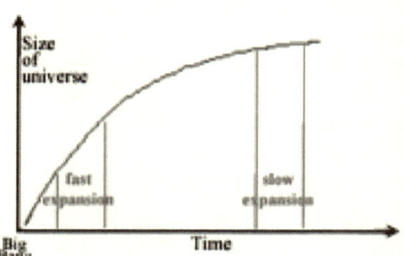

Reference Frame

Now that we have established the basics of system-building and the unified systematic hierarchy, we can apply these to the next stages of the universe's development in time. As most of the gamma ray photons created by the Big Bang have yet to interact with anything, and have been stretched by the continuing expansion of space-time into microwaves, the CMB provides a universal and absolute reference point for the measurement of physical time.

As our galaxy is, and always has been, essentially at rest with respect to the CMB, and there is essential agreement between our time scale and that of the Universe.

One Logos, Two Realms

The emergence of the Logos in the Abstract Realm preceded the Origin Event of the Big Bang.

$$O \xrightarrow{Logos} + \begin{cases} \uparrow \xrightarrow{Logos} P\ universe \\ \\ \downarrow \xrightarrow{Logos} S\ universe \end{cases}$$

The Logos separated Nothing into two abstract pixels with a quality that, for simplicity, we can equate with Left action and Right action. We will later discuss what is being twisted. This 'something out of nothing' is a similar dynamic to the vacuum creating a virtual particle pair (such as an electron and positron pair that do not have a pixel of action between them), i.e., or the reverse of the equation, $(+1)+(-1)=0$. The separated Left and Right pixels, obeying the Logos, developed in a complementary manner into two complementary realms, the S universe and the P universe.

The Left pixel developed, obeying the Logos, and resulted in the P universe, the hot Big Bang and the development of the physical universe. This development is quite well understood from the merest fraction of a second after the very start of the process up to the present day, and this we will focus on. The complementary S universe was only detected in the last decade of this writing, and we will later discuss what little is known to science about it, and offer a précis of what other disciplines have inferred about it.

Structure in Spirit Realm

While both realms are substantial, there is no requirement that they have similar structures—in fact, we might expect them to have quite different structures. The physical world involves matter clumping together under attractive gravity into relatively small aggregations called galaxies, stars, planets, etc.

What is the structure of the spirit world? The current techniques of detecting dark energy are too crude to tell us much, but they indicate that it is not in clumps but fills all of space. The little communication that has occurred between the spirit world and the physical world suggests that the structure of spirit world involves vast layers, extensive and endless substantial planes that are not wrapped around little planets.

This speculation about the spirit realm is from *The Great Divorce* by C. S. Lewis, about a bus trip from the outskirts of Hell to the edge of Heaven. First, consider the vision as the aeronautic bus soared up away from the miserable bus stop in Hell:

> We were now so high that all below us had become featureless... I got the impression that the grey town filled the whole field of vision... astronomical distances... millions of miles away, Millions of miles from us and from each other..."[151]

Then, arriving in the at the edge of Heaven:

> A cliff had loomed up ahead. It sank vertically beneath us so far that I could not see the bottom, and it was dark and smooth. We were mounting all the time. At last the top of the cliff became visible like a thin line of emerald green stretched tight as a fiddle-string. Presently we glided over that top: we were flying above a level, grassy country through which there ran a wide river.... I got out... I had the sense of being in a larger space, perhaps even a larger *sort* of space, than I had ever known before: as if the sky were further off and the extent of the green plain wider than they could be on this little ball of earth. I had got 'out' in some sense which made the Solar System itself seem as an indoor affair.[152]

A similar sense of the vastness of the substantial spirit world is from *Life in the World Unseen:*

> Space *must* exist in the spirit world. Take my own realm alone, as an example. Standing at the window of one of the upper rooms of my house I can see across huge distances whereon are many houses and grand buildings. In the *distance* I can see the city with many more great buildings. Dispersed throughout the whole wide prospect are woods and meadows, rivers and streams, gardens and orchards, and they are all occupying space, just as all these occupy space in the earth world. They do not interpenetrate any more than they interpenetrate upon the earth-plane. Each fills its own reserved portion of space. And I know, as I gaze out of my window, that far beyond the range of my vision, and far beyond and beyond that again, there are more realms and still more realms that constitute the designation *infinity of space*.[153]

Assuming that the dark energy of modern cosmology is in the quanta of the spirit world, and that this negative energy has a structure (and this is an open question in astronomy) arranged in vast layers that cross the vastness of space, we can combine this with the physical world, ending up with a composite picture of "one universe: two substantial realms" that looks something like layers stacked upon layers.

In this view, the vast empty intergalactic physical vacuum is not an unnecessary "waste of space" but room for the substantial spirit world to develop and fill. A house in the highest spirit realms could share space with a supercluster of galaxies. They would not inconvenience each other, being on different sides of spacetime, one extending in the spiritual metric, the other in the physical metric.

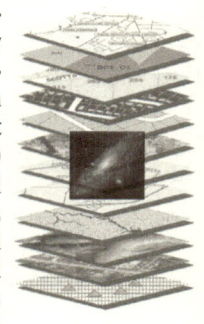

The universe is expanding and the positive energy of physical matter is getting even more spread out. The dark energy component of the cosmos is not like this, it just keeps on growing as the spiritual metric expands. This makes sense for an eternal spirit world: it just keeps on getting bigger. This is not the case for the physical world; it has an expiration date. In 60 billion years or so, all the fuel in all the stars in the physical world will have run out and the galaxies will go dark and life, as we know it, will become impossible. We can assume that God has foreseen this and is confident that humans will have figured out how to trigger our own Big Bangs and make new universes with them. The spirit world, on the other hand, will just keep on getting bigger into eternity.

To conclude, we have expressed the internal truth of two substantial realms in the language of modern quantum physics in a plausible, if not necessarily correct, way.

Sir John Templeton, who dedicated his fortune to a Foundation for the reconciliation of theology and science, and who passed away in July 2008, espoused what he called a "humble approach" to theology. Declaring that relatively little is known about God through scripture and present-day theology, he once predicted that "scientific revelations may be a gold mine for revitalizing religion in the 21st century."[154] We can hope that he was correct.

The Physical Universe

The pixel of Left (or 'false vacuum' or 'inflaton' as it is variously called) was at the Planck temperature ($\sim 10^{35}$K) and had a negative pressure. Driven by this negative pressure, this pixel entered a period of inflation during which it doubled itself with everyPlanck tick of time.

Quantum pixels

We have noted the difference between a discrete situation and a continuum when we briefly looked at the types of infinity in *Volume Two*. We saw how a discrete situation, such as the countable fractions between 0 and 1, can look just like the continuum of the uncountable irrationals between 0 and 1 when the precision is not too great.

Whatever the abstract construct at the start, the physical world that emerged from it is most decidedly of the discrete variety. Things that appear continuous to us, and were considered so in classical science, only appear such because our precision of observation was too rough to notice. This is similar to how the screen of my computer looks white to me, but under a magnifying lens is actually seen to be equal numbers of red, blue and green pixels all ablaze together. Again, similar phenomena occur in science when nature appeared to be continuous—e.g, space, time, energy, existence, etc.—until we developed the technology capable of such precision and resolution that the discrete nature of most things was revealed.

The discrete pixels are called quanta (*sing.* quantum) in science, and as Max Planck was the first to notice this in the quantum of action (existence), the units are named after him.

In natural units, all pixels have a size of 1 with all larger measures being integer multiples of this. In human units, the pixels are so inordinately

Measure	Unit	Value
Existence	Planck's Constant	$6.6\times10^{-34}kg$ sec $= 4.1\times10^{-15}eV$ sec
Time	Planck Time	5.4×10^{-44} sec
Space	Planck Length	$1.6 \times 10^{-35}meter = 4\times 10^{-33}inch$

small that the numbers would be huge. The speed of light is 1pL each 1pT in natural units of space and time, while it is 299,792,458 meters each second in human-scale units.

The Hot Big Bang

The initial pixel of spacetime, driven with Left, doubled each Planck tick of time for a time period that is estimated to have lasted for 10^{-35} seconds which, while brief by human standards, is 10^9 quantum ticks. And the number of pixels doubled each tick. The final number of ticks is roughly $2^{10^9} \sim 10^{33,333,333}$. As the current volume of the visible universe has $\sim10^{30,000}$ cubic Planck units in it, we are only able to see a fraction of the entire physical universe that is, though invisible, the same age as our visible piece of it.

This exponential expansion phase in the universe's history was brought to a close when the separation of fundamental entities with a color charge braked the inflation and converted the potential energy of expansion into the positive energy that we call the Hot Big Bang. This energy appeared as an intense mangling of spacetime pixels into every kind of particle and antiparticle, fermion and boson. The average energy was 10^{28}eV, the Planck temperature of 10^{32}K so everything was moving at light speed, and the universe was an ultra dense fluid of bosons and fermions with an equal amount of left and right, in which the original Left was only an insignificant asymmetry to the whole.

The pixels of spacetime, throughout this time retained their entanglement, and have done so up to the present. This global entanglement is observed today in the absolute nature of inertial mass and angular momentum. The Logos determines that this entanglement also has a local aspect that attracts, which we call gravitational mass.

In the early universe, the mass density was enormous, as was the gravitational attraction, but equal in all directions; it was *isotropic*. It resisted expansion, further slowing down the inflation of the universe.

As the universe expanded, the temperature fell, and one by one all the matter-antimatter froze out and all that remained was mainly photons, neutrinos and anti-neutrinos, with the original Left emerging as a 1 in 100 billion number of electrons, protons and neutrons.

The first system building was when the density and temperature were not sufficient for free quarks and gluons, and they condensed into hadrons and ended up in the two nucleons, the proton and neutron. (During the first minute, the 11-minute half-life of the neutron was long enough not to alter the equality of protons and neutrons.) By the time it was cool enough, however, for the deuterium nucleus to be stable, the balance had shifted towards the protons, and all the neutrons ended up in the helium nucleus. Both of these system building interactions were governed by the Logos.

Cooler still, and the electrons and anti-electrons annihilated leaving behind the 1 in a 100 billion excess of left-handed electrons, a diluted version of the original Left, along with the quarks.

The neutrinos and antineutrinos had stopped interacting once the average spacing became much larger than the weak force radius and, while a few of them did get close enough for matter-antimatter untwisting, most did not. As neutrinos can travel through light years of solid lead without ever interacting, and the antineutrinos are a lot less densely packed than lead, the 200 billion neutrinos and antineutrinos are around to this very day. They, like the photons, have had almost all of their kinetic energy reduced by the expansion of spacetime. All they have left is their rest energy which, while only a few eVs, mounts up into the billions, and these relic neutrinos are the best candidate of the dark matter in the universe that is ~10 times as great as that of the nucleons in the universe.

Unlike the early Hot Big Bang where the braking of the exponential inflation had jangled all of spacetime in every possible way, spacetime was much calmer now, with energy locked up in discrete quanta and there was the beginning of a separation of matter and the vacuum. While the vacuum was still being disturbed by the passage of photons and neutrinos, this became less and less as time went by until we have the placid vacuum today of the Voids in which spacetime is hardly ever disturbed.

The pixels of this 'perfect vacuum' have an internal wave aspect that is usually cancelled out by all the others around. When this cancellation is disturbed by energy passing through, the result is the local effect we call gravity. The internal aspect of the spacetime pixel also has an excited, but not twisted, state. This high-energy state is called a Higgs Boson, and it has no spin, no charge, and no color. A search is currently on to detect a Higgs, and the current lower bound for its rest mass energy is ~160 GeV, about 160 times the energy in the gluon field inside a proton. If it has a quantum of action, it is a real boson with a very great mass; if it does not it is a virtual boson and, not actually disturbing the pixels, has no mass at all.

Just like the other three forces, gravity can be described as coupling with virtual bosons, though not with a spin of 1. The local aspects of gravity (the disturbance of the pixels of spacetime entanglement) appear as local coupling with massless spin-2 bosons called gravitons, while the global aspects of gravity, the entanglement of all, appears as a global coupling with the massless, spin-0 Higgs. This was explored by Richard Feynman before the Higgs entered the scientific lexicon. While spacetime is discrete, at a low enough resolution it can be treated as a continuum, and the mathematical tools dealing with the smooth continuum are well-developed. This was pioneered by Einstein who described both the global and local aspects of gravity as a bending of spacetime in his epochal General Theory of Relativity.

Clumping

As noted, the early universe was isotropic, but grainy. This graininess was magnified by inflation so that at the end of the Hot Big Bang there was a slight variation, just 1 part in 10,000, between different areas of the early universe. This slight anisotropy began in the neutrinos, the dark matter, and these gravitating clumps and strings of dark matter were the seeds around which the regular matter later clumped together around at start of superclusters of galaxies.

Gravity, however, did not come into its own as a system builder until the universe had cooled sufficiently for the electromagnetic interaction to convert the universal plasma of electrons and nuclei into neutral atoms of hydrogen and helium. This is called the 'recombination period' (a bit of misnomer inasmuch as there had only been a plasma up to this point in history).

At this point, the photons became 'decoupled' from the matter and travelled freely hereafter. It is these photons that, having all been stretched by the expansion of spacetime, fill the universe today with the CMB. The CMB is isotropic to one part in 10,000, which is how we know of the slight bumpiness in the early universe. It was only after this decoupling from the overwhelming number of photons that clouds of ~75% hydrogen-1 (protons) and 25% helium-4 could start to condense around the dark matter.

The gravitational instabilities in these clouds fragmented them into galaxies and then into stars. The gravitational potential energy of the in-falling and colliding atoms was converted into kinetic energy, and the clouds started to warm up and emit thermal radiation. This radiation pressure opposed the gravitational collapse, but had little impact on the collapse at first.

As this collapse continued, the atoms re-ionized back into a plasma, and when the core temperature and pressure reached high enough values, the protons started to fuse together as a deuterium (hydrogen-2) nucleus (with the emission of a positron that annihilated with an electron into gamma rays) and so on to helium-4. The star ignited.

The radiation pressure outwards was now as great as the inward pull of gravity, and the star reached an equilibrium where the two were balanced. It took up a position on the *Main Sequence, MSS*—the linear relation of mass, temperature, luminosity and lifetime of 99% of all stars—that is determined solely by its mass accumulated before the increasing heat blew the rest of the natal cloud of gas away.

As massive stars have a greater inward pull, they need to be at a higher temperature to reach equilibrium, and they blaze with an intense violet-white light for a brief, by stellar standards, profligate 10,000 years of brilliance that ends in a great cataclysm. These are the O-class stars on the high end of the MSS.

Small mass stars, on the other hand, reach the same equilibrium but generating a miserly red light for hundreds of billions of years of their dim lifetime. These are the M-class stars. Our medium-sized Sun is a G-class star that will prudently shed its beneficent light for another 20 billion years or so.

As hydrogen and helium are poor couplers with photons, massive clouds could condense before the radiation pressure blew the outer layers away, so the first generation of stars were very massive, of the O class on the MSS. Before the stars ignited, the universe had been dark since all the gamma photons of the Big Bang has lost so much energy opposing the expansion of the universe, that they were now infra red, and so this period is called the Dark Age. This period of darkness ended with the ignition of the first stars, and visible light reappeared in the universe. (It should be noted that all the billions of stars in the billions of galaxies, over all 13 billion years of shining, have added but a tiny fraction to the number of photons in the CMB.)

The Main Sequence

The equilibrium between gravitational collapse and radiation inflation is reached at the same temperature for stars of the same mass powered by hydrogen-to-helium thermonuclear burning at the core. This relation of temperature and mass is the MSS bracketed by the massive O-class stars radiating intensely in the X-ray-UV-violet white at the high end of the sequence, and low-mass. M-class stars dimly radiating in the red at the other end.

The amount of energy needed to power the intense radiation of massive stars is such that they run out hydrogen to burn very much more rapidly than the low mass stars, at which point they leave the MSS.

A blue-white O-type star with a mass 100 times that of our sun, has a surface temperature of 30,000K and emits 100,000 times as much energy as the sun. It runs through its hydrogen in only 10,000 years before leaving the main sequence. A red M-type star with a mass just one tenth of the sun's dribbles out just a ten thousandth of the sun's energy and can last over ten trillion years.

The 1st Generation of Stars

The First Generation were O-type stars and, within the first million years, had used up all the hydrogen in their cores and moved off the MSS. The reduced radiation pressure allowed the core to collapse until the temperature became high enough for helium to start to fuse. The energy released by this process, and all the subsequent stages, liberates only a fraction of that released by proton fusion, and so the lifetime remaining is proportionally much less.

At the bloated out surface, the increased amount of energy is spread out over a larger area, so each square centimeter will be cooler. The surface will have a red color because it is so cool and it will be much farther from the center than when young and on the MSS. Despite its cooler surface temperature, the red giant is very luminous because of its huge surface area. An example is of a red supergiant is Betelgeuse, a pink star that can be seen with the naked eye. If it were placed at the center of our solar system, all of the planets out to Jupiter would be inside its bloated surface.

The beryllium-8 nucleus formed out of 2 heliums is very unstable, but it lasts long enough for a third helium to add on, creating a carbon-12 nucleus. When the core runs out of helium, it heats up until the nuclei fuse into oxygen, then neon, sodium and magnesium, then silicon and phosphorus, and finally into iron and nickel.

This is the end of the line, as iron has a minimum binding energy, and no more energy can be released by further fusion. At this point, the star has an onion layer structure, with the lighter elements around the core. The collapse continues and the core reaches a temperature and pressure at which reverse beta decay becomes possible—absorption of electrons by protons—and neutrinos are produced in great abundance. These leave the core at light-speed thereby removing energy and cooling the core. The core and then the outer layers commence collapsing. The release of gravitational potential energy heats the entire star, and it explodes as a supernova. The star shines out with the brilliance of 100 billion regular stars, and the elements heavier than helium (all called *metals* by cosmologists) are scattered, adding themselves to the primordial hydrogen and helium.

Star Death

Stars have a lifetime that starts when the core of the collapsing cloud of hydrogen/helium reaches a high enough pressure and temperature for it to 'ignite' the thermonuclear fusion of hydrogen and the star enters the MSS. It stays on the MSS for as long as it takes until the core hydrogen is all used up. The helium 'ash', along with the primordial helium, then ignites and the metals start to be formed in quick succession up to iron/nickel.

The final stages of a star depend on its mass, and can end in a whimper or a bang. For stars about the size of the sun, the core never reaches the extreme conditions necessary for the heaviest metals to be produced. So the thermonuclear reactions cease somewhere between the carbon and silicon stage. At this point, the star is no longer able to create the radiation pressure to oppose gravitational collapse, and star shrinks. The outer layers heat up and prolong the collapse, but the star inexorably shrinks into a "white dwarf" which is at a high temperature but has a small surface area, so is much less luminous than a star at the same temperature that is still on the MSS. At this point, another factor comes into play—the fermion nature of the electrons in the star's plasma.

As mentioned earlier, the quantum probability of two fermions being in the same state is zero; it is impossible. As the volume of the white dwarf shrinks, the electrons reach a "degenerate" state in which they are on the verge of being forced to enter into the same state. The impossibility of this happening prevents any further collapse. As the white dwarf continues to cool, the volume remains a constant, held up by the 'degeneracy pressure' of the electrons, as it slowly cools through yellow heat to red heat and ends as a black dwarf. The mass is still that of the sun, but the volume is about the size of the earth's. The universe is too young for even the first generation of G-type stars to have turned into black dwarfs. This is the 'ice' ending to a star. (As our sun will end its life as mainly carbon, I like to think our sun will end up as a great big diamond.)

For stars much bigger than the sun, the core does turn into iron/nickel and the star goes supernovae. Much of the star is ejected to fertilize the primordial hydrogen for the next generation. The force of this ejection compresses the remaining core and smashes through the electron degeneracy. Rather than be forced to do the impossible, the electrons take the only alternative route and combine with the protons to create neutrons.

If the mass of the core remaining after the ejection is not too great (~2 solar masses) the neutrons, being fermions, also have a degeneracy pressure, and refuse to do the impossible and share a state. This can stop any further collapse, and the result is a neutron star that retains its size as it slowly cools of from the millions of degrees at formation as a gamma star through X-ray, UV, white, yellow, red stages towards blackness.

The neutron star has the mass of the sun in a volume the size of a comet. The density of this 'neutronium' is that of the atomic nucleus. The neutrons are stable because it is the proton/electron, in such a situation, that has the most free energy.

If the remaining mass is too great for the neutron degeneracy to resist the intense gravity, the collapse continues and a 'black hole' is the end result. To understand a black hole, we need the concept of escape velocity.

Black Holes

We have already discussed the interconversion of kinetic and potential energy in the simple pendulum in a gravitational field. A similar thing happens when a bullet is shot straight up in the air. As it rises in the gravitational field of the earth, it slows down until it comes to a momentary halt. All the kinetic energy of the bullet leaving the gun has been sapped away and stored as potential energy. It then falls towards the earth with increasing speed until (ignoring the frictional heat lost to the air) it hits the ground with its original speed. The energy is now all kinetic and, when it hits the ground, this is randomized as heat.

The consequence of the gravitational disturbance of physical energy is attractive, and this applies to all forms of physical energy, including that of a photon. Just like a rising bullet, the gravitational field drains energy from a photon. Since a photon is always moving at light speed, it does not slow down, the energy is drained from the photon. It is redshifted as its wavelength is stretched, its period increases and its frequency decreases.

Gravity is, however, an extremely tiny effect. Two positive protons will, for instance, repel each other with a force that is a billion trillion, trillion, trillion times the attractive force of their mutual gravitation as a consequence of their physical energy.

For example, the earth and a human body both have equal numbers of positive and negative charges, and the virtual photons are all well-confined. The electromagnetic interaction between the two is minimal, and even a minuscule imbalance is quickly restored by a flow of current, such as a static spark or lighting stroke.

Unlike electromagnetism, however, which has an equally powerful repulsion between like charges, physical energy is only attractive, and does not have an expansive effect. (We will encounter an expansive form of gravity when we get to the section that explains why we are using the term 'physical energy' rather than just plain energy.) Gravitation is cumulative, and even an infinitesimal value can amount to a large value if it is multiplied by a big enough number. Most of the mass-energy of an atom is in the blaze of gluons that is the nucleons, and it is this energy en masse which gives rise to the gravitational mass of a composite body.

The human body is a unified wave confining and giving form to ~30,000,000,000,000,000,000,000,000,000,000 quarks, electrons and an even greater number of gluons and photons. The earth has 2×10^{22} as many of them confined. Each one of these quanta of energy is gravitationally attracted to all the others, and the consequence is the force of gravity that keeps us on the earth's surface.

Sir Isaac Newton, while entertaining no hypothesis about the internal workings of gravity, derived these very useful approximations about the external nature of gravity that are all still most useful to this day:

1.   The mass of an extended rigid body can be treated as localized at a point within the rigid body, its 'center of mass' which, for an isotropic sphere, is coincident with its geometric center.

2.   The attractive force between two extended bodies is proportional to the product of their masses.

3.   The attractive force between two extended bodies is inversely proportional to the square of the distance between their centers of mass.

The surface of the earth, and thus the position of the center of mass of the human body, is ~6,400 km from the center of the earth. The tiny force of attraction between all those zillions of quanta sums up to our 'weight' that holds us to the earth's surface, a force of 9.8 newtons for each kilogram of mass.

In an airplane flying at 10 km (33,000 ft), we are at a greater distance from the earth's center, and the force is proportionately reduced. The the proportional change is very $$9.8 \times \frac{(6,400)^2}{(6,410)^2} = 9.77$$ small, however, and a person of my weight of 170 lbs. only weighs 8 ounces less at this altitude than on the ground.

Even on the International Space Station, at an altitude of 370 km, the force is still 8.76 and I would weigh 152 lbs. It $$9.8 \times \frac{(6,400)^2}{(6,770)^2} = 8.76$$ is by no means a gravity-free environment, the weightless experience is a product of 'free-fall' and is akin to the weightless feeling momentarily experienced on a roller coaster.

At the distance of the Moon, the force of attraction is reduced to just 0.003 newtons/kilo. The Moon is 1/80th as $$9.8 \times \frac{(6,400)^2}{(376,400)^2} = .0003$$ massive as the earth, but the surface is just ¼ of the distance to the center, so the surface gravity of the moon is 1.62 newtons, about ⅙th that of earth's, not $1/80$th.

The sun is 332,950 times the mass of the earth, but its visible surface is 109 times as far from the center, so the surface gravity of the sun is just 28 times that of the earth. Just as a bullet fired from the earth loses kinetic energy rising against the pull of the earth, so a photon leaving the sun's surface loses energy and is red-shifted, but only by about one millionth of its energy (and the red shift on leaving the earth is proportionately even smaller).

A bullet fired from the earth's surface is slowed until it stops, and then starts to fall back. The faster the bullet rises, the higher it will climb. But as the bullet rises, the force of gravity falls off. At a high enough initial velocity, the bullet will rise so far that the gravity is too weak to stop it so it never comes to a full stop but continues to climb. This initial velocity is called the 'escape velocity' (EV) and is a useful measure of the gravity gradient of a

body	EV (km/s)
Earth	11.2
Moon	2.4
Sun	618
White dwarf	5,200
Neutron star	100,000
Black hole	299,792

body. The earth's EV is 11 km/sec while a bullet from an M16 rifle travels at 10 km/sec and would be almost fast enough to never fall back again if air resistance didn't sap much of its energy.

It would, however, easily escape the Moon because the EV is only 2.4 km/sec. At the sun's visible surface, the EV is 618 km/sec, but the concept can be applied to any distance from the center. At the 'surface' of a sphere as large as the earth's orbit, the escape velocity is down to 42 km/sec, so even though the bullet can escape from the Moon, it cannot shake free of the sun.

The surface of a white dwarf has an EV of ~5,000 km/sec, and a photon rising against this gravitational field experiences a significant redshift.

The surface of a neutron star has such an intense gravitational field that the EV is >100,000 km/sec, or one-third the speed of light. When a neutron star implodes, all the mass collapses through a surface, the event horizon, at which the EV is the speed of light. A photon attempting to escape from the event horizon has an infinite red shift that is indistinguishable from no photon at all. Nothing, not even photons, can escape from the event horizon. It emits no light, hence its name.

Structure of a Black Hole

No one has ever seen inside an event horizon, even in a thought experiment, so we can only theorize. Some think that the mass collapses into the infinite density of a point, a singularity. Quantum physics, however, suggests something that does not involve an infinite quantity since it connects separation of particles with energy. The closer two entities get, the higher is their kinetic energy and thus the temperature.

As the collapsing star breaks through the neutron degeneracy pressure, rather than do the impossible, they first dissolve into a quark-gluon plasma. As they are compressed ever closer, the temperature reaches the stage at which average energy is capable of creating electron/positron pairs and then nucleon/anti-nucleon pairs. At even higher temperatures the electromagnetic and weak interactions unite as the kinetic energy soars above the rest mass of the weak bosons and so on back to the earliest stages.

These are the last stages of the Hot Big Bang, and as the temperature rises in the collapsing core, the stages reappear in reverse order. The outward pressure of this recreated primordial plasma prevents further collapse. Just how many stages are necessary to stop the collapse depends on the in-falling mass. The very largest Black Holes—greater than the mass of a billion suns—probably have to go all the way and recreate a few pixels of false vacuum at the very center whose enormous expansive pressure holds up the layers above.

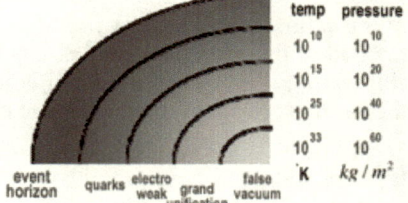

2nd Generation stars

There are only three possible endings for a star that has exhausted its fuel:

1.   White dwarf stars, which cool off into black dwarf stars, held up by electron degeneracy pressure.

2.   Neutron stars, which start at an ultrahigh temperature, but will eventually cool to darkness, held up by neutron degeneracy pressure.

3.   Black holes, which emit no light from their event horizon, held up by a reversal of the Hot Big Bang, a small 'crunch' as it is technically known.

The stars of the first generation were massive (>100 suns) and in their death throes that scattered metals into the primordial gas, turned into black holes.

At the center of the galaxy where the stars were closely packed (for stars, that is) these black holes merged into the central black hole with a mass greater than a billion suns. The energy liberated in this merging of black holes formed the active core of a quasar which, being so bright, can be seen across a distance of 10 billion light years.

Our galaxy went through this stage until everything settled down ~10 billion years ago. The black hole that lives in the center of our galaxy is named *Sagittarius A** (pronounced "Sagittarius A-star"). It is ~26,000 light-years from Earth and its event horizon is measured to be about 14 million miles across. This black hole would fit inside the orbit of Mercury and is estimated to have the mass of ~4 billion suns.

The rest of the 100 billion stars in our Home Galaxy, the Milky Way, rotate roughly in a plane about this central point with an orbital period of ~250 million years. We can consider this a galactic year. Our earth is ~4 billion earth-rotation years old, so we can consider our home planet to be a youthful 16 year-old on the appropriate timescale.

All of the processes just described—from galaxy formation to 'metal' production in stars—occur under the direction of the Logos, and the end result was that the clouds out of which the second generation of stars formed had a smattering of the 'metals' in them. The 'metals' are much more efficient at coupling with photons, and the second generation tended to have smaller masses than the first generation. They were still large, however, and by about 5 billion years ago, they had also gone supernova and further enriched the interstellar medium.

While only about 1%, there was still enough 'metals' around that when the third generation of even smaller stars ignited and blew away the outer hydrogen and helium, there was enough 'metals' remaining to form dust, which amalgamated into planetesimals and then into planets. Also about this time, it has only recently been discovered, the heretofore unknown form of energy, called "dark energy" (not dark matter) began to show its influence. The third generation of stars to emerge were enriched with metals. Our sun is a third-generation star and has a significant enrichment in the 'metals' at 71% hydrogen, 27.1% helium, 1.3% carbon and oxygen, and less than 0.7% all the other 'metals'.

A small fraction of the collapsing gas did not fall into the sun. The metals remained in orbit and combined into molecules, grains and planetesimals, while the hydrogen and helium was blown away by the solar wind. Such aggregates condensed to form the proto-earth about 5 billion years ago.

A collision with another small planetesimal created the Moon—the splash condensed at half the distance it is now). The original crust of the proto-earth was like that of Venus, all of one piece, but this impact cracked the shell into tectonic plates that are shifting around to this day. The Moon also stabilized the axis of the Earth's rotation at a ~23° inclination instead of it wandering all over the place as happened to the planet Mars.

While the water molecule is small, it is chemically attracted to minerals and aggregated with them. Along with the water contributed by the comets during the heavy bombardment phase of the earth's formation, the water condensed on the cooling Earth into oceans that were saturated with reduced, soluble iron in the ferrous state. The

Moon raised enormous tides which eroded the land and created great beds of porous clay in the cracks in the ocean floor. These faults between the tectonic plates circulated seawater through the crust as black and white smokers perfusing they clay beds. The intense UV light and lightning drove many organic reactions in the dense atmosphere of carbon dioxide, ammonia and nitrogen.

The history of the universe up to this point has evolved under the direction of the Logos, particularly that aspect we earlier described as the Perfect Wave. Many waves combine over time into a wave with unique properties. As the science describing the formation of 'rogue waves' and 'perfect storms' is in its infancy, this period is as yet only imperfectly described.

But one thing is known for certain. The end result of this 13-billion-year history was a womb perfect for the origin of life to occur with relative rapidity (a few hundreds of million years). As we shall see, the influence of the Moon was crucial to the rapidity of the origin of life. As very few planets can be expected to have such a large moon, we can expect that life was kick-started on the Earth, and has only slowly developed on the moonless majority of planets in the *Goldilocks Zone*—being neither too close or too far from the sun to allow for liquid water. It is to be hoped that most of these planets will have at least developed photosynthetic bacteria and an oxygen atmosphere by the time we get to them.

The emergent properties that were inherited from the Logos made the earth a perfect womb for the emergence of life, the transition from systems characterized by chemistry to a new set of systems that were predestined to be the subsystems of Life. Perhaps slighting the stars that incubated all the 'metals', the early earth is our first example of an eden that provides a womb that is 'just right' for the origin of systems with a much higher level of sophisticated properties that are inherited from the Logos.

THE PROKARYOTES AND *VITAL UNITS* OF LIFE

Under the direction of the Logos, the system-building of simple systems becoming the interacting subsystems of more sophisticated systems continued as the earth cooled and liquid water accumulated.

In the scientific hierarchy of 'systems of study,' physics and chemistry deal with systems that are decidedly not living systems, while biology deals with systems that are decidedly alive. The gap between the two is the province of biochemistry.

The systematic hierarchy of living systems that emerged has, at its very lowest level a system that does not have a given name, so we will call it the *vital unit*. We will define a *vital unit* as a volume of water confined by a surface with portals, whose analog form and function is determined by stored digital information.

The simple living systems called bacteria are a single vital unit. These are called the *prokaryotes*. Sophisticated living systems have many, even trillions, of interacting vital units as their basic subsystems. These are the *eukaryotes*, the fungi, plants and animals.

We will start our exploration of the systematic hierarchy of living systems at the very bottommost level, with the biochemistry common to all vital units.

Fundamental biochemistry

Biochemistry, the study of the chemical subsystems of life, is the discipline where the great division of the sciences into two ways of viewing the world occurs.

Worldview One

This is the classical view of the world as matter in motion. All systems have only an external aspect that needs to be taken into consideration. Physics pre-1900, chemistry pre-1950, and all current biology, genetics and on up to the brain sciences are classical Worldview One. The math is all real numbers and $i=\sqrt{-1}$ rarely appears in the descriptions.

Pedagogically, all of science up through high school is strictly Worldview Two and the math is in real numbers and $i=\sqrt{-1}$ is strictly for the pure mathematicians and a mystery to be introduced only to the few at the university level. Quantum science would be so much more accessible if complex numbers were introduced in early schooling somewhere after fractions.

The concept of probability in this view is strictly classical probability, which is founded on the axiom of independent assortment. It is not considered a fundamental aspect of reality but rather an admission of incomplete understanding.

Classical probability excels in situations where independent assortment applies. An example is the kinetic theory of heat and the Gas Laws, which precisely relate the pressure, temperature and volume of a perfect gas. In a perfect gas, the atoms do not interact with each other except to bounce off each other as perfect Newtonian spheres. The helium in a ballon is an example of a perfect gas.

If, however, the atoms interact in any way—they are sticky—small corrections have to be made to the perfect gas laws. When two water molecules bounce off each other, for example, their interaction creates a tug opposing the rebound. Energy is removed from kinetic motion and the rebound is much slower than it would be in a perfect gas. At temperatures lower than 100°C, there is insufficient kinetic energy to separate the molecules, and the gas laws are ignored as the water liquifies.

A molecule of glucose added to liquid water will be pushed about on a 'random walk' through the volume of water. This is the kinetic diffusion of a glucose molecule through water that obeys classical probability—the distance travelled from the starting point is proportional to the thermal velocity and the square-root of the number of collisions along the way. In this 'Brownian motion' through the water, the glucose molecule will sweep out a volume that is proportional to time.

Worldview Two

The modern sciences of physics and chemistry share a worldview in which all systems have an internal wavefunction (complex numbers) that gives an external form (real numbers) to a set of confined and interacting subsystems. All the math of physics and chemistry is in complex numbers and $i=\sqrt{-1}$ appears in all the fundamental, and highly accurate, equations of these Worldview Two sciences.

Probability, in this worldview, is a fundamental aspect of reality on a par with location in time and space. This quantum probability is not classical but comes in two varieties, both of which are founded on dependent assortment, not independent assortment.

Quantum probability determines, over time, what happens in space; in bounded situations such as systems where the Law of Large Numbers applies, the internal form to the quantum probability completely determines the external form of the system.

Unfortunately, all the way up to high school level, the teaching of physics and chemistry is classical and deals only with the external aspect and not the internal, which is mentioned only in graduate school.

Biology and the higher sciences share a worldview that is strictly classical. The internal aspect so central to Worldview One is utterly ignored and is no longer considered a part of the picture.

Schizoid Biochemistry

The switch in the consensus from the external-only Worldview One to the internal/external Worldview Two falls uncomfortably within biochemistry. There, while the internal wave aspect is accepted as important in the catalytic activity of enzymes, by far the most important concept in the rest of biochemistry is that of the 'lock-and-key' as two molecules externally fit together as 3-D jigsaw puzzle pieces.

This specific 'binding' of two molecules that neatly fit together is found throughout biochemistry. Examples are:

- The specifically folded chain of amino acids in the active form of a protein
- A specific substrate bound to its enzyme
- A steroid molecule bound to its receptor
- A strand of DNA bound to its complementary RNA
- A calcium ion bound to a muscle protein
- A virus bound to the surface of its host-to-be.

Internal quantum probability plays no role in all these examples of the lock-and-key motif. All follow the same basic steps:

1. The thermal motion of the molecules brings them into contact
2. The two molecules bind together
3. The bound complex goes through a conformal change in shape and function
4. The two molecules can stay together, or changes occur and the molecules separate and go their separate ways

Unified Science

It is the intention of this work to illustrate how internal quantum probability plays a central role in all the sciences. Inclusion of the internal at every level of the science hierarchy, as we shall see, solves a lot of problems that do not follow independent assortment and classical probability.

We shall call a science 'unified' if it considers the internal quantum probability aspect of matter to play a primary role in the systems under study. In this view, the sciences of physics and chemistry are unified sciences, biochemistry is partially unified, while biology and on up are not unified sciences. The goal of this work is to include the internal aspect in all the sciences and see what results from the extension of the worldview.

As befits the scientific, bottom-up approach to a systematic hierarchy, we shall start at the very bottom with the external concept of lock-and-key binding so central to biochemistry.

Lock and key

The lock-and-key motif is most simply demonstrated in the study of enzymes.

The 'lock' is a large protein enzyme. This is a linear polymer of amino-acid monomers that have 'folded' into a precise form—roughly spherical in the simple, globular enzymes—as the amino acids fit together as a 3-D jigsaw puzzle.

Almost the entire surface of this amino-acid globule has the property of H-bonding nicely with the surrounding water molecules except for a patch that does not. This water-hostile patch is the 'binding site' where the key is inserted into the lock.

The 'key' is a specific small substrate molecule—say D-glucose—that perfectly fits into the binding site. The bound complex now has an entire surface that perfectly H-bonds with the surrounding water molecules. Even a very similar molecule such as L-glucose is the wrong key and will not fit.

The lock and correct key come together as their thermal motion brings them together. When they bump into each other on their random, classical walk through the water, the substrate slots into the binding site of the enzyme; the key is inserted into the lock. Only the precise substrate will fit perfectly, the very similar molecule L-glucose will not do the trick.

When the key is inserted, the entire protein undergoes a conformal change in which an internal section close to the substrate becomes the 'active site.' This is 'turning the key' and the lock is activated. The active site is now positioned next to the substrate.

This is where current biochemistry makes the switch to a unified view. The internal quantum probability wave aspect of the substrate and active site blend together into a new probability form. The external alters to reflect this new form and the external form of the substrate takes up this altered form of the metabolite product. No longer fitting neatly into the binding site, the metabolite is ejected and they go their separate ways.

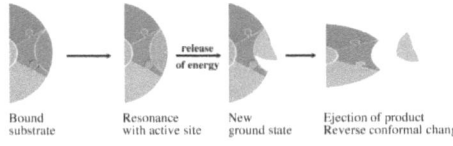

The only difference between the classical view of enzymatic and a unified view is in the very first step, the probability that the substrate will bump into the binding site on its thermal random walk through the water.

The classical view is that of independent assortment. The probability will depend on the random walk of the substrate molecule coinciding with the random walk of the enzyme macromolecule. The volume swept out over time will depend on the thermal velocity. The thermal velocity is inversely proportional to the square-root of the molecular weight. A substrate-glucose molecule has a molecular weight of 180, while that of hexokinase, an enzyme with a binding site for glucose, has a molecular weight of 100,000, so it moves only at 1/20th the speed of the glucose and can be taken as stationary to simplify the discussion.

Using a 2-D illustration, the volume of the water, W, is 100, the volume of the enzyme is 9, and the volume of the substrate is 1. The ratio of the stationary enzyme volume to the water volume is $E/W = 9/100$. If the substrate has a speed of 5, it will sweep out a volume of 5 in 1 second. The probability that they will coincide by independent assortment is 9/100 : 5 which is 9/20 : 1 or about 45%.

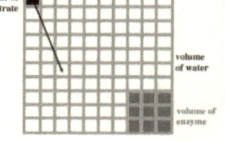

The volume of a glucose molecule in very small, $\sim 5 \times 10^{-23}$ cm^3 and that of the enzyme is just a few magnitudes or so greater. This is their cross section for interaction if only the external aspect of the enzyme and substrate are taken into consideration.

For a reaction to occur, the total cross section of all the enzyme macromolecules plus the total cross section of all the substrate molecules must be commensurate with the volume of the water they are moving through with thermal motion.

External Cross section

In typical milli-molar enzyme experiments, the volume of solution is usually in the cubic-centimeter range. There are trillions of enzyme molecules and substrate molecules wandering about and their combined cross section is greater than this so they do not, so to speak, need to waste time in finding each other. The probability is high that many encounters will happen each second and the reaction will proceed at a reasonable rate.

At extreme dilutions, however, the total cross section for bumping into one another becomes less than the total volume, and they now have to spend time waiting for an encounter, and the rate falls off exponentially with further dilution. The cusp point in the reaction rate is called the kinetic limit and is a measure of the cross section of the substrate and enzyme.

The classical view predicts that the cross section for interaction will be commensurate with the external size of the two molecules. The substrate and enzyme follow independent assortment and only influence each other when in local proximity. The cross section volumes of substrate and enzyme are known, and it is

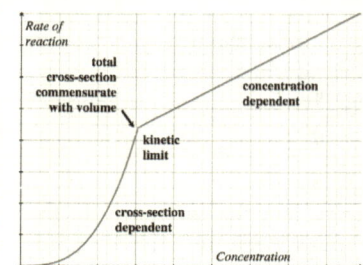

simple to calculate the expected concentration at which the total cross section volume will fall below the solution volume and the kinetic limit will begin to dominate the reaction rate. As the external cross section is so small, the kinetic limit is calculated to be at a relatively high level of ultra dilution.

Excited States

Internal Cross section

In the unified view, the substrate molecule is a system of interacting subsystems in the ground state of the internal wave aspect. The wavefunction extends outwards in a set of un-occupied states, and the system can be excited into these states.

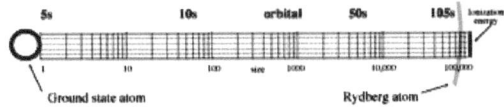

The same thing applies to the enzyme, its subsystems are in the ground state and they can also be excited into the wavefunction that extends outwards.

The very first step in any interaction is the overlap of the internal aspects of the two systems. The resonance between the two determines the probability of external coupling with subsystems—the internal interaction is primary; the external coupling is secondary. The cross section for the internal aspects of the substrate and enzyme to overlap is very much greater that their external cross sections would suggest.

If, in the instant in which the substrate fits into the binding site but before the active site swings into action, a photon of just the right energy hits the complex, the substrate will jump to a higher excited state. A larger photon of just the right energy will kick the substrate into an even higher excited state. This type of behavior is exhibited on a smaller scale in Rydberg Atoms.

Rydberg Atoms

The size of an unexcited atom is $\sim 10^{-10}$ meters, possessing a volume of $\sim 10^{-30}$ cubic meters. A low-energy photon of just the right energy can be absorbed and an electron will jump to a higher orbital, say from the 4s to the 5s orbital. This excited atom has a slightly larger volume than the ground state atom. The photon is rapidly ejected and the atom returns to the ground state.

If an impinging high-energy photon is at or above the ionization energy, however, the electron shoots off at high speed and the atom becomes a positive ion that is smaller than the ground state atom.

By the careful use of tuned lasers, photons of an intermediate energy can kick the electron to orbitals that are fractionally less than the ionization energy, e.g., from the 5s to the 105s orbital. Such a *Rydberg Atom* is somewhat stable—its chemistry can be examined—but the electron soon ejects a cascade of photons as it jumps back towards the ground state.

The size of such a Rydberg

Atom is $\sim10^{-5}$ meters with a volume of $\sim10^{-15}$ cubic meters, a volume that is a million billion times as large as the ground state volume, and is approximately the size of a bacterium. This illustrates how non-local the wave aspect of a system can be, and it should be recalled when we discuss internal waves embracing a mere million or so atoms.

A substrate at a distance far from the enzyme but close enough for their internal aspects to overlap significantly is akin to the excited electron in the Rydberg Atom—it ejects a cascade of photons and jumps towards the ground state. Rather like the tractor beam of science fiction, the enzyme reels the substrate in from a distance.

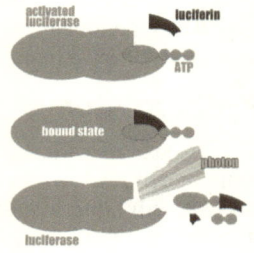

Experiment: Internal or external

Unlike many other disciplines, science insists that experimental protocols have to be devised that can decide if it is the external cross section or the internal cross section that is involved in the very first step of an enzyme and substrate interaction.

If I now had the access to a modern laboratory that I had in my callow youth, I would perform the following experiment to distinguish between the two views using the glowing tubes of greenish lights that appear at many after-dark celebrations. Inside the main tube is a thin-walled glass container that must be crushed to start the contents glowing.

The mix contains the substrate luciferin, a small sulfur-containing molecule, and the 550-aminoacid enzyme, luciferase activated by ATP. The enzyme adds its energy to the substrate and the excited luciferin drops to its ground state with the release of a yellow-green photon. This is the green glow which lasts until either all the ATP or luciferin is used up. This is how fireflies generate light on a summer's evening.

Modern single-photon detectors are quite capable of noting the interaction of a single luciferin molecule with a single activated-luciferase. When solutions of one luciferin molecule and one luciferase are mixed together, the time it takes for the two to bind can be measured. Given a fixed final volume, repetition will give the average time it takes for the two molecules to find each other and bind. Altering the final volume will give an experimental measure of the cross section for the luciferin and luciferase to interact.

This can be compared with the external cross section which is readily calculated from their sizes when in crystal form. Lacking access to such a laboratory, however, we will look to the literature and what experimentalists have already noted about ultra-dilute enzyme reactions.

The Literature

As early as the 1980s, violations of the thermal diffusion kinetic limit expected in enzymatic interactions had been recorded, but not dealt with. A textbook from that era gives it a single sentence, "The action of an enzyme in bringing together two different substrate molecules is more rapid than the rate of combinations which result from chance collision of molecules."[155] No explanation is offered.

Such unexplainable rates were also noted in 2002:

"The [rates] of the enzymes superoxide dismutase, acetylcholinesterase, and triose phosphate isomerase are [above the kinetic limit]... In these cases, there may be attractive electrostatic forces on the enzyme that entice the substrate to the active site. These forces are sometimes referred to poetically as Circe effects."[156]

The collective wisdom of the Wiki in 2013 also notes the violation of the classical, external expectation:

"Some enzymes operate with kinetics which are faster than diffusion rates, which would seem to be impossible. Several mechanisms have been invoked to explain this phenomenon. Some proteins are believed to accelerate catalysis by drawing their substrate in and pre-orienting them by using dipolar electric fields. Other models invoke a quantum-mechanical tunneling explanation, whereby a proton or an electron can tunnel through activation barriers, although for proton tunneling this model remains somewhat controversial."[157]

In simple enzymes, the protein is large and the substrate is small. In other situations, the roles are reversed and it is the protein—thousands of atoms—that is small in relation to the 'substrate' it binds to, such as DNA with billions of atoms:

"A number of vital biological processes rely on fast and precise recognition of a specific DNA sequence (site) by a protein. How can a protein find its site on a long DNA molecule among [a billion or so] decoy sites?"[158]

The binding process is impossibly fast if taking only the external size of the molecules, and only sensible in this age if the internal aspect of an extended quantum probability is included in the picture.

While these few quotes googled from the literature are just hints and intimations, they do give me confidence that the ultra dilute luciferase experiment just outlined would show that the internal cross section was involved in the interaction (though one can never be absolutely sure of experiment) and not support the classical expectation of an independently assorting external cross section.

Influence of the internal aspect

The natural question that now arises is, "Just how far does the influence of the internal aspect extend beyond the external aspect?" The answer to this could be measured directly in the luciferase experiment. Lacking this, we can get a rough estimate by looking at the relationship between two of the main molecules that are the basic subsystems of life. The first is the very smallest molecule of all, with a molecular weight of only 18, the water molecule that abounds in great numbers in all life. The second are the much larger and less numerous molecules, the proteins with molecular weights in the hundreds of thousands.

Water structure

Over 75% of the subsystems of a living system are water molecules, and the structure of this water is directed and controlled primarily by proteins. The structure of bulk water is a reflection of the structure of the water molecule, and this is a reflection of the internal wavefunction.

The oxygen atom has two singlet electrons while the hydrogen atom has one singlet electron. One oxygen and two hydrogen atoms satisfies the valence requirement of paired electrons in filled orbitals in the water molecule. The wavefunction of the covalent bond is such that the electrons spend more than 50% of the time about the oxygen and less than 50% of the time about the hydrogens. The net result is that molecule is polar, with a net negative charge on the oxygen atom and a ½-positive charge on each hydrogen.

This separation of charge on the water molecules makes them very sticky, they are strongly attracted to each other. This is why water molecules are only a perfect gas in steam well over 100° C. The geometry of this charge separation is fully satisfied in a structure when each negative O atom is near to two ½-positive H atoms, and vice versa. The H atom vacillates between the two O atoms as a hydrogen-bond. These requirements result in the open structure of solid ice. In plain water, however, the molecules can only settle into this low energy ground state at 0° C. Above this temperature, the kinetic energy constantly disrupts this stable structure and the water is liquid.

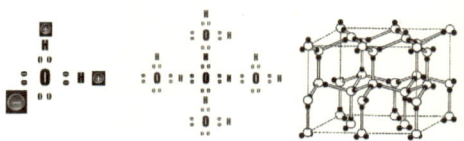

Warm ice

There are macromolecules, however, that can generate a powerful wavefunction in which the ice structure of the surrounding molecules is stable at 20° C and above, a wavefunction in which the stable structure is no longer disrupted by the thermal energy.

Proteins are excellent examples of such water-organizing molecules, and a familiar example is the 'active ingredient' in Jello™ dessert, i.e., the protein 'gelatin.' This is created by methods best-not-discussed of breaking down the large, insoluble proteins of animal cartilage into smaller, soluble proteins.

The contents of a Jello™ packet are mainly sugar along with only about one gram of the dried gelatin protein (which is why the package demurs it is "not a good source of protein"). It is dissolved in two cups (500 grams) of hot water which, when it cools to room temperature, turns into a solid. The majority of the water molecules are now in the ice-like structure even though the temperature is well above the freezing point of sugar water. The confinement of the water is not perfect—the jello is much softer than ice—and jello left at room temperature will slowly weep water as the molecules escape from the wavefunction of the gelatin.

In a jello, one gram of protein is structuring 500 grams of water. The molecular weight of gelatin is ~300,000 and that of water is 18. In solid jello, for each molecule of protein present, there are

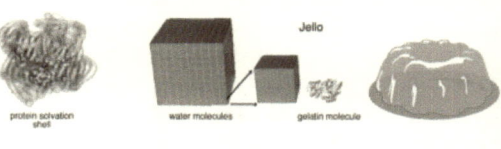

~8,000,000 water molecules, all of them being provided a wavefunction in which they can settle into the stable ice-like structure.

Another biological, if less appetizing, example is to watch a drop of blood flow from a small wound and then, in seconds, turn into a jello-like clot, triggered by a small amount of inactive protein in the fluid blood flipping to a gelatin-like active form and sealing the breach.

In a terminology that we will find useful, we can say that a small number of proteins are 'generating' a wave in which a huge number of water molecules are 'resonating'. This is akin to the role of a small number of musical instruments structuring the large amount of air in a concert hall during the performance of a symphony.

Each gelatin macromolecule can be considered the center of a sphere of eight million water molecules. The influence of the internal aspect of this small protein is magnitudes greater than its external aspect. In earlier terms, we would say that the cross section of the gelatin molecule for interaction is its global internal aspect and not its local external aspect.

The Vital Unit

Now eight million water molecules is a large number of molecules. The smallest systems that are unequivocally alive are the bacteria. A typical bacterium, with a mass of $\sim 10^{-15}$ grams, contains on the order of ten million molecules of water. So just two gelatin molecules are quite capable of generating a wavefunction that allows all the water molecules inside a bacteria-sized container to fall into the stable state. The water is a solid gel.

This bacteria-sized container of water is significant in understanding the basic living system, what we shall call a vital unit. The bacteria-sized containers of the 10 million or so water molecules in a living bacterium is the water-impassible bi-lipid membrane that is an essential component of any living system.

The Bi-lipid Membrane

Lipids are based on the glycerol molecule with its three hydroxyls. Two of these hydroxyls are ester-bonded with two organic acids having long linear non-polar hydrocarbon chains, about twice the length of those in gasoline, that are utterly incapable of forming H-bonds with water molecules. The third hydroxyl is bonded to a variety polar groups, often containing phos-

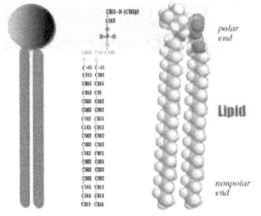

phate, and very capable of forming H-bonds with water molecules. The molecule has two ends, one hydrophilic (*water loving*)—the phosphate, and one hydrophobic (*water hating*). All soaps and detergents have these opposite ends.

The stable state for all concerned is when the polar end of a lipid is in water and the nonpolar end is not. Both ends are in a stable state in a double layer of lipids, with the polar ends of both layers facing the water, and the nonpolar ends facing each other in between.

The bi-lipid membrane is very ancient inasmuch as it occurs in every living system without exception. For all but the slightly different lipids used in the most ancient lineages of bacteria, the basic structure is identical. In the bi-lipid membrane, the lipid chains are stacked inside with the polar ends on either surface facing water on both sides of the membrane.

Water structuring wavefunctions do not penetrate through a bi-lipid membrane. What the water structure is on one side has no influence on what it is on the other. A drop of water coated with such a bi-

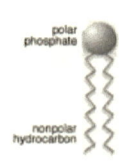
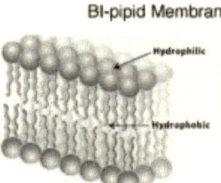
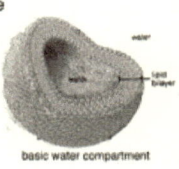

Bi-pipid Membrane

layer, dropped into cooling jello, will remain liquid. The internal waves generated by the gelatin cannot penetrate into the compartment. They are completely blocked and the water inside the compartment remains liquid.

It is customary to consider the cell as the basic unit of life, and the bacteria as having particularly simple cells. The eukaryote cell in this view is basic, and is a view that we think should be reconsidered.

All eukaryote cells—animal, plant and fungi—have an external boundary of bi-lipid membrane just as the bacteria do. The vast difference between the two is that while the water-containing compartment of a tiny bacterium is simple, the much larger eukaryotes are all divided into subcompartments by internal bi-lipid membranes.

These eukaryote subcompartments are of the same magnitude as the bacteria and, in fact some of them are unequivocally descended from bacteria. Each of these subcompartments has portals that connect it to other compartments or to the environment outside the cell.

Molecules, other than proteins, do have a role in controlling the water structure of living organisms. We earlier mentioned that the Rydberg limit was about the size of a bacterium. All living things are composed of compartments of about this size that confine water structuring within a *bi-lipid membrane*. The prokaryotes (bacteria) have an outer lipid membrane and are on the order of this size, while eukaryotes (animals, plants and fungi) are composed of ~10,000 compartments, defined by internal lipid membranes, that are also about this size. The nucleus is also divided into this size subcompartments, but here the work is by proteins and nucleic acids, not lipids.

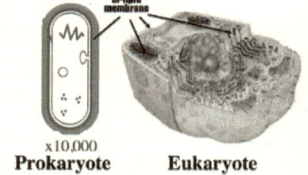

x10,000
Prokaryote   **Eukaryote**
simple compartment   *sub-compartments*

In the systematic hierarchy of the inorganic realm, the atom is the fundamental subsystem. We propose that the fundamental 'atom' of living systems is a volume of water commensurate with bacteria of ~10 million molecules confined by a bi-lipid membrane. Since we will often refer to it, we will call such a volume of confined water a vital unit. Bacteria are then single vital units. Eukaryote cells are 'vital molecules' whose interacting subsystems are vital units.

In the next section, we discuss how proteins generate wavefunctions that structure the water of a vital unit and, in the section after that, how

System	Interacting subsystems	Examples
Vital molecule	Vital Units	Eukaryotes: plant, animal cell
Vital Unit	$\sim 10^7\ H_2O$ + bi-lipid membrane	Procaryotes: bacteria

the analog wavefunctions generated by proteins is recalled from digital memory stored using nucleic acids.

Proteins

Proteins are often subsystems of higher structures—such as membranes—but we start by discussing simple proteins that perform their role in isolation. Proteins have a set of emergent properties that are preeminent at manipulating and controlling the analog forms of molecules, such as the ice structure of water, and the analog form of single molecules, such as in the metabolism of glucose.

All living systems use the same set of 20 'natural' aminoacid monomers that are linked together as protein polymers. Proteins have a wide variety of chemical properties and can basically do all the possible chemistry of carbon-containing compounds in an aqueous environment at regular temperatures. Proteins are master chemists; they can do chemical manipulations still impossible in the laboratory.

All aminoacids have a similar structure with one end a strong hydrogen donor, the other a strong hydrogen acceptor. All of the natural aminoacids monomers of protein chains are linear, except for proline (technically an *iminoacid*), which puts a kink in any chain. In between these complementary polar ends are attached twenty different radical groups with a variety of chemical reactivity, symbolized by R. This radical part of each aminoacid gives each of the 20 a unique set of overlapping chemical reactivities.

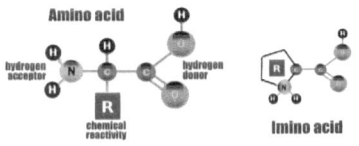

A protein chain is created by linearly connecting the complementary ends of two aminoacids by the elimination of water, creating a peptide bond. Both the oxygen and nitrogen atoms in the peptide bond are still quite capable of hydrogen bonding. At one end of an aminoacid polymer is a free amino group, while at the other is a free acid group.

Unlike most of the chemical manipulations in living systems, the wavefunction for connecting aminoacids with peptide bonds is not provided by a protein but by a nucleic acid assisted by proteins.

Protein structure

Proteins are long linear chains of amino-acids linked by peptide bonds. The order of the aminoacids in the chain is called the *primary structure* of the protein, and is read from the amino to the carboxylic end. These are the first 200 of the 525 aminoacids in the primary structure of human pyruvate kinase, using the one letter code of biochemistry.

```
  1  MSKPHSEAGT AFIQTQQLHA AMADTFLEHM CRLDIDSPPI TARNTGIICT
 51  IGPASRSVET LKEMIKSGMN VARLNFSHGT HEYHAETIKN VRTATESFAS
101  DPILYRPVAV ALDTKGPEIR TGLIKGSGTA EVELKKGATL KITLDNAYME
151  KCDENILWLD YKNICKVVEV GSKIYVDDGL ISLQVKQKGA DFLVTEVENG
```

A chain can link to itself or another by a disulphide bond between two cysteine aminoacids. This is the *secondary structure* of the protein. The much smaller hormone, insulin, has two chains with 1 intra- and 2

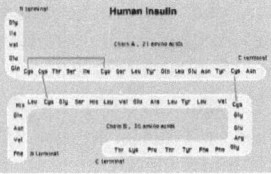

inter- disulphide bonds as its secondary structure.

The folding of the secondary structure into the active folded form with its active sites for water and substrate is called the *tertiary structure* of a protein. Some active proteins contain more than one folded protein, and they way they fit together with a surrounding water shell is called the *quaternary structure*. Hemoglobin has four protein subunits—with two primary structures—each with a heme group attached.

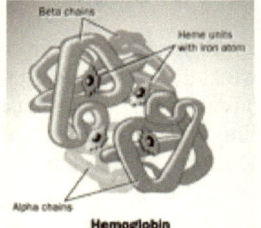

Hemoglobin

The way that the unified wave of the complex changes when a substrate binds to one, two, or more active sites is called *allosteric* conformal change. Haemoglobin is actually a rather poor absorber of oxygen. When one oxygen binds, the whole wave alters and mono-oxyhemoglobin is a much better absorber of oxygen. The wave shifts with another oxygen and di-oxyhemoglobin is much, much better at absorbing oxygen. The wave shifts with another oxygen and tri-oxyhemoglobin is most avid at absorbing oxygen. It is this allosteric effect on the unified wave that allows hemoglobin to pick up oxygen where it is plentiful in the lungs and release it where the concentration is low.

This most useful property comes from the Logos; it is not accidental. Given a primary sequence, a particular form is received from the Logos which, when expressed in particle density, determines which waves are generated and what substrates will resonate with them.

Finite set of forms

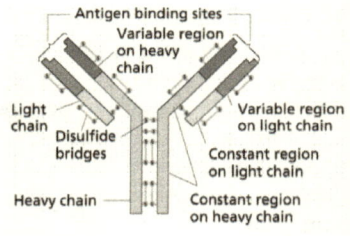

How many such distinct forms does the Logos hold? The workings of the immune system suggest, as follows, that this is a finite number of about 100 billion. The immune system recognizes any non-self macromolecule, the antigen, and generates a barrage of a protein, the antibody, specifically tuned to bind the antigen. Antibodies are 'Y'-shaped with a two-pronged variable end, where the binding occurs, and a constant region which, when the binding occurs, flips from an inactive form to an active form. This active form is recognized by a protein that initiates the destruction of the antigen-antibody complex.

All such antibodies to any macromolecular non-self form are made by the lymphocytes—with each lymphocyte making only one antibody,— all with the same variable region.

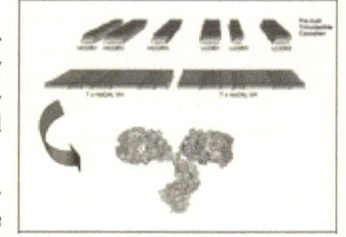

While still in the womb, cells in the bone marrow proliferate into immature lymphocytes. Every one ends up programmed to make a different variable region by being given a permutation of a small set of digital 'cassettes.'

Each one of the 100 billion different permutations are given to one of the 100 billion immature lymphocytes that are generated in the marrow. This army can make an antibody to any macromolecule, which is why we can say that the Logos has a finite number of forms of about 100 billion. The immature lymphocytes then travel to the thymus where they display their particular antibody for testing by a thymus cell. If its antibody happens to be tuned to a substrate the thymus cell can make, the thymus cell orders the immature lymphocyte to self-destruct. If the antibody passes this test, the lymphocyte is ordered to mature and is released back to the blood.

There it circulates for years with its antibody poking out of its surface. The army of mature lymphocytes is now capable of recognizing any of the 100 billion macromolecular forms from the Logos that are not made by the body itself. When this elimination of self-forms misses one by mistake, an autoimmune disease develops.

Protein Active Form

Each amino acid in a long chain has a set of chemical reactivities that must be satisfied either by interactions within the chain or by interactions with the surrounding water. Some of these chemical 'needs' come from the peptide bonds linking the chain together.

In a long chain of amino acids, the hydrogen bonding activity of the polar O and N atoms in the peptide bonds can either be satisfied by H-bonding with water molecules or with polar atoms elsewhere in the same chain or in another chain.

This self H-bonding of the polar peptide bond creates two well-studies motifs that appear in the structure of many different proteins.

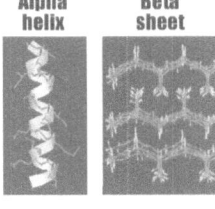

Alpha helix   Beta sheet

One is the alpha helix, in which the aminoacid chain coils as a spring; the other is the beta pleated-sheet in which parallel chains are arranged in layers.

The chemical reactivity of an amino acid in a chain includes the tendency to form an alpha helix or a beta sheet. Those good at being in a helix tend to be poor sheet makers, and vice versa. The kink-creating proline is good at neither and tends to disrupt both structures when it is present.

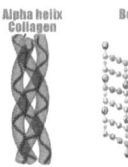

Alpha helix Collagen   Beta sheet Silk

Some proteins are mainly of aminoacids that are very good at helix creating or sheet creating. An example is the protein that makes up collagen, which is rich in aminoacids that are good helix builders. (Incidentally, it is collagen that is broken down in gelatin.) The coils are then supercoiled giving the composite great stretching ability. Another is silk that is rich in sheet building aminoacids. It is this sheet structure that gives silk its extraordinary strength.

Depending on their geometry, all amino acids have an alpha ability—we can assign it 0 for alpha-disruptive, 1 for a mild alpha, and 2 for a strong alpha tendency—and a beta ability—also 0 1 or 2. A list of the 20 will include this in the reactivity and it will look something like this: A long chain of aminoacid B would have a strong tendency to helix while a chain of aminoacid C would have a strong tendency to sheet.

Aminoacid	REACTIVITY	
	Alpha	Beta
aa A	1	1
aa B	2	0
aa C	0	2

Reactivity

The various side chains add to the set of reactivity for the aminoacid, except in glycine with a single hydrogen which adds little to the tally. The side groups on the aminoacids are chemically active, which is probably why protein, unlike other vital macromolecules, is not stored in an inert form to any great extent in living systems. There are 20 universal amino acids used in living systems, each with is own set of chemical reactivities, a few of which are illustrated. Proline always puts a kink in the chain and is used to terminate alpha helixes, composing 17% of collagen fibers.

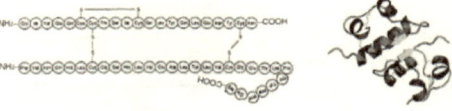

The side groups can be acidic, basic, reducing, oxidizing, H-donor/acceptor, hydrophobic hydrocarbons, etc.—all in different amounts—pretty much covering all the reactions possible in water solutions. This set of 20 different arrays of chemical reactivities in the natural aminoacids is important in the transition of an aminoacid chain from the linear form of its synthesis to the folded active form of the protein.

Each living system has a specific set of proteins that it employs. Some of these sets are so central to life that they are in constant use, the 'housekeeping' proteins, while others in the repertoire are only called upon during certain occasions.

Each of the proteins in this repertoire has a specific and exact 'primary' sequence, the linear list of aminoacids in the chain. The assembly of each protein chain always starts at the free amino end of the chain and concludes at the free carboxyl end.

Each of these linear aminoacid chains in the repertoire then, interacting with the surrounding water, rapidly folds into a 3-D shape with a surface and an interior. The shape that each chain folds into is precisely the same for each primary sequence. This precisely-folded aminoacid chain is the active form of the protein. Many chains spontaneously fold just interacting with water, others need the participation of other proteins called chaperonins. In either case, the end result is a precisely-folded, active protein.

One of the milestones in 20th-century biochemistry was the elucidation of the primary and secondary structure of the small, but important, protein called insulin. It contained two aminoacid chains—one with a proline kink—linked by covalent disulphide bonds between cysteines. X-ray crystallography later revealed the 3-D structure of the precisely folded active insulin protein with its internal coils and sheets.

In the ground state, folded protein, all the chemical requirements of the aminoacids in the chain are satisfied. Examples are the hydrophobic side groups that all cluster in the interior and the alpha-beta requirements of the peptide bonds. Those with an aromatic ring like to stack up together inside, the bent proline does not want to be stressed and the sulfurs like to be paired up. H-acceptors need to be near H-donors in the interior or water at the surface, while H-donors want to be near H-acceptors in the interior or water at the surface. Positive charges must be near negative charges, and so on for all the chemical reactivities.

In the folded state, every chemical requirement is met except for those surrounding the binding site. Every one of the chemical requirements of each aminoacid in the chain is fully met in the precise folding of the active form except at the binding site. The H-bonders at the surface, excluding the binding site, are all perfectly spaced for ice-like bonding with the surrounding water, which it participates in structuring.

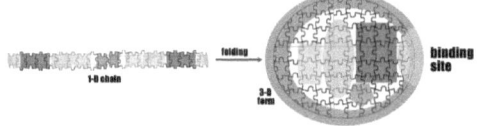

Protein Folding

So each aminoacid ends up in its correct place in the active form in which all its chemical reactivity is balanced. This is akin to connecting all the pieces of a completed jigsaw puzzle to a thread and then jiggling the jigsaw into its component pieces. Protein folding is then putting all the pieces back together again.

The lock-and-key concept of classical biochemistry, which does not take the internal aspect into consideration, has a major problem with random thermal collisions precisely folding a long chain. The problem is there are far, far more ways of incorrectly folding than there are of correctly folding. This problem—and it is known as the Protein Folding Problem—of finding the least energy configuration by exploring all possible configurations is known as the *traveling salesman problem*, TSP.

The task in the TSP is to find the shortest route when a set of towns have to be visited. When there are just a few in the set, exploring all possible routes—the brute strength approach—is feasible. The number of possible routes increases as the factorial, however, and this approach becomes unfeasible for larger sets. The number of pixels of spacetime in the visible universe is minuscule compared to factorial two hundred, $200! = 10^{100} \times 10^{100} \times 10^{100}$. The problem is that chains of aminoacids 500 or more long are quite common.

n	n!
5	120
10	3.6 million
25	15 (trillion)^2
50	3 (trillion)^5
100	>google
200	>(google)^3
500	>(google)^{11}

Allosteric proteins

Even worse, the addition of a simple calcium or phosphate ion can cause the whole chain to fold in a completely different way into another precise form of a protein with a completely different set of properties.

The concentration of calcium ions in solution is firmly controlled in all living systems, and is usually very low inside cells and higher outside. Many proteins are to be found in cells in their folded, but inactive form. They remain inactive until the cell membrane lets calcium ions, in response to some external stimulus, flood into the cell interior. The ions attach to the inactive proteins which all flip into their active form with a set of coordinated consequences, such as a muscle cell contraction.

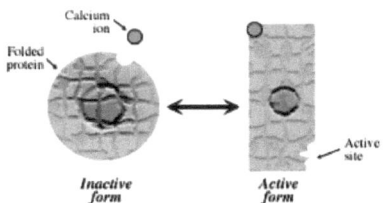

The calcium is quickly cleared from the

cell and expelled back to the outside, the ions disassociate from the proteins and they all, in coordination, flip back to their inert forms. The cell is now ready to respond to the next stimulus.

Such 'allosteric' enzymes often work in cascades. A tiny influx of calcium activates a few thousand molecules of an enzyme that phosphorylates another inactive protein. The addition of the phosphate causes this inactive form to flip to an active form. This active protein then processes thousands of another protein to activate them. By such a cascade, a few calcium ions can alter the activity of huge numbers of proteins.

This change of protein form with the influx of calcium can be forceful, as we routinely experience in the contraction of our muscles, powered by the breakdown of adenosine triphosphate, ATP, the ubiquitous carrier of energy in all living systems. The head of the protein myosin has a bound ATP in an inactive form that is inserted in a ladder-like actin protein. When a calcium ion appears—allowed into the cell by a nervous impulse—the myosin breaks the ATP and changes the protein form that, like a cog, ratchets the actin forward a notch. It is the coordinated stimulation of this conformal change in myosin fold-
ing that is the contraction
force of a muscle

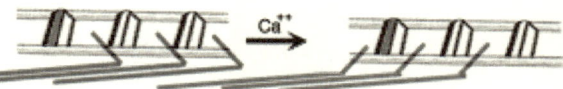

External only

Both chains of aminoacids folding into precise form and allosteric flipping between forms need to be explained. Classical biochemistry looked to the external aspect of the aminoacids and the thermal motions of the chain bringing them into external contact to explain both folding and flipping.

The speed of thermal motion is known, and the number of random collisions the aminoacids in the chain have with each other and with water molecules can be estimated. From this, the time can be estimated for a 500-aminoacid chain to fold precisely if only the collisions of
the external aspect are taken into
account, as in classical biochemistry. A similar calculation estimates the time for the precise
reconfiguration of the chain
when a calcium ion arrives. The
chart gives typical examples.

Classical biochemistry	Estimated time	Measured time
Folding of 500-aminoacid chain	> trillion years	>> second
Flipping form on addition of Ca	> trillion years	>> second

The magnitude the discrepancy between the estimated time for folding and flipping, considering just the external aspect, and the observed time of flipping and folding in experiments is immense. This is why it is called the called the protein folding *problem*.

Including the internal aspect

There are enzymes of ~5,000 aminoacids that are reversibly denatured by heat. Heat a solution of the active folded enzyme above a certain temperature and the enzymatic activity disappears since the proper folding is disrupted by the thermal energy. When the solution is cooled, the enzyme activity returns as all the trillions of enzyme molecules spontaneously refold back into the same precise configuration. Not only does this not take eons, as the lock-and-key concept predicts, it happens so quickly that the process is difficult to observe.

This problem of explaining protein folding with classical concepts is akin to the problem that physics had when attempting to explain the structure of the atom using classical concepts. For, in a universe that obeyed classical concepts, the powerful electric attraction between the proton and electron should pull the two tightly together, just as two powerful magnets cling together so that separating them is difficult. Yet the electron does not cling tightly to the proton, it remains at a distance that is ~10,000 times the diameter of the proton. What could possibly be holding the electron up there against the powerful force of attraction? Nothing in the classical panoply would suffice.

It was only when classical concepts were abandoned and the internal wave aspect was taken into account that the extended structure of the atom began to make sense. Following this clue, we will see what protein folding looks like when the internal aspect is taken into account.

A classical solid is always local and never nonlocal. A classical solid is never in two places at the same time, and needs to traverse the spatial separation between them to get from one place to the other. Classical biochemists consider the atoms in an aminoacid chain to be classical solids that are always local and never nonlocal. They might admit that the atoms were made of things, such as electrons and photons, that had decidedly nonlocal behaviors, but that this was a subtlety that could be ignored when considering something as large as a 500,000-aminoacid chain.

But it is this very ignoring of the non-local, non-solid internal aspect of the atoms that creates the protein folding problem. We earlier saw that the system wavefunction of the helium atom totally confines all its subsystems, none of them—electron, virtual photon, quark and gluon—escape or enter the atom in normal circumstances. The wavefunction itself is confined, and is everywhere zero beyond the bounds of the atom.

The helium atom does behave as a classical solid, a Newtonian massy sphere, and is a perfect gas when all other atoms are not. In its thermal collisions, it behaves exactly as would two colliding billiard balls. Only when the thermal kinetic energy is minimal near absolute zero do helium atoms all abandon their classic perfection and, with utter abandon, merge their internal aspects into a single wavefunction, and the atoms are united as a superfluid—superconducting quantum fluid—that has properties that are decidedly nonlocal.

To a lesser extent, all the other noble gases, such as neon, are classical solids like helium that have a system wave that perfectly confines the subsystems. None of these elements has a role in living systems, however, and all the other atoms are characterized by a system wave that does not perfectly confine the subsystems; they have chemical reactivity. In certain situations, the atoms used by living systems are nonlocal and do not behave at all like a classical solid.

Not-solid atoms

Two examples of the atoms in aminoacids behaving non-locally and not as classical solids in everyday situations are the nitrogen atom in the ammonia molecule and the carbon atom in the diamond atom.

Earlier we discussed internal nodes in standing waves, where an electron was 50% in both lobes but never in between the two lobes. A similar situation is found in the ammonia molecule—three hydrogens bonded to a single nitrogen atom—where thermal energy has kicked it into a slightly excited state.

The nitrogen atom has five electrons in the valence outer shell of electrons. In ammonia, three of these couple a covalent bond with the three hydrogens, and the remaining two are in a lone pair that the ammonia molecule can share with others. Its core subsystems that do not participate in coupling are the two inner electrons, seven protons, seven neutrons and innumerable photons and gluons of the nucleus that are all firmly confined.

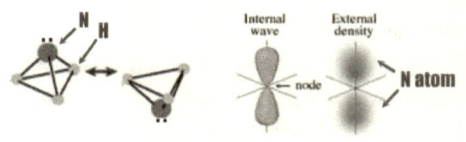

The slightly-excited ammonia molecule resonates at a constant frequency between two configurations; the core nitrogen can be above the plane of the three hydrogens, or it can be below it. The lone pair is on the top of the molecule in one configuration, at the bottom in the other. The core nitrogen vibrates between these two states at a constant frequency while the H atoms remain stationary.

The internal wave has a node in the H-plane and the probability of the nitrogen core being there is exactly zero. This is similar to the p-orbital we have already discussed—the electron is 50% in either lobe and never in between them. An electron behaving strangely is one thing, but the nitrogen is substantial matter, at least in the classical view. This substantial atom is 50% in one place, 50% in another place, but never in between them. All 40-odd core subsystems and the lone pair are 50% in one lobe and 50% in the other, but none of them is ever in between the two lobes. Teleportation perhaps, but the technical term is *tunneling* between the two lobes. This is an odd tunnel; it has zero length but its two ends are in different places.

The nitrogen atom is clearly capable of non-local behavior that is decidedly not what a solid, massy local atom is supposed to do. Carbon atoms can go in quite a different direction from being a local, solid atom when zillions of them blend their individual internal aspects into a single entity. The wavefunction of a diamond, a giant molecule of just carbon. This wavefunction

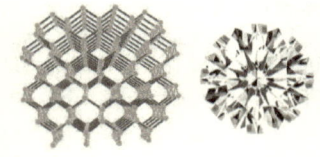

so fiercely confines its subsystems that diamond is a super-solid and is the hardest known solid. In a natural diamond, while most atoms are C-12, a few are C-13, C-14 and N-14, which spoils the overall perfect resonance. When pure C-12 diamonds are created, they are expected to be harder than natural diamonds.

This non-local behavior of carbon atoms, like that of nitrogen, is not what local, classical solid atoms are supposed to do. If carbon and nitrogen atoms can behave in a non-local fashion in one molecule, there is no good reason to expect that every atom in a long aminoacid chain cannot do likewise.

Protein Folding

Just as a single nitrogen can tunnel, all we are suggesting is that all the atoms in all the aminoacids in an extended chain ca also tunnel from the excited state to the ground state without being anywhere in between. The linear chain of core atoms and coupling electrons and surrounding water is in an excited state, just like a Rydberg Atom but on a larger scale. This excited state then rapidly sheds a cascade of energy as it jumps to the ground state, and becomes a precisely-folded protein surrounded by structured water.

A measure of the energy difference between the ground state and the first excited state is the transition temperature below which reversible enzymes are in the active ground state, while above that temperature, they are inactive and in an excited state. This is a sharp transition point where the average kinetic energy at that temperature equals the energy of the first excited state.

The probability density of the atoms changes from the ground state to the excited state, a non-local change in the wavefunction, without atoms having to move as solid massy spheres would move in a local fashion. On cooling, the macromolecule sheds energy and falls into the ground state, the precise form it had before the warming episode.

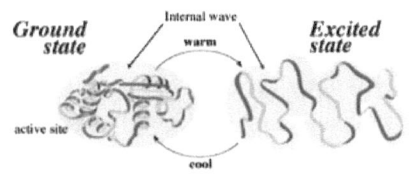

This transition from an excited state to the ground state occurs quickly. This can be observed in the distinctive color of sodium atoms when kicked into the first excited state by the thermal energy of a flame. The excited atoms, all in the same state, quickly jump back to the ground state, releasing a flood of identical, yellow photons. The same color is created in street lights using electric energy to excite the sodium atoms (Na). In my high-school chemistry class, I spent many a pleasant hour identifying elements by their various colors. Copper (Cu) gives a green flame, rubidium (Rb) a crimson, etc.

Each photon liberated has 1 unit of the action but differently distributed between the energy and time period of the electromagnetic oscillation in the oriented boson. In natural units, the energy, E, and time period, T, are the reciprocals of each other. The Rb's red photons have the least E and most T, the Cu's green photons have the most E and least T, while the Na's yellow photons have an intermediate value.

$$E = \frac{1}{T}$$
$$ET = 1$$
$$T = \frac{1}{E}$$

This rapid transition to the ground state applies to the natal aminoacid chain in its unfolded, extended linear form. It is in one of the excited states and the macromolecule sheds energy and falls into the ground state, which is the precise form of the protein. All the aminoacids are in stable surroundings except for those in the active site. The jigsaw has a mold to fall into, so to speak.

The addition of a calcium puts the chain in an excited state, and it sheds energy and jumps to the new ground state, the allosteric twin of the original form. As jumps between quantum states occur with great rapidity, the folding and flipping of an aminoacid chain can be expected to

Unified biochemistry	Estimated time	Measured time
Folding of 500-aminoacid chain	>> second	>> second
Flipping form on addition of Ca	>> second	>> second

happen very quickly. When the internal aspect is included in the picture, the protein folding problem disappears. When the internal aspect is included in the picture in a *unified* biochemistry—one where both the internal and external aspects are taken into account—the estimated and observed times are no longer in violent disagreement. This is just how physics was forced to accept the wavefunction as being real, since it kept the atom from collapsing. Protein folding, for the opposite reason, will eventually force all biochemists to include the internal aspect in all the foundations of living systems. It is only at this point that biochemistry can be called a unified science.

Sense of smell

That the internal wave aspect cannot be ignored in macro systems—and this is the prevailing view—but plays a role at every level is supported by the science of smell. This was originally explained by the lock-and-key concept, i.e., the odor molecule would fit into a receptor and trigger a response. The problem was that odor molecules whisk by so rapidly that there is no time to stop and bind.

A competing theory proposes that the odor molecule and the receptor protein resonate together momentarily on an internal level without the molecule actually stopping, an example of an excited state that can bring about a change in the protein structure and register a signal. Support for the resonance view over the locking view is growing, as noted in a recent review article:

> How does the sense of smell work? Today two competing camps of scientists are at war over this very question. And the more controversial theory has just received important new experimental confirmation.
>
> At issue is whether our noses use delicate quantum mechanisms for sensing the vibrations of odor molecules (aka odorants). Does the nose, in other words, read off the chemical makeup of a mystery odorant—say, a waft of perfume or the aroma of wilted lettuce—by "ringing" it like a bell? ...
>
> The predominant theory of smell today says: No way. The millions of different odorants in the world are a little more like puzzle pieces, it suggests. And our noses contain scores of different kinds of receptors that each prefer to bind with specific types of pieces. The finding represents a victory for the vibration theory....[159]

Water structuring

In the methane molecule, one carbon and four hydrogens, the internal wave is such that electrons are 50% with the carbon and 50% with the hydrogens. This is a nonpolar covalent bond. In the water molecule, one oxygen and two hydrogens, the internal wave is such that the electrons are 60% with the oxygen and only 40% with the hydrogens. The hydrogens, so to speak, each have a 10% positive charge and the oxygen has a 20% negative charge. This is a polar covalent bond.

One of the fundamentals that influence the wavefunction is a separation of charge, the ground state will either spread out the charge as much as possible, or unite it with an equal and opposite charge. This is accomplished in a water molecule having two positive hydrogens from other molecules attached to the negative oxygen, and its positive hydrogens attached to two negative oxygens on other molecules.

The stable H-bond is very directional, the coupling H is on a straight line, any asymmetry of the 180° line is not stable. In the three dimensions, the ground state of H-bonding is open hexagonal rings. This is the structure of ice when all the H-bonds are in their ground state.

The great difference that H-bonding creates is seen in the difference between the boiling point of methane and water. The boiling point is when all intermolecular interaction is overcome by the thermal energy and the molecule flies free as a gas. The molecular weights of the two molecules, 16 for methane and 18 for water, are similar. Methane has a boiling point of 109° K (degrees above absolute zero) while water has a boiling point of 373° K, a difference of 264° K. This difference is due solely to the H-bonding capacity of water. The attraction that water molecules have for each other can be directly observed as the considerable surface tension at a water/air interface.

When a crystal of sodium chloride dissolves in water, it does so because each ion becomes surrounded by the appropriate end of the water molecule—the positive sodium by the negative oxygen end and the negative chloride by the positive hydrogen end. This is the solvation, or hydration, sphere that spreads out the charge to such an extent that the attraction holding the two ions together in the crystal is neutralized and the ions move into solution. The hydroxyl in methanol has H-bonding capacity, and it is also soluble in water with an H-bonded solvation shell.

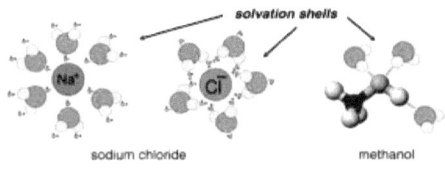

sodium chloride   methanol

In living systems, proteins are all surrounded by water and they also have a solvation shell about them (even if, for other reasons, they are not in solution). In fact, many of these water molecules should really be considered subsystems of the protein since they are a constant part of the folded structure. This also happens when ions come out of solution and form a crystal. Both sodium and chloride ions abandon their water companions, but this does not always happen. When zinc and chloride leave solution they usually take four molecules of water with them into the crystal, the 'water of crystallization'.

This applies to the ground state of a folded protein. The pattern of electric charges and H-bonding must allow the surrounding water to take up the stable ice-like configuration. The ground-state internal wave of the protein is structuring this water shell as much as it is structuring the aminoacid chain.

Vital Unit water

In a vital unit, each small volume contains proteins that generate a resonance that structures the surrounding water—solid, liquid or in between. This water, which is ice-structured by the internal waves generated by proteins, is akin to the air in a concert hall being pressure-structured by the external waves generated by musical instruments. Each particular state of the water is the harmony of all their waves.

The proteins also generate waves that resonate with their particular substrate. This is like harps, trumpets, etc., all playing at the same time in the harmony. Just as a listener has no problem distinguishing the

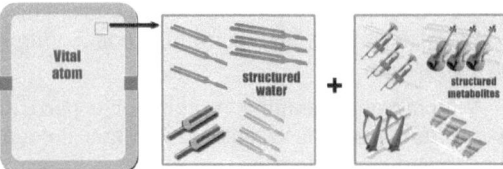

sound of the tuning forks, the sound of the violins, or the sound of the xylophones, the substrates have no problem distinguishing their particular protein 'instrument' playing in the harmonious resonance of the vital unit.

All living systems are 75% water, and the structure all of this water is firmly controlled at all times in the healthy organism. Almost all of this structuring is accomplished by proteins although the other vital macromolecules do play a role. As 75% of a living system, we can say that 75% of the collective wavefunction generated by proteins is structuring this water. The other 25% of the collective wavefunction being generated is structuring all the other molecules in the 'symphony' that is the healthy cell. This water-structuring role can be preeminent, such as it must be in growing embryos that can be greater than 95% water, with everything else being less than 5%.

While proteins structuring water in a harmoniously-unified thriving vital unit is more sophisticated than musical instruments structuring air in the performance of a harmoniously-unified symphony, the analogy is useful and informative in generating the following summary.

Generator and resonator

In a symphony, a small number of massive, multimolecular instruments generate waves that structure a large number of light-weight resonating air molecules.

In a living system, a small number of massive proteins generate internal waves that structure a large number of light-weight water molecules.

Confinement and portals

In a concert hall, the resonating air molecules are confined by the surrounding soundproof walls (it is this lack of resonance that makes an outdoor symphony sound so flat). In this confining barrier, however, there are portals that allow for interaction with the outside, the passage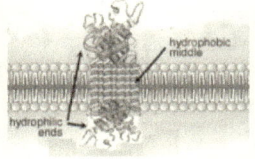
through these doors being highly controlled, such as in Carnegie Hall, and involve an expenditure of money.

In a vital unit, the resonating water molecules are confined by the water-impassible bi-lipid membrane. In this confining barrier, however, there are portals that allow for interaction with the outside, the passage through these portals being highly controlled by proteins that span the membrane, and involve an expenditure of energy.

Lipids are occasionally enzymatically attached to proteins whence they radically alter the internal wave. There is a whole class of proteins that have a hydrophobic middle with two hydrophilic ends. They span the membrane and perform dozens, if not hundreds, of roles in connecting what is going on on either side of the membrane.

A membrane-spanning protein for a particular small molecule—the ligand—has a binding site at its outside end that recognizes and binds the ligand (just as enzyme and substrate do).

The ligand-protein flips its conformation with a variety of consequences:

- The ligand is now on the inside where it is released. This is 'passive' transport and the ligand can pass out just as easily so the internal and external ligand concentrations will be similar. If an ATP is split during the conformal change, this is 'active' transport and is inwards only, so the internal concentration can be much higher than the external.

- The inside end becomes an enzyme. An example is the binding of a messenger molecule, such as insulin, to the outside end with the inside enzyme that generates a 'second messenger', which creates a cascade of consequences. An example is the inner end becoming an adenyl cyclase that converts ATP into cyclic-AMP (cAMP). Almost every living system uses cAMP as a second messenger even if the consequences that ensue can be very different.

If a single vital-atom bacteria is like a single symphony, then the compartmentalized eukaryote is like a vast multiplex where the soundtracks of the many movies playing do not interfere too much with each other. The main difference is that all the activities of each vital unit in the vital molecule are coordinated and unified, a topic we will discuss in detail when we eventually move from the simple prokaryotes to the eukaryotes.

Analog form and digital information

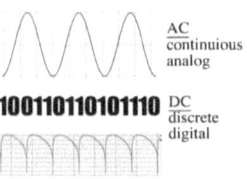

The final, and most useful, analogy between a symphony and a bacteria is the relation between analog form content and digital information content, which we will abbreviate as AC and DC. The key difference between the two is while AC involves smooth, continuous transitions, DC involves abrupt, all-or-nothing, discrete transitions.

A simple example of AC form is a sine wave and a simple example of DC information is a binary string. Both the pressure waves in a concert hall and the wavefunctions generated by proteins in a vital unit are examples of complicated analog waveforms. The analog forms, in both cases, are generated according to digital information. In a symphony, the music being played is determined by the score distributed to the musicians before the start. Each instrument translates its part of the DC score, generating an analog waveform that it contributes to the harmonious whole. If a different score is distributed, an entirely different analog form is generated.

There is a subtle difference between this method of generating analog waves from digital information and the method used in living systems. A musical instrument/performer is capable of generating a large number of different sound waves, and either generating or not generating them according to the digital score. The score and instrument are separate entities, and the score can easily be changed to a different one.

In living systems, the score and instrument are one and the same entity. The score itself is the generator of the analog wave, a wave that is generated continuously and never turned off. The wave only ceases when the protein is destroyed—and most proteins present in a vital unit are in a constant turnover—or it undergoes a conformal change and starts generating a new wave. Most proteins have at least two conformal states, and some have a dozen or more conformal states.

Compared to a piano, a particular aminoacid chain has a small set of waves it can generate, and then only continuously.

The primary structure of a protein wave generator is digital, it is an all-or-nothing situation of a particular aminoacid being present in the chain. Each digital primary structure is an instruction to generate a particular wave. Biochemists have a single-letter code for aminoacids, so the primary structure, starting at the free-amino end, looks like a paragraph with no spaces between words. Each primary structure is a digital instruction regarding how to generate a specific waveform.

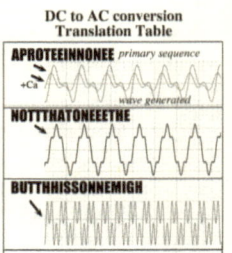

This is a 'translation table' that associates each primary structure with generating a specific analog wave, a DC to AC conversion.

A symphony changes the AC wave by changing the DC score, not the instruments. A living system changes the AC wave by changing the DC score and the wave generators. Most living atoms have a limited set of suites of proteins, of symphonies they can play—these are called the states of a bacterium.

Bacterial states

A bacterium has set states that that depend on what is traversing its bi-lipid portals to the outside world, states that can be as different as Beethoven and the Beatles. A partial listing of these various states is:

Abundant state

Everything that is required is freely available from the outside. The suite of proteins that is deployed in this situation generates a wave in which metabolism is directed to grow and divide.

The members of the suite rise and fall as the stages are run through in cycles—the movements of the symphony—as the cell grows and divides. The suite of proteins is a suite of DC instructions that are generating this analog wave. This is the digital Grow and Multiply score or, in computer lingo, the *GrowMultiply* program.

In an abundant environment running this program, a bacteria can divide in this way every 20 minutes, three times each hour. Assuming the food does not run out, in one day a single bacterium can multiply into 2^{72} bacteria, or about a billion trillion of them.

When running the *GrowMultiply* program, the vital unit makes a lot of housekeeping proteins and enlarges to a maximum size. Sensing that the compartment is growing to a size beyond the capacity of the wave, the cell switches to the *Divide* subroutine of asexual multiplication.

1. The cell reaches maximum size and the water-metabolic wave changes to the first movement of the *Divide* symphony.

2. Proteins and RNA copy the long-term digital information stored on DNA. The double helix is split into its two strands. Each strand is then complemented to create two double helixes identical to the original. The second movement of the *Divide* symphony begins.

3.   Proteins physically separate the two duplexes and constrict the cell membrane between them. The final movement movement of the *Divide* symphony begins.

4.   Proteins pinch off the cell membrane resulting in two separate compartments, two small but identical clones of the original mother cell.

Collegiate state

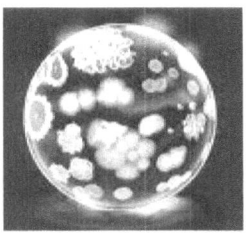

When a few bacteria scattered on a petri plate of nutritious jelly are left to incubate overnight, each cell multiplies into a 'bacterial colony' with a characteristic form and color.

The suite-of-proteins DC program that is running in each identical bacterium is Collegial, and this directs resources into recycling and maintenance rather than multiplication. The environment is mainly clonal sisters.

Scarcity state

An isolated bacterium in a so-so environment will use the differential inputs of its portals to detect gradients. The suite of DC proteins deployed will then divert resources into its motile ability so that it moves up a gradient of molecules it likes, and down a gradient it dislikes.

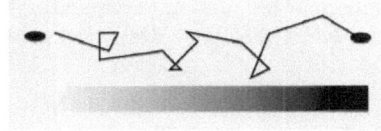

Famine state

In a situation when the environment is totally hostile and core metabolism of ATP and glucose fragments starts to falter, the DC program being run is changed to Famine.

The suite of proteins now generates a wave that diverts all remaining resources to drastically altering water and metabolism in an asymmetrical cell division: one half self-destructs, while the other settles into a compact, dehydrated form, a spore with a very minimal metabolism.

If the symphony of life is liquid change, that of the spore is frozen stasis. Bacteria can remain in this frozen-life state for a remarkable length of time—the longevity of a dry dust of anthrax spores is legendary—and only a Phillip Glass symphony can rival such an endlessly unchanging chord as in Famine.

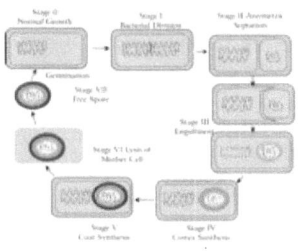

This dormant form includes a stash of ATP cached with the portal-detectors. When their binding sites signal both water and food, the conformal change generates a cyclase activity that turns the ATP into cAMP, the second messenger. This is the signal to switch the DC score and the spore awakes as the *GrowMultiply* program starts up.

Computer Programs

I have a friend who is a music composer. On his ever-versatile Mac he has what he calls 'an orchestra in a box.' It converts a DC score directly into an AC wave form that is an excellent facsimile of a recording by the Berlin Philharmonic.

"You want 100 first-violins? 20 Renaissance horns? 10 timpani? 30 harps? No problem." So the segue from music scores to computer programs is a smooth one.

In the computer world, the food-and-water detector of the spore is akin to a port that can be in state 1 or state 0, with the 0 state being absence of cAMP, and the 1 state being the presence of cAMP.

In a computer programming language, the Famine program would be an idle, do-nothing loop with an exit:

```
IF PORT = 0 THEN RUN FAMINE ELSE RUN GROWMULTIPLY
```

To conserve resources, this program need only be run once an hour by the digital processing system we have yet to discuss. All living systems have such digital programming subsystems, and much of housekeeping metabolism is run by such routines.

To understand the manipulation of digital information in living systems, we have to study another type of polymer—the nucleic acids, RNA and DNA. Just as the proteins are the master manipulators of analog form, the nucleic acids are master manipulators of digital information. RNA is also capable of a small, if primal, subset of the analog manipulations proteins are capable of. An example of this is the RNA that provides the wavefunction for the peptide-bond linking of aminoacids into a long chain.

In order to discuss the nucleic acids, however, we need to discuss yet another class of monomer-polymer, the carbohydrates, that are a component of nucleic acids.

Metabolic relations

Carbohydrates and carbohydrate-fragments are central to metabolism, all the other polymer-classes spin off from this center. The housekeeping core of all living systems, from bacteria to Man, are all remarkably similar.

The flow of atoms in this core metabolism centers on the carbohydrates, which can either be transformed into the other monomers or broken down to carbon dioxide and water, an oxidation whose energy is trapped in the high-energy bonds of ATP, adenosine triphosphate, one of the nucleic bases. This is an irreversible reaction. Plants and some bacteria use the energy of sunlight-photons to reverse this step and transform carbon dioxide and water into carbohydrate and free oxygen. 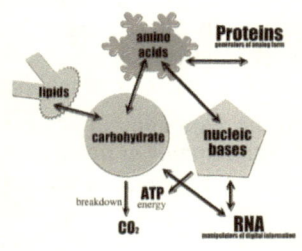 From the carbohydrate, along with a few inorganic ions, a plant can synthesize all the other types of molecules.

Carbohydrates can also be reversibly transformed into lipids or, with reduced nitrogen added, into all 20 natural aminoacids. Fragments of these aminoacids are fused into rings to generate the four natural bases of RNA—adenine and guanine with two rings, and uracil and cytosine with one ring. A nucleic base attached to a carbohydrate (ribose) attached to a phosphate is the basic monomer out of which the RNA polymer is constructed.

A note about DNA. For a host of reasons we will later discuss, and central to the consensus about origins called the 'RNA World', DNA is actually just one of the many varieties of RNA. All the subunits of DNA are created by altering the subunits of RNA to reduce their polarity. The polar free hydroxyl on the ribose is replaced by a nonpolar hydrogen, and a nonpolar methyl group—a drop of oil—is added to one of the four nucleic bases. In the DNA polymer, all the polar hydroxyls along the backbone of RNA are replaced by nonpolar methyl groups—severely reducing the polymer's ability to H-bond with water—and one-quarter of all the bases have a small hydrogen replaced by a large methyl group—further enhancing the hydrophobic tendency of the ground state wavefunction. The 'deoxy' of DNA is a recognition of its drastic loss of polar attributes.

While most varieties of RNA are facile with water and fold up into a wide variety of analog forms with a wide range of activities, oily DNA folds into only one analog form, the famous double helix.

This oily variety of RNA has one preeminent property—it is a remarkably stable macromolecule. While the carbohydrates, lipids, proteins and RNA of insects trapped in amber in the dinosaur age have all disintegrated, DNA that has remained stable for 100 million years is still there, its digital information almost intact.

If the many digital manipulations by all the other varieties of RNA are likened to the active CPU of a computer, then the passive DNA is as a hard drive to which digital information can be written, duplicated, read and translated. DNA is a store of digital information that is stable down a lineage through deep time.

In most of the following, we will be discussing just RNA, but before we can do that, we have to understand its carbohydrate component. So, our discussion will progress from carbohydrate to RNA, and then finally to the manipulation of digital information.

Carbohydrates

All the manipulations we are about to describe are performed by protein catalysts. A simple 'catalyst' acts by providing waves in which the activation energy of the transition state is lowered. The thermal energy is then sufficient to excite the transition state and the reaction rapidly proceeds. The product, which no longer fits, is expelled and another substrate takes its place.

An example is hydrogen peroxide which has more free energy than water and oxygen. It is metastable because it has an activation energy beyond the reach of thermal energy. So a peroxide solution is quite stable. Put it on a wound, however, and the enzyme 'peroxidase' goes to work lowering its activation energy. The solution bubbles furiously as the peroxide decomposes into water and a singlet oxygen (which disinfects the wound by its reactivity). A sin-

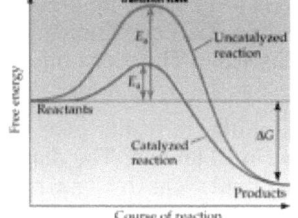

gle peroxidase molecule can do this for over a billion peroxide molecules a second. A reaction can also go 'uphill' if an ATP supplies the energy that is in the products.

We will be using the term carbohydrate generically for polymers, monomers and monomer fragments. All carbohydrates are based on the motif H-C-OH, which looks like a carbon atom inserted into a water molecule, hence the name. They are classified by the number of carbon atoms: the glycerol bit in lipids being a linear

3-carbon molecule, the ribose bit in nucleic acids being a 5-carbon ring, and the glucose in sugar being a 6-carbon ring. Since the hydroxyls at different locations have different properties, the carbohydrates have a consistent convention for numbering the carbon atoms.

The hydroxyls of two monomers form a bond, eliminating water, and link up into carbohydrate polymers studded with hydroxyls that are excellent H-bonders with water or other molecules. In the illustration, the two glucose molecules are linked by a 1-4 elimination of water creating a maltose, a disaccharide. The most common organic compound on Earth is cellulose, a large polymer of glucose all linked in the same way. The unused hydroxyls on adjacent chains can also form cross-links between chains.

Most of the carbohydrates in metabolism have one or more phosphates ester-bonded to the hydroxyls. These phosphates are all stripped away and recycled in creating all of the end products except RNA, which has a sugar-phosphate backbone. It is this abundance of phosphate that is the acid in nucleic acid.

Central metabolism

In the lock-and-key classical picture, a metabolite moves through the system by thermal motion. The substrate molecule bumps into the first enzyme, is processed, then released. This first intermediate metabolite molecule moves about in thermal motion until it bumps into the second enzyme, is processed and released as the second intermediate. This moves about by thermal motion until it bumps into the third enzyme, is processed, and released as the third intermediate, and so on until the final step.

The view of metabolism that includes the internal wave is not so haphazard. Each intermediate is in an excited state with respect to the next enzyme in the sequence; it moves to the ground state of being bound to the enzyme from wherever it is within the lipid-bound compartment. There are no concentration gradients within the compartment because the internal wave ignores spatial separation on this scale. The metabolites are structured by internal waves generated by proteins, just as the water molecules are structured.

Either way, the steps by which molecules are transformed into other molecules are well-known. The diagram is the sequence of manipulations by enzymes wherein a 6-carbon glucose molecule is activated by the addition of a phosphate, flipped into fructose, and another phosphate added. The molecule is then split into two identical halves. The phosphates are then removed and, after one more transformation, the glucose molecule has been transformed into two molecules of 3-carbon pyruvate. This small molecule can be considered a central molecule in metabolism since so many metabolic paths lead onwards from it.

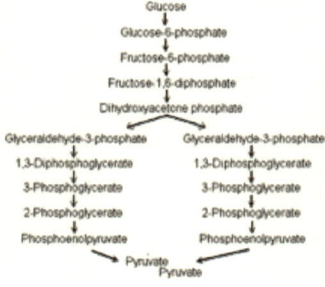

Pyruvate can go into lipids or aminoacids, or it can end up as alcohol or lactic acid in anaerobic respiration. In respiration using oxygen, the pyruvate is fed into a cycle of carbohydrates, starting with citric acid, that splits off carbon dioxide, feeds energy into ATP, and returns back to citric acid.

Krebs cycle

Pyruvate is one of the central molecules in the core metabolism common to all life. At the very core of metabolism is the Krebs cycle (aka the citric acid cycle) whose intermediates are important beginning points for the assembly of other molecules such as aminoacids and nucleotides. Running in the direction illustrated, the 3-carbon pyruvate first gives up a hydrogen to the carrier, NAD, and then releases a molecule of carbon and forms a thioester with a molecule called Coenzyme A. This transfers the 2-carbon fragment to the 4-carbon oxaloacetate (OAA) to form the 6-carbon citrate.

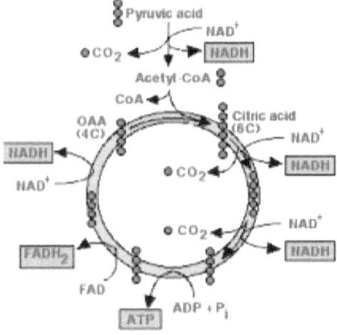

This progressively gives up hydrogen to the hydrogen carriers—the nucleotides NAD and FAD—loses two carbon dioxide molecules, phosphorylates ADP to ATP, and ends up as oxaloacetate to start the cycle again. The activated hydrogens liberated can be used by all organisms to drive chemical transformations and, in aerobic organisms, are fed into the pathway that combines them with oxygen powering the addition of phosphate to ADP generating ATP. Running in this direction, the cycle creates energy and releases carbon dioxide.

The cycle can also run in the opposite direction, given a source of ATP and activated hydrogen when it incorporates carbon dioxide into more complex molecules. A more complicated cycle of carbohydrates, but in reverse, is used by plants to fix carbon. The cycle starts by combining phosphorylated-ribulose with a carbon dioxide molecule. It ends by turning three of them into one of the phosphorylated 3-carbon precursors of pyruvate plus a ribulose to start the cycle over again. The carbohydrate cycle of this 'dark process' is driven along by using the high-energy products created by the light-capturing mechanisms of photosynthesis. From this 3-carbon carbohydrate, core metabolism can create any of the other molecules of life, including the 5-carbon-ribose that is a subunit of RNA.

Polysaccharides

The table sugar we are most familiar with is a disaccharide, a linking of a 6-ring glucose and a 5-ring fructose. There are plenty of hydroxyl groups which are mild H-bonders: sugar is very soluble in water, and structures water somewhat; it gets viscous and sticky. The molecular wave embraces a small shell of structured water.

The most abundant polysaccharide is the cellulose that all plant cells construct for support. This is innumerable glucose units linked one way. Starch is

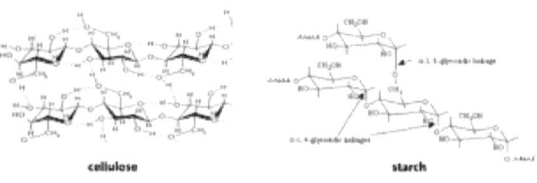

cellulose   starch

also made of glucose, but they are linked a different, and branching way. Unlike cellulose, it is soluble in water and forms a sticky glue (or, more palatable to adults, a thick sauce when corn starch is used in cooking).

The very different and well-designed properties of cellulose and starch come from the Logos and are expressed in the unified, macromolecular chord of the quantum wave.

Saccharides often adorn proteins, put there by enzymes, and they tune the waves that are generated. Some of the molecules constructed by manipulation of simple structures from core metabolism are large indeed and have remarkable properties inherited from the Logos. One of these is based on a ring called a porphyrin (pronounced POR-fur-in). It is a sophisticated wave with the form of 'aromatic' rings in an aromatic ring.

The porphyrin ring is adorned with a variety of 'handles' that allow it to link to other molecules. The hole in the center is filled by a metal ion coordinated by the four nitrogen atoms. The most familiar porphyrin is haemoglobin (not used in bacteria), which has an iron in the center and four rings attached by their handles to a protein complex. It provides a wave that resonates with oxygen. In high concentrations, oxygen binds it to the iron; in low concentrations, it releases the oxygen. This is how blood carries oxygen from the lungs to the rest of the body.

Chlorophyll has a magnesium atom in the center, and its handles are also connected to a protein complex. A red or blue photon can merge its wave with the delocalized electrons, kicking them into an excited state, a state where they are localized where they can be syphoned off onto the top of a chute of 'electron-transport' proteins. As they fall to the ground state down the chute, their energy is captured in ATP, which can be used to turn carbon dioxide into glucose. This is how the sun provides all the energy for life.

Most large molecules in living systems are polymers of small monomers, and the largest molecules are the nucleic acids that can contain billions of atoms.

Ribonucleic Acid, RNA

The RNA polymer is composed of nucleotide monomers all linked together by ether bonds between a sugar and a phosphate. There are only a few nitrogenous bases making up the natural nucleotides used by all living systems, these bases being synthesized from amino-acid fragments. The bases are linked to ribose which is linked to phosphate.

Nucleotides

The nucleotide, adenosine triphosphate, ATP, is the energy supplying cofactor in a host of enzymes, and energy is supplied by the breaking of the high-energy phosphate bonds. ATP is composed of a nitrogenous base, adenine, linked to the 1'-hydroxyl of ribose, linked via the 5'-

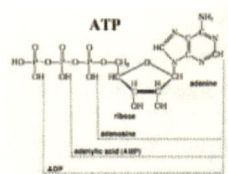

hydroxyl to a polyphosphate.

Many other nucleotides and nucleotide-like molecules are used as cofactors in metabolism. Consider the following two examples:

1. Cyclic-AMP has the phosphate at 5' linked back to the ribose at 3, and is a ubiquitous second messenger.
2. Replacing the final phosphate in ATP with a ribose attached to a nitrogenous base forms vitamin-B. This is nicotinamide adenosine dinucleotide (NAD) that excels at conveying hydrogen atoms between molecules. NAD is used as a cofactor by almost as many enzymes as ATP.

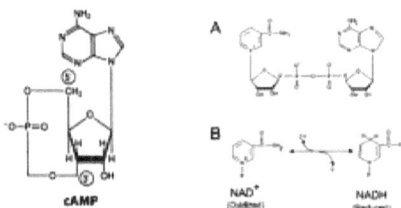

RNA Folding

The chemical activities of the various side groups along the RNA, while not as diverse as in protein, have plenty of hydrogen bonding capacity, and an RNA molecule folds up into a precise form in water just as do proteins. The DNA variant of RNA replaces the free 2'-hydroxyl on the ribose with a hydrogen atom, becoming a sugar called deoxyribose. This DNA is also made more hydrophobic by adding a methyl group to the ~25% of the bases that are uracil, adenine's partner. The base is now called thymine.

In proteins, the folded aminoacid chain binds sympathetically with water. The geometry of the H-bonding groups at the surface is just right for H-bonding with water. At the binding site, just about any type of molecule can be accommodated by the surrounding aminoacids.

The ribose-phosphate backbone of RNA excels at binding with water. The geometry of the H-bonding in the nitrogenous bases makes water-bonding awkward, but H-bonding with a complementary base is extremely easy. In the folded form, the H-bonding in the interior is very often between complementary bases, not with water. The 'substrate' of a nucleic acid is often another nucleic acid. In nonpolar DNA, all the interior bonding is entirely between complementary bases; none of the bases need to bond with water.

There are four natural nucleic bases with either one or two rings and capable of making two or three hydrogen bonds. The names come from early chemistry and are not particularly informative.

	1 ring	2 rings
2 H-bonds	Uracil (thymine)	Adenine
3 H-bonds	Cytosine	Guanine

In an RNA polymer, one end has a free 5'-hydroxyl on the ribose while the other end has a free 3'-phosphate on the ribose. Synthesis of all varieties of RNA always starts at the 5'-end and concludes at the 3'-end, and base-to-base binding is always between ribose-phosphate chains going in opposite directions.

Base-to-base bonding always follows these two rules for *base pairing*:

1. The sum of the rings in the two nucleic bases is 3.

2. The difference in bonding number of the nucleic bases is 0.

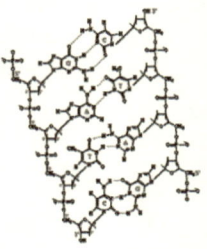

These requirements are met by the two pairs of complementary bases that sum to three rings and are commonly referred to by their initial letters, the two-bonding A–U(T) pair and the three-bonding G–C pair. Enzymes that exchange an amine group on adenosine with an oxygen, generating inosine, can switch between 2-bond and 3-bond pairs.

The 'binding site,' so to speak, of an RNA chain is another RNA chain in which the sequence of bases is complementary going in the opposite direction. Like a zipper, the complementary bases bind and align the two chains, the G-C triple bond being stronger than the A-U double bond.

In either case, the distance across the complementary is a constant three rings.

In a DNA double helix, there are two zippered chains that cling tightly together. In other varieties of RNA, the chain folds back on itself, as in the small tRNAs where a single chain creates loops as complementary sections zipper together.

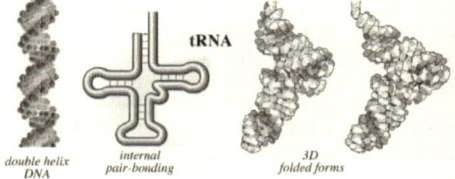

tRNA

double helix DNA · internal pair-bonding · 3D folded forms

The folding of an RNA/DNA chain is exactly like that of proteins, except that to the internal complementing of chemical activities is now added the complementing of digital information on the pattern of nucleotide H-bonds. It has already been noted that there is an inadequacy of the lock-and-key concept to explain the rapid fashion by which nucleic acids find and unite with their exact complements. In DNA, all the internal structure is base-pairs.

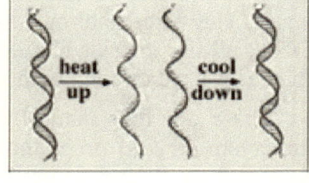

heat up · cool down

If DNA is heated above a certain temperature, the thermal energy becomes sufficient to disrupt the double helix and the two strands separate. When cooled below that temperature, the DNA 'anneals' and the two strands very rapidly align and bind back together again as a double helix. This is exactly the same process as in the reversible denaturation of enzymes by heating and cooling.

In unified science, the double helix is the ground state, while the two denatured strands of DNA are in an excited state of the extensive wavefunction. In unified biochemistry, the rapid annealing of RNA and DNA is to be expected; it is not a puzzle.

Terminology

Modern technology has accumulated a lot of wisdom about the manipulation of digital information. We would like to apply this wisdom in unraveling the way that RNA manipulates digital information in living systems.

In modern technology, digital information can be stored and disseminated in a wide variety of physical forms. Magnetic domains, electric charge, electric current, reflective pits and radio waves are a brief sample. Digital information is readily transformed from one physical form onto another many times, but the digital information always remains unchanged.

Whatever the physical form,—N or S magnetism, reflective or non-reflective, etc.—the digital information is notated in the same way, as a string of zeros and ones, i.e., in the binary code. All that wonderful stuff that digital computers, phones, routers, TVs, etc. do is a highly-sophisticated manipulation of 0s and 1s responding to input from the environment.

The problem is that manipulation of digital 0s and 1s is highly arithmetical, and A, U(T), G and C are not suitable symbols for these mathematical operations. It will make the application of binary-code wisdom to living systems much easier if we use numerical, not alphabetical, symbols for digital information. Binary code will not do because it cannot handle the relations between the four bases.

We will use the four integers, 0, 1, 2 and 3 instead of letters to name the four bases in RNA. The chemical difference between uracil and thymine is irrelevant to the digital information and need not be noted.

	1 ring	2 rings
2 H-bonds	2	1
3 H-bonds	3	0

Logic operations

The mathematical manipulation of digital information is also a systematic hierarchy, where simple operations are concatenated into more sophisticated operations. At the very lowest level are the simple logic gates. They are physical structures that take a digital signal input and generate a digital output. The simplest NOT gate turns an input of 1 into an output of 0, and an input of 0 into an output of 1. It simply inverts the binary digits, and a simple circuit in a transistor easily does the binary inversion.

More sophisticated logic gates take two inputs and, depending, have a single output. Examples are gates AND, NOR or XOR that take two inputs that result in a single output. The NAND gate, for instance, always outputs a 1, unless both inputs are 1 and then it outputs a 0. A transistor with a few connected AND and NOT gates on it will do this.

NAND gate

Input A B — Output

A	B	Output
0	0	1
0	1	1
1	0	1
1	1	0

The logic operation that a gate performs on two inputs of binary information is shown in a 'truth' table in which 1 is true, 0 is false. While it seems an unpromising manipulation of two digits, sets of connected NAND gates can do every kind of logical manipulation possible to a digital input. Everything that a computer does can be done by sets of NAND gates alone.

Using the numerical notation, the manipulation of digital information at the base of the systematic hierarchy can also be expressed as a truth table. As a reminder that direction is important, we will use • to signify the free hydroxyl at the 3' (*three-prime*) end of the chain.

TC	
•N	•|N-3|
0•	•3
1•	•2
2•	•1
3•	•0

An example is the logical operation of threes-complementing, TC, a number in which the output is the size of the number after 3 is subtracted, $TC(N\bullet) = |\bullet N{-}3|$, and the TC of this returns the original, $TC(TC(N\bullet)) = N\bullet$.

This logical operation can be applied to a string of bases, $N_1 \, N_2 \, N_3 \ldots N_n\bullet = \mathbf{N}\bullet$ and performing the operation twice, TC^2, duplicates the original string $\mathbf{N}\bullet + TC^2(\mathbf{N}\bullet) = 2 \, \mathbf{N}\bullet$. To make a copy of digital information, the TC logical manipulation is performed twice on the original, creating two identical copies.

Barring a few exceptions, such when a poly-adenosine primer is assembled, all RNA and DNA is generated by the TC logical operation—the enzyme *transcriptase* transfers digital information from DNA to RNA, while the enzyme *reverse transcriptase* copies it from RNA to DNA and, in one of the set of analog operations that RNA 'rybozymes' are capable of, information on RNA is copied onto RNA. These are the physical structures that perform the logical operation and these enzymes are equivalent to the circuits on a transistor that perform the NAND operation.

The digital information on DNA is duplicated by performing a TC on both strands of the double helix, which are pried apart for the operation. The enzyme helicase separates the strands and two DNA polymerase enzymes assemble the complementary strand. As nucleotides are only added to the 3' end of a growing polymer, the leading template strand is simple to assemble, while the lagging template is complemented in sections which are then linked together by a ligase. The end result is two identical DNA helixes.

$$1N\bullet \, 2\bullet N + TC(1N\bullet) \, TC(2\bullet N) = 1N\bullet \, 2\bullet N \; + \; 1N\bullet \, 2\bullet N$$

This gives two copies of the digital information. After n repetitions of this there will be 2^n duplicates of the original digital information. After 100 duplications, the number of copies would be 2^{100} or about a billion, trillion trillion.

Another example of a logical operation on digital information in living systems is the transformation of an adenosine to inosine.

The truth table of the operation is simple but, as shown, performing a TC twice changes a 1 in a digital string to a 0. The logic operation is the exclusive subtraction of 1 from the number, $XSB(1) = 0$, while leaving the other numbers unchanged. This is an example of performing this sequence on a string of bases $0321\bullet$

N	XSB	TC	TC
0•	0•	•3	0•
1•	0•	•3	0•
2•	2•	•1	2•
3•	3•	•0	3•

$$XAD(TC^2(0321\bullet)) = 0320\bullet$$

We can simplify the notation. As we have made clear, the digital information on codon and anticodon is the same except for it being in complementary form. We can signify this identity of DL by simplifying the notation and just signify the DI and the direction it is going by a •.

So the complement of $0123\bullet$ is $\bullet0123$; the numbers remain unchanged, and just the dot moves. The threes complement of a number is generated by moving the dot from one end to the other. This is in deference to the tradition that mRNA is the codon while DNA and tRNA are anticodons (though the reverse, like electricity, would make more sense).

So the passing along of DI in protein synthesis entails that the numbers do not change, just the location of the dot. At the end, it signifies the codon, so the numbers equal the usual bases. At the start, it signifies the anticodon, so the numbers signify the threes-complement base. Flipping from T on DNA to A on mRNA to U on tRNA is cumbersome and adds no new digital information, just the chemical nature of the particular media it is on which is quite irrelevant.

Media	DI	AW
DNA	**•0123**	
mRNA	**0123•**	
tRNA	**•0123**	
aminoacid	Bezier generator	

While the exploration of the systematic hierarchy of sophisticated operations founded on the NAND logic has given us the iPhone, the mathematical exploration of the systematic hierarchy of operations founded on the TC and XSB logic operations is minimal, and there are probably more such basic operations that have yet to be discovered.

What is well-known, however, is the connection between digital information, DI, on RNAs and the digital information in proteins, their primary structure. As discussed, this is the sequence of aminoacids in a linear polymer, and it is this DI that determines the automatic folding of the chain into the active form of the protein. This folded form, in turn, determines the analog wave, AW, that the protein generates to add to the harmony in a vital unit, the sequence being represented as:

DI-RNA → DI-protein → AW-generated.

To make this connection, we first have to describe a few of the many RNA varieties.

An RNA World

It is becoming clear that DNA, which has heretofore played the central starring role in biological theory, has only a supporting—if essential—role comparable to that of the somewhat passive role of a hard drive in a PC computer. All the really important action—including getting life started in the first place—is performed by RNA with a role comparable to the central processing unit (CPU) of a computer that reads and writes digital information to the hard drive as well as manipulating it, combining it with input from the environment (the User) before outputting it as analog information to a screen or loudspeaker.

We will now briefly summarize the varieties of RNA that are well-understood and then just mention those that have been so recently discovered that their function is still being deciphered.

Messenger-RNA

One of the simplest is an RNA whose sole function is similar to that of digital film cassette that carries digital information from one location to a distant location. Digital information stored on a DNA strand is threes-complemented as a strand of mRNA is assembled.

Each section of DNA that is transcribed into a particular stretch of RNA is called a gene. Some genes create mRNA, others create all the other types of RNA. The mRNA is conveyed to a ribosome which then uses the digital information to assemble proteins.

In the monomer-bacteria, the mRNA is now ready for its role in protein synthesis. In all eukaryotes—animals, plants and fungi—the mRNA is processed by another variety of RNA that excises specific, long stretches of bases, called *introns*, and only the remaining *exons* get used to assemble a protein.

Transfer RNA

The small transfer RNAs are generated by threes-complementing DNA onto RNA in the same way, except that now the strand is chopped up into the 61 varieties of tRNA. "However, many cells contain fewer than 61 types of tRNAs... A minimum of 31 tRNAs are required."[160]

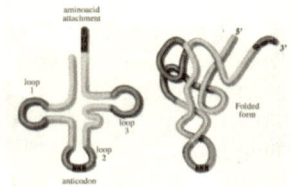

Each tRNA folds into a particular, if similar cloverleaf form, two of which we saw earlier, with three loops separated by paired-base stem sequences. Every tRNA has an identical 3' ending sequence of 001• (cytosine, cytosine and adenine with its free hydroxyl) and each tRNA has an unpaired triplet of bases, NNN•, exposed to water called the anticodon. An adenine in the third 'wobble' place in the anticodon is often XSBed to an inosine, so its complement is cytosine, not uracil, XSB(NN1) = NN0. This reduces the number of essential tRNAs that must be assembled.

There are 20 enzymes in all living systems, one for each of the natural aminoacids, with two binding sites, one for its particular aminoacid, the other for an ATP. When both are present, the active site splits the ATP and adds AMP to the carboxyl end of an amino/imino acid. This activated aminoacid remains bound, and the protein undergoes a conformal change, creating a third binding site. The substrate that fits this site is the shape, and sometimes the anticodon as well, of a particular tRNA. Some enzymes have a binding site that can accommodate six different tRNAs. In this way, just 20 enzymes can process up to 61 different tRNAs.

When the appropriate tRNA is bound, another active site swings into action transferring the aminoacid from the AMP to the free 3'-hydroxyl on the terminal adenine of the tRNA. The tRNA is now 'charged' with an aminoacid. This linking of RNA and aminoacid is the basis for transferring DI on an RNA form to DI on a protein primary structure. This is akin to a computer transferring DI from magnetic domains on a hard drive to DI in the CPU as electric potentials in silicon gates.

Every single one of the natural aminoacids is now associated with at least one anticodon on a tRNA, serine being the champ with six different tRNAs it can be attached to. There are 64 different triplet anticodons. Three of them are are not expressed on a tRNA; they are reserved for simple commands such as *Start* and *Stop*. The other 61 are assigned an aminoacid in a 'translation table' called the Universal Triplet Code. All extant living systems use the same assignments, some with a few very minor variations.

Translation Tables

In the systematic hierarchy, the universal triplet code plays the same role in living systems as the universal ASCII code plays in computer systems. Even the earliest computers did not deal with single bits, they dealt in constant-sized blocks of multiple bits. Four bits were used at first and the block was called a nibble. This was limited in scope, and the industry quickly settled for blocks of eight bits, called a byte, as it could comfortably encode the roman alphabet. This is the '8-bit' level of sophistication in the computer systematic hierarchy.

There are 2^8 different bytes, and these 256 suffice to encode the symbols on a keyboard in what is called the ASCII code. The initial 128 of these (0-127, 00000000-01111111 binary) encode a few simple commands (such as *carriage return*) and the common numbers and letters. The latter 128 bytes (128-255, 10000000-11111111 binary) encode more exotic symbols as in the illustration.

0	0011 0000	@	0100 1111	`	0110 1101
1	0011 0001	P	0101 0000	n	0110 1110
2	0011 0010	Q	0101 0001	o	0110 1111
3	0011 0011	R	0101 0010	p	0111 0000
4	0011 0100	S	0101 0011	q	0111 0001
5	0011 0101	T	0101 0100	r	0111 0010
6	0011 0110	U	0101 0101	s	0111 0011
7	0011 0111	V	0101 0110	t	0111 0100
8	0011 1000	W	0101 0111	u	0111 0101
9	0011 1001	X	0101 1000	v	0111 0110
A	0100 0001	Y	0101 1001	w	0111 0111
B	0100 0010	Z	0101 1010	x	0111 1000
C	0100 0011	+	0110 0001	y	0111 1001
D	0100 0100	b	0110 0010	z	0111 1010
E	0100 0101	c	0110 0011	.	0010 1110
F	0100 0110	d	0110 0100	,	0010 1100
G	0100 0111	e	0110 0101	;	0011 1010
H	0100 1000	f	0110 0110	?	0011 1111
I	0100 1001	g	0110 0111	!	0010 0001
J	0100 1010	h	0110 1000	*	0010 1010
K	0100 1011	i	0110 1001	(	0010 1000
L	0100 1100	j	0110 1010	)	0010 1001
M	0100 1101	k	0110 1011	}	0010 1001
N	0100 1110	l	0110 1100	space	0010 0000

For example, when shift-A is pressed on the keyboard, the byte 1000001 (65) is sent to the computer which looks it up in a table of analog shapes. In the early days, these shapes were all 12-Courier, which looked like this. The analog shape A is sent to the graphic processor which displays this shape on the screen.

In a similar way, the digital information in living systems is manipulated in sets of three bases called a codon. There are $4^3 = 64$ different codons from 000 to 333. The codon on tRNA provides a lookup table for the elements of analog form, the aminoacids in a protein. The array of tRNAs is a translation table from digital triplets to the elemental generators of analog form, the aminoacids.

$Q \longrightarrow$

Elements of form

The simple 8-bit computers used a table of bitmapped shapes. It was useless to try to alter their size since the curves would all come out lumpy, not smooth at all.

Nowadays, sophisticated systems generate the analog form of a font in any size for screen or printer by storing shape information as a set of Bezier curves; a simple mathematical length, L, plus a curvature, C, at either end.

Each relative length and two curves contributes to the overall analog shape of the letter. The letter 'm' in the illustration involves over 30 Bezier points separating the 30-odd curves it takes to create the analog form of a Times New Roman 'm.' As it takes 3 bytes to describe each curve, the letter 'm' takes about 100 bytes of storage. Simple letters, like 'c' and 's' require less information, while complex letters, such as Q and E, require more. As each component of each curve in this scheme can have 256 different values, there are ~65,000 different Bezier curves.

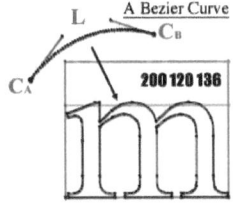

L   A Bezier Curve
C$_B$
C$_A$
200 120 136

If I select a block of 12-point text in the document, and tell the computer to make it 24-point, all the computer has to do is multiply the L value of each letter by 2 to generate the requested change, just the first byte is altered and 24-point type is displayed.

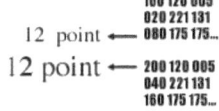

12 point ← 100 120 005
020 221 131
080 175 175...

12 point ← 200 120 005
040 221 131
160 175 175...

Amino acids are like Bezier curves in multi-dimensions. Each adds its contribution of curve to the final shape of the system form when folded. This is the curve to the internal wave aspect.

While there are only 20 different Bezier-amino acids, the description now involves thousands of different points, not just dozens, each contributing to the final form of the active folded protein. In the computer system, the curves are variable in length, while in the primary sequence, each amino Bezier takes up a fixed space.

The computer system takes the opposite tack to that of a living system in distribution of numbers, but the end result is the same in that a huge number of forms

Element	Curves	Points	Length	Describes
COMPUTER SYSTEM A letter	**Bezier**	**Many**	**Few**	**Variable**	**Form**
LIVING SYSTEM A protein	**Amino acid**	**Few**	**Many**	**Fixed**	**Complement**

can be accurately described. Again, they take opposite tacks in that the Bezier curve description of the letter 'm' describes the form of the letter, while the aminoacid description of a glucose-binding enzyme describes the resonant complement to the glucose form and the required surrounding water structure.

The Triplet Code

Each digital anticodon on a tRNA is thus linked to an aminoacid-Bezier, a tiny generator of form. All these aminoacid-Beziers, when linked in a chain, contribute to the final folded form of the protein and the wavefunction it generates to structure water and metabolism. The role of tRNA is a set of adaptors with DI-carrying nucleotides at one end and AW-generating aminoacids at the other

For historical reasons, the triplet on the tRNA is labelled an anticodon but the lookup table linking RNA and aminoacid is always given as the codon. As the digital information on a codon is the same as that on an anticodon, it does not really matter.

The table has the 1st-position in the codon as rows, the 2nd-position as columns, and the 3rd-'wobble' position at the 3'-end added to complete the 64 possible codons. Three of the codons are reserved for *Stop* command to mark the end of a chain. A methionine codon always marks the start of a chain (and is often later removed before folding).

1st		2nd			
		0	1	2	3
	0	000• GLY 001• 002• 003•	010• GLU 011• 012• ASP 013•	020• VAL 021• 022• 023•	030• ALA 031• 032• 033•
	1	100• ARG 101• 102• SER 103•	110• LYS 111• 112• ASN 113•	120• MET 121• ILE 122• 123•	130• THR 131• 132• 133•
	2	200• TRP 201• [end] 202• CYS 203•	210• [end] 211• 212• TYR 213•	220• LEU 221• 222• PHE 223•	230• SER 231• 232• 233•
	3	300• ARG 301• 302• 303•	310• GLN 311• 312• HIS 313•	320• LEU 321• 322• 323•	330• PRO 331• 332• 333•

The other 61 are distributed among the 20 aminoacids. The code is mathematically *degenerate* since multiple codons are assigned the same aminoacid. There are patterns to this assignment table, one being that the 3rd position is often irrelevant, eight aminoacids have a box all to themselves, while six split a box in exactly the same way, serine doing both at once.

Another pattern is that when only the stronger, 3-bonded guanine and cytosine pair are in the 1st and 2nd positions, they code for two pairs of complementary 'contributions' to the analog form of a protein:

00N• codes for glycine, which is a small simple amino acid with a strong tendency to participate in structures such as coils and sheets.

33N• codes for its complement, proline, which is a large ring 'imino' acid that adds a distinct bend to an amino acid chain and strongly disrupts structures such as coils and sheets. Glycine and proline are a complementary pair.

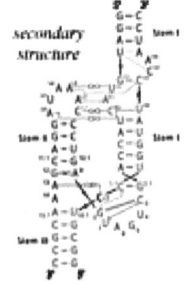

03N• codes for arginine, a large polar molecule with multiple H-bonding capacity.

30N• codes for its complement, alanine, a small nonpolar molecule with zero H-bonding capacity.

It is such patterns of codons and properties that have convinced many that, like the 4-bit nibble that preceded the 8-bit byte in computer technology, the very earliest versions of the RNA to aminoacid was a 2-bit binary codon that preceded the 3-bit triplet codon in the sequence of steps that occurred during the origin of life.

We have now discussed two types of RNA: the mRNA is a long molecule complemented from DNA, and a set of tRNA charged with an aminoacid   and marked with an anticodon triplet of exposed bases. The final RNA we need to complete the picture is the ribosomal RNA, rRNA, that brings the two together in a ribosome. This is a major example of RNA manipulating analog forms just as protein enzymes do.

Ribozymes

One of the reasons why it is thought that the RNA world preceded that of protein and DNA is that RNA, like protein, can generate waves that can perform many simple metabolic steps. In particular, it excels at manipulating nucleotides and nucleic acids. Unlike the specialized DNA and protein, RNA is capable of both analog and digital manipulations.

A ribozyme (from ribonucleic acid enzyme, also called RNA enzyme or catalytic RNA) is an RNA molecule possessing a well-defined tertiary structure that enables it to catalyze a chemical reaction. They can comprise one or more chains of RNA.

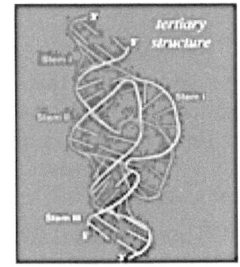

While this is an emerging field, it has already been established that ribozymes can: Catalyze the hydrolysis of one of their own phosphodiester bonds (self-cleaving ribozymes); Catalyze the hydrolysis of bonds in other RNAs; Catalyze the amino-transferase activity of linking aminoacids on tRNA into protein; Catalyze their own duplication and synthesis out of nucleotides.

As in a protein, the primary structure of a ribosome is the linear sequence of ribonucleotides, the secondary structure is how the chains loop and connect by complementary binding, while the tertiary structure is the final active form with its shell of water and compensating ions, usually magnesium.

Every living system has complexes of ribozymes that, with a suit of proteins as assistants, are the sites of protein synthesis. All prokaryote bacteria use the same basic ribosome for assembling proteins, while all eukaryote plants, animals and fungi use larger and more sophisticated ribosomes (except in their organelles, the mitochondria and chloroplasts, where the simpler, bacterial ribosomes are still in use. These organelles are self-contained and their duplication is kept quite separate from that of the rest of the eukaryote cell.

Ribosome

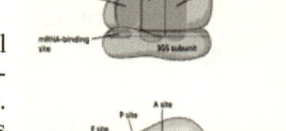

The ribosome is a complex of ribozymes that do all the chemical manipulations that turn the digital instructions on mRNA into the primary structure of proteins. The ribozymes are assisted by a dozen or so proteins that tune the ribozyme activity.

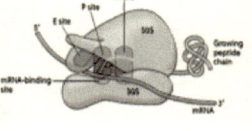

The rRNA is complemented from DNA, processed, and assembled into two complexes of different sizes. These two unite around a mRNA at the starting methionine codon. The rRNAs in prokaryote vital units and eukaryote vital molecules are similar in size, with the latter augmented with an extra RNA.

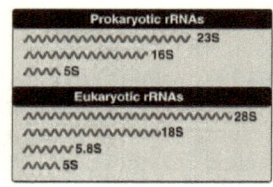

The larger subunit has three active sites: the A site binds loaded-tRNA, The P site binds the previous tRNA with the attached growing peptide, and the E site where the now-unloaded tRNA ends up and is ejected for reloading. The smaller subunit has one site where mRNA is bound.

The subunits bind and unite at the required methionine codon at the 5'-end of an mRNA. The ribosome then behaves as a complicated machine that has a repetitive clockwork-like mechanism. The mRNA is ratcheted through one triplet codon at a time, the mechanism being powered by GTP, rather than the more usual ATP. A t-RNA with an anticodon that matches the newly-exposed codon is bound. The aminoacid chain is split from its tRNA and linked to the new aminoacid. The old tRNA is released to be recharged by its enzyme.

Stripped of its time element, the ribosome aligns the codons on the mRNA with anticodons on the loaded-tRNA and then links the aminoacid chain together. This is basically how the digital information on RNA is transformed to digital information as the primary sequence of aminoacid-Bezier wave generators.

The ribosomal RNA links all the aminoacid-Bezier AW generators into an aminoacid chain. The linear chain of aminoacids is liberated when one of the *End* codons is encountered. While the analog form of the *Start* Met-codon is necessary for the assembling of the two rRNA subunits together on the mRNA, the analog form of the *End* codons is antithetical and the two subunits dissociate and go their

separate ways. The linear combination of all the aminoacid-Beziers gives a specific form to the wavefunction of the ground state. The chain shakes off energy and folds, as discussed, into the final active form specified by the ground state.

This is the manipulation of DI by RNA reading it from memory on DNA and converting it into an AW to join the harmony prevail-

		AW aminoacid	AW protein
DNA	**mRNA**	**tRNA**	
DI anticodon	DI codon	DI anticodon	

ing in the vital unit. From DNA to mRNA to an array of tRNAs, the flow is DI that switches from codon to anticodon at each step. The array of adaptor tRNAs is also an array of Bezier AW simple generators. These are linked into a primary structure which folds into a protein AW generator that structures water and substrates.

The digital score has become analog music. Other than requiring the initial Met-codon, the ribosome is indifferent to the sequence on the mRNA or what protein it is making. A human ribosome will freely translate mRNA from a bacteria, a radish, or an extinct dinosaur, into bacterial, radish, or dinosaur protein.

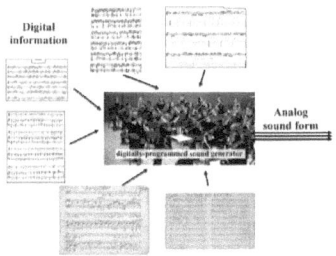

Programmed Generator

In a symphony orchestra, an instrument such as a piano can produce any music at all. A piano is a universal analog wave generator, and the actual waves it generates are digitally-programmed depending on what musical score it is supplied with.

In living systems, the analogue to the musical instrument is the ribosome that does all the manipulations to turn a digital mRNA into a protein, a 'note.' It is a universal generator and will turn any stretch of mRNA into protein. This structure was also inherited from the Logos before the basic living system emerged and has changed little over 3 billion year

Many ribosomes can be at work on the same mRNA, following each other at a short distance, in what is called a polysome. In this way, thousands of a particular protein can be made from just one mRNA. The mRNA has a set lifetime, however, as enzymes that hydrolyze mRNA patrol the cell and are constantly reducing the mRNA population. The turnover of mRNA is such that the 'score' being played can be changed in minutes.

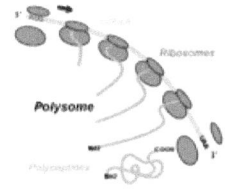

A single mRNA molecule can give rise to over 100 identical proteins, all of them generating the same wave. This number of proteins is quite sufficient, recalling the abilities of gelatin, to completely alter the wave in a vital unit.

This is exemplified by the bacteriophages, the viruses that need bacteria to multiply in. When a single molecule of virus RNA is injected into a bacterium, within minutes all the resources of the vital unit are redirected into constructing new virus particles. A single molecule of viral RNA can control the symphony being played in a vital unit, and this applies to the natural state as well as pathological states.

In summary, this is the way that digital information as DNA is transformed into analog waves structuring a vital unit, and so determines the interactions that it is capable of. This is exactly what happens when a CD player reads digital information as a series of reflective pits on the CD (the DNA) and outputs a DC current that is either on or off, read as a sequence of bytes of binary information (the mRNA). The adaptor is a DC/AC transformer (the tRNA) that takes a binary number and converts its value into a voltage, Vt, at that moment (the amino acid) that drives the amplifier for the loudspeaker (the protein) that generates the analog sound wave. The sampling rate determines how long each Vt lasts—a 1/1,000th of a second is not unusual—and this gritty voltage that drives the loudspeaker is blurred into a smooth-sounding wave by the inertia of the loudspeaker membrane.

This flow of DI to AW is called the 'reading' of stored digital information.

$$DI_{DNA} \xrightarrow{READING} DI_{codon\ mRNA} \longrightarrow DI_{tRNA} \longrightarrow AW_{aa} \longrightarrow AW_{protein}$$

$$DI_{CD} \xrightarrow{READING} DI_{DC\ byte} \longrightarrow DI_{DC/AC} \longrightarrow AW_{Vt} \longrightarrow AW_{loudspeaker}$$

The reading direction, from DNA to protein, is reasonably well understood and is foundation for the next level of sophistication in the hierarchy of life, the level that is the object of study in genetics.

Unified Genetics

I live in a society where the question, "Did you back it up?" is usually a portent of disaster. Our whole digital technology is based not only on the reading of DI but also on the writing of DI in the reverse direction. Analog form is converted into digital information and stored. This is the writing direction from analog form to digital information.

The process of turning the glorious analog form of a Beethoven symphony into the linear pits that give CDs their particular rainbow shine is basically just the reverse of of the reading process.

Analog waves have a property that is utterly absent in digital information. Analog waves are holographic, the structure of the wave is nonlocal, and a small sample is all that is needed to comprehend the entire wave. This is obviously not the case for digital.

The structured air in a symphony hall is immense compared to the area of an eardrum, yet this tiny sample of the wave is all that is required to appreciate the performance. Just like the eardrum, the sensor in a microphone takes a tiny sample of the wave and turns the analog pressure-wave in time into an analog electric current which is input to an AC/DC converter.

The AC/DC converter measures the amplitude of this wave, which can be plus or minus, each moment at the sampling rate. This measure is output as a binary number byte. The linear stream of binary bytes is written onto the master CD as a stream of linear pits in the surface.

$$DI_{CD} \xleftarrow{\quad WRITING \quad} DI_{DC\ byte} \longleftarrow DI_{AC/DC} \longleftarrow AW_{Vt} \longleftarrow AW_{microphone}$$
$$DI_{CD} \xrightarrow{\quad READING \quad} DI_{DC\ byte} \longrightarrow DI_{DC/AC} \longrightarrow AW_{Vt} \longrightarrow AW_{loudspeaker}$$

The master CD can then be duplicated for distribution. The writing operation is not just for long-term, unchanging DI; it is also used for short-term, changeable DI manipulation. An example is the writing of the manuscript on a MacBook Pro.

As you will have probably noticed, I revel in illustrations, diagrams, graphs, etc. These are created on a small suite of programs: *Safari* for googling, *Adobe Elements* for diagrams, *Grapher* for graphs, *Magic Number* for calculations, and *MathType* for equations. When I am working, these programs with their multiple windows, plus Mail and WiFi, are all running at the same time, creating a digital structure that is many gigabytes in size. I use Apple *Pages* to assemble all these disparate elements together, and it remembers what I have done so I can backtrack at any time. This enormous structure of digital information is written to the hard drive as a temporary file, as short-term memory.

I can jump easily from one part of the digital structure to another. The CPU just calls up that section of DI from active memory on the hard drive into its working memory on its silicon registers, replacing what was there. If I switch from *Pages* to *Elements*, the appropriate part of the DI structure is read into active memory. I now draw a circle filled with the red. The changes I make are sent to the graphic processor for display, and also written to the hard drive, updating the virtual image on the disk. Everything I alter is displayed and written to active memory.

In the old days, a dog knocking the electric plug out of its socket, plunging the screen to blankness, would be a disaster. But not for a modern Mac. It restarts, and one by one, starting with the *Finder*, all the programs reload with all their windows just as they were before the crash.

Other examples could be given, but the main point is that the Writing operation—AW to DI—is just as important as is the Reading operation—DI to AW—in our sophisticated manipulation of digital information in this computer/Web world.

Fundamental Dogma

Knowing what we know about how digital information and analog wave are bi-directional in modern technology, it is surprising that the foundations of classical genetics *embrace the Reading direction but totally reject the Writing direction*. This rejection is known as the Fundamental Dogma of Genetics, which states that information only flows from genotype (DI on DNA) to phenotype protein (AW as bodily form), never from phenotype to genotype.

In this view, where 'writing to disk' is not allowed, the only way that DI in the DNA can ever change is by random, accidental mutations—alterations in the physical DNA that randomly alter the DI. Such mutations can be as simple as a single base being changed by a copying error (the probability of which is very low, but not zero) or a radioactive carbon detonating or as complex as whole sections of DNA being duplicated, flipped, or moving from one location to another.

These random changes in the physical DNA—like running a magnet over a hard drive—can radically alter the DI in the genotype. This is reflected in alterations in the analog wave, alteration in the phenotype. These new phenotypes are sorted out by 'natural selection', i.e., those that fit well in the world around them get to flourish and multiply, while those that do not fit well simply perish and disappear.

This foundation of read-only classical genetics is called the modern synthesis of Darwinism and molecular biology. I know it sounds weird in our digital age, but that is the nub of modern evolutionary thinking.

Reading and Writing

As any programmer working on the iPhone will affirm, this is not an efficient way to work with digital information. You need to have the writing direction down first, then the reading direction. A unified genetics will have a biochemistry that links the analog wave of life, in all its diversity of forms recalled from the digital information on the DNA, with a Write direction as well as the Read direction, and the Write coming before the read.

Taking a lesson from CD technology, we would expect that, if there was a writing direction, then it would look similar to the reading direction in reverse.

One such reverse section of the reading process in living systems is already well-established. The enzyme *transcriptase* complements a DNA strand to assemble an RNA strand, while the aptly-named enzyme *reverse transcriptase* complements a strand of RNA to assemble a strand of DNA (which can be complemented to form a double-helix).

$$DI_{DNA} \xleftarrow{\text{reverse transcriptase}} DI_{RNA} \longleftarrow DI_{tRNA} \longleftarrow AW_{aa} \longleftarrow AW_{protein}$$

$$DI_{DNA} \xrightarrow{\text{transcriptase}} DI_{RNA} \longrightarrow DI_{tRNA} \longrightarrow AW_{aa} \longrightarrow AW_{protein}$$

Transcription:

In transcription, under the control of an enzyme complex, a double helix stretch of DNA is unwound and one strand is used to assemble a complementary strand. The only difference to DNA copying being that ribo-, not deoxyribo-, nucleotides are linked together in the sugar-phosphate backbone. Only one of the strands, the *reading strand*, is transcribed by the enzyme. This 'messenger RNA' (mRNA) then departs from the DNA and the helix reforms.

Reverse Transcription:

In transcription, DNA is used to make RNA. In reverse transcription, under the control of a *reverse transcriptase* enzyme, a stretch of RNA is used to make a complementary strand of DNA. This DNA is then complemented, as in duplication, and a DNA double helix is formed. The study of this process is still in its infancy, as is an understanding of the function of the ~60,000 *reverse transcriptase-like* genes found in the human genome.

Once in DNA, the DI can be passed on down a lineage (with a little tweaking) through deep time. The illustration, for instance, is the primary structure of the same core housekeeping protein found in all contemporary living systems whose lineages diverged billions of years ago. Yet, the

DI for the primary sequence on DNA has hardly altered in all that time. This is an example of cellular long term memory; i.e., once something is truly learned, it is never forgotten.

Reverse mRNA?

What is not known is whether a stable linear array of tRNA anticodons can be used to assemble a linear sequence of codons on a single RNA.

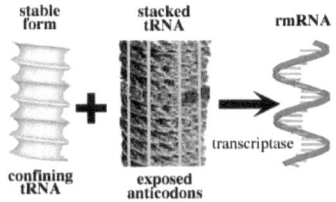

stable form   stacked tRNA   rmRNA

confining tRNA   exposed anticodons

A stable array of tRNAs in a straight line is obviously impractical, but how about the possibility of spiraling around in a stacked helix of tRNAs with ends pointing to the central axis and the anticodons exposed on the outside in a linear spiral?

Starting at one end, a transcriptase can move along the spiral assembling a single reverse-mRNA that is the complement of the tRNA spiral. Information about the analog form that is confining the tRNA into the helix is now converted to linear digital information on an rmRNA. Such a situation would add a second section to the Write operation that is also a reverse of the Read operation.

$$DI_{DNA} \xleftarrow{\text{write}} DI_{RNA} \longleftarrow DI_{tRNA} \longleftarrow AW_? \longleftarrow AW_.$$
$$DI_{DNA} \xrightarrow{\text{read}} DI_{RNA} \longrightarrow DI_{tRNA} \longrightarrow AW_{aa} \longrightarrow AW_{protein}$$

If this is so, we need an analog form that provides the stable confinement of tRNA while being complemented.

Sampling the wave

In the example we used from technology, it was not an entire wave that was involved, just a sample. Due to the holographic nature of waves, this sample reflected the entire wave.

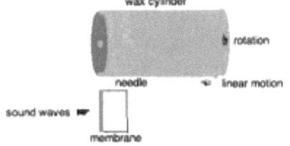

wax cylinder

rotation

needle   linear motion

sound waves

membrane

The great American inventor, Edison, was the first to exploit this aspect of waves and put it to practical use when he invented the first method of sound recording. The method involved a thin membrane that moved back a forth with the change in air pressure. This vibration of the membrane was the sample of the wave in the air. Attached to this membrane was a light needle that made a scratch on a wax-coated cylinder. This cylinder was put into two motions. It rotated at a moderate, constant speed, and it also moved slowly from left to right. In quiet air, the needle inscribed a smooth helix about the cylinder. Placed in the room with a resonating tuning fork, the needle scratched out a sine wave in the wax.

The wave on wax was transformed into a wave on metal and the system was then run in reverse. The wavy scratch on the metal drum was rotated and a needle was placed at the start of the wave. The needle was attached to a membrane that vibrated as the metal wave passed under the needle. A sound wave was generated.

In this purely analog-analog situation, the writing process samples a wave in 3-D air changing with time, into a 1-D wave in wax changing with distance along a line. The reading process just reverses the steps, converting the wave in wax into a wave in air.

The waves in a vital unit are not structuring pressure but the spatial structure of water and metabolites. The most important of these is the water structure standing wave followed a close second by the structuring of ATP and other core metabolites.

What we require for an AW to DI converter is a structure that can sample these important waves and convert them into DI about the wave. This requires an adaptor RNA that has both analog and digital 'ends,' the reverse of the Read direction, where a loaded tRNA has an analog end in the attached aminoacid. As aminoacids are not involved, to make the scenario plausible, we will need to examine some of the less well-known types of RNA to see if any have an analog side to them.

Types of RNA

Almost every week a new type of RNA is discovered with a new set of properties to add to the dozens of RNA types already uncovered. This is a list of some of the recently established varieties of RNA. Only the first three (unshaded) are included in the classical 'read-only' worldview of read-only Darwinism, as eloquently propagated by Richard Dawkins *et al*.

mRNA	Codes for protein	srpRNA	Membrane integration
rRNA	Translates mRNA	snoRNA	Base modification
tRNA	Link to aminoacids	smyRNA	mRNA splicing
miRNA	Gene regulation	teloRNA	Telomeres on DNA
piRNA	Chromosome stability	siRNA	Gene regulation
gRNA	mRNA modification	xistRNA	Chromosome inactivation
rnpRNA	tRNA maturation	aRNA	mRNA translation
yRNA	DNA replication	lncRNA	various

For some time, it has been known that RNA information is easily written into DNA information by the activity of *reverse transcriptase* enzyme (RT), but this was thought to be significant only for RNA 'retroviruses' such as HIV. This insignificant role for RT in living systems made it difficult to account for the 500,000 or so different variants of RT found in the human genome. Richard Dawkins discounted this as yet more Junk DNA—affirming this at a time when it is now known that this DNA, which is never translated into protein structure, is transcribed into RNA at low, but significant rates unnoticeable by earlier, cruder methods.

If Writing of digital information has a role in unified molecular genetics, we can expect that these 500,000 versions of RT have 500,000 different roles and are also transcribed at a rate currently below detection. One variety of RNA called *small interfering* RNA (siRNA) is double stranded—two RNA chains locked together as in DNA—and is used by genetic engineers to silence genes..

Yet another variety of RNA that is only now being explored is *long noncoding* RNA (lncRNA), which must be a contender for the RNA CPU manipulating digital information since it has such a wide gamut of activities:

> Evidence has accumulated showing that (lncRNA) play a significant role in a wide variety of important biological processes, including transcription, splicing, translation, protein localization, cellular structure integrity, imprinting, cell cycle and apoptosis, stem cell pluripotency and reprogramming, and heat shock response.[161]

Riboswitches

To conclude, one new type of RNA activity has only recently been documented that fits the requirements for an AW to DI connector, i.e., the *riboswitch* RNA. This is a type of RNA that structures water and has a binding site for a substrate. This RNA generates a wave that attracts and binds its substrate in a nonlocal way just as a protein enzyme does. Just like an enzyme, a riboswitch flips its form when the substrate is bound, causing a conformal change in anything it is attached to:

> Riboswitches bind cellular metabolites and control gene expression: Segments of RNA, typically embedded within the 5'-untranslated region of a vast number of mRNA molecules, have a profound effect on gene expression through a previously-undiscovered mechanism that does not involve the participation of proteins. In many cases, riboswitches change their folded structure in response to environmental conditions (e.g. ambient temperature or concentrations of specific metabolites), and the structural change controls the translation or stability of the mRNA in which the riboswitch is embedded. In this way, gene expression can be dramatically regulated at the post-transcriptional level.[162]

This riboswitch 'sampling' of the wave in vital units is an area of research that has opened a door to a heretofore unexpected field of study connecting analog form with digital information. Much has been established about what happens, much is yet to be learned about how it happens:

> Riboswitches are structures that form in mRNA and regulate gene expression in bacteria. Unlike other known RNA regulatory structures, they are directly bound by small ligands. The mechanism by which gene expression is regulated involves the formation of alternative structures that, in the repressing conformation, cause premature termination of transcription or inhibition of translation initiation. Riboswitches regulate several metabolic pathways including the biosynthesis of vitamins (e.g. riboflavin, thiamin and cobalamin) and the metabolism of methionine, lysine and purines. Candidate riboswitches have also been observed in archaea and eukaryotes. The taxonomic diversity of genomes containing riboswitches and the diversity of molecular mechanisms of regulation, in addition to the fact that direct interaction of riboswitches with their effectors does not require additional factors, suggest that riboswitches represent one of the oldest regulatory systems.[163]

This generation of precisely folded forms in water solution is exactly like those generated by Bezier-aminoacids in a protein, except now there is Bezier-RNA generating the wave. So, an assemblage of RNAs resonating in the wave of the vital unit is not at all impossible.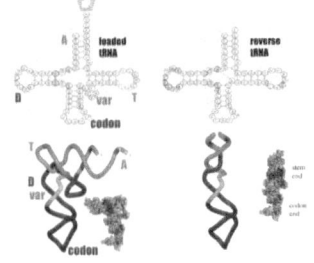

My candidate for such an RNA, seeing that aminoacids are not involved, is a variant of the tRNAs, a variant that preceded the familiar ones. All tRNAs have the same basic form, a perfectly symmetrical cloverleaf tuning fork that is just what might be expected of a good resonator except for the variable section, a kink that spoils the symmetry about the codon and stem.

Before the tRNAs appeared, they could have been preceded by a set of RNAs without the kink. The 3'-triplet for adding an amino acid could be absent but a new internal adenine-uracil bond could be added, straightening the final shape, which is estimated in the illustration. This symmetrical shape is just what a good resonator should look like, a tuning fork. Such a molecule of reverse transfer-RNA (rtRNA), would resonate with analog waves as well as carrying a digital byte of information and could well stack into a form with the codons linearly arranged on the outside for a transcriptase to move along.

Unfortunately, I do not find such an RNA with google, and do not have a well-equipped laboratory. The number of such rt-RNA molecules would be expected to be small, just a few of each kind in an aggregate.

There may be a few of such aggregates, for as we know from our technological transition from mono to stereo, just two sensors taking their samples collects another level of analog information, such as phase difference. Two microphones are used to sample the analog air wave, one left of center, the other to the right. The analog wave from the left is converted into a binary stream, as is that from the right, at the sampling rate. The two streams of binary digits is merged into a single stream—a computer has no problem separating them if the first digit and subsequent odd bits are the left channel, the second and all even bits the right channel.

When the separated channels are converted into analog current that drives a loudspeaker on the left and one on the right, the recreated sound wave in the lounge is a lot more like the original sound in the concert hall. So we can expect at least two stacks of r/tRNA sampling the wave in a vital unit, perhaps even more channels for 3-D wave samples. Even so, the numbers of reverse transfer RNA molecules can be expected to be small. I shall try to keep up with RNA and see what emerges in the literature. Leaving this to basic research, we turn to looking at what use such an AW to DI adaptor-RNA could be put to.

State Sensor

In our picture of a vital unit, a minimal living system, we have a large number of water molecules that are resonating to the unified standing wave generated by the small number of proteins. In terms used in physics, there is a 'water field' with a value for the local 'ice-like' structure of the water molecules.

Almost all vital units are in thermal equilibrium with the physical environment and are subject to change in temperature with the weather. The ice-like structuring of water is sensitive to temperature, and the water-wave of a vital unit is sensitive to temperature. A sensor that provided a digital readout of this would be a most useful input to the RNA for digital output that determines the state of the vital unit. In those animal vital units that are kept at a constant temperature, the water-wave would not change with temperature, a drone note that anchors every chord like that of a bagpipe.

Every protein generates its particular water 'note' to contribute to the symphony. Many proteins also generate from a binding site a wave for ATP to resonate with and bind with. All these harmonize in a standing wave that resonates with ATP, and there is a constant ATP-field as well as a water field. In the symphony analogy, the water wave is like the percussion section setting the basic beat, while the string section is the ATP wave that dominates the melody. A sensor that generated a digital output about the state of this ATP standing wave would also be a useful input to the digital processing that controls the state of the vital unit.

It is estimated that in a bacteria growing and multiplying in an abundant environment, there are ~1,000 different metabolites being processed in the vital cell.[164] Each and every one of these 1,000 metabolites has its own field—i.e., the orchestra is composed of 1,000 different types of instruments. The state of some is more significant than others; they dominate the melody, so to speak. The water and ATP standing waves are examples of such major fields. The NAD and pyruvate fields are also major fields, as are others.

Sensors that generated a digital output about the state of all these major fields would also be useful in managing the state of the cell.

In our discussion so far we have dealt with the water in living systems as if it was plain. It is not. The water in each and every vital unit has a concentration of ions in it—simple atoms or molecules with an electric charge. The pH is a measure of the ratio of OH^- / H^+ concentrations, and the pH of the water is strictly maintained. In eukaryotes this is 7.4, a small preponderance of hydroxyl and slightly alkaline. The concentration of the other ions is always strictly controlled and, as in the case of calcium ions, is sometimes altered as a global signal for change in the state of the vital unit. In an animal, for instance, the concentration inside the cell is lower than that outside, except for potassium which is the reverse. Note that calcium is rigorously excluded from animal cells.

ION	INSIDE	OUTSIDE
K^+	139	4
Na^+	12	145
Cl^-	4	116
HCO_3^-	12	29
Mg^+	0.8	1.5
Ca^+	0.0001	1.8
PO_4	1	0.9

Proteins not only structure water with the wave they generate, they also structure the ions. There is a field for each ion, and the phosphate level is known to be a major one: "Cellular metabolism depends on the appropriate concentration of intracellular inorganic phosphate (Pi). Pi starvation-responsive genes appear to be involved in multiple metabolic pathways, implying a complex Pi regulation system in microorganisms and plants."[165] A digital sensor for this can be added to the list.

On a personal note, after studying biochemistry at Sussex University, I was a scientist with Eli Lilly looking into the antihistamine-resistant side of asthma. As this was basic research, I delved into the second-messenger role of cyclic adenosine monophosphate (cAMP) and its guanosine-analogue, cGMP. They seemed to have complementary actions, in many situations where cAMP did something when raised in concentration, then cGMP did the same when lowered in concentration, and vice versa. These two small molecules have a global effect on metabolism that is as drastic as an influx calcium. RNA is known to be sensitive to them both, and a digital sensor for cAMP and cGMP can also be added to the list of major fields that are monitored by digital sensors.

Analysis of wave

The human ear converts an analog sound wave into a digital stream, but it does so in quite a different way than our technology. We have seen how simple waves, sines and cosines, can be combined to form every other wave, even a sharp saw-tooth.

It was Fourier who made this mathematically-sound by proving that this equation could describe any wave to an accuracy, limited only by the number of included terms. $AW = a\sin 1x + b\sin 2x + c\sin 3x ... + q\sin 100x + ...$

Each sensor in the ear is tuned to resonate with only one of these component waves, say sin(100x), and the strength of its resonance is a measure of the amount of the sin(100) wave, the coefficient, Q. The sensor has a dedicated line to the brain along which only information about the sin(100) is ever transmitted. The digital representation of the number Q is sent down the line to be received as digital information about sin(100x).

Each sensor is tuned to a different component, so the brain receives in massively-parallel form a current Fourier Transform of the current analog sound wave. The brain puts the sines back together and we hear the music, not a lot of sine waves. $a\sin 1x + b\sin 2x + c\sin 3x ... + q\sin 100x + ...$ = Music

The slightly different folded forms of the putative r/tRNAs are tuned so that they also perform the equivalent of a Fourier Transform on the sampled wave of a vital unit. One is responsive to the water-field, others to all the major fields, ATP, NAD, phosphate, etc.

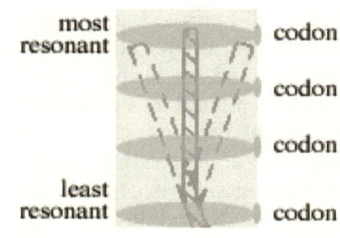

The amount of resonance each sensor has in its field determines its position in the resonating stack of r/tRNA, most resonant at one end, least at the other.

A reverse mRNA transcribed from this would be a digital report about the state of the vital unit that is perfect for input to an IF/ELSE digital processing decision. This aspect of RNA is only now beginning to be explored, so we will need to fall back on an illustration of how such a digital sensor could influence a vital unit.

Recalling memory

To illustrate the possibilities of digital sensors, we shall use a system of just three sensors, colored according to the wave they resonate with. The blue RNA sensor with anticodon •000 resonates with the fundamental water field and provides the basic reference point. The red sensor with •111 resonates with the cAMP field, while the green sensor with •222 resonates with the cGMP field.

The three sensors resonating with their fields in a vital unit will arrange themselves in order from least resonant to most resonant, and the transcriptase will then create a rmRNA with the anticodons as templates.

There are four possibilities we will consider in rela-

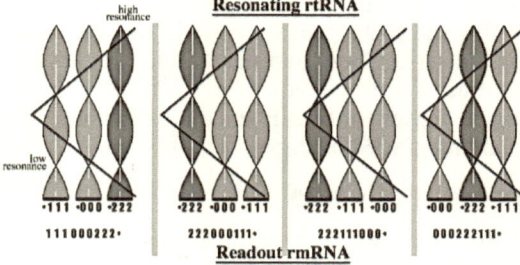

tion to the water sensor:

1.   Red cAMP is less and green cGMP is greater than the water resonance. The rmRNA is 111000222•

2.   Red cAMP is greater and green cGMP is less than the water resonance. The rmRNA is 222000111•

3.   Both are less than the water resonance. The rmRNA is 222111000•

4.   Both are greater than the water resonance. The rmRNA is 000222111•

1.   **111000222•**
2.   **222000111•**
3.   **222111000•**
4.   **000222111•**

There are four different outputs (ignoring the other possible two for the moment) that the digital output can be.

The digital sensors have an output of DI on 'readout' RNA that reflects the state of the wave in the vital unit. This DI can then be an input to IF/ELSE digital computing.

In a simple computer before the days of point-and-click, a crude calculator program would display a screen of choices and wait for an input from the keyboard, a port into the computer, which is assigned to the variable X.

```
Press
1 to add
2 to subtract
3 to multiply
4 to divide
```

The CPU of the computer runs a simple MS Basic routine that calls upon four different subroutines depending on the value of the variable. The computer ignores any key that is not 1, 2, 3, or 4 and just waits as it constantly scans the digital input for a match. If there is one, the CPU looks up the physical address of the DI on the hard drive and reads it into active memory.

Each segment on a hard drive is numbered, and this number is the physical address of the DI stored in that segment. The RUN command in the above program is an instruction to look up in a table the physical address of the named programs, copy the DI stored at that location on the hard drive into active memory, and then follow its instructions.

```
IF     X = 1 RUN ADD
ELSEIF X = 2 RUN SUB
ELSEIF X = 2 RUN MUL
ELSEIF X = 3 RUN DIV
RETURN
```

Naturally enough, living systems do not use MS Basic for such IF/ELSE processing but RNA does use a similar set of addresses for processing. The DI on DNA has a digital address in the promoter region that precedes each segment of DNA that is to be transcribed into RNA. The RNA polymerase binds to this *promoter* region with a strand of RNA that is complementary to that promotor region. With this complementary strand, the polymerase binds to the promoter and then initiates the assembly of RNA at the initiation site.

This complementary strand binds to the promoter sequence on the non-template strand, so the 'address' on the template DNA is the same as on the strand.

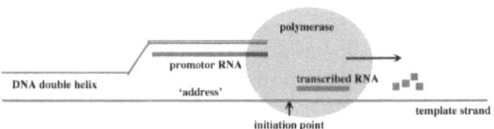

In our simple vital unit there are four stretches of DNA storage, each with a digital address. In the vital unit, each of these four stretches codes for a single mRNA that can profoundly alter the state of the vital unit. We have already seen in the viruses, that a single RNA molecule can suborn the mechanism of the vital unit into making more viruses.

The DI is only read from the DNA onto RNA, however, when the readout rmRNA binds to the promoter address—they are threes-complements of each other.

1.   In times of abundance, both the cAMP and cGMP sensors are less resonant than the water and the readout RNA is 222111000•. This binds to the address •222111000 and the polymerase binds and the mRNA transcribed. This is the program *GrowMultiply* that generates the analog wave of growing and multiplying.

2.   Bacteria excrete a small amount of cAMP. When they are close together, the external cAMP level soars and this alters the internal cAMP. The cAMP resonator is now greater than the water. The readout RNA is 222000111• and the address it binds to is •222000111. The mRNA transcribed is the program *Colony* that generates the analog wave of uniting as a characteristic colony.

3.   When food is getting scarce, the cGMP alters and the sensor is greater than water. The readout is 111000222• and the called address is •111000222. The mRNA transcribed is the program *Search* that generates an analog wave that channels resources into motility.

4.   When there is famine, both the cAMP and cGMP fields cause more resonance than water in the sensor. The readout is 000222111• and this binds with address • 000222111, and the transcribed mRNA is the *Spore* program; the vital unit shuts down.

•111000222 --- Search
•222000111 --- Colony
•222111000 --- GrowMultiply
•000222111 --- Spore

Feedback loop

Earlier, we described how the digital output of DI in the Read direction as mRNA determined the overall standing wave of the vital unit, its state. To this we can add a suggested feedback in digital form of the state of the vital unit. Digital information about the cell moves in the Write direction, from r/tRNA to rmRNA to DNA to determine what is transcribed as mRNA.

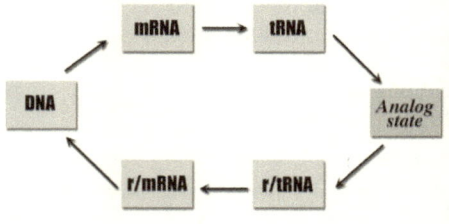

We can refer to all DNA that has such an address as the digital 'repertoire' of a vital unit. Only DNA that has an address can be transcribed into RNA, although the address itself is not transcribed into RNA. Very complex systems, such as eukaryote cell, have a great deal of DNA involved in these addresses that is not transcribed, and is probably part of what used to be called 'junk' DNA.

The repertoire includes all the DNA that is called on to make protein as well as DNA that is called on to assemble all the other types of RNA involved in calling up subroutines and the like.

Challenge and response

In our simple illustration using a triple sensor, there were two possible digital readouts that we did not assign a match on the DNA. What happens if these unassigned rmRNAs are the digital readout. Here we encounter the basic level in the systematic hierarchy of learning and memory.

a. **111222000·**

b. **000111222·**

If a readout rmRNA does not find a match, it is copied by RT into a blank address to store the result of the learning process that follows. The vital unit now calls up, in order, on its digital repertoire to see if it is a useful response to the input. In the case where both cAMP and cGMA were less than the water but their order is reversed, copying the *GrowMultiply* program into the blank address and transcribing the DNA works just fine, so the *GrowMultiply* program is now called up when the cAMP and cGMP are both less than the water sensor.

In the final unassigned readout, both cAMP and cGMP sensors are greater than water. This rmRNA is copied into DNA and the orderly exploration of the repertoire begins as the overall state falls below optimum. The *GrowMultiply* program fails, as does *Search* because the bacterium is caught in a matrix and cannot move. Before the last resort of *Spore* is called upon, the vital unit can start mixing and matching some of its subroutines, the protein modules it has available.

Proteins are modular assemblies. The same module is found in many different proteins and performs the same way. Many proteins, for instance, have a module that binds ATP with very similar primary sequences. This ATP module is in a string with other modules on the mRNA for the entire protein. When the module splits an ATP, its conformal change is transmitted to another module to drive an energy-absorbing catalytic change to the substrate of another module.

This ATP module appears in many other enzymes driving a host of very different reactions. New proteins can be generated by mixing modules, and simply requires tacking their mRNA together and copying it back to DNA.

Let us return to our stressed-out vital unit with its blank address. The vital unit now assembles new combinations of modules to see if it can bring things back to normal. The mRNA for the modules is copied into the DNA at the blank address, and its single mRNA transcribed and translated into protein.

One of the proteins it generates by mixing modules raises the normally low rate of digestive enzyme export. These convert the large molecules of the restraining matrix into small food monomers and the vital unit springs back to health, the cyclic nucleotides fall, and the *GrowMultiply* program restarts.

The vital unit has now learned a lesson. It has gained a bit of wisdom about how to handle this situation when it occurs. The sequence of modules that did the trick is now the program *Digest* with an address on the DNA that matches the readout RNA when both cyclic nucleotides sensors resonate more than the water sensor.

a. **·111000222** --- Search

b. **·222000111** --- Colony

c. **·222111000** --- GrowMultiply

d. **·000222111** --- Spore

e. **·111222000** --- GrowMultiply

f. **·000111222** --- Digest

The repertoire of responses with a three-field RNA sensor has now expanded. This is basically how the store of digital wisdom is expanded.

Unified Genetics

This diagram summarizes the discussion of a unified genetics that includes the internal wave aspect in the picture. All the digital information in this feedback loop is in the pattern of base-pairs on DNA.

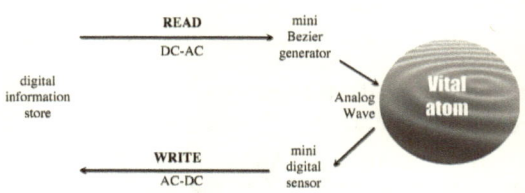

Living systems also have another level of digital information that is not stored in base-pair patterns but the pattern of methyl groups added onto bases along the DNA molecule.

The only analogy I can draw from computer technology is pressing the little language button in my menu bar. The keyboard input—Richard—is the same, but the analog shapes on the screen are utterly different. If I choose Greek, it is Pιψηαρδ, if I choose Arabic, it is زيصحارد, while if I choose Hebrew, it is ובישרג and it types from right to left. The same input but a very different output.

The computer does this by changing lookup tables. RNA would do it another way. It would block all the addresses of all the other languages by dabbing a spot of paint to make them unreadable. Now, only the addresses of the chosen language are available for calling as output. Difficult on a hard drive, but the result is the same: an input can have a different output depending on this second level of writing. Methylation of the bases along the DNA strand makes it even more hydrophobic and coil up even tighter, thereby making it less unavailable than usual. (The transition from RNA-only to DNA+RNA probably involved a similar mechanism.)

Since this level of digital information is written using the genetic level as substrate, this feedback loop is called Epigenetics.

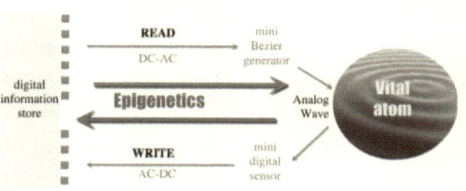

In the Great Debate over a designed or a random universe, classical genetics with its read-only concept of digital information has been the scientific support for randomness with its Fundamental Dogma. This 'foundation of sand' is inevitably shifting. A sign of this is to be found in *Nature*, the pre-eminent science journal, and reprinted in *Scientific American*, purveyor of mainstream science to the general public:

> A student referring to textbook discussions of genetics and evolution could be forgiven for thinking that the 'central dogma' devised by Crick and others in the 1960s — in which information flows in a linear, traceable fashion from DNA sequence to messenger RNA to protein, to manifest finally as phenotype — remains the solid foundation of the genomic revolution. In fact, it is beginning to look more like a casualty of it...

The review also went on to debunk another concept propounded by Richard Dawkins, a major promoter of random-chance-and-accident Darwinism, that of Junk DNA. This considered that the 95% of human DNA that was not read into mRNA, rRNA and tRNA to be meaningless 'selfish' DNA that was just useless baggage:

…Starting in 2003, ENCODE researchers set out to map which parts of human chromosomes are transcribed… Last year, the group revealed that there is much more to genome function than is encompassed in the roughly 1% of our DNA that contains some 20,000 protein-coding genes — challenging the old idea that much of the genome is junk. At least 80% of the genome is transcribed into RNA.[166]

It is already known that, just like a single RNA virus that suborns a cell, some human DNA is only briefly transcribed into RNA and then shuts down for the rest of a lifetime. The most astonishing example of this is the SDR on the human male Y-chromosome. This is only transcribed into RNA for a few hours in a 6-week old embryo. It then shuts down and is never transcribed again. Yet the RNA generated in that brief burst is sufficient to divert the default female path of development into that of the male. An embryo with a Y-chromosome lacking this comparatively short stretch of DNA will develop into a female, a genetic glitch that has troubled sex-divided sports testing.

As the sensitivity of the methodology improves, it will probably turn out that almost 100% of DNA is transcribed into RNA at some point in a person's lifetime.

As our technological example of virtual memory suggests, it is quite possible that the DNA of somatic cells is used for long-term memory. This possibility cannot be explored until DNA sequencing is applied to various tissues of the same individual, the glia cells of the brain being the most likely to need DNA for long-term storage of digitally-encoded memories.

Epigenetics

Unlike most major advances, epigenetics as a field of study had its origins not in observation of simple systems, but in the observation of the most complex system of all, human beings. It was found that the grandsons of grandmothers who had survived the famine during the siege of Stalingrad had significantly lower life expectancy than the control group. This was true even though the postwar experience of the parents and children in America were very similar.

The only known biological thing that grandsons inherit only from their grandmothers is their sole X-chromosome with its DNA of base-pairs. If the fundamental axiom of classical genetics were true, that reading but not writing only ever happens, then all these hundreds of disparate X-chromosomes must have undergone the exact same random change in the DNA to be passed on to the grandchildren.

However, this is exactly the opposite of what is expected of randomness when tossing 100 coins that all came up heads would cause extreme suspicion of the coins' fairness. But something was definitely being passed down the generations on the X-chromosome. The fundamental axiom of classical genetics trembled as it became clear it was incorrect: writing digital information happens alongside the reading of digital information.

Thus, in the last decade, this is how the science of *epigenetics* got its start, and it promises to be at least as significant as genetics, probably more so. While the science is still in its infancy, it will come as no great surprise, seeing what we have already discussed, that RNA has a role to play in epigenetics as well as in genetics:

[Epigenetics] could be mediated by the effect of small RNAs. The recent discovery and characterization of a vast array of small, non-coding RNAs suggests that there is an RNA component, possibly involved in epigenetic gene regulation. Small interfering RNAs can modulate transcriptional gene expression via epigenetic modulation of targeted promoters.[167]

While most research into epigenetics is currently in the animal realm because of its implications for aging and disease, it has also been established as a major player in the prokaryotes, the bacteria and simplest of vital units:

Like many eukaryotes, bacteria make widespread use of post-replicative DNA methylation for the epigenetic control of DNA-protein interactions. Unlike eukaryotes, however, bacteria use DNA adenine methylation (rather than DNA cytosine methylation) as an epigenetic signal...regulates the cell cycle and couples gene transcription to DNA replication... Switching between alternative DNA methylation patterns can split clonal bacterial populations into epigenetic lineages in a manner reminiscent of eukaryotic cell differentiation... DNA methylation plays important roles in the biology of bacteria: phenomena such as timing of DNA replication, partitioning nascent chromosomes to daughter cells, repair of DNA, and... regulate cell functions involved in RNA stability, mRNA translation, or protein turnover. However, the underlying molecular mechanisms remain to be identified.[168]

Bacterial DNA is turned off by adding a methyl to all the adenines in a stretch of DNA while eukaryotes do it to the guanine. The transition is probably a clue to what happened during the prokaryote-eukaryote leap in sophistication.

While epigenetics has a great future, it is still in its early days and most of the work has been done with eukaryotes. We will be discussing them in a later chapter.

We earlier outlined the history of the physical realm, starting with the Big Bang and ending with the formation of the planets. Our historical narrative now continues with that of the planet Earth. The 4-billion-year history of the earth is divided into Eons, and the first was the Hadean when the Earth was molten and subject to a tremendous bombardment of comets and the like, which raked the surface. There was no liquid water and there were certainly no living systems. The Hadean Eon was hellish and abiotic.

The Earth cooled, accumulated a liquid ocean and, within 100 million years, the fossil record shows clear signs of abundant bacterial life. The transition from an abiotic world to one teeming with life is the topic of the next chapter.

To summarize the view of evolution presented here. If the read-only, external-only, materialistic view of evolutionary development can be captured in the phrase, *Survival of the Fittest*; then this write and read, internal and external, theistic view of evolutionary development can be captured in the phrase, *Wisdom of the Ancestors*.

THE ORIGIN
OF LIFE

Biogenesis is the study of the origins of living systems. Classical biogenesis is based on the concept that natural law directly influences the external aspect of systems. A unified biogenesis is based on the more sophisticated view of modern science where natural law determines the internal aspect of systems which determines over time the external aspect of systems.

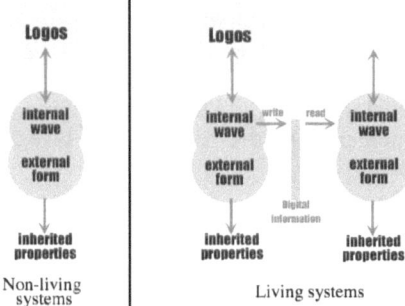

The Logos is an abstract systematic hierarchy that determines indirectly the structure and emergent properties as the physical hierarchy is constructed over time. The external form resonates with the Logos and the emergent properties are directly inherited from the Logos. If the system is disrupted, it no longer resonates with the Logos and the properties are lost. All the properties of the helium atom that we discussed are lost if the thermal energy disrupts it into independent electrons and nuclei.

We have also discussed how living systems can read digital information and convert it into analog waves. If these waves resonate with the structures in the Logos, then the emergent properties are expressed, even though the Logos was not involved.

We have also seen that there is writing of digital information in living systems, even if it is as yet poorly understood.

Putting all this together, we can say that the very first living system, the first vital unit was a system that resonated with the Logos and inherited a set of emergent properties that allowed it to sample its analog wave and write it to digital information. This was duplicated and used to recreate multiple copies of the analog wave that also resonated with the Logos and indirectly inherited the same set of emergent properties as the very first 'parent' system.

Non-living systems

Living systems

First Life

To be a contender for first life, we need inanimate molecules that can: (1) Sample analog waves (2) Output and store digital information (3) Recreate the analog wave by reading the digital information (4) Duplicate the digital information.

It should be quite clear from what we have discussed why the consensus is that the only molecule with this set of emergent properties is RNA, and that it was an RNA world in the beginning.

Systematic edens

Living entities form a systematic hierarchy, just as the inanimate realm does, and the basic principles apply:

1. A system has an internal wave aspect that determines over time the confined forms of interacting subsystems that are coupling with their subsystems. This is the external form of the system.

2. If the confinement of subsystems is not perfect, the system can overlap its internal wave with that of another similar system so that they couple with their valence subsystems and interact. These are the external forms of the interactions of the system and gives it a set of characteristic properties not possessed by its individual subsystems.

Up until the advent of human beings, everything that happened during the history of the physical realm was a direct reflection of structure of the Logos. Each step in the expression of this abstract structure was, in turn, highly probable given that a few requirements were met:

1. Systems had emerged from a previous step.

2. Their circumstances are such that they can interact.

3. There is sufficient time for the Law of Large Numbers to make the external form of interaction be that of the internal form of interaction.

These requirements for system building are that there be sufficient time and circumstances for the interacting systems to stumble upon the probable form of a higher system that is waiting there but empty in the Logos.

Borrowing the concept of an Eden where everything was just right for the origin of man, we shall use the concept of an eden in describing system building. An eden for a system is where all the requirements are met, i.e., the subsystems are in an environment that is conducive to them and their interactions. The systematic development of the inanimate realm is a sequence of edens that is just right for the emergence of the next, higher level. The chart lists the sequence of inanimate edens.

We can expect to find a similar sequence emerging in the system building in living systems. So, the first step towards understanding the system building of life is the emergence of an RNA eden in which all its subsystems are in an environment suitable for interaction.

The unfolding of the Logos that led to the environment in which RNA emerged is only poorly understood at present. The only difference between the classical and the unified view is that the sequence of steps is inherent in the Logos in the unified view while it is random in the classical. Both views necessitate a sequence of steps and an environment in which RNA subunits can naturally occur.

One recent development is the possibility that the first RNA subunit did not emerge by linking base, ribose and phosphate together, but by the assembly of quite different fragments:

[One ring bases] and their respective nucleotides have been prebiotically synthesized by a sequence of reactions that by-pass free sugars and assemble in a stepwise fashion by going against the dogma that nitrogenous and oxygenous chemistries should be avoided. In a series of publications, The Sutherland Group at the School of Chemistry, University of Manchester have demonstrated high yielding routes to cytidine and uridine ribonucleotides built from small 2 and 3 carbon fragments such as... glyceraldehyde-3-phosphate, cyanamide and cyanoacetylene. [169]

History of Earth

In our discussion of the unfolding of the Logos over time, we got as far as the third generation of stars that formed from collapsing clouds of interstellar H and He that also contained 1% of all the other elements as a dust. The dynamics of the formation of our Sun and planets have yet to be fully understood but 99% of the material ended up in the sun (or was ejected when the sun ignited) and 99% of the angular momentum ended up in the planets.

The inner rocky planets—Mercury, Venus, Earth and Mars—are mainly dust, while the gas giants—Jupiter, Saturn, Neptune and Uranus—are mainly H and He. To my mind, even though it was written over 60 years ago, the answer that George Gamow supplies to his question, summarized in his diagram, resonates with a unified perspective of standing waves in the dynamics of the collapse:

> It is clear that, in such heavy traffic, numerous collisions must have taken place between the individual particles, and that, as the result of such collisions, the
motion of the entire swarm must have become to a certain extent organized. In fact, it is not difficult to understand that such collisions served either to pulverize the "traffic violators" or to force them to "detour" into less crowded "traffic lanes." What are the laws that would govern such "organized" or at least partially organized "traffic"? [170]

The Moon

One thing that is firmly established about the final step in the aggregation of the proto-Earth is that it involved a collision with a Mars-sized planetesimal, their combined splash coalescing in a few years into the Moon, then at only half the distance it is today.

This presence of our large satellite Moon is of great importance in making the Earth an eden for life for (at least) the following reasons:

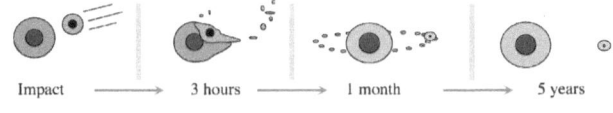

1. The Earth's axis was tilted by 23° from the normal and most of the angular momentum of the Earth-Moon system ended up in the rotation of the Moon.

This made the system very stable and the axis of the Earth, unlike that of Mars, has kept a constant orientation.

2.   The crust of the Earth was shattered into fragments allowing the heat of the Earth's interior, stoked by radioactive decay, to drive the motion of tectonic plates with the gradual release of internal energy as ongoing volcanic activity. Venus, of similar size but without a moon-fractured crust, had no such gradual release and the buildup of internal heat eventually escaped as global eruption that remelted the entire surface.

3.   The Earth cooled and water from vulcanism and icy comets condensed into the oceans. The Moon raised tides in these oceans that were enormous compared to those seen these days. The tidal effect is directly proportional to the cube of the distance. When the Moon was at only half the distance, the tides it raised on the Earth were $2^3 = 8$-times as great in amplitude as those

today. As noted earlier, the energy of a wave is the square of the amplitude, so the primordial tidal waves had 64 times the energy of those today and coastal erosion was 64 times as efficient. Huge amounts of sediment were deposited in the oceans, creating the great China-clay beds we mine today for our pottery and industry. The photo is a remnant of one atop a mountain after centuries of mining.[171]

The Moon-contributed stability, volcanic tectonic plates, and great beds of clay gave the Earth-Moon system the Logos-inherited properties that, as we shall see, were just-right to be the eden for life to emerge and flourish.

The Eden for Life

While there is no current consensus as to how life actually emerged, a lot is known about aspects of the early earth that probably contributed to the momentous advent of life. I shall mention just a few of the emergent properties on the early earth that are most likely contributors to the origin of life.

Following the pattern prescribed by the Logos, three great systems interacted together as the subsystems of the eden for first life to emerge, two of which are indirectly attributable to the Moon:

1.   Surface of the Earth; the oceans and atmosphere illuminated by the young Sun

2.   Tectonic vulcanism

3.   Sedimentation of great beds of clay minerals

We will deal with each in turn, then see how they interacted together as the eden for first life.

Earth's Surface

Water in its liquid state is the most important ingredient for life. In all living systems, water molecules are the main resonator subsystems that take up the form of the internal waves created by the generator subsystems. Just as an orchestra playing in the absence of gaseous air molecules cannot create a symphony, the generators cannot create a living system in the absence of water molecules.

The Earth is in what astronomers call the *Goldilocks zone* about the Sun. It is not too close to the Sun that would otherwise result in a surface temperature above the boiling point of water; it is not too distant from the sun, otherwise resulting in a surface temperature below the freezing point of water. It is just the right distance for liquid water to exist alongside solid water and atmospheric water. Neither closer-in Venus at $+900°F$ nor further-out Mars at $-110°F$ is in the Goldilocks zone and consequently do not have oceans.

A very important property of water is that it is an excellent solvent. We can use common salt, sodium chloride, to exemplify this. An ionic solid like sodium chloride is not really composed of NaCl molecules. There is no sharing of electrons; the sodium loses an electron completely to a chlorine atom. The result is two charged spheres since both have their outer electrons in the noble gas configuration. The spheres stack as closely as they can to minimize their repulsion and maximize their attraction. For sodium chloride, this a cubic array, and salt crystals composed of zillions of atoms exhibit this shape. These ionic bonds are very strong and it takes a temperature of $1,074°K$ ($801°C$; $1,474°F$) before salt will melt, and only at $1,686°K$ will the ions be liberated as a free gas of ions, a plasma.

If a crystal of salt is dropped into water at room temperature, however, the ions dissolve into free ions surrounded by water molecules. There is less free energy in solvated ions than in the crystal, so sodium chloride is quite soluble and 36 grams of it will dissolve in 100 grams of water.

At the other extreme is dicalcium triphosphate, $Ca_3(PO_4)_2$, where the free energy of the tightly bound ions is much less than when they are surrounded by water, and this ionic compound is quite insoluble in water as only 0.002 gram of it will dissolve in 100 grams of water. Living systems often use the powerful interaction of calcium with phosphorous as a signal coordinating the generator subsystems within a cell.

At the time of the origin of life, the newly formed oceans were not pure water. The great tides raised by the Moon had dissolved various salts in the water. It was in this milieu that living systems emerged. The 'symphony of life' has as its resonating subsystems the water molecule and a small number of ions. Ignoring the rare ions, this solution is approximated by mixing four salts together, as listed in the chart.

salt	%
NaCl	0.8
KCl	0.02
Na$_2$HPO$_4$	0.14
KH$_2$PO$_4$	0.024
pH	7.4

It is this solution that life emerged in, and it is roughly the ionic concentration in the cytoplasm of all living systems to this day. All cells go to a good deal of trouble to maintain this concentration of ions.

Small Organics

The atmosphere of the very early earth would have reflected the gas cloud out of which the earth accreted, composed mainly of hydrogen and helium. We have already noted that the speed at which the molecules of a gas are moving at a given temperature goes down as the molecular eight increases. Both the hydrogen molecule and the helium atom are small, at 2 and 4, so they were moving fastest of all. Since their speed was greater than the escape velocity, over time almost all of these two gases escaped from the atmosphere.

This is called the Hadean era of the earth's history. It started at Earth's formation about 4.6 billion years ago, and ended roughly 3.8 billion years ago. This was a period of heavy bombardment, including the impact that created the Moon, and intense volcanic activity.

Based on today's volcanic evidence, the resultant atmosphere would have contained 80% water, 10% carbon dioxide, 7% hydrogen sulfide, and smaller amounts of ammonia, nitrogen, carbon monoxide, methane and inert gases. Free oxygen would be entirely absent.

Atmosphere

The tremendous tides raised by the early Moon ensured that the ocean and atmosphere were in equilibrium. There was no free oxygen, and hence no ozone layer, so the UV from the sun could penetrate into the lower levels of the atmosphere. Lightning was also commonplace as the weather back then was awful.

The cataclysmic formation of the Moon probably stripped away most of the earliest atmosphere, but the rain of comets brought with them water and a great number of small organic molecules.

The energy input by lightning and UV fragmented the molecules in the atmosphere by knocking out hydrogen atoms, creating free radicals with unbalanced electrons. These free radicals then combined with each other, creating more complex molecules.

$$H_2O \rightarrow HO-$$
$$NH_3 \rightarrow H_2N-$$
$$CH_4 \rightarrow H_3C-$$

The classic Miller-Urey experiment, as illustrated, recreated the primordial atmosphere and conditions, and subjected it to an electric discharge.

At the end of one week of continuous operation, Miller and Urey observed, by analyzing the cooled water, that as much as 10-15% of the carbon within the system was now in the form of organic compounds. Two percent of the carbon had formed aminoacids, including 13 of the 22 that are used to make proteins in living cells, with glycine as the most abundant. The Miller-Urey experiment inspired many experiments in a similar vein. It was found that aminoacids could be made from hydrogen cyanide (HCN) and ammonia in a water solution, and significant amounts were formed of the nucleotide adenine which is one of the four bases in RNA and DNA.

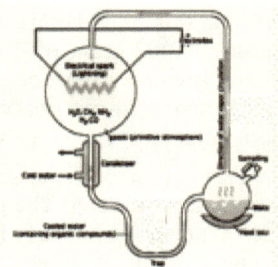

glycine

adenine

Ocean

The primordial ocean was saturated with iron in its reduced, ferrous state Fe^{+2}, which is soluble in water. The ferrous ion can easily give and receive electrons as it reversibly changes into the ferric state Fe^{+3}. In this way, it can transfer electrons (each accompanied by a H^+ ion) and act as a simple reversible mediator between chemical interactions.

The intense UV light striking the oceans was absorbed by ferrous ions, which changed into ferric ions. The liberated electron unites with a H^+ to create a very reactive hydrogen atom that can participate in a host of chemical reactions and even drive chemical reactions 'uphill.' The generation of excited electrons provided the primitive Earth with reducing potential and the abiotic synthesis of reduced raw materials such as H_2 and HCN.

$$Fe^{++} + UV \rightarrow Fe^{+++} + e^-$$

This was a crude echo of the photosynthetic splitting of water, and the liberated oxygen precipitated out as an insoluble ferric/ferrous mixed oxide called magnetite. This is how the *banded iron* strata were deposited—the main source of iron ore today, $[FeO][Fe_2O_3]$.

One reaction that many think played a central role is the condensation of a sulfhydryl, the sulphur analog of an alcohol, e.g., CH_3SH, with a carboxylic acid, e.g., CH_3COOH, to form a thioester. The thioester linkage formed is "energy rich," releasing between 7.5 and 8.5 kcal/mol upon hydrolysis. The significance of this is that this is sufficient energy to form pyrophosphate from two inorganic phosphates, and pyrophosphate is the energy store used to drive many chemical reactions in an uphill direction.

Thioesters still play a key role in the *housekeeping* metabolism common to all life, and drive the formation of ATP directly in *substrate-level* phosphorylation.

Tectonic Vulcanism

The crust at the bottom of the oceans is naturally saturated with seawater, and the heat escaping from the inner mantle heats this water and sets it into convection as the hot water rises and the cool water from the ocean sinks through the crust. The upwelling of heated water can be channeled into two types of outlets called *smokers*—one black, one white.

Black smokers

The faults between tectonic plates are sites where hot magma can come close to the surface. When these areas are underwater, the result is *black smokers*, where seawater percolating down is heated and returned to the surface acidified and saturated with minerals, particularly sulfides. When the superheated, mineral-rich water leaves a vent and mixes with the cold ocean-bottom water, it precipitates a variety of minerals as tiny particles that make the vent water appear black in color and, over time, this precipitation creates a chimney-like structure. This is why these sulfide chimney structures are called black smokers.

The sulfides are energy rich and can be used as a source of free energy to power chemical changes. To this day, black smokers host a prolific biota that is powered by high-energy sulfides, rather than the light of the Sun, as is the case for almost all other life.

White smokers

White smokers refer to vents that emit lighter-hued minerals, such as those containing barium, calcium, and silicon. These vents also tend to have lower temperature plumes. They are not powered by magma coming close to the surface as tectonic plates spread apart, but by the reaction of the newly-surfaced rock with seawater. The water chemically reacts with the rock, forming hydroxide minerals like serpentine.   The reaction liberates heat, hydrogen, methane and ammonia. These break through the seafloor as the white smokers. Rather than the dense precipitate of iron sulphide, the minerals that precipitate out in white smokers form lacy and porous networks, the pores being of bacterial dimensions. The reaction of

methane with hydrogen is catalyzed by iron sulphide crystals, and builds organic molecules while releasing energy. While carbon dioxide is very stable, it reacts with free radicals in the vents to form acetyl thioester, an activated form of carbon dioxide. Carbon dioxide will spontaneously react with this to form pyruvate, a simple CHO molecule that is at the center of core metabolism in all organisms

These alkaline hydrothermal vents also continuously generate acetyl thioesters, providing both the starting point for more complex organic molecules and the energy needed to produce them, since acetyl thioesters can incorporate phosphate, as acetyl phosphate, which can, like ATP, transfer its high-energy phosphate to other molecules. Microscopic structures in such alkaline vents show many interconnected compartments that are thought have possibly provided an ideal womb for the origin of life.

Lost City is a small forest of such white smokers in the mid-Atlantic ocean of about 30 chimneys made of calcium carbonate 90 to 180 feet high, with a number of smaller chimneys. The outflow from these white smokers contains a variety of hydrocarbon molecules:

> Radiocarbon evidence rules out seawater bicarbonate as the carbon source for [hydrogen and carbon monoxide] reactions, suggesting that a mantle-derived inorganic carbon source is leached from the host rocks. Our findings illustrate that the abiotic synthesis of hydrocarbons in nature may occur in the presence of ultramafic rocks, water, and moderate amounts of heat.[172]

Sedimentary Clay Beds

There was intense weathering on the early earth and the sediment settled in the oceans as immense beds of clay. Clay has many interesting emergent properties, including a wide variety of catalytic properties that are exploited today in the chemical industries.

A typical clay particle is normally very small, <2μm, which results in the presence of very large surface areas. Clay minerals are composed of silicon, aluminum or magnesium ions or both, and water. Iron can be a substitute for aluminum and magnesium, and potassium, sodium, and calcium are often present in abundant quantities as well. Chemical solutions perfused through clay beds undergo many useful transformations. The catalytic properties of clays are still being explored, but a wide range of chemical ability has already been documented in a recent review:

> Clays exhibit specific features such as high versatility, wide range of preparation variables, use in catalytic amounts, ease of set-up and work-up, mild experimental conditions, gain in yield and/or selectivity, low cost, etc., which may be very useful tools in the move towards establishing environmentally friendly technologies. Furthermore, the possibility of upgrading these materials by the [separation of layers] opens new and interesting perspectives, also considering possible shape selective advantages. Recent catalytic applications of cationic and anionic clays in organic or fine chemistry (acid- or base-catalyzed reactions, Diels–Alder reactions, reactions using metallic nitrates, etc.), environmental catalysis ([sulphide and nitrate] oxidation) and energy exploitation (partial oxidation of methane) are discussed as very promising research subjects with a wide range of possible future developments.[173]

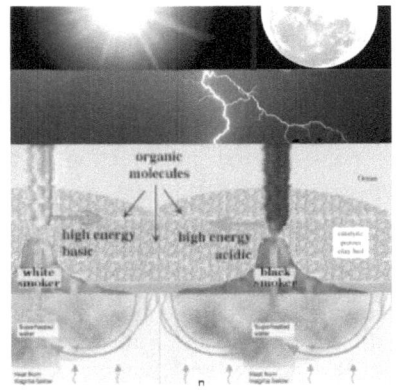

The primordial clay beds were, therefore, places where diverse chemical interactions between molecules could occur, and it has been proposed that the catalytic properties of clay were involved in creating life's precursors.

Womb of Life

The primordial earth, with all the properties just enumerated (and probably many more that are yet to be understood) that were inherited from the Logos, was an eden in which all the subsystems were present for the system-building interactions that led to the first living systems. All of these emergent properties came together in what can only be called a womb in which this momentous event could occur.

This womb had three main components:

1. Ocean water saturated with iron and small organics is circulated through clay beds deposited on the crust driven by the heat exiting the mantle.

2. The bed has both black and white smokers that add high-energy organics to the perfusate and establish gradients of both temperature and pH in the clay bed.

3. The catalytic properties of the clay allow many transformations to occur and, where the conditions are just right as prescribed by the Logos, allow for nucleotides to emerge.

While there is no consensus as to the exact details, we can surmise that a bed of clay was deposited over a tectonic fault and that it was perfused by both black and white smokers. The small organic molecules that enriched the cold ocean water were drawn through this bed and, after being heated by the magma below, entered the clay as high-energy thioesters and phosphates.

This circulation of water through the catalytic pores in the clay (with the smokers playing the role of a heart) and the temperature and acidity gradients that were set up, allowed a multitude of chemical reactions to occur and a 'chemical ecology' to be established expressing a plethora of Logos-derived properties.

One aspect of this prebiotic chemistry in its clay bed eden is that the properties that it inherited from the Logos had a distinct chiral nature. This is similar to the very first eden of all, the eden for the emergence of matter—the Hot Big Bang of matter and antimatter in essentially equal amounts. The one-in-ten-billion deficit of antimatter fermions—a right ½-twist along the complex time axis—over matter fermions—a left ½-twist along the complex time axis—was sufficient when the universe cooled to form all the 100 billion galaxies in the visible universe (not to mention that the physical realm is probably magnitudes times greater, though not infinite).

The eden for matter was chiral, with a fundamental preference for left over right on the internal, complex level. The eden for first life was also chiral, with a fundamental preference for *right* sugars and *left* amino acids.

Left-right asymmetry

The light is a transverse wave—the axis of waviness is at right angles to the direction of travel (while sound is a longitudinal wave where the axis of waviness is parallel to the direction of travel). In unpolarized light, the axis is at any, and all directions, around the direction of travel. In *plane* polarized light, there is just a single axis of waviness.

If such polarized light is passed through water, the axis will be unchanged. If it is passed through a solution of an "optically active" compound, however, the axis is rotated. This does not involve any transfer of energy, just the configuration of the chemical compound. The rotation can be to the right (clockwise) or to the left (counterclockwise). This rotation can be measured using a simple experimental setup.

Molecules involving carbon atoms are optically active when there is an 'asymmetric carbon', i.e., a carbon atom in which all four of the tetrahedral bonds are connected to different sets of atoms. In this situation, it is possible to distinguish between two isomers that are identical except for being mirror images of each other. The asymmetric carbon is said to be 'chiral.'

The glyceraldehyde (GAD) molecule that occurs in core carbohydrate metabolism has an asymmetric carbon

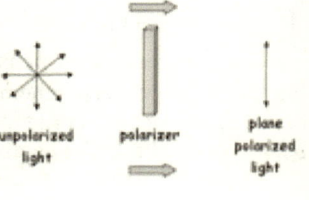

with its two mirror-image isomers that are equally and oppositely optically active. The central carbon atom of GAD is attached to a hydrogen, a hydroxyl, a formalde-hyde radical, and a methanol radical. As this is the simplest sugar, the isomer that rotates light to the right is called D- (*dextro-*) and to the left, L- (*-levo*). All molecules in metabolism can be sourced to this molecule, so it is used to define the chirality of organic molecules. All the aminoacids that reflect the L-form of GAD are assigned a chirality of L (even if they rotate light to the right) and those based on the D-form of GAD have a chirality of D (even if they are now left rotators). Similarly, all the many carbohydrates used by life can be classified as D or L sugars.

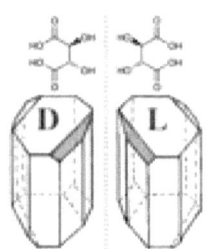

The crystals that these isomers form in the pure state are also mirror images of each other, as is illustrated by the two crystals of D- and L-tartaric acid, which played a key role in the exploration of optical activity.

Currently, there is nothing that is known that would suggest that all of the sugars and aminoacids that could have entered into the womb of life were not balanced in their chiral forms. That there were an equal number of both the D and the L forms around.

Something happened during the gestation of life, however, and the universal ancestor of all life emerged constructed solely of L-aminoacids in all its proteins, and D-sugars in all its nucleic acids. As far as I am aware, there has been little work done on proteins constructed of D-aminoacids or L-sugar nucleic acids, so the underlying reason for this chirality is unknown.

If the properties are identical, then the first to emerge stole the show. If their properties are subtly different, as might be expected given the chirality of the Logos, it was because the proposed solution within the Logos to the challenge involved only the subsystems with a specific chirality.

Challenge and Response

Inanimate History

Moving back to the big picture, there is a subtle difference between the view of history based on the principles of a unified science and those of current classical science.

Our science is now confident that we have a pretty accurate description of the history of the physical realm from the Big Bang origin of matter fermions to the emergence of the eden for life on the planet Earth some nine billion years after.

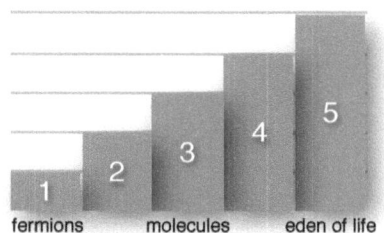

Sophistication

In the classical picture, this development was a fortuitous accident in a universe that just happened to have the precise natural laws that allowed for it to happen. That one universe should be blessed with such a fortuitous set of natural laws is to be expected at random in a multiverse of a trillion {*forty* trillions} trillion other universes each with their random selection of natural laws.

That is the classical perspective. The unified perspective views the history of the physical realm—composed of twisted components of four complex dimensions—as being the stepwise expression of abstract structures in the Logos—composed of a hierarchy of complex dimensions.

A similar stepwise expression of the Logos occurs in living systems:

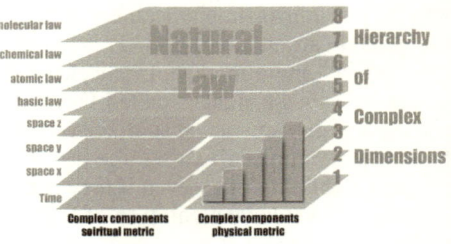

1. Prokaryotes. Vital Units with a wide variety of emergent properties that allow them to populate and thrive in almost every environment found in and around the surface of the Earth.

2. Eukaryote protists. Poly vital units, single plant, fungi and animal cells. A higher level in the sophistication of their emergent properties, such as the amoebas, yeasts and the denizens of ponds, such as euglena and paramecium. They are almost as ubiquitous as the bacteria.

3. Multicellular eukaryote plants. Wide variety of emergent properties, autotrophic so can live on sunlight, carbon dioxide and water plus a few minerals.

4. Multicellular eukaryote fungi. Heterotrophic recyclers of dead plants and animals.

5. Multicellular eukaryote animals. Active analog muscles and calculating digital nervous system.

This development over time is called evolution; living systems have developed over time into a systematic hierarchy of ever-increasing sophistication. The emergence of a new level of sophistication does not happen instantaneously; it is a process involving time. By its very nature, a process has a beginning, a middle and an end. Any process in system building can be divided into three stages:

1. *Formation.* All the necessary systems are gathered together in a suitable environment.

2. *Development.* The systems interact with each other and explore their possibilities.

3. *Completion.* The interaction of the systems resonates with a form in the Logos, and they become subsystems of a more sophisticated system, a step up in the systematic hierarchy.

An simple example is the system building step when atoms first emerged in the universe out of a plasma of free electrons and protons, which happened about one million years after the Big Bang.

The early universe was utterly hostile to any atom of hydrogen or helium at all. For every electron and proton, there were 100,000,000,000 high-energy gamma photons that instantly disrupted any liaison if they ever encountered each other. All that energy had a gravitational attraction that opposed the residual inflation of the universe, slowing the expansion of spacetime. This drained energy from the gamma photons and they became X photons, then UV photons, then visible light photons, then IR photons and, by our era, they became microwave photons, the CMB.

1. *Formation.* During the last stages in this cooling, the energy of the still-as-abundant photons became too small to disrupt helium and then hydrogen atoms.

2. *Development*. Opposite electric charges moved towards each other, while equal charges moved apart, still interacting with photons.

3. *Completion*. There were an equal number of positive and negative charges that emerged from the Hot Big Bang, and the electromagnetic force is so strong that all found a mate as subsystems of a hydrogen atom or a helium atom.

Each of these three stages is also a process in time, so each of the three also has a start, middle, and end. A process can be broken into a series of of subprocesses.

This also applies to the system-building in the hierarchy of living systems, and here the classical read-only view and unified write-read perspectives diverge.

Consider a vital unit that is multiplying in an environment rich in items A and B. The vital unit has the digital information to generate an analog wave to utilize A as food but not B. As the supply of A diminishes as multiplication proceeds, there is *selection pressure* to use B as food.

In the classical view, the response to this selection pressure is to wait until the random changes in the current DI come up with a digital sequence that happens to generate an analog wave for utilizing B.

In the unified view, the vital units shuffle their resources and test them out one by one, recording what works, discarding what does not.

For even simple systems, there are a large number of things that can happen, i.e., the things that are possible given the circumstances. To describe such large sets of possibilities, scientists have developed the concept of a *phase space* and the history of a set of interacting systems as having a *trajectory* in that phase space.

Phase space

Each system has a *degree of freedom* for each of the things it can do, e.g., moving in the three spatial dimensions. Expressing an interaction it is capable of is another simple example.

In a phase space, every degree of freedom of the system is represented as an axis of a multidimensional space. A phase space may contain a huge number of dimensions. For instance, a gas containing many molecules may require a separate dimension for each particle's x, y and z positions and momenta, as well as any number of other properties. For every possible state of the system, or allowed combination of values of the system's parameters, a point is plotted in the multi-dimensional space. This trajectory, which is a succession of points in phase space, is a representation of the system's state evolving over time.

In summary, the phase diagram represents all that the system can be and can do, and its shape can easily elucidate qualities of the system that might not be obvious otherwise. An example is a pendulum bob swinging to and fro in an arc. The actual trajectory of the bob in actual space is variable, and it is momentarily motionless at either end while moving rapidly through the center of the arc. In a phase space where position and momentum are the degrees of freedom, however, the motion is a simple constant circle. If the pendulum is frictionless and swinging in a good vacuum, the circle is a constant. If energy is lost to friction, the trajectory is a spiral that winds slowly to a stop at the center.

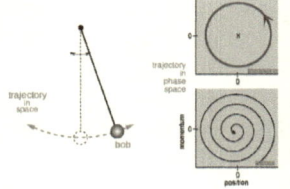

As it is technically impossible to illustrate thousands of orthogonal dimensions, we will restrict our diagrams to just two representative dimensions. Although this is a loss, there are still useful trajectories that can be illustrated in just two dimensions. For instance, in a system such as helium whose internal wavefunction firmly confines all of its subsystems, the trajectories of all of the subsystems in phase space are confined within a volume. The only degrees of freedom a helium atom has in everyday circumstances, i.e., excluding ultracold and ultrahot environments, is movement in the three spatial dimensions. The confined volume of subsystem phase space is moving as an entity in the helium atom phase space. System building involves a confined volume of subsystem phase space and the extension of a new set of dimensions to describe the system phase space.

The incomplete subsystem confinement of most system waves, other than perfect helium, are the coupling possibilities of the system, its valence couplers, and each extends a degree of freedom to the new phase space. The core subsystems, like helium, are firmly confined to a volume of phase space. In the stepwise expression of the Logos over time, each step can be broken into the three stages:

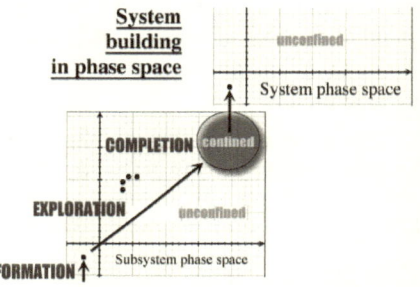

1. *Formation.* Emergence of systems from previous levels into an eden provided by the Logos in which the systems thrive and explore their interactions

2. *Exploration.* The systems explore the possibilities of their interactions, contributing growth and development of the environmental eden they are in. They have a trajectory in phase space.

3. *Completion.* The interactions of the subsystems resonate with the Logos and the resultant internal wave confines the systems as subsystems of a higher system. Their partial confinement is the coupling ability of the higher system, which emerges as the confined form is established.

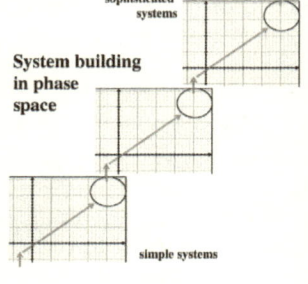

This emergent higher system is now ready to participate with others in the formation stage of the next step in system building as they explore the phase space of the degrees of freedom extended by the emergent interactions.

For inanimate systems, this does not add much to the picture of historical system building but, for living systems, we have the added aspect of learning and digital memory. The learning is about interactions with others and the environment, and the environment is an expression of the Logos. The living system is learning about the Logos and committing it to digital memory.

In the classical view, the trajectory through phase space in response to selection pressure is random and only a description. In the unified view, the trajectory through phase space already exists in the Logos and negative selection pressure can be viewed as a opportunity, a challenge with a prize.

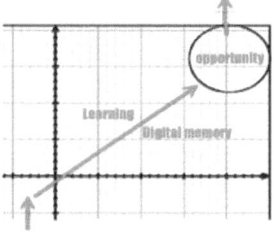

There are no living systems that exist forever and, given a population of living systems that are unable to multiply, it will decline in numbers over time with a certain half-life. If the multiplication rate of the living system is greater than the average half-life, then the population will increase. So, the movement through phase space is a lineage moving through time, the leading edge increasing and the tail end fading out.

The trajectory through phase space is a lineage of living systems accumulating *wisdom*, knowledge about the Logos, along the way.

As long as the benign natal eden endures, the lineages that remain the same do just fine. If, however, the eden disappears as the Logos-guided development of the environment proceeds, then the opportunity is more of a challenge to be solved. Not so much later in the evolution of life, free oxygen was added to the environment and sufficiently entered into the clay beds to completely alter the possible chemistry, and the eden for life disappeared. Only those lineages that solved the challenges of life outside of the clay bed eden survived to leave descendants. This is akin to the classical concept of selection pressure: a lineage adapts or it becomes extinct. The only difference is that learning and writing digital information is a lot faster than waiting for random errors to generate the requisite information in the modern synthesis of genetics and evolution.

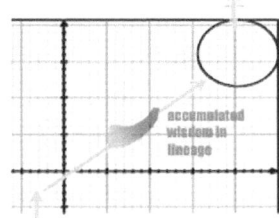

In the clay-smoker womb of life, the challenge is not to get swept away by the water flow—selection pressure for water confinement and control—and how to synthesize the molecules of life independently of the abiotic sources—selection pressure for metabolic control.

Kerogen

As any organic chemist will testify, it is all too easy for molecules to link up into a disorganized tangle of connected atoms that coats the experimental glassware. Such amorphous *kerogen* probably coated and filmed the channels in the backwaters of the clay-bed wombs. It could have been such films that the first vital units used to confine water and an environment in which phospholipids could be assembled.

The transition from a chemical ecology to the emergence of the first living systems took a few hundred million years, and much remains to be understood. What is most remarkable is that all the evidence is that one, and only one, lineage solved the challenge of independent life and emerged as the ancestor of all life on Earth. This lineage is the *Last Universal Common Ancestor* (LUCA) from which all known living systems are descended.

Last Common Universal Ancestor *LUCA*

The subsystems that entered the womb of life were an inanimate heterogeneous mix of activated small molecules. What emerged, a hundred million or so years later was the Last Universal Common Ancestor (LUCA) of all life on earth. LUCA was a living system with all the properties necessary for life independent of the clay-bed womb.

The properties of the LUCA lineage is the set of properties that all living things have in common plus some that were lost along the way. These include:

1.   An outer coat very similar to a kerogen of cross-linked polymers, i.e., something like a coat of mail. All prokaryotes have one but the lineage that led to eukaryotes discarded it along its trajectory in phase spaces.

2.   A bi-lipid membrane that confined the vital unit of water within the protective coat

3.   The core carbohydrate metabolism of small molecules

4.   The set of nucleotide-related prosthetic assistants in metabolism—ATP, NADP, etc.

5.   DNA for deep-time storage of digital information down a lineage

6.   The universal triplet code for converting digital information into analog waveforms

7.   The large and small RNA subunits of the ribosome for assembling proteins

8.   Using hydrogen ion gradients, *proton-motive force*, to recharge ATP

We have already discussed these properties except the last. This is the basis for many sophisticated processes that also use proton-motive force to link ADP and phosphate together with the high-energy bond of ATP.

Proton-motive Force

One of the challenges facing early life was the control of acidity, the concentration of hydrogen ions, the pH of the vital unit of water. The molecules of life work well only at or near a neutral pH—if too high or too low, all sorts of unpleasant things start to happen.

In order to thrive in an acid environment, energy must be expended to drive the H-ions against the concentration gradient. This energy is provided by a portal protein breaking ATP down to ADP and P and ejecting an H-ion. This ability to control pH must have been discovered early on in living systems.

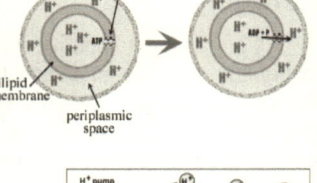

This is a reversible process. If excess H-ions outside are allowed to flow inwards, ADP and P are united into an ATP. If H-ions are pumped out through the membrane by some alternative route, then they can be allowed to flow back inwards through the portal with the generation of ATP. The H-ions can be pumped out again to repeat the process.

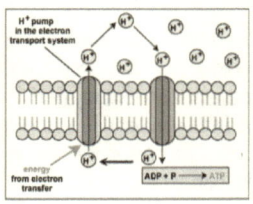

The LUCA lineage had discovered how to create such a current of protons and use it to generate ATP by using the electrons released by the stepwise breakdown of carbohydrates to pump H-ions out through a membrane, using a sequence of electron carriers and an electron acceptor, such as ferrous and/or sulphide ions.

Photosynthesis and Respiration

This proton-motive or *chemiosmotic* force was later adapted for trapping the energy of light. The energy of absorbed photons is used to drive the electron transport of H-ions into the periplasmic space. A water molecule is the electron acceptor with the liberation of free oxygen.

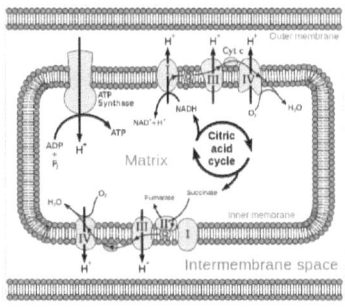

This free oxygen, in turn, became the electron acceptor in respiration, mastered by the ancestor lineage of the mitochondria as summarized in the Wikipedia diagram.[174]

The cyclical breakdown of pyruvate to carbon dioxide releases energetic electrons that fall down a chain to oxygen and drive H-ions into the periplasmic space. The return flow of these H-ions is used to generate ATP.

LUCA Ribosome

The ribosome on which digital information was turned into analog waveforms was central to the ability of the LUCA lineage to use proteins to control water and metabolism. The ribosome was so central to the process that it has altered surprisingly little over the 3 billion years it has been passed on down the three great lineages that radiated into phase space from the LUCA as the eubacteria, the archaebacteria and the eukaryotes.

The ribosome subunits in the two prokaryote lineages are the same size, but the RNA sequences are different. The eukaryote ribosome is similar to, but larger than, the archaebacterial ribosome. In contrast, the eukaryote lipid membrane is eubacterial, which complicates unraveling the lineage that led to the eukaryotes. Other than this, the little-changing ribosome is very useful for outlining the basic family tree of living organisms.

The illustration shows the similarities among small subunit ribozymes for cellular life forms (bacteria, archaea, and eukaryotes). Each branch is labeled with the name of a representative member of that group, and the length of the branches corresponds to the degree of difference in the rRNA sequence. The changes down all these lineages have been minimal over the billions of years since the

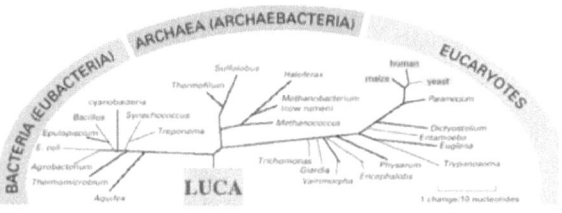

LUCA emerged. On this scale, plants, and fungi—here represented by maize and yeast—are our close cousins, while E. coli is not.

The eden in which the lineage that led to the LUCA had all the ingredients for RNA to thrive in. But this eden in which life emerged was only a small locality in a much larger abiotic earth, and it is quite possible that only one locality, a womb in the eden, had the mix of conditions that were just right for RNA to flourish and explore its phase space as RNA mastered the use of amino acids to manipulate analog waveforms.

While the womb had everything necessary, the rest of the eden was supportive but lacking in one or another of the essential subunits needed for RNA.

The challenge was to explore the vast phase space opened up by proteins and, one by one learn how to create these essential subsystems from other substances in the environment. The core metabolism in current use by all living systems contains relics of the course of this exploration. We have already mentioned that the early ocean was saturated with iron and it seems that electron transfer to ferrous ions from the abundant hydrogen sulphide was first harnessed to drive the proton-motive force of generating ATP. Many of the archaebacteria use this method to this day, and many enzymes important in electron transfer have an iron-sulphur cluster at their active site.

As the clay-bed eden involved extremes, it is thought probable that the bi-lipid membrane of LUCA was of the archae-bacterial type, inasmuch as they thrive in the most extreme environments. Unlike the bi-lipid membrane enclosing eubac-teria and eukaryotes in which the two layers are free to slide over each other, the archae-lipids are cross-linked, and the membrane is much sturdier as a consequence, but stiff at low temperatures.

While it was suitable for clay-bed temperatures, it was unsuitable for ocean temperatures, so it was a lineage that discovered a more fluid membrane that graduated from the clay eden with the resources to thrive and explore this new phase space in the ocean. At least one archae-lineage also learned how to make fluid lipids—perhaps by horizontal gene transfer of the necessary digital information from the eubacteria—and joined them in exploring the world ocean.

The LUCA was a vital unit that used digital information to generate analog waves to control water and metabolism in a way akin to how a music score is transformed into the symphonic control of the air pressure in a concert hall.

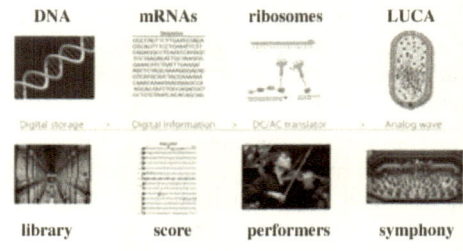

The LUCA was probably more akin to the archaea because they survive in environments similar to those found on the young earth—hot springs, sea vents releasing sulfide-rich gases, boiling muds around volcanos, etc. They use mineral energy to drive their metabolism, i,e., sulfides, etc.

Procaryote Radiation

The fluid membrane opens up a whole new phase space to explore and learn about. The iron and hydrogen sulphide in the ocean activated by UV light provided a source of energy at first, but the exploration soon discovered *pigment* molecules that absorbed light and elevated electrons to levels that could drive the proton-motive ATP cycle with hydrogen sulphide as electron accep-

tor. These pigment molecules had conjugated double bonds whose delocalized electrons resonated with photons, just as in an antenna, and absorbed their energy. These excited elec- 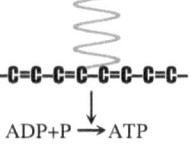 trons were then passed down a cascade to drive the proton-motive force.

Eventually, such exploration led to the discovery of chlorophyll as perfect for an antenna. The magnesium's electrons are delocalized over the entire body of the molecule. These delocalized electrons absorb a photon and jump into the excited state. The shape of this excited wave is localized near the

long hydrocarbon chain by which the body is embedded in the phospholipid membrane. This tail is adjacent to other, simpler conjugated molecules, and the excited electron is passed on down a cascade to drive proton-motive force.

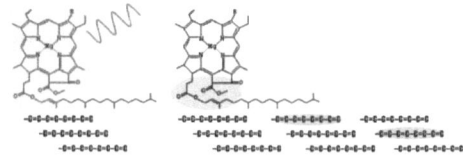

When a hydrogen atom in the body of chlorophyll-A is replaced by a methyl group, as in chlorophyll-B, the resonant frequency is altered and a different color of light is absorbed. It was soon discovered that if an excited electron from A was passed to B and another photon absorbed, then the electron had sufficient energy to strip hydrogen from a water molecule and attach it to NAD, with the liberation

of free oxygen. The activated hydrogen of NADH could then be used to reduce the plentiful carbon dioxide to carbohydrate.

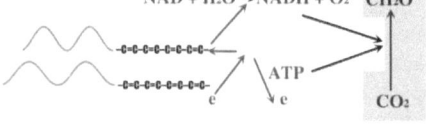

This photosynthetic double activation by light totally liberated living systems from any abiotic molecules for food, and was perfected in the green photosynthetic eubacteria that flourished.

The oxygen released was at first absorbed by *sinks*. These oxygen sinks included the ferrous iron that saturated the ocean, which was oxidized into insoluble ferric oxide that precipitated out as the great iron ore beds we mine today. Hydrogen sulphide and ammonia were oxidized in the UV irradiated atmosphere.

photosynthesis   Light

$$CO_2 + H_2O \xrightarrow{} CH_2O + O_2$$

respiration

$$CH_2O + O_2 \xrightarrow{} CO_2 + H_2O$$

ATP

Eventually all the sinks were full and free oxygen appeared in the atmosphere, a major change in the environment over a few million years. Adjustment to free oxygen was a major challenge, and exploration of phase space eventually established a successful lineage that could thrive in the nighttime by reversing the photosynthetic cascades to use oxygen to convert carbohydrate into ATP, carbon dioxide, and water. This is *oxidative respiration* and is the great cycle of carbon dioxide and water in life powered by light photons.

The LUCA is estimated to have lived some 3.5 to 3.8 billion years ago when the atmosphere contained no free oxygen, and all metabolism was anaerobic. There is evidence that living systems emerged in the period of 500 million years after the earth was formed, the LUCA some time after that, and the prokaryotes diversified after that for the next 1.5 billion years.

Prokaryote Division

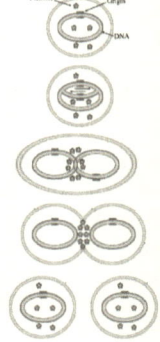

The asexual division of a prokaryote into two independent vital units is relatively simple. The unified internal wave changes as the protein AC generators change in number through a series of steps. The DNA double helix is in a single loop attached to the membrane at the *origin* stretch of DNA.

The DNA replicase enzyme-complex unwinds the origin DNA and duplicates it in both directions. As the two copies of DNA elongate, the tubulin proteins (ancestral to the tubulin of the eukaryotic cytoskeleton) collect in a ring about their connecting point, pinch in the membrane and cell wall, and complete the membrane and cell wall that separate the two daughter cells.

The next eden

Cyanobacteria use water, carbon dioxide, and sunlight to create their food, the byproduct of this process being free oxygen. At first, this oxygen was rapidly reduced by the omnipresent ferrous ions in the seawater, but as the ferric iron precipitated out, free oxygen started to linger around. This was a major challenge, as oxygen is reactive and disruptive to an orderly metabolism.

At least one lineage learned the solution to this challenge presented by the Logos in the current state of the environment. The lineage explored phase space by using a duplicate of the electron cascade from photosynthesis and running it in reverse, letting the electrons released by carbohydrate breakdown fall down the cascade, generating ATP by proton-motive force, and ending up on oxygen. Two hydrogen ions then add themselves, generating a molecule of water.

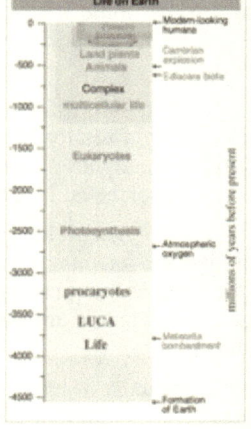

This lineage could thrive in the light of day, splitting water to liberate the oxygen and adding the hydrogen to carbon dioxide as carbohydrate. The lineage could equally thrive in the dark of night when carbohydrate could be split into carbon dioxide and the hydrogens added to oxygen regenerating water. This was the origin that developed into the oxidative respiration used in most eukaryotes.

A layer of mucus often forms over mats of cyanobacteria. In modern microbial mats, debris from the surrounding habitat can become trapped within the mucus, which can be cemented together by calcium carbonate to grow thin laminations of limestone. These laminations can accrete over time, resulting in the banded pattern common to stromatolites. The dome-shaped morphology of stromatolites is the result of the vertical growth necessary for the continued infiltration of sunlight to the organisms for photosynthesis.

Stromatolites were abundant on the planet by three and a half billion years, a billion years after the formation of the Earth. The earliest stromatolite of confirmed microbial origin dates to 2.72 billion years ago, while a recent discovery provides strong evidence of microbial stromatolites extending as far back as 3.4 billion years ago. In the primordial ocean, the photosynthetic and respiring bacteria thrived and utterly changed the Earth as the concentration of oxygen in the atmosphere and seawater inexorably increased over

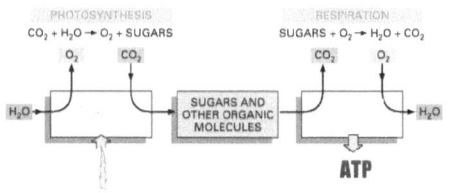

millions of years. Colonies of these bacteria grew below the lethal UV depth in the shallows and, as sediment slowly settled onto them, they grew towards the light. As these *stromatolites* slowly extended upwards, a mix of sand and organisms, the light intensity in the interior fell and exploration allowed growth to continue, albeit slowly, in dim light.

Deep in the interior of a stromatolite where the light was absent was the eden in which the next level of sophistication emerged in a lineage of archaebacteria with fluid membranes. They abandoned the now useless photosynthesis and turned to a lifestyle as a recycler, as a heterotrophic scavenger. All a lineage had to do, really, to respond to this lightless challenge was to learn to feast by exporting its degradative enzymes to the exterior. These secreted enzymes would dissemble the dead and dying photosynthetic bacteria drifting down from the sunlight far above.

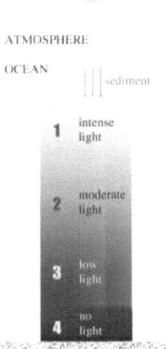

This lineage open up, and radiated into, a phase space in this stromatolite eden and explored and learned the system-building steps that led to the sophisticated multi-vital-unit eukaryotes.

THE EUKARYOTES

Stromatolites only form these days in shallow sea water unsuitable for most other life. In other locations they would be grazed away before they had a chance to form.

In the early ocean, stromatolites were by far the most sophisticated type of system around. They lasted tens of millions of years and provided the eden in which prokaryote lineages could explore their interactions and the system building that led to the eukaryotes.

The mix of founder bacteria that initiated the stromatolite diverged into different varieties, simple races, that specialized as they explored and learned how to thrive in the different zones within the stromatolite.

1.   At the surface, especially the top, the useful light and disruptive UV are intense. Here, races specialize in handling RNA damage and membrane oxidation.

2.   Below is a zone of moderate light and low UV, and these races do well in moderate conditions.

3.   In the low-light zone, the challenge is efficiency and making the most of every photon.

4.   The response to the challenge of darkness is to abandon photosynthesis and rely on respiration alone and the recycling of debris drifting down from above.

Logos derived, emergent properties

All bacteria secrete a cell coat to protect the delicate lipid bilayer. The cell coat is commonly made of murein, a rigid complex of linked sugars and amino-acids that surrounds and protects the delicate outer bi-lipid membrane like a coat of chain mail.

In prokaryotes, the primary function of the cell wall is to protect the cell from internal turgor pressure caused by the much higher concentrations of proteins and other molecules inside the cell compared to its external environment. The simple bacterial cell wall seems to be a relic of much earlier metabolic diversity in that instead of the usual L-amino acids it incorporates D-amino acids. The wall is located immediately outside of the cytoplasmic membrane. Peptidoglycan is responsible for the rigidity of the bacterial cell wall and for the determination of cell shape. It is relatively porous and is not considered to be a permeability barrier for small substrates.

Loss of cell wall

One of the first steps along the pathway to eukaryotes was a lineage abandoning this confining wall and learning how to manage without it.

Much much later, the lineage divided as one line developed a protective wall of cellulose, a branched mega-polymer of glucose, which led eventually to the plants. Later still, the wall-less lineage divided yet again as one line developed a protective coat of chitin, a branched mega-polymer of N-acetyl-glucosamine, a derivative of glucose. In terms of structure, chitin may be compared to the polysaccharide cellulose. It is the main component of the cell walls of fungi. The lineage that did not readopt a protective exoskeleton that defined its shape developed instead a sophisticated internal cytoskeleton to define its shape and was the ancestral lineage that founded the animal kingdom.

But for many ages in the protective stromatolite eden, the ancestral lineage of all eukaryotes did without an external cell wall. Exploration of the scavenger phase space that opened up in the stable darkness found that the sturdy outer coat could be abandoned without harm. This allowed the digestive enzymes that were secreted to freely hydrolyze the surroundings and generate monomers that could be absorbed.

Powerful digestive enzymes, however, are dangerous to have around, especially in proximity to the digital store on DNA. At least one lineage learned the advantages of dividing but not duplicating DNA, so only one compartment had

DNA in it. The extra DNA-free compartments could contain all the heavy industry, so to speak, with one compartment reserved for the delicate computer systems, the RNA CPU and DNA data stores.

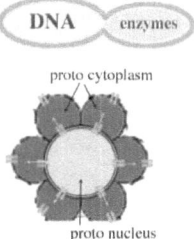

All that would be needed is cell division, without DNA duplication, and incomplete separation. This was the beginnings of the nucleus in which the DNA is strictly segregated from the rest of cell metabolism.

Exploration of the advantages of many, diverse compartments continued. The path through   phase space involved learning how to control the transfer of material between compartments. These discoveries included protein rails, the precursors of filaments as guides, and protein motors with loads to ratchet along these filaments. The filaments could transfer in both directions, so information on the state of the industry could flow back to the computer area. These were the simple beginnings of the cytoskeleton.

Further exploration learned the advantages of further specialization and, before the eukaryotes diverged, dozens of specialized compartments had been added as the basic eukaryote structure. The diagram illustrates are just a few examples of some of the compartments found in all plant, fungal, and animal cells.

  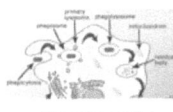

Golgi body   endoplasmic reticulum   mitochondrium   phagocytic vacuoles

Emergent properties

At least one lineage eventually made the leap in sophistication from prokaryote to eukaryote. This involved learning how to assemble a number of interrelated subsystems that differentiate the sophisticated eukaryotes from the simpler prokaryotes.

Very few of the transitional forms were able to survive outside the stromatolite eden and there are only a few *primitive* eukaryotes still around today. The list of levels of sophistication that emerged in the lineage that left the womb and radiated into the wider world is extensive and, not surprisingly, it took a long stretch of time to learn it all and a lot of extra DNA to store all the digital information accumulated down the lineage that was needed to generated all the different analog forms that were required.

Much is still to be elucidated regarding the details of this development, but it seems that only one lineage accomplished all of the steps with the ability to leave the stromatolite eden and explore and populate the larger world. It did not take the descendants long to learn that stromatolites were an excellent food supply, and the stromatolite were grazed almost to extinction. Stromatolites now only survive in waters almost all other life forms find too hostile to bother with.

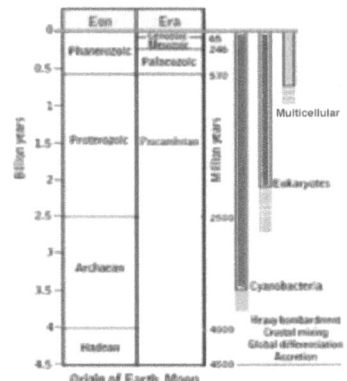

The list of developments along the line to the eukaryotes that left the stromatolite womb

includes:

1. Phagocytosis
2. Cytoskeleton
3. Centrosome
4. The assembly line
5. Chromosomes
6. Nucleus and Nucleolus
7. Sophisticated digital processing
8. Domestication of prokaryotes
9. Mitosis
10. Sexual reproduction

We will now discuss each of these developments separately, although they are often intermixed with each other.

Phagocytosis

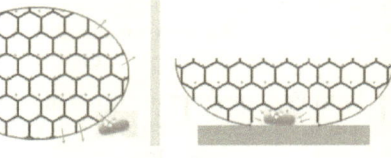

In any other environment, being without a protective wall was a lethal experiment but, in the stromatolite eden, it had a great advantage that the scavenger was in intimate contact with the dead cells drifting down into darkness from the light above. The cell could grow without being limited by its surface area that could be irregular. It could learn how to envelope fragments of dying cyanobacteria against a surface and 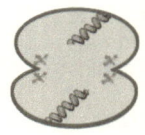 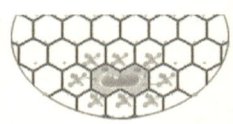 digest them in a pocket, rather than just wastefully secreting digestive enzymes into the environment as bacteria do.

Experimenting with the proteins used to divide a prokaryote into two, variants were developed that closed off the pocket so that the prey was in a *vacuole* into which the enzymes were poured to digest the prey into its monomers. The lineage learned about phagocytosis from the Logos and flourished. It was safe and protected. The lineage further explored the phase-space possibilities provided by the Logos.

Cytoskeleton

Rather than control its shape and form with an external cell wall, the eukaryote lineage controlled its shape with an internal cytoskeleton. As the number and variety of vital units increased, so did the cytoskeleton become more elaborate. Each vital unit, like a prokaryote, could be in a number of states. The cytoskeleton controlled the state of each vital atom, switching them about according to the digital information patterns stored on DNA that were expressed in the cytoskeleton. All eukaryotes have this structure:

> The apparently formless background of the cytoplasm of the cell... has been shown to have an internal structure when examined by high-voltage electron microscopy.... A microtravecular lattice, an irregular 3-D lattice of very slender protein threads.... A 3-D spider web in which are suspended the ... organelles ... enclosed by membranes which effectively partition the cytoplasm into specific compartments.... The contents within a compartment may be quite different

from the environment in the general cytoplasm or inside other [compartments].[175]

The details of the development of the cytoskeleton common to all eukaryotes are currently unknown, but the lineage must have explored the phase space of possibilities opened up by structural proteins, such as the microtubules. These are mega-structures in which the system wave embraces large proteins as subsystems.

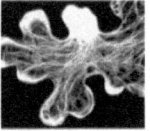

During this period, a lineage developed the three main kinds of cytoskeletal filaments found universally in eukaryotes: the microfilaments, intermediate filaments, and microtubules. The cytoskeleton provides the cell with structure and shape. Cytoskeletal elements interact extensively and intimately with cellular membranes at attachment points.

The filaments can be thought of as the ropes that pull and the posts that push in the erection of a circus tent. The structural elements can also transmit and receive images from one end to the other; they are both sensory and motor.

It is the interplay of these that gives shape to the wall-less ancestors and allows them to develop phagocytosis. The fibers of the cytoskeleton control the state of the vital unit they are attached to.

The Centrosome

The state of the cytoskeleton controls the states of the vital atoms, and the control of the cytoskeleton is centralized in the centrosome.

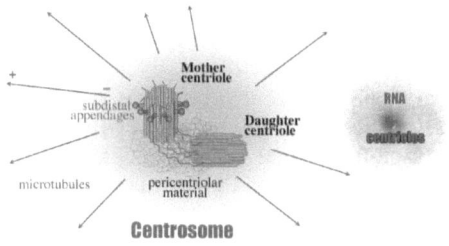

All of the cytoskeletal fibers have an *output* plus end attached to a vital unit, while all the *input* negative ends converge at the centrosome.

The bulk of the centrosome is a halo of RNA and attendant proteins, the pericentriole, and at the center are the *mother* and *daughter* centrioles.

In the period of the cycle called *interphase,* the cell is not preparing to divide, and it is the mother centriole that is in charge while the daughter just tags along.

The mother centriole converts the digital information on the RNA into the activity of the microtubules that control the states of their attached vital units. Digital information is converted into the analog form of the many vital units of which the cell is composed. The centriole translates digital information into analog form—it is a DC/AC converter. This is the role of the ribosome at a higher level of sophistication.

A novel RNA was detected in the centrosomes.... This RNA was named centrosomal RNA (cnRNA); five different cnRNAs were described. During the sequencing of the first... it was discovered that the transcript contained a conserved structure—a reverse transcriptase domain. In a 2005 study, we speculated about several possible mechanisms for determining the most important functions of centrosomal structures and referred to one of them as an "RNA-dependent mechanism".... The presence of a reverse transcriptase domain in this type of RNA, together with its uniqueness and specificity, makes the centrosome a place of information storage and reproduction.[176]

As the workings of the centrosome are still currently a topic of investigation, we can only suggest an analogy for the connection between the digital information in RNA and the state of the cytoskeleton and, hence, the state of the attached vital units.

Consider a kite with eight struts radiating from a core that adjusts the length of each strut as indicated by a sequence of eight digital numbers sent up the kite's cord from a handset. Each sequence of digital numbers results in a different analog form to the kite, just two of which are illustrated.

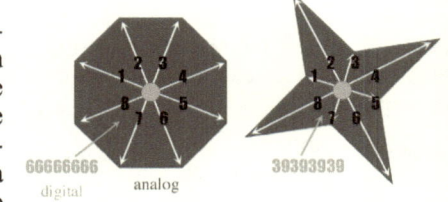

This is an example of digital information being translated into specific analog forms, another DC/AC converter.

In the lineage leading to the eukaryotes, this type of digital control was perfected. The fibers of the cytoskeleton not only control the state of the many vital units, they can also bodily transport whole units about within the cell and perform many other functions that are only now being explored.

The centrosome is the microtubule organizing center of the cell as well as a regulator of cell-cycle progression. The centrosome we have described is thought to have evolved only in the animal lineage of eukaryotic cells. Fungi and plants use other similar structures to organize their microtubules. A centrosome is composed of two centrioles, barrel-like structures made mostly of tubulin. They are at right-angles to each other and surrounded by a star-shaped 'aster' of RNA and many proteins. The proteins are responsible for microtubule nucleation and anchoring. Each centriole of the centrosome is based on a nine triplet microtubule assembled in a cartwheel structure.

Eukaryote form

The primary cilium, a long skinny cellular extension found in most cells, grows off one of the centrosomes. This primary cilium is probably involved in some sensing mechanism. In fact, in some cells (such as the rod and cone cells of the eye), the primary cilium is where most of the cellular sensory apparatus is located.

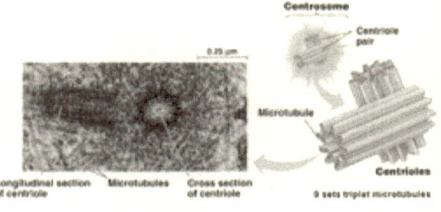

The centrosome receives RNA from the nucleus and adjusts the cytoskeleton to the requested form. As eukaryotes can change form quite rapidly, we can assume that a constant stream of RNA arrives at the centrosome from the nucleus to balance the breakdown of old messages. This is the way that ribosome output on the mono unit level is regulated. The ribosome and centrosome are the DC/AC converters on their respective levels of sophistication.

The pattern is, as always, systems of organized interacting subsystems. Ribosomes organize aminoacid polymers to generate the internal AC wave of a prokaryote-level mono vital unit. Centrosomes organize vital units to generate the internal AC wave of a eukaryote-level poly vital unit.

The Golgi Assembly Line

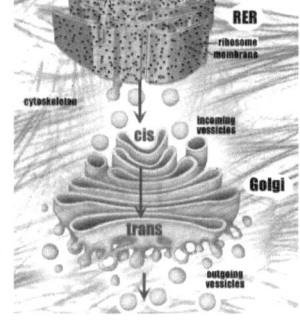

The exploration of the phase space in the Logos resulted in the specialization of vital subunits for centralizing the bulk manufacture of proteins. They developed into a set of vital units that acted as an assembly line.

Those at the start of the assembly line are called the rough endoplasmic reticulum (RER). The walls of the RER are studded with ribosomes that are busy translating mRNA into proteins—the mRNA originating from the nucleus. The proteins that accumulated inside the walls are regularly budded off as vesicles which the cytoskeleton conveys to the Golgi Body.

The Golgi is a set of stacked, flattened vital units that, like an assembly line, has a starting point—the *cis face*—and an ending point—the *trans face*. A vesicle arriving from the RER merges with the cis face of the Golgi. Its contents are passed from compartment to compartment being specifically modified along the way. Finally reaching the trans face, the products are sorted and packed into vesicles, an address label affixed and the vesicle budded off for the cytoskeleton to deliver to its intended destination.

Cells synthesize a large number of different proteins in the RER. The Golgi is integral in modifying the proteins from the RER—such as attaching a carbohydrate to specific places on the primary structure, thus adjusting its folding and properties. The Golgi is also involved in the transport of lipids around the cell, as well as the generation of lysosomes—recycling units filled with digestive enzymes to which the cytoskeleton can transport worn out organelles or ingested prey for breakdown and reuse.

The Golgi sort, package, and address the vesicles for transport by the cytoskeleton to the many different destinations within the cell. This last function is similar to that a post office; it packages and labels items which it then ships to different parts of the cell.

The efficiency of such an assembly line with its linear sequence of specific actions is to be seen today in the computerized manufacture of a car, where each robot performs its function in turn.

As the number of sophisticated analog forms learned and accumulated down a lineage increased, the digital information needed to store and manipulate these analog forms also increased. Two major innovations in digital technology occurred in the ancestral eukaryote lineage. The innovations were probably interwoven, but we will deal with them separately.

The first innovation was librarian-like, it involved the compact storage of huge amounts of digital information and an efficient way of reading a required piece of information. This is the chromosome with DNA stored on reels of histones.

The second innovation was akin to the computer leap in digital sophistication from the 8-bit world of ASCII and MS-DOS to that of the 16-bit computer of Windows, in which ASCII is only a subset of the information manipulated. This advance involves *post-translation* manipulation of mRNA before it is translated into proteins. The triplet-code *exon* sequences are embedded in not-triplet code *intron* sequences. They are copied together from DNA store to RNA and the introns are then stripped out. The remaining exons are linked together as an mRNA for transport to the ribosomes for translation into protein.

A similar stripping of higher-level code to create a simple ASCII text file happens in modern computers. While this manuscript is full of high-level information—inset graphics, multiple variations on fonts, etc.—if I choose to save my document as a *Text Only* file, all this high-level information will be stripped away and a simple ASCII file saved to disk; a file identical to those of the 8-bit era of technology.

Chromosomes and Histones

While both prokaryotes and eukaryotes use DNA for deep-time storage of their inheritance of digital information, they manage their DNA store in very different ways. We have already mentioned that the defining difference is that eukaryotes have multiple compartments, with the DNA stored in one compartment, the nucleus, and metabolism occurring in all the other compartments.

In the prokaryote vital unit, the DNA is a single loop attached to a protein anchor embedded in the cell membrane at a specific location called the nucleoid at about the midpoint of the cell. Close to this anchor is the unique origin for the initiation of DNA replication. Starting at this origin, the DNA is duplicated in both directions to generate two copies.

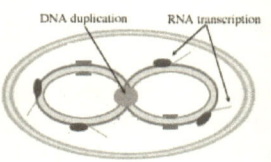

The DNA is loosely confined to the nucleoid by *packing proteins* which, when the state switches to spore, effect the condensation and dehydration of the DNA for stability over deep time. Other than this, the prokaryote DNA is always in the same state, and bacteria in the abundant state duplicate the DNA and transcribe RNA all at the same time.

The much larger amount of DNA in a eukaryote nucleus is divided into a number of highly-structured chromosomes. The DNA double-helix is wrapped twice about a nucleosome, a reel composed of eight histone proteins, and locked in place with a ninth histone. These histones are among the

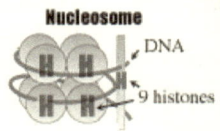

most conserved of all proteins, and they have only slightly diverged down countless lineages since their discovery by the ancestor of plants, fungi, and animals.

The DNA is unwrapped from its nucleosomes for RNA transcription and then rewound when no longer required. The DNA is also unwrapped from the nucleosomes for duplication at multiple bidirectional sites along the chromosome and the fragments are then joined together. Unlike prokaryotes, duplication and transcription of DNA do not occur at the same time. There are three basic states that eukaryote DNA can be in:

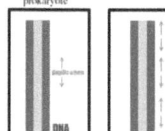 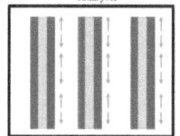

1. *Growth.* The DNA is partially condensed and attached to the nucleus membrane. The unwound stretches are free to be transcribed into RNA. Each chromosome has its own volume inside the nucleus.

2. DNA *Duplication.* There is no transcription of RNA and all the DNA is unwound from the nucleosomes for copying.

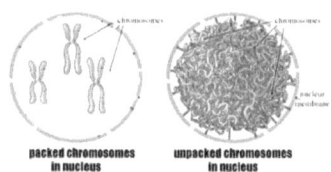

3. *Cell Division.* The DNA is entirely wrapped on nucleosomes, and the spools are stacked and compacted into the condensed chromosomes. These have caps of repetitive DNA, called telomeres, and a somewhat central place specialized for cytoskeleton attachment called the centromere.

The DNA in each human cell is about 6 feet long— this is by no means a record—and this entire length is condensed on scaffolding proteins into a ~3 trillionths of an inch. As this condensation usually happens after DNA duplication, there are two identical *chromatids* attached at their centromeres ready for cell division (yet to be discussed). When the eukaryote cell is not preparing to divide, the DNA strands each have their own volume in the nucleus in which the DNA is strung from attachments and is in a partially open and partially packed state.

packed chromosomes in nucleus   unpacked chromosomes in nucleus

In certain situations, the DNA of a chromosome remains condensed and inactive. A mammalian example of inactivation of an entire chromosome is the X *dosage* difference between a male and female.

A male has one X chromosome in every cell that was received from his mother. All the cells of a female start with two X chromosomes, a maternal and a paternal. About 6 weeks into gestation, however, every cell in the embryo condenses one of its X chromosomes—apparently at random—and it becomes a Barr Body that gets duplicated but remains condensed down its lineage.

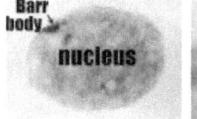 

Barr body ⬎

nucleus

 This mosaic of maternal and paternal active chromosomes is visible in the tortoiseshell cat which is always female. One color coming from mom's X and the other from dad's.

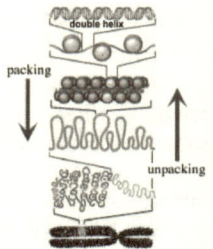

Prior to cell division, the DNA is duplicated and packed neatly into two chromatids—the two DNAs—attached together at the centromere.

In the next steps, the reels are tightly stacked together, the stacks folded, and folded again, to form the condensed chromatids attached at the centromere. The proteins that participate in this condensation are descendants of the packing proteins that herd the DNA in the prokaryote nucleoid. In the *interphase* state of the cell (i.e., when the cell is not preparing for division), most of the DNA is kept wrapped on stacked nucleosomes except when sections are unwound for RNA transcription.

The Nucleus and Nucleolus

All the DNA digital store of a eukaryote is contained in the membrane-bound nucleus, and much of its content of RNA is concentrated in the nucleolus. The nuclear membrane consists of two lipid bilayers—the inner nuclear membrane and the outer nuclear membrane. The outer nuclear membrane is contiguous with the membrane of the endoplasmic reticulum, and the space between the two membranes is contiguous with the contents of the endoplasmic reticulum.

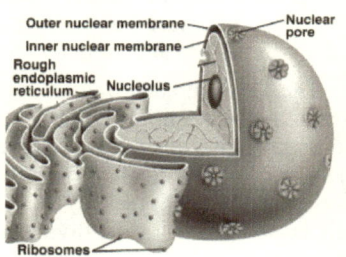

The inner nuclear membrane encloses the nucleoplasm, and is covered by the nuclear lamina, a mesh of *nucleoskeleton* proteins (similar to the cytoskeleton outside the nucleus) which stabilizes the nuclear membrane, as well as being involved in chromosome function and gene expression. The inner and outer membranes are connected by nuclear pores that penetrate the membranes. The nuclear membrane is punctured by thousands of these nuclear pore complexes—large hollow proteins with an inner channel.

The main structures making up the nucleus are the nuclear envelope, and the nucleoskeleton (which includes nuclear lamina), a meshwork within the nucleus in charge of transportation within the nucleus. The interior of the nucleus does not contain any lipid membrane-bound subcompartments, but the contents are not uniform, and a number of subnuclear bodies exist, made up of unique proteins, RNA molecules, and specific parts of the chromosomes. The largest of these is the nucleolus that takes up to about 25% of the nuclear volume.

In the prokaryote, almost all the RNA transcribed from DNA is translated into protein primary structure. The opposite is true in eukaryotes, where the majority of the RNA transcribed from DNA never makes it out of the nucleus to instruct a ribosome. The majority of RNA remains in the nucleus and is involved in the manipulation of digital information. This is akin to the central processing unit (CPU) of a modern computer where masses of digital information are manipulated and only the results are sent to the output ports of the CPU. The output ports of the nucleus are the nuclear pores through which the results of the digital manipulations are exported via skeletal proteins: the flood of rRNA , tRNA and mRNA all creating the protein activity directing each vital unit, and the sporadic centro-RNA to program the centrosome and the overall state of the cell.

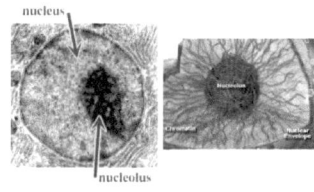

Chromatin is DNA not in the highly compacted form of the chromosome. The nucleolus is a knot of chromatin. The functions of the nucleolus are currently under investigation, but one thing that is established is that it controls the production of ribosomes for export out of the nucleus. It also has all the characteristics expected of a CPU involving RNA manipulations:

> The nucleolus also contains proteins and RNAs that are not related to ribosome assembly and a number of new functions for the nucleolus have been identified. These include assembly of signal recognition particles (SRP), sensing cellular stress and transport of HIV messenger RNA.[177]

The last item is very suggestive. The RNA that redirects the metabolism of a cell into making virus particles is managed by the nucleolus. This suggests that the nucleolus also manages the top level RNA, the current program called up to run the cell in non-pathological situations. It is becoming clearer in the literature that the nucleolus is in charge of determining the cell cycle and cell differentiation.[178]

The SRP are RNA/protein complexes that attach to proteins as they are coming off ribosomes and are the addresses used by the cytoskeleton to deliver the protein to its correct destination.

In some organisms, particularly plants, when two nuclei are combined into a single cell during hybridization, the nucleoli compete for control. The RNA genes of one nucleolus are suppressed and not generally transcribed, though reactivation of the suppressed RNA genes may occasionally occur. This selective preference of transcription of RNA genes is termed nucleolar dominance.

The connection between sensing stress in a prokaryote and the writing or reading of digital information of DNA in response was discussed earlier. As the nucleolus has been established as a sensor of stress in the eukaryote cell[179], we can expect that a similar dynamic will be discovered in the nucleolus.

I suggest that all this is very CPU-like manipulation of digital information generating the output of the nucleus and thus the state of the cell. The exact location and functioning, however, of the digital CPU in the nucleus will need to await more detailed elucidation.

Sophisticated digital processing

It should be quite clear from all the above that the eukaryote system is a whole new level of sophistication in which the prokaryote system is but a subsystem. While the ASCII-like triplet code suffices admirably for the simple prokaryote, it is incapable of encoding the sophistication of the eukaryote cell.

We again have an illustrative example from the realm of computer technology. The first computers using the ASCII code were only capable of the very rudiments of word processing, the reading and writing the digital information of simple text. The screen and printer had only one font for output, the monospaced *Courier* font that mimicked the fixed space of the manual typewriter—and looked `like this`.

The document processing of my MacBook Pro is much more sophisticated, it can deal with so much more than plain Courier text—even unusual text **like** *this*. If, however, I choose from the menu the **Plain Text** command, all the extras are stripped away from the ASCII, and the display is, again, like this.

This increase in sophistication of computers was due to the input of human ingenuity. The increase in sophistication from prokaryote to eukaryote was due to input from the higher levels of the Logos as the exploration of phase space step-wise learned to manipulate digital information in a systematic hierarchy of increasing sophistication.

Digital systematic hierarchy

In the computer world, there developed over decades a systematic hierarchy founded on the the byte of the ASCII code.

The first step in sophistication was to manipulate digital data two bytes at a time—the realm of 16-bit computing. The ASCII code of eight bits never changed, but now the second byte could be used to code for single character properties such as *italic,* ***bold italic,*** and ***bold italic underlined***. The next step was to a 32-bit system in which the ASCII was still embedded, as ever, but the third and forth bytes could be used to code for thousands of colors.

The Mac I am writing on can run in 64-bit mode but accommodates older software that runs in a 32-bit environment. This manuscript I am writing takes up 14,000,000 bytes when it is stored on the hard drive, but almost all of that is not in ASCII code but in instructions about what to do with that ASCII which encodes about 100,000 words in about 100,000 bytes of ASCII. Only about 5% of the digital information is in

1ST BYTE	2ND BYTE	3RD BYTE	4TH BYTE
ASCCI	italic	Colors, fonts	
ASCCI	bold	Colors, fonts	
ASCCI	underline	Colors, fonts	
data	instructions		

ASCII which is on the same order of the 1% of the digital information in human DNA that is in the triplet code and is translated into protein. Like early computers where almost 100% of the digital information is in ASCII code, almost all of bacterial DNA is in Triplet Code.

If a disk utility is used to examine the bytes of information stored in the normal file, it will be found to consist of ASCII bytes separated by non-ASCII multibyte stretches of binary information. In the plain text file, it will look exactly like that in the earliest computers, a sequence of ASCII codes.

All computers have basic 'housekeeping' that runs all the basic functions, such as keyboard input, screen display and reading and writing to digital storage, the Disk Operating System, DOS or *System*.

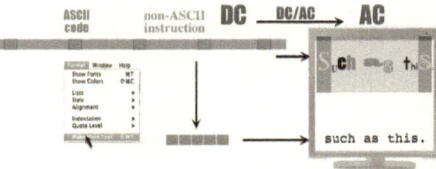

Simple Computers

On the 4-bit machines, the system that became the standard was CPM, which could hardly handle wordprocessing. This was quite rapidly replaced by 8-bit systems that could handle one byte of eight bits at a time, and ASCII became the standard. The standard emerged because of interaction, you could only read another file if the same coding was used. In living systems, the equivalent stage was the 2-bit codon.

Wordprocessing could now handle content well, but to indicate any kind of formatting you had to insert codes into the text marked off with reserved characters. A phrase such as "He *said* that E=mc^2 was bunk" had to be entered as "He {i}said{i} that E=mc{sp}2{sp} was bunk." You could do it, but it wasn't easy. Forget about color. This was the world of MS-DOS, 400K disks for storage and plain text files.

The simple living systems called prokaryotes (bacteria) use a similar system on a single strand of DNA. Almost all of its inherited digital information is codons that code for amino acids, the equivalent of plain text files in MS-DOS.

Sophisticated Systems

For computers, the step-up to 16-bits soon settled on the standard called Windows. Now there was space to mingle data and simple instructions. No more insertion of a code, now *italic* could be turned on and off at the click of a button.

On my 64-bit computer, the ASCII code is still there taking up 1 byte with the other 7 bytes allowing millions of colors, scores of different fonts, and a plethora of different languages. With a few clicks I can now write ιν γρεεκ, صيباري ني, ん ジャパネセand יקנרוק' (*in Greek, in Arabic, in Japanese and in Hebrew,* respectively), in colors and sizes as easily as in English.

While the 8-bit computers struggled as word-processors, my 64-bit Mac can play a plethora of roles with ease: publish books, play music, videos and TV, take photos, do international videoconferencing, Photoshop people out of photos, etc.

Introns and exons

All the more sophisticated eukaryotes—everything else except the bacteria—use the equivalent of Windows, although the method of mixing data and instructions is different. All the data in eukaryotes, the codons for amino acids, are broken into long strings of bases, called exons, separated by long stretches of codons that are not translated, the introns. These untranslated stretches contain instructions, such as how to splice out the introns and link up the exons correctly into an mRNA where the stored digital information is converted into the correct primary sequence for a protein. The spliced-out intron mixes with the RNA pool for a while, informing it, at the very least, that such and such an mRNA was recently generated.

The human dystrophin gene is an extreme example of this preponderance of introns. The RNA transcript is the largest known to date, spanning 2,200,000 bases on the X chromosome that takes 16 hours to transcribe into 2,200,000 bases of

RNA. The gene size is mainly accounted for by huge intronic regions. After splicing out 2,186,000 bases of RNA introns, the 79 exons that remain are spliced into an mRNA of just 14,000 bases for translation by a ribosome. There

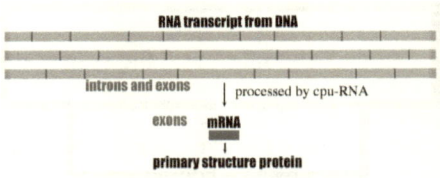

it codes for a protein of 3,500 aminoacid residues that is essential to proper muscle functioning. Comparison of intron sequences of the human and mouse genes revealed that they are extremely conserved in size and that a similar fraction of total intron length is represented by repetitive elements. It seems like there is something that is important to protect from mutation—which suggests that cells control when and where changes occur in the DNA.

Boilerplate mixing

In recent years it has become apparent that most proteins have a modular structure, and the same module can appear in a large number of different proteins. Each module has a specialized function, such as binding ATP or sticking a zinc finger into DNA. A protein can have a dozen or so modules and unite all their disparate abilities into the precise ability of the protein. If correct, the modules would be the primary subsystems of proteins. The immune system does something similar when it permutates a set of modules to generate almost all the shapes a molecule can be before testing them to eliminate the self-shapes.

Sophisticated living systems can do even more with modules, they have the equivalent of 'boilerplate' text. Both prokaryotes and eukaryotes translate the codon-equivalent of a plain ASCII text recalled from disk into an analog output from the

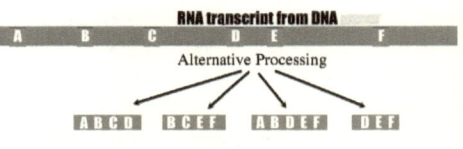

ribosome. The difference is that the mRNA that is read from the DNA is not processed before translation by the bacteria; in all other life, it is processed as the introns are excised before the RNA gets sent to the ribosome. Depending on the state of the RNA pool, a transcribed mRNA can be spliced into a variety of different mRNAs for translation.

This is akin to picking the first paragraph from boilerplate text:

     1.   Dear Sir
     2.   Dear Madam

Then picking the second paragraph:

     3.   Thank you for your check…
     4.   We are concerned that we have not received your check…

Just as in computers where 8-bit was simple and 64-bit is sophisticated, the prokaryotes running data-only ASCII never got very sophisticated, while all eukaryotes use the sophisticated method of mixing data and instructions.

Domestication of Prokaryotes

The control of life-units was so sophisticated at an early stage of the exploration of the phase space provided by the Logos that two very crucial events occurred in history—the domestication of the mitochondrion and, later, the chloroplast.

Phagocytosis has its dangers, e.g., a bacteria that is engulfed as food might resist being digested and, instead, infect the cell and kill it. In this way, the cell becomes food for bacterial multiplication.

An intermediate standoff situation can also occur. The bacteria enter the cell and end up in a digestive vacuole, but the bacteria is neither digested nor does it multiply. Such situations occur today, as with the tuberculosis bacterium. Most people who are exposed to TB never develop symptoms, since the bacteria can live in an inactive form in the digestive vacuoles of the lung cells. But they are constantly trying to escape control and spread from cell to cell. If the immune system weakens, such as in people with HIV or elderly adults, TB bacteria can become active. In their active state, TB bacteria cause death of tissue in the organs they infect. Active TB disease can be fatal if left untreated.

The oxygen generated by the photosynthetic bacteria in the stromatolites was not a problem at first because it quickly diffused away and was mopped up by the ferrous ions in the ocean. Some of the oldest known rock formations, formed over 3,700 million years ago, include banded iron layers. The formations are abundant around the time of the great oxygenation event, and become less common after 1,800 million years ago. The total amount of oxygen locked up in the banded iron beds is estimated to be perhaps twenty times the volume of oxygen present in the modern atmosphere. Banded iron beds are an important commercial source of iron ore.

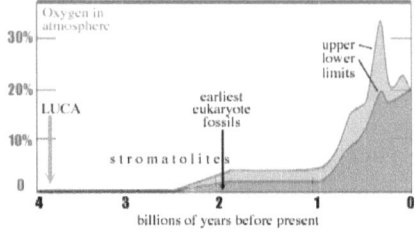

Oxygen is an active gas, and it can cause many problems if it is more than a trace. It can form free radicals, such as oxygen and peroxide, that readily interact destructively with the molecules of life.

Something similar to the tubercle bacterium happened in one lineage of early eukaryotes. It ingested a bacterium that had perfected the process of aerobic respiration, the reverse operation of photosynthesis.

$CO_2 + H_2O$ + energy — photosynthesis —> $CH_2O + O_2$

$CH_2O + O_2$ — respiration —> $CO_2 + H_2O$ + energy

This bacterium was able to use the oxygen that was accumulating in the atmosphere and ocean to efficiently 'burn' food for ATP production instead of relying on the inefficient fermentation process used up until then by scavengers, which is what our ever-so-distant ancestors were in the stromatolite. The polite word is heterotrophs in contrast to the primary producers, the photosynthetic autotrophs.

This latest-model bacterium, with its superior skills inherited from the Logos, ended up in a stalemate with the eukaryote ancestor. It could prevent itself from being digested, but it but could multiply unchecked.

It turned into a partnership as planned in the Logos. The eukaryote prospered: the bacterium soaked up all the otherwise toxic oxygen and excreted excess ATP. The bacterium prospered: the phagocyte proved a safe environment with an abundance of pyruvate to eat. It was a match made in heaven.

The phagocyte cautiously learned how to domesticate the bacterium, to let it grow and multiply, to attach it to the cytoskeleton, to govern its states. This was perfected in a lineage, the universal ancestor of eukaryotes-with-mitochondria (which is almost all of them) as the bacterium life unit was domesticated and became an integral, and essential, subsystem in sophisticated living systems.

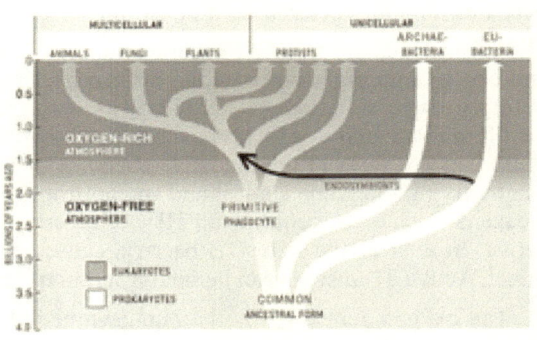

Many of the mitochondrial genes we eventually transferred to the nucleus, with the proteins synthesized in the RER and transported to the mitochondria by the cytoskeleton. Presumably, long-distance control was inefficient for some crucial proteins and the mitochondrion retains a set of genes for the mRNA, rRNA and tRNA necessary to translate the few genes into protein. These are of the prokaryote varieties, and the mitochondrion divides just as a prokaryote does.

The mitochondria of all living things is descended from this lineage.

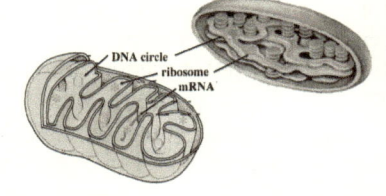

Later, a similar thing happened with a photosynthetic cyanobacteria. It was engulfed by a phagocyte with mitochondria, and it was also domesticated to became the ancestral chloroplast.   Only a few of the genes were transferred to the nucleus, and the chloroplast has almost a full complement of prokaryote genetics.

This eukaryote lineage abandoned scavenging for a living and, following the plan in the Logos, became an autotroph that developed into the single-cell and multicellular plants.

These adopted bacteria are known generically as endosymbionts. Both the mitochondria and the chloroplasts, unlike most of the vital units, have a double bilipid membrane. All the DNA, ribosomes, tRNA and mRNA are within the inner membrane (as would have been the case for the ingested and tamed prokaryote). The inside of a chloroplast also has stacks of light-absorbing plates containing the chlorophyll.

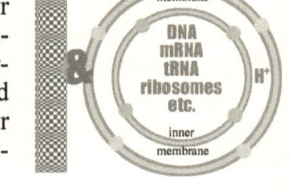

In both endosymbionts, the space between the inner and outer membrane is used for generating the proton-motive force. Both membranes have their specific portals that allow communication between the inner and outer space and the vital units surrounding it. The outer membrane is attached to the cytoskeleton so that endosymbionts can be moved as necessary.

The control of these endosymbionts is total. An example is the mammalian ovum, which has an abundance of mitochondria ready to power the development of the zygote. For decades, however, these mitochondria are all in the "off" state, as the dormant egg waits for release and possible fertilization.

During the mitotic division discussed in the next section, the centrosomes apportion the endosymbionts along with all the other organelles between the mother and daughter cells, and not always equally since it depends on what program they are running.

Eukaryote Division

The structure of the eukaryote cell is complex, so dividing into two clones is not as simple as in the prokaryote. We have used the instructive analogy of a vital unit to a symphony. The process by which a eukaryote divides is *mitosis* and a useful analogy is a ballet. The digital score for a ballet includes music and, in addition, a whole new set of instructions about arms, pirouettes, solos, duets, and corps behavior.

In an analogous way, the digital score stored on DNA is transformed by the DC/AC centrosomes in a beautiful choreography that has delighted and intrigued cytologists since mitosis was first observed. Mitosis occurs in a sequence of precisely organized steps, the very first being the duplication of the centrosome.

Centrosome duplication

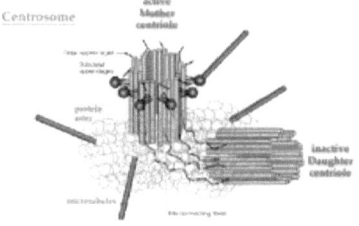

A eukaryote has just one centrosome in a cell. The two centrioles at the heart of the centrosome are not equivalent. One is the 'mother' centriole that is reading the mRNA tape and controlling the many vital units. The other is a subordinate, 'daughter' centriole that has a passive role until cell division.

The two centrioles come apart within the aster. The daughter cell becomes active, and the two centrioles read the same digital instructions conveyed to them on RNA. A new centrosome seems to extend at right angles off each old centrosome in a way reminiscent of semiconservative replication of DNA. These are grand-daughters, and the end result is a mature mother-granddaughter centrosome and an immature daughter-granddaughter centrosome.

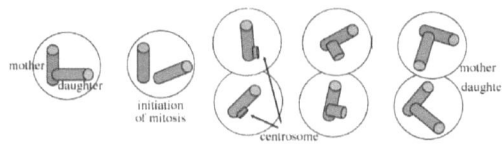

The immature centrosome now fully separates and is taken on what is called a 'kissing tour' of half the cell run by the mother centrosome. During this tour, the attachments of the life units to the mother centrosome are transferred over to the daughter centrosome.

After this transfer of responsibility over half the cell, the mother and mature daughter move to opposite poles to perform the amazing duet of cell division, or mitosis, learned from the Logos by the universal ancestor of all 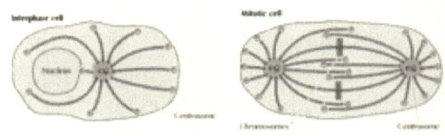 eukaryotes (protists, plants, fungi, and animals). The digital memory of how to do this has been passed on to all eukaryotes.

The next step involves the duplication of the DNA store and its complete condensation into a pair of chromosomes attached at their centromeres. The nuclear membrane is disassembled and a *spindle* of thick cytoskeleton forms between the separating centrosomes. The chromosomes are aligned on the *central plane* of the spindle and attached to it by their centromeres. The centromeres split and the duplicate chromosomes are pulled apart by the spindle to opposite ends. A new nuclear membrane is assembled around the chromosomes with the centrosome remaining outside.

The spindle is disassembled as the cell is pinched in two as new membrane is assembled at the central plane. The end  result is two daughter cells, identical clones except that one has the original mother centrosome while the other the original daughter centrosome.

Spindle

During cell division, the gross manipulation of structure is visible under a microscope. Together, the two centrosomes run the same program that occurs in well-defined stages.

As the cell enters mitosis, the dynamics of microtubule assembly and disassembly change dramatically. First, the rate of microtubule disassembly increases about tenfold, resulting in overall depolymerization and shrinkage of the regular microtubules. At the same time, the number of microtubules emanating from the centrosome increases by five- to tenfold. In combination, these changes result in disassembly of the regular microtubules and their replacement by outgrowths of large numbers of short microtubules from the centrosomes.

Right before the physical cleavage separating the two cells occurs, the mother centrosome migrates to the cleavage furrow and gives it a final 'kiss' of farewell. This event signals the final separation of the two cells into mother and daughter. After the completion of cell division, the active centriole in the daughter cell (originally the inactive daughter) becomes a mother centriole in the next round of cell division.

Enumerating a lineage

When a bacterium divides by asexual reproduction into two, there is little to distinguish the two resultant cells. They are both called daughter cells.

This is not true in eukaryote multi-unit life. One cell gets the mother centriole, and the other cell gets the activated and educated daughter centriole, now an active mother centriole with attached inactive daughter.

This allows us to digitally record the lineage of each cell by noting which gets the mother centriole. A simple method would be to add to the lineage record a C-G pair (0) if it gets the mother and an A-T pair (1) if it gets the daughter at each cell division.

After 3 rounds of cell division, there are 8 cells each neatly labeled with a binary number from 0 to 7 with the labels being 3 bases in length. It takes roughly 45 rounds of cell duplication to get from a single zygote to the 50 trillion-or-so cells of a mature human being.

Using this method, the label on each cell is now 45 bases long, and each cell has a unique numerical label that, in binary code, ranges from 0 to 35,184,372,088,831, or $2^{45} - 1$. After 45 rounds, and assuming no cell death, every cell has a label that is 45 bases long. There is a cell that has the original mother centriole and comes from a lineage solely of mothers, with label:

00000 00000 00000 00000 00000 00000 00000 00000 00000

and another cell that comes from a lineage solely of daughters, with label:

11111 11111 11111 11111 11111 11111 11111 11111 11111

This universal method of keeping track of cell division has not yet been detected in living systems even though it has obvious utility. A similar, crude method of keeping track of cell division is in the telomeres, the caps at the ends of the DNA helix in a chromosome.

Telomeres

The telomeres are buffers capping the ends of the chromosomes, and are consumed during cell division and replenished by an enzyme, the telomerase reverse transcriptase, that copies RNA into DNA.

Telomere length varies greatly between species, from approximately 300 to 600 base pairs in yeast to many kilobases in humans, and telomeres are usually composed of arrays of guanine-rich, six-to-eight-base-pair-long repeats. Eukaryotic telomeres normally terminate with a 3' single-stranded-DNA overhang which is essential for telomere maintenance and capping. The telomeres consist of hundreds of repeats of the base pattern 221000 (TTAGGG).

At each cell division in cells lacking telomerase reverse transcriptase (which is the majority of human cells), the telomere loses a few of these repeats and the telomere shortens. When it gets short enough, the division of the chromosomes becomes impossible.

The telomere shortening mechanism normally limits cells to a fixed number of divisions, and animal studies suggest that this is involved in aging on the cellular level and sets a limit on lifespans. Telomeres protect a cell's chromosomes from fusing with each other or rearranging—abnormalities which can lead to cancer—and so cells are normally destroyed when their telomeres are consumed. Most cancers are the result of 'immortal' cells which have ways of evading this programmed destruction.

The precise labeling of cells with a numerical label is probably important in the regulation of cells at the organ level of sophistication. A liver cell from a whale, a man, and a mouse are identical in all respects. They originate in the embryo in a similar fashion from a lineage of cells that goes, respectively, through 50, 43, or 36 rounds of cell division to create the 2^{50} cells of the whale liver, the 2^{43} cells of the human liver, or the 2^{36} cells of the mouse liver.

Hack a slice off the livers, and the identical cells will vigorously multiply until the original number is restored—the whale cells stop when there are 2^{50} of them, the human cells stop when there are 2^{43} of them, and the mouse cells stop when there are 2^{36} of them.

This would easily be accomplished by a register in the DNA heritage that translates as 50 in the whale, 43 in the human, and 36 in the mouse. The only difference between a whale, a human and a mouse liver cell is the content of this register; call it SIZE. The development and maturation of the liver is the instruction to multiply until the cell lineage label is SIZE bases long, each cell with a unique numeric label from 0 to $2^{SIZE} - 1$.

Healing is just maintaining the array of numbers. If one goes missing, cells duplicate and assign it the missing number. To accomplish this, each cell in the array is constantly sending a numeric message to its neighbors on a snippet of RNA, "I am #n." When the message is not received by its neighbor, #n+1, it duplicates and the daughter is assigned the label, n.

The daughter #n, in turn expects to get the message, "I am #n-1" and, if it does not, it in turn duplicates and assigns its daughter the #n-1 label. When the two edges of the wound meet, conciliation of numbers occurs. If this is perfectly accomplished, there is seamless healing; if it is not, 'scar tissue' accumulates until the mismatch is sorted out.

It has always been known that plant cells do not completely separate when they divide, there are tiny channels of cytoplasm through the thick cellulose walls called plasmodesmata connecting them. A typical plant cell may have between 103 and 105 plasmodesmata connecting it with adjacent cells. It had long been thought that animal cells lacked these connections, but it now turns out that they have them as well. Membrane nanotubes, as they are called, are transient long-distance connections between cells that can facilitate intercellular communication. They can also contribute to pathologies by directing the spread of viruses. Recent data have revealed considerable heterogeneity in their structures, processes of formation, and functional properties, in part dependent on the cell types involved.

Like the pilli of bacteria with their transfer of digital information during conjugation, these are channels of digital information transfer as well as other types of material transfer such as entire organelles.

Sex

All prokaryotes multiply by asexual reproduction. By making two identical copies out of one. An asexual lineage is a clone of many copies of a single founder and this is called *vertical* transmission of digital information. System building involves steps and, for an asexual lineage, all the steps must be discovered alone by that lineage.

Stripped of all detail, all sexual processes are distinguished by a *horizontal* transfer of digital information between different lineages, usually, but not always accompanied by multiplication as sexual reproduction.

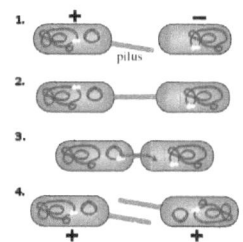

Many prokaryotes indulge in conjugation, the horizontal transfer of DI between different, but related, clones that does not involve reproduction. A *plus* bacteria extends a thin tube *pilus* that pierces a *minus* bacteria. DNA polymerase copies some DI of the plus and injects it into the minus through the pilus. As the DI usually includes the instructions on how to make a pilus, both are now of the plus 'sex.'

The DI for antibiotic resistance is stored on such mobile DNA and the ability can spread horizontally between lineages. In this way they get some of the advantages of sexual horizontal transfer. We can illustrate the advantage of sex with a system having a set of simple attributes, R, expressed from Level 1 in the Logos. A lineage can explore phase space and learn over time from the Logos how to use these simple attributes in two different, more sophisticated ways, B or G. With all three attributes present, a fourth quite different attribute emerges, i.e., a new level of sophistication, Y*.

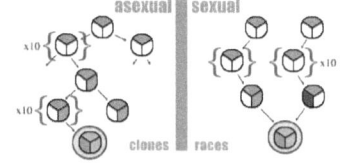

If it takes, on average, 10 generations to learn either G or B, it will take a lineage of asexual clones 20 generations to learn both ways of doing things and make a jump in sophistication. Two sexual lineages, however, can develop into two races, one with RB and one with RG. A sexual union between the two races can then discover the new level of sophistication in just half the time. Sexual reproduction with horizontal transfer speeds up the emergence of levels of sophistication.

Origin-Division-Union

Sexual reproduction illustrates the philosophical principle of development that Unification Thought calls Origin-Division-Union. We have already encountered this in the leap in sophistication from fundamental fermions and bosons to atoms and molecules. The maelstrom of the Big Bang (origin) separated electric charge (division) into negative electrons and positive quarks, whose colors ended up as positive protons and neutral neutrons. All evidence suggests that, for every electron in

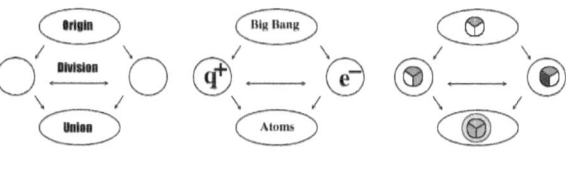

the universe, there is a proton. When the universe cooled, the positive and negative charges recombined (union) in neutral atoms. This step, in turn, was the formation stage for the emergence of the elements, and a whole new level of sophistication in the Logos was expressed as their chemistry. Sexual reproduction allows the formation-stage origin of a new species to develop separate lineages of specialized races that can come together in a quantum leap in sophistication.

At some point in the evolution of the eukaryotes, a lineage discovered the advantages of having two sets of chromosomes that could slightly differ from each other, adding a layer of fine-tuning to digital manipulations.

The digital score of mitosis was partially duplicated to create the process of *meiosis* that allowed for the alternation of generations, one with a single set of chromosomes (haploid) and the other with two sets (diploid). Meiosis turns a diploid cell into four haploid cells, while the fusion of two haploid cells creates a diploid cell.

Meiosis

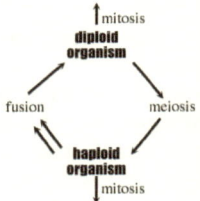

One of the great advantages of meiosis is that the diploid pairs of chromosomes are connected up and then the DNA is duplicated. There are now four chromatids attached together as a *tetraplex*. Sections of DNA can then be spliced together in different combinations following epigenetic digital information recorded by the lineage. It is at this tetraplex stage that the chromosome rearrangements occur that result in speciation.

The centrosome then performs two rounds of mitosis-like spindle separation of the four chromatids into four haploid cells each with one chromatid.

In the simple, single-cell eukaryotes, both the diploid and haploid cells multiply by mitosis, and the cells are very similar. The alternation of generations is often in response to a change in the environment.

Edens and Evolution

So far we have discussed a sequence of edens that emerged under the guidance of the Logos, where all the conditions were just right for extending the systematic hierarchy. Edens in which system building occurred as simple systems interacted to form higher systems of greater sophistication. The creation of an eden involves analog waves coming together as a perfect wave, just as ocean waves sometimes combine as a rogue wave with exceptional properties. The main difference between system building in living and nonliving systems is that living systems memorize analog waves as digital information and recall the analog waves as needed.

The sequence of edens, starting with the Big Bang origin of space, time and the fundamental entities, we have described so far is:

1. *Atom eden*. The expansion and cooling of the universe allowed system building of neutral atoms of hydrogen and helium.

2. *Element eden*. Gravitational condensation into stars, and H and He burning, generated all the other elements to be scattered in supernovae.

3. *Molecule eden*. Emergence of third-generation stars facilitated the accretion of the rock-and-water planet Earth with its moon.

4.   *Macromolecule eden.* Driven by tectonic smokers, RNA, and digital manipulation, proteins and analog wave control as basic living systems emerged.

5.   *Life eden.* The control of water and metabolism in a lipid confined volume was mastered and the LUCA left the smokers and radiated into the ocean.

6.   *Eukaryote eden.* With the mastery of photosynthesis and colony formation, stromatolites emerged where a lineage could shed its armored coat and explore the possibilities of scavenging. The eukaryotes radiated into the ocean.

The next step that occurred, about 500 million years ago, was that single-cell eukaryotes had learned from the Logos how to aggregate into simple multicellular plants, fungi, and animals. The eden for this step in system building was probably the ooze that coated the ocean floor, but the details are still to be worked out.

Evolution

The unfolding of the Logos as a series of levels of sophistication is *evolution*. We shall use this word as in the dictionary meaning of:

1.   Development, advancement, growth, rise, progress, expansion, unfolding; transformation, adaptation, modification and revision.

We shall **not** use it in the sense of its secondary definition:

2.   Darwinism, natural selection of random variation.

We can roughly distinguish two types of evolutionary development depending on their outcome, on the increase in the level of sophistication attained.

Mega evolution

These involve major transitions in the sophistication of living systems. These are the steps upward that are preceded by the Logos constructing a distinctive eden. These steps in mega evolution include:

Sophistication		Logos-derived emergent properties
Pre-life	0	Simple organics, clay catalysis, iron e^- transfer, high-energy bonds
RNA life	1	Water control, reading and writing digital memory
Prokaryote life	2	Triplet coded proteins, metabolism, ATP, phospholipids
Eukaryote life	3	Multiunit, organelle domestication, centrosomes, mitosis and sex

These mega-evolutionary steps which opened new vistas of phase space are the great leaps in evolution in which the Logos must provide an eden in which all the ingredients for system building are present together in a stable, safe environment.

Micro-evolution

In contrast, microevolution does not involve a leap in sophistication but in the exploration of the phase space opened up by the Origin event.

This microevolution exploration of an unoccupied phase space (the advance is often driven by the decay of the eden of the Origin event). Two examples of transitory edens are:

1.   The emergent prokaryotes altered the chemistry of their smoker-clay eden with their oxygen and mopping up of organic chemicals.

2.   The emergent multicellular eukaryotes grazed away the stromatolite edens of the eukaryotes.

The radiation of the prokaryotes from the LUCA in the smoker-clay eden into the world ocean, the radiation of the eukaryotes from the stromatolite eden into the ocean, and the radiation of the insects onto dry land are all examples of microevolution.

The evolution of computers from the 1980s Mac Plus with 128kB of RAM and 400kB disks to the MacBook Pro with 2,000,000kB of RAM and a 1,000,000,000kB disk is an example of microevolution. The two computers use the same basic architecture, except one is small and simple, while the other is still small but very sophisticated.

Multicellular Life

All of the living systems we have discussed so far are microscopic and invisible to the naked eye. The realm of prokaryotes and single-celled eukaryotes was unknown to the ancients, and was only uncovered by the invention of the microscope.

It is only when systems emerged in which the single cell was a subunit, that the multicellular life seen by the naked eye emerged.

It would seem that the evolutionary path of mono-vital-unit prokaryote to poly-vital-unit eukaryote is similar to the path of mono-eukaryote to poly-eukaryote. The former was a scavenger in the debris at the bottom of the stromatolite, the latter a scavenger in the ooze at the bottom of the ocean:

> It seems likely that an early step in the evolution of multicellular organisms was the association of unicellular organisms to form colonies. The simplest way of achieving this is for daughter cells to remain together after each cell division. Even some prokaryotic cells show such social behavior in a primitive form. Myxobacteria, for example, live in the soil and feed on insoluble organic molecules that they break down by secreting degradative enzymes. They stay together in loose colonies in which the digestive enzymes secreted by individual cells are pooled, thus increasing the efficiency of feeding (the "wolf-pack" effect). [180]

This was probably the earliest exploration of multicellular phase space. This is a rough outline of the macroevolutionary steps starting with the simplest levels:

Sophistication		Logos-derived emergent properties
Multicellular life	4	Tissues, differentiation, plants, and fungi
Tube animal life	5	Nervous system, worm brain, muscles
Segmented life	6	Organs, segmentation wings, legs, insects
Deuterostome life	7	Gills, jaws, kidney, heart, gut brain
Amniotic egg life	8	Land adaptation, amphibians, dinosaurs, reptile brain
Mammalian life	9	Warm-blooded, milk, care of young, family brain
Primate life	10	Tribes, social awareness, clan brain
Hominid life	11	Upright, clans, fire, pidgin language, hunting weapons, tribe brain

Sophistication		Logos-derived emergent properties
Human life	12	"I Am" self-awareness, infinite creativity and potential

There is an interesting parallel between the systematic structure of the realm of chemical systems with that of the realm of living systems. Both start at level zero, with systems that have zero chemistry—protons and electrons—or systems that have zero life—RNA and protein molecules.

Level One of both chemistry and life have multiple subsystems of level zero interacting to form the basic unit. In chemistry these are the atoms of the elements; in life these are the vital units.

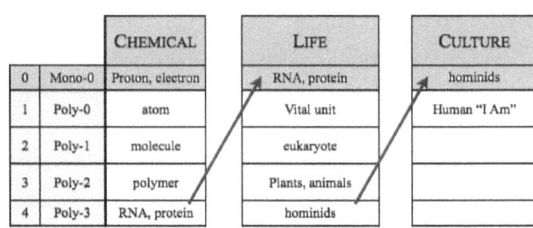

		CHEMICAL		LIFE		CULTURE
0	Mono-0	Proton, electron		RNA, protein		hominids
1	Poly-0	atom		Vital unit		Human "I Am"
2	Poly-1	molecule		eukaryote		
3	Poly-2	polymer		Plants, animals		
4	Poly-3	RNA, protein		hominids		

The systematic hierarchy is constructed starting with chemical activity, and proceeding through molecule, polymers, and then the sophisticated biopolymers, RNA with its digital processing, and proteins that generate analog forms to manipulate many small molecules, starting with water.

The combination of RNA and protein is level zero in the hierarchy of living systems. A systematic hierarchy stands on this foundation. We will define that troublesome word *mind* to be the measure of the sophistication of the set of emergent properties, determined by the Logos at the Origin and memorized for the descendants, of the system wave that is confining all the electrons and nucleons it is composed of. Sophistication refers to the kinds of systems the wave is manipulating and organizing. By this definition:

The mind of a prokaryote is capable of organizing chemicals into a unified whole.

The mind of an eukaryote can do all this, and also organize vital units into a unified whole.

The mind of a plant can do all this, and also organize sensory images into a unified whole.

The mind of an animal can do all this, and also organize motor images into a unified whole.

The mind of hominid can do all this, and also apply concepts in mastering fire and vocalizing simple language.

The hominids are level zero out of which the human has its origin.

The great leap is to the ability to name objects, to associate motor images (the idea in your mind) with distinct aspects of the environment. The stream of language you hear in your mind is exactly the same as if you were speaking aloud, except the destination is different. Naming and recognizing objects leads to the recognition of self and non-self, the "I Am" of self-awareness.

Cell differentiation

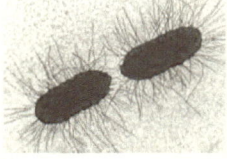

In all but the very simplest of multicellular organisms there is cell differentiation and, although the genotype is the same, the external phenotype is not. During cell multiplication, the somatic lineage diverges into same-cell tissues. The simplest seaweeds, for example, have a holdfast, i.e., a tissue that excels at attaching to and gripping rock, and the rest is photosynthetic fronds.

Differentiation requires that information be passed between cells. Much of this information flow is known to be in the analog form of chemical hormones that fluctuate in concentration and gradients.

We can speculate that digital information is also involved. We have already outlined a simple numbering system for a somatic lineage based on where the mother centrosome ends up in cell division. After three rounds of division, each of the eight cells that result is numbered 0 through 7 in digital store with address ###.

This numbering allows for a digital control of differentiation with a simple program, such as:

```
IF ### = 0 THEN RUN holdfast
IF ### = 1 THEN RUN holdfast
IF ### = 2 THEN RUN holdfast-blade
IF ### = 3 - 7 THEN RUN blade
```

Support for ways that the RNA-CPU can keep track of digital numbering in multicellular eukaryotes is, for example, provided by the mammalian liver. Once a mechanism is perfected, it persists essentially unchanged down through a lineage. The mammalian liver cell was perfected in the ancestral mammals, and a mouse liver cell is indistinguishable from an elephant liver cell.

The mature mouse liver contains ~10^9 of these cells, while a mature elephant liver contains ~10^{14} of these cells. The precursor cell of the somatic lineage in a mouse goes through 30 rounds of division, while in an elephant it goes through 47 rounds of division. They then stop dividing. The relative sizes of most mammalian organs is the same, which suggests that there is a digital store of what scale the mature body is to be.

If either mature liver has a lobe excised, the cell division resumes and the liver regenerates to the normal size. This suggests that the array of cells are numbered, and when a section is excised, the array reestablishes itself. Something is keeping track of numbers. The simplest method would be the passing of digital information about who the neighbors are. This suggests that RNA is passed between cells, a phenomena not yet observed, but supported by the recent discovery of micro channels connecting cells together. These connections are reminiscent of the pili through which bacteria swap digital information. In normal circumstances, for example, cell #26 receives "my # is 25" and "my number is 27" from adjacent cells, and is content. If, however the array is damaged and the "my # is 27" is no longer received, the cell divides and assigns the daughter cell #27.

As differentiation is still a topic of debate, the strategy for growth and healing is as yet unclear. But all three of the main eukaryote lineages— plant, fungi, and animal—found ways of digitally storing analog patterns of differentiation and recalling them as required.

We will not go into great detail as these topics are well-covered in the literature. The only difference here is that each advance is learned and recorded in unified science, while such an advance is considered a fortuitous accident in the current mainstream view. To further reduce the length of this work, we will just outline the evolutionary steps learned in plants and animals, and not deal with the fungi.

Multicellular Systems

The ocean is so hospitable to plants that the only real advantage of multicellularity is the holdfast—once you find a good spot, you get to stay in it. So, plants in the oceans are really sophisticated seaweeds, with air pods for buoyancy and sexual spots for reproduction. The forests of sea kelp are habitats for a wide variety of organisms.

Unlike the ocean, dry land posed a series of challenges that we mastered as the plants radiated over the land, starting with the mosses radiating into the moist zones around the land. The next big advance was vascular tissue that could distribute food and water around the plants.

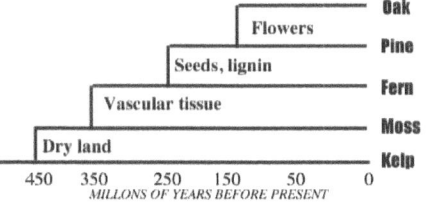

The discovery of lignin, which is used to strengthen cellulose, allowed for height and closeness to the light, and the seed for water-less reproduction allowed for a massive proliferation of simple trees over the land, whose remains are the great coal beds we mine today. It is thought that the recycling ability of the fungi was temporally stymied by lignin, a very tough molecule to degrade, allowing great swamps of undigested plant material to accumulate. The final innovation was the green broadleaf that was jettisoned in the fall, and later developed into a wide range of colored flowers.

Ocean Animals

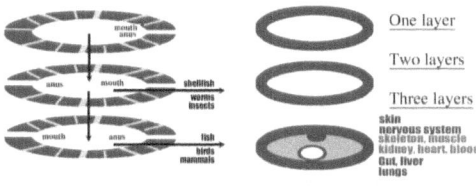

The animal lineage went through a much more complex series of developments in the ocean before attempting the colonization of dry land.

The very simplest animal was a sphere of cells with a hole in it. Prey could be engulfed into the proto-gut, digested, and the indigestible remains ejected. The hole acted as both mouth and anus. The next innovation was a second hole; the prey was engulfed by the first hole and ejected from the second. The exploration of *protostomic* phase space with mouth first, anus second, later radiated into the shellfish, one type of worm and the insects.

In ways that are difficult to explain except as digital program shift, one protostome lineage flipped things around so that the first hole was used as the anus and the second hole as the mouth. This *deuterostomic* lineage radiated into the other type of worm, the fish, reptiles and mammals.

The very simplest of animals had only a single layer of cells. This has a limited phase space, and the most complex extant animals are the sponges. A second layer of cells was added by outside cells slipping through the mouth and coating the inside. These *germ layers* are called the outer ectoderm and the inner endoderm. This is also a limited phase space, and the most complex *diploblastic* extant animals are simple worms.

The lineage that explored developing a third layer, the mesoderm, opened up a vast phase space, the exploration of which led to the emergence of all animals more sophisticated than flatworms. The ectoderm developed into the skin and nervous system, the

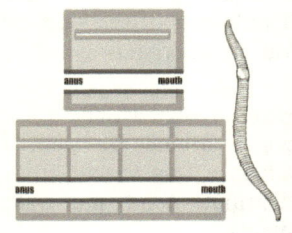

mesoderm into bone, muscle, heart, and blood, and the endoderm into gut, liver, and lungs. Both protostomes and deuterostomes independently discovered the versatility of a mesoderm third layer but, reflecting the mouth-anus flip, vertebrates and insects have opposite spinal cord/gut configurations.

So far, we have discussed two examples of a phase space opened up by duplication and association: eukaryotes as poly-prokaryotes, and animals as poly-eukaryotes. A third example is the lineage that explored making attached duplicates of itself—the exploration of segmentation by both protostomes and deuterostomes, then differentiation of the segments. The mouth-end segments developed into the head with a centralization of the nervous system and sensory organs. Middle segments developed legs, and the hind segments developed into the tail. The spatial order of segments is reflected in the spatial order of the Hox genes on the chromosome, an aspect of digital processing that is under investigation.

The millipede is an extreme example of duplication of segments, while human segmentation is only preserved in the spinal vertebrae that protect the spinal column.

The digital control of segmentation and differentiation is uses a set of Hox genes. The Hox genes have been remarkably conserved down through the ages, and fly and chicken Hox genes can be interchanged and still function. While the genes are the same, sophisticated organisms have many sets of Hox genes to control local differentiation. The fruit fly has two clusters of Hox genes, while a human has four.

A simple circulatory system developed, with a separation between the water locally bounding a cell—the lymph—and mobile water that distributed oxygen, food, and wastes globally—i.e., the blood. Muscles developed to circulate the blood and developed into a heart in stages from a simple muscular tube to a chambered pump.

Muscles and a skeleton they could move developed. An external skeleton of chitin emerged in the anthropoids and insects, while an internal skeleton of cartilage emerged in the chordates. Up to this point in history, dead animals were soon degraded and left little impact in the fossil record. Discovery of how to calcify cartilage and construct bone changed that situation, since bones are sturdy enough to be occasionally fossilized.

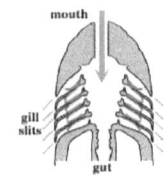

The oxygenation of the blood was accomplished by allowing oxygenated water to enter the mouth and, passing over gill beds of fine blood vessels, it was expelled out of gill slits in the sides so as not to dilute the gut contents. The gill arches were supported by cartilage or bone.

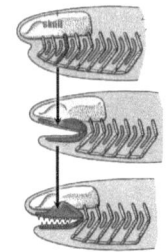

The phase space opened up in the Logos by the advent of the gill arches was enormous. Exploration dropping the first two arches and developing the third arch, generated lineages leading to the cartilaginous sharks and the bony-jawed fishes.

The arches still play a role in human development, ending up modified into bits that eventually ended up as the tiny bones of the inner ear, the jawbone, and the voice box. The radiation of the bony fishes filled the oceans with all kinds of fish that thrive to this day.

One adventurous lineage explored surviving the potentially-fatal circumstance of being stranded by tidal retreat, by gulping air and burrowing into mud—the lung fishes. One branch started using their four fleshy fins to crawl about—the coelacanths that were long thought extinct, but were discovered in the depths of the ocean off the coastlines of the Indian Ocean and Indonesia. Further exploration of the opening phase space of the moist edges of dry land resulted in the advent of the early amphibians.

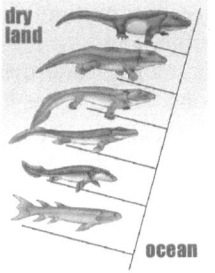

Land Animals

The colonization of the land by both plants and animals required that both discover how to reproduce free of the constraint of limited water. The basic process was the same—learning how to generate analog forms and sample them for digital storage. This digital information could be used to regenerate the analog form or be copied and passed on down a lineage. As digital information accumulated, lineages became more sophisticated.

The simple mosses and amphibians remained dependent on environmental water. Just as the seed was protected by a coat, the eggs of reptiles were protected by an egg-shell.

The higher plants discovered flowers and fruits to protect reproduction, while the mammals developed the womb and mother's milk.

Only the animals, however, explored the phase space of muscular movement coordinated by a nervous system. While the muscles of mice and men are similar, it is in the nervous system that great leaps occurred as higher and higher levels of the Logos resonated with physical systems.

THE NERVOUS SYSTEM

The cells in multicellular animals communicate with each other in an analog fashion by varying the concentrations of small molecules in the lymph fluid that they are sharing. The ectoderm cells that developed into neurons extended thin protoplasmic filaments, driven by the cytoskeleton and the centrosome, to use these chemical signals to communicate over a distance.

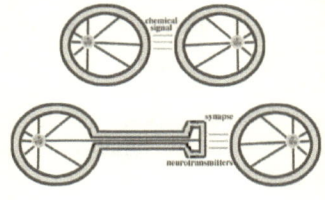

The final step was polarizing the bi-lipid membrane, so that signals from other neurons depolarized the membrane at many input filaments, the dendrites, which if sufficient, sent a depolarized signal down a single output filament, the axon. These connections created neural nets that could process signals, before output to an effector cell.

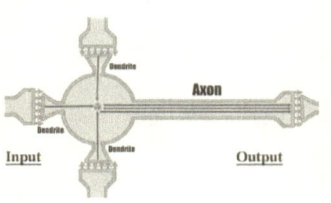

The basic pattern is the same for human neurons as it is for jellyfish neurons: There is a cell body that extends many short input filaments, the *dendrites*, and a single long output filament, the *axon*. The axon ends either on a dendrite or a regu-

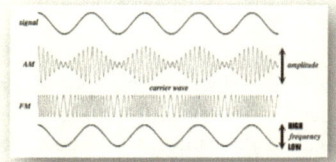

lar cell. A stimulus arriving down the axon activates the dendrite, which can be either a plus or minus type. If the sum of the plus and minus activation of the dendrites is greater than the threshold value, a stimulus is sent down the axon to activate the dendrites it is connected to.

Waves of stimulation spread over neural nets, and they are *frequency-modulated* (FM) waves, not amplitude-modulated (AM) waves.

FM and AM waves

A radio station works by imposing a very low-frequency sound wave on a high-frequency *carrier* wave. The carrier wave is modulated with the sound wave. The radio receiver strips away the carrier and only allows the sound wave to enter the amplifier and loudspeaker.

The first, and simplest type of modulation is to vary the amplitude of the carrier wave in time with the amplitude of the signal. This AM radio had a problem with static—stray waves generated by many diverse sources, such as thunderstorms and air conditioners.

The problem with static was solved by modulating the frequency of the carrier wave. A heterodyne receiver combines this modulated wave with a set frequency wave, and the variation in frequency, not amplitude, carried the signal wave.

In AM radio, the frequency is constant and the amplitude varies. In FM radio, the amplitude is constant and the frequency varies. Neurons use FM; the amplitude of a signal down the axon is constant and it is its frequency that varies.

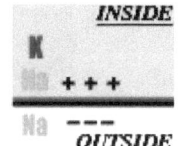

We have already discussed proton-motive force driven by an imbalance of charge across a bi-lipid membrane. The membrane of a neuron in the resting state also has a charge imbalance but, in this case, it is because ATP is used to accumulate potassium ions so the inside of the cell becomes positive and the outside is negative. Like a battery, at rest the membrane of the neuron cell body is primed with electrostatic energy.

When a plus dendrite is stimulated by the incoming axon, a positive charge ripples outwards across the outer cell membrane. When a minus dendrite is stimulated, a negative charge ripples outwards, and the plus and minus ripples interfere with each other.

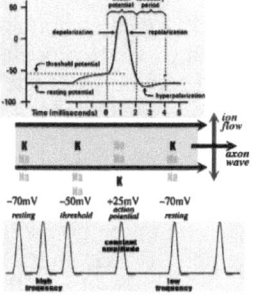

The axon has a root in the cell membrane. If the ripples at this root sum up to 20% positive, it initiates an action potential down the axon. The membrane depolarizes and sodium ions flood out of the cell, and the membrane now reverses polarity. Potassium ions flood out of the axon and the membrane peaks at a positive value. The membrane repolarizes as sodium and potassium are pumped in opposite directions. The rise in the action potential triggers the next bit of membrane, to depolarize, and the spike is propagated down the axon with a constant amplitude but varying frequency.

We earlier described an electromagnetic photon as having two orthogonal components—the electric and the magnetic—waving orthogonal to the direction of travel. The wave down an axon is similar, it has two *parallel* components—the sodium- and potassium-ion concentrations—that also wave orthogonally to the direction of travel down the axon.

The action potential lasts two milliseconds After this, for another two milliseconds, the membrane is hyper-polarized, and quite unresponsive to stimulation. The resting state of ion imbalance is restored after another four milliseconds. The nerve is now primed and ready for the next spike to arrive.

Analog neuron nets

Neurons deal in analog information. An analog pattern of inputs to the dendrites creates an analog pattern of waves down the axon. An example of the simplest of neural nets is the nervous system of the single-opening, mouth-anus hydra. This has sensory cells on the surface connected by a neural net to muscle fibers in the hollow body. Touch a hydra and it im-

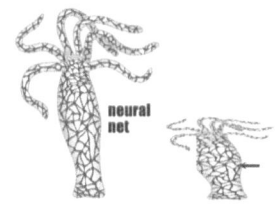

neural net

mediately contracts, a simple example of a reflex arc. We humans still use such simple analog circuits in the reflex that immediately removes our finger from a hot stove without involving anything sophisticated in our nervous system.

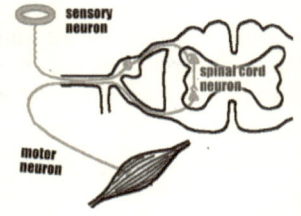

These neural nets also occur, with much greater sophistication, in all animals and humans. A single neuron can receive input from up to 10,000 different axons arriving at its dendrites. A neuron can ramify the end of its output axon and synapse with up to 10,000 different dendrites.

The human nervous system is estimated to comprise 85 billion neurons that connect together in the brain with 1,000 trillion synapses, i.e., a quadrillion. Each neuron creates a tiny electrical wave when it fires, and the firing of billions of neurons combine to create electrical waves that can be measured by electrodes attached to the outside of the skull.

Brain waves

Brain waves are created by the synchronized firing of neural nets—otherwise there would be no waves large enough to register. The science of electroencephalography (EEG) studies these brain waves, which are a global measure of synchronized neural nets firing in the brain. The waves measuring this global analog aspect of the brain can be divided into five types, characterized by their time period, that are associated with five different states of the brain:

1. *Delta waves* have the lowest frequency, ½ to 3 cycles per second (cps) and the highest amplitude, ~100 microvolts (μV). These waves are characteristic of unconsciousness and deep, dreamless sleep. When the delta waves appear with others in the conscious state, they seem to be associated with intuition and day-dreaming.

2. *Theta waves* have a higher frequency (4 - 7 cps) and a smaller amplitude (~50 microvolts). They seem to represent the subconscious mind, and are prevalent while drifting off to sleep, in dreaming sleep, and during meditation and peak creative experiences.

3. *Alpha waves* have a midrange frequency (8 - 14 cps) and amplitude (~30 μV). They are seen in the relaxed state, and also while daydreaming and in the recall of sensory experiences.

4. *Beta waves* have a frequency of 15 - 38 cps and amplitude of ~30 μV. They are associated with alert, normal consciousness and logical, analytical thinking, and conversation. High-frequency beta waves are associated with stress, anxiety, and inner conflict. Low-frequency beta waves are associated with a clear, alert and creatively focused mind.

5. *Gamma waves* have the highest frequency (38 - 100 cps) and the smallest amplitude (~10μV), which makes them difficult to study. They are seen in peak performance—both physical and mental—high focus and concentration,

and in transcendental experiences. When they do occur, they involve waves of synchronization over large parts of the brain.

We earlier discussed the analog forms generated by folded proteins in terms of the sum of multiple individual Bezier generators, the aminoacids in the primary structure. The wavefunction structuring water and metabolism in a vital unit is the composite symphony of all the proteins generating their waves. In a similar way, the neurons of the brain generate a Bezier wave unit that combines into the wave of the neural net, and the overall wave of the brain is the combination of all the nets together.

The analog form generated by a protein is recalled by RNA from digital memory in DNA. The proteins specialize in generating analog form, and the RNA specializes in digital information. If the same pattern repeats itself, then the analog forms generated by specialized neurons are recalled from digital memory by cells specialized for digital storage and manipulation. The open question is: Where does this digital aspect reside?

Digital Memory

The digital aspect of memory is only now starting to be explored but there is one thing that might provide a clue: we can expect that the ability to store, recall, and manipulate digital information increases with the sophistication of animal mental ability, from worm to human.

As mentioned, the human brain contains ~85 billion neurons specialized in analog wave generation. They are a 10% minority, however, as the are embedded and surrounded by ~850 billion glia cells, which are the stem cells that give rise to the neurons in development.

The current consensus is that all that is of primary important in brain functioning occurs in the neurons and the synaptic strengths in neural nets. The iconoclastic *Roots of Thought: Unlocking Glia*[181] by Dr. Koob makes the case for neurons being directed and controlled by the glia cells.

He considers this glia programming of the neurons to be analog, conveyed bursts and waves of calcium. I consider this a gross effect, and that RNA is passed between neuron and glia, conveying digital information. Unfortunately, I cannot google anything about RNA transfer between these cells—like the centrosome, only single molecules need be involved as, noted earlier, a single molecule of RNA is quite capable of completely altering a cell's state. So, we will just précis in the following section the case that Dr. Koob makes for glia analog control of neurons.

As might be expected, the ratio of glia controllers to neuron generators increases as one moves up the hierarchy of animal sophistication. In the jellyfish, only a small fraction of the cells in the simple nervous system are glia, while the rat's nervous system has twice as many glia as neurons. In the cortex where the highest functions are performed, the human brain has 35% more glia cells than the chimp brain does.

Sophistic.	glia	neuron	ratio
Jellyfish	1%	99%	0.01
Worm	10%	90%	0.1
Rat	60%	40%	2
Chimp	80%	20%	4
Human	90%	10%	9

The neurons are rather like DNA, i.e., they both have a single function which they perform excellently—storing digital information or generating analog forms, respectively. The glia cells are akin to RNA in that they can perform a wide variety of functions. We have earlier listed the dozens   of emergent properties possessed by RNA, from digital information manipulation to analog form generation. The glia cells have almost as wide a spectrum of properties:

1.   Ependymal glia—lining the brain ventricles, stem cell for neurons and other glia
2.   Schwann cells—myelination of axons in periphery
3.   Oligodendrocytes—myelination of axons in white matter of brain
4.   Müeller cells—glia in eye
5.   Velate cells—glia in nose
6.   Epithelial glia—surround brain blood vessels
7.   Microglia—response to infection
8.   Bergmann glia—astrocytes in cerebellum
9.   Astrocytes—most abundant cell in human cortex

The astrocyte seems the most likely candidate that digitally programs neurons since the neurons cannot function without them:

> An astrocyte is a self-sufficient, self-replicating cell signaling to itself content-edly. Neurons have no reason to exist except to support astrocytes. Mature neu-rons cannot function alone, whereas mature astrocytes have no difficulty exist-ing without neurons. When placing mature neurons in a Petri dish, they are un-able to survive without astrocytes. Astrocytes are perfectly content without neurons.[182]

It is the astrocyte network over the which the Ca^{++} wave passes, and the cere-bral astrocytes are connected to other astrocytes and neurons by *gap junctions* that are similar to synapses.

Astrocytes are known to store information and are involved in short- and long-term memory. They control neuron firing and control the addition, modulation, and deletion of synapses between axons and dendrites on neurons.[183]

In the wiring of the developing brain, the ependymal glia surrounding the ven-tricles send out radial extensions toward the outside. The cell body then moves outward along this extension and, as it returns inward, divides into two cells. The daughter cell moves outward along the radius and becomes a neuron, while the mother cell returns inward to the ventricle.[184] The neuron climbs the path estab-lished by its mother glia (and it can be expected that the mother glia also ends up with the mother centrosome).

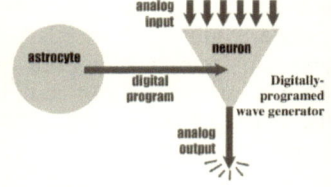

In our view, the digital program, provided by the glia and running at the top-most level of the neuron, determines its output given a certain pattern of inputs. The variably programmed neuron recognizes, we might say, the analog form of the input in the output it sends down the axon and its many connections.

An example of such variable programming is the development of the visual cortex. The level of the cortex that is programmed to recognize vertical lines is divided into narrow stripes with alternate inputs from the left and right eyes. In normal development, the strips are of equal size. If, however, the eyelid of the right eye is sewn shut at birth—the experiment was on cats—the right-strips are vestigial and the whole layer becomes programmed to respond to the left eye. If the eyelid of the mature cat is opened, the right eye is oblivious to vertical lines and the cat just does not see them.

Universal Computer

Returning to modern computer technology to illuminate the digital aspect of living systems, there is the concept of the universal computer that can run any program. This concept seems commonplace nowadays, in that a digital computer can be programmed to emulate any number of devices, but it was only established by Turing a half century ago. Depending on the program, a computer can emulate a TV with full color and sound, a typewriter, a graphic design studio, a symphony manipulating sampled sounds, a film editing studio, and many other functions via a burgeoning variety of useful apps.

This is the *universal* aspect of modern computers, in that any number of different programs can be run on the same computer. There is a hierarchy of universal computers running in living systems.

The simplest is the ribosome. This will take any stretch of RNA and turn it into a protein. A ribosome will happily translate even a nonsense stretch of RNA made synthetically in the lab. It was this that allowed for the first step in uncovering the Universal Triplet Code when a synthetic RNA made only of uracil was translated by ribosomes into long strings of peptide-linked phenylalanines, and **UUU** was established as its digital code.

A step-up in sophistication is the universal computer that is the centrosome. The variety of digital programs received from the nucleus is reflected in the variety of cell shapes, movements, and behaviors of the eukaryote cell. The most elaborate program being that of mitosis and meiosis.

Finally, we reach the universal computer that is the neuron that runs whatever program it receives from the astrocytes. It is flexibility in neuronal programming by astrocytes that gives the brain a remarkable ability to alter and adjust:

> Neuroplasticity occurs on a variety of levels, ranging from cellular changes due to learning, to large-scale changes involved in cortical remapping in response to injury. The role of neuroplasticity is widely recognized in healthy development, learning, memory, and recovery from brain damage. During most of the 20th century, the consensus among neuroscientists was that brain structure is relatively immutable after a critical period during early childhood. This belief has been challenged by findings revealing that many aspects of the brain remain plastic even into adulthood.[185]

The centrosome uses the ribosome level to direct protein activity. The neuron level uses the centrosome level, with its connection to each dendrite and each ramification of the axon end, to direct synapse behavior.

Serial and parallel computers

At this point, the digital technology of modern computers fails to assist in understanding that in living systems, because current technology deals with *serial* computers while living computers run in *parallel*—massively parallel, to be precise.

An excellent example is provided by the construction of the Mandelbrot Set (MS) which, like matter and natural law, involves complex numbers. As discussed earlier, the Mandelbrot Set illustrates the behavior of a complex number, z_0, under the iteration: $z_{n+1} = (z_n)^2 + z_0$

This iteration generates a series of complex numbers, the *Julia* set that, when plotted on the complex plane, give the orbit of the z_0 under the operation.

A serial computer has a single CPU that draws the MS in the following way:

```
Read the complex coordinates corresponding to the pixel top
  left pixel of the screen.
Calculate the Julia set for that number, say up to n = 1,000.
If the magnitude of a result:
Wanders off beyond 2 to infinity, the pixel is left white as
  the number is not in the MS.
Remains bounded around the unit circle, the pixel is in-
  structed to turn black as the members of the MS are those
  with bounded Julia Sets.
Read the complex coordinates of the next pixel and repeat.
```

The sophisticated CPU rapidly examines each pixel in turn, one after the other, and instructs the display according to what happens to the series in the Julia set. This is the basic principle of the serial computer established by Turing. The MS appears line by line as the MS is displayed. If there are one million pixels on the screen, the illustration is of the MS when it gets to the 600,000th pixel.

A massively-parallel computer has one million simple and slow CPUs, one for each pixel. Each CPU can send a message to any of the pixels, but the pixel only turns black if it receives many signals. Each CPU knows its coordinates, and proceeds to calculate the coordinates of each member in the Julia set of its number. It sends a signal to the pixel corresponding to each result (so every pixel receives at least one signal because it is the starting number). As each simple CPU plods through its Julia set, the corresponding pixels are sent a signal. As the million Julia sets are being slowly calculated, the MS slowly emerges, fuzzy at first but with increasing sharpness as all the Julia sets merge and combine with each other. The illustration is a low-resolution example after a few minutes, but the characteristic shape of the MS is already emerging from the overlap of Julia sets.

Human Mind

The question, "What is my mind that expresses itself in my thoughts and body?" has attracted the attention of philosophers—with varying success—and scientists—who tend to duck the question. To contribute to the discussion, we shall now attempt a 'theory of mind' by applying the principles of a unified science we have established.

In unified science, there are only two basic components to material systems—(1) a set of external fundamental entities and (2) this set of fundamental entities is confined and directed by the internal wavefunction.

So far, we have used the convention of dealing with systems as things-in-themselves. Things such as carbon atoms, glucose molecules, RNA macromolecules, bacteria, cells, organs, etc. Again, every single one of them is the same—a set of fundamental particles moving over time to reflect the form of the internal wavefunction. The external aspect does what the internal aspect tells it to do.

This is as true for a human as it is for an atom, the only difference being the number of external particles and the intricacy of the structure to the internal wavefunction.

When this intricate internal structure resonates with an internal structure in the Logos, a set of emergent properties is expressed in the external structure and function of the system. When it does not resonate, the emergent properties are absent and the system is described as broken or diseased.

The internal analog waveform generated on waking has a resonance with the Logos, giving rise to the emergent property of self-awareness of the "I Am" object and its complement of other "I am not" objects. The mind is an emergent property of the intricate structure of the internal wavefunction, not the consequent movement of valence electrons and core nuclei and confined electrons.

This "I Am" structure in the Logos is a hierarchy of complex dimensions so we can assume that the human mind is also a hierarchical construct in an abstract poly-dimensional complex space.

The awake "I Am" with its thoughts and sense of self is the center of an environment. While walking through the woods, I see, hear, smell, and touch the world around me. As each complex dimension has two components—the real and imaginary, magnitude and amplitude—a 3-D world has six components. In order to model the external world, six components must be kept track of. We shall assume that it is no coincidence that the cortex of the human brain has six distinct layers.

It is in these six layers that the two aspects of awareness are projected onto these six layers. Like the massively-parallel method of generating the Mandelbrot Set, each neuron is contributing a tiny ripple to the overall wavefunction that is the mind:

1. The self. This is the 'I am' that inhabits the body. Every recorded memory is being expressed in this analog wave, but at a low level that we are unaware of, i.e., the subconscious. All are contributing a murmur to the sense of self, singing softly "I remember" and it takes little effort to magnify these and bring them into resonance with the "I Am".

2. This abstract entity is embedded in another abstract construct in complex space that represents the environment. It can also wander into self-generated constructs. I can walk through a forest but be totally unaware of the trees as I go over yesterday's argument and what I *should* have said.

As the Creator Parent also has the "I Am" awareness, we can assume that within God's abstract structure there is also a hierarchy of complex dimensions. For pure mathematicians, the implications are that the emergent properties uncovered in the exploration of poly complex dimensions will be intriguing and delightful. Eventually, the existence of God will be seen as inevitable as is the unique prime factorization of the integers. It is inconceivable that it could not be true as proved.

Systematic hierarchy

As with everything we have discussed so far, both the human mind and the nervous system that generates it are a systematic hierarchy. While the experience of a healthy human is that the mind is a single, unified entity and that the environment is also singular, exploration of the brain reveals that aspects of experience are processed in disparate areas of the brain. Each sense has an area it projects onto, and is also broken down to different areas. For example, in the visual cortex, there are areas that respond to color, areas that respond to grey, areas that respond to vertical lines, areas that respond to horizontal lines, curved lines, solid areas of color, etc.

This is called the *binding problem* in brain science. How are all these disparate bits of detail combined into the unified experience of a healthy human being. If we consider each area generating a wave that combines into a unified wave, this is no longer a problem. A cello and a piccolo locally generate waves that contribute to the unified whole that is the symphony. A protein locally generates a wave that contributes to the unified whole that is the water structure and metabolism of a vital unit. The same basic principle applies.

Externally, while the human nervous system is spread throughout the body, all the higher functions are localized, particularly in the brain. But the organ inside our skull is not the only location in the body, we have other concentrations throughout our body, in particular, in the gut and around the heart. They look very different from what's in our heads, but they do the same type of work – regulating the system they are attached to in ways that are responsive to our surroundings.

Gut brain
solar plexus

The brain in our gut, the "enteric nervous system" manages every aspect of digestion, from the esophagus to the stomach, small intestine, and colon. This solar plexus in the abdomen is situated behind the stomach and contains many ganglia distributing nerve fibers throughout the viscera. There is also the heart, which has a complex intrinsic nervous system that is sufficiently sophisticated to qualify as a "little brain" in its own right. The heart's brain is an intricate network of several types of glia and neurons, like those found in the brain proper. Its elaborate circuitry enables it to act independently of the cranial brain – to learn, remember, and even feel and sense. The lowest part of the cranial brain stem is also a part of the gut brain.

While the simplest of animals have basically a gut brain, this is the basic level of the systematic hierarchy of all animal nervous systems. The lineage that led to humans added layer upon layer as the sophistication increased. The systematic hierarchy, with each added layer working through the lower layers, is observed in the human brain structure.

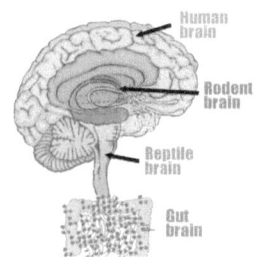

At every level, the basic principle is at work, i.e., analog forms are recalled from digital storage and expressed as internal waves that organize and unify the external particles as a well-functioning body.

EVOLUTION AND EPIGENETICS

The current scientific theory of evolution is called the *modern synthesis* that unites Darwin's concept of survival of the fittest with the 'read-only' digital aspect of modern genetics and its random mutation and alteration.

The analog forms generated by this randomly-altered digital information in the genotype on the DNA are expressed in the phenotype of the body. If a form happens to increase survival and procreation, the DNA enters the gene pool. If it does nothing, it also enters the pool. If it is deleterious, it is purged from the gene pool by the death of the phenotype.

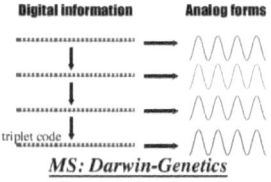

MS: Darwin-Genetics

As we have seen, the workings of a modern computer require that there be an ability to read and write digital information. Both read and write are crucial to the sophisticated manipulation of information.

In unified science, learning about the Logos and writing it to memory for later recall, underlies evolutionary development exploring a phase space. This requires that both writing and reading are possible at every level.

While learning requires memory, memory does not imply learning. The computer is an example. It has a systematic hierarchy of memory, from active memory that is constantly changing, short-term memory in buffers and registers, medium-term memory in virtual images, and long-term memory when data is written to disk.

With this in mind, we can expect to find that living systems have various levels of memory from short-term active memory on RNA to deep-time storage on DNA. In the early days of evolutionary thought, Charles Darwin became associated with the concept of *random variation* underlying evolution, while Jean-Baptiste Lamarck was associated with *accumulated learning* underlying evolution. Lamarckism implies that writing to digital memory must exist alongside reading from digital memory.

Memory
Active
Short-term
Medium-term
Long-term

While this *writing to disk* is absent in the Fundamental Dogma of genetics, the new and burgeoning science of *epigenetics* is explicitly exploring this aspect of the writing of digital memory in living systems. In a unified science, the theory of evolution is a postmodern synthesis of Lamarck and epigenetics.

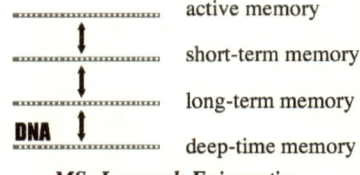

MS: Lamarck-Epigenetics

Epigenetics

While epigenetics is now so well-established as to have a recent Nova episode[186] devoted to it on PBS television, it is probably a field that is unfamiliar to most people. In many ways, it can be considered the reemergence of Lamarckism in a much more sophisticated form.

> "For years, genes have been considered the one and only way biological traits could be passed down through generations of organisms. Not anymore. Increasingly, biologists are finding that non-genetic variation acquired during the life of an organism can sometimes be passed on to offspring—a phenomenon known as epigenetic inheritance. An article ... in the July issue of The Quarterly Review of Biology lists over 100 well-documented cases of epigenetic inheritance between generations of organisms, and suggests that non-DNA inheritance happens much more often than scientists previously thought."[187]

The 'central dogma' of molecular biology is that there is a one-way flow of information from the genotype—the genes and DNA sequence—to the phenotype—the proteins and the result of protein action—i.e., the development and eventual form and function of the body. It is upon this central dogma that the whole of Darwinism is constructed since, as there is no 'back-flow' of information from the body to the genome, the only changes allowed in the genome are random mutations, random rearrangements, and other such random occurrences for natural selection to go to work on.

This dogmatic assertion, so fundamental to Darwinism, is clearly up for revision. Note that the preeminent proponent of materialistic Darwinism, Richard Dawkins, assumes in his many works that all is now understood of the basic principles of evolution premised on random mutation and variation.[188] As he does not, however, mention epigenetics even once in any of his writings, by this fact alone he is condemned to have only a partial view of the truth; the classic mistake of the blind man confusing his odiferous grasp of the elephant's tail with the whole beast. Richard Dawkins is not unique in this respect; this premature assumption of complete knowledge happened to many elder statesmen in physics just a century ago:

> "It seems that every so often, a fairly large group of scientists begin to assert that science is just about complete, that the vast unknown is gone, and that all the really major research can stop because we now know everything except the details. For those who fall under the spell of this sort of belief, be aware that a similar belief seemed to have taken hold at the turn of the last century. This was just before Relativity and Quantum Mechanics appeared on the scene and opened up new realms for exploration.... 'The more important fundamental laws and facts of physical science have all been discovered, and these are now so firmly established that the possibility of their ever being supplanted in conse-

quence of new discoveries is exceedingly remote.... Our future discoveries must be looked for in the sixth place of decimals.'- Albert. A. Michelson, speech at the dedication of Ryerson Physics Lab, U. of Chicago 1894. 'There is nothing new to be discovered in physics now. All that remains is more and more precise measurement' - Lord Kelvin, 1900."[189]

Just as in physics—where the advent of relativity and quantum mechanics punctured this 'we know it all' attitude—so the advent of epigenetics has the potential to puncture the biological 'we know it all' attitude prevalent in current Darwinism as exemplified by Richard Dawkins in all his writings. One can only feel sorry for Dawkins as the dustbin of history is not a comfortable place for one so arrogant.

The first hint that the one-way "central dogma" of Darwinism was wrong came when it was noticed that the identical genetic defect in the human genotype had very different effects on the phenotype depending on whether the faulty gene was inherited from the mother or the father.

> Even though both parents contribute equally to the genetic content of their offspring, a developmental process called genomic imprinting sometimes leads to the exclusive expression of specific genes from only one parent. This process was first described in 1984, when two laboratories discovered a mark, or 'imprint,' that differentiates between certain genes on the maternal and paternal chromosomes and results in the expression of only one copy of those genes in the offspring. The genes in imprinted areas of an organism's genome are expressed depending on the parent of origin.[190]

This phenomenon was eventually traced to a pattern of chemical alterations—methylation of the cytosine bases—imprinted on the structure of the DNA. Here the DNA was acting as the substrate for a layer of information to be written on. This has nothing to do with the base sequence itself—the genetic code—it is defined as a level of epigenetic information impressed on the genetic level.

Epigenetics and Lamarckism

Even more dramatic examples that violated classical Darwinism were soon uncovered.

> "Toward the end of World War II, a German-imposed food embargo in western Holland—a densely populated area already suffering from scarce food supplies, ruined agricultural lands, and the onset of an unusually harsh winter—led to the death by starvation of some 30,000 people. Detailed birth records collected during that so-called Dutch Hunger Winter have provided scientists with useful data for analyzing the long-term health effects of prenatal exposure to famine. Not only have researchers linked such exposure to a range of developmental and adult disorders, including low birth weight, diabetes, obesity, coronary heart disease, breast and other cancers, but at least one group has also associated exposure with the birth of smaller-than-normal grandchildren. The finding is remarkable because it suggests that a pregnant mother's diet can affect her health in such a way that not only her children but her grandchildren (and possibly great-grandchildren, etc.) inherit the same health problems.

"In another study, unrelated to the Hunger Winter, researchers correlated grand-parents' prepubertal access to food with diabetes and heart disease. In other words, you are what your grandmother ate. But, wait, wouldn't that imply what every good biologist knows is practically scientific heresy: the Lamarckian inheritance of acquired characteristics?"[191]

In this case, the epigenetic information involved chemical tagging of the his-tones, the protein 'spools' on which the foot-long DNA molecules are wrapped around to keep them manageable. This is an image of how histones and DNA combine:[192]

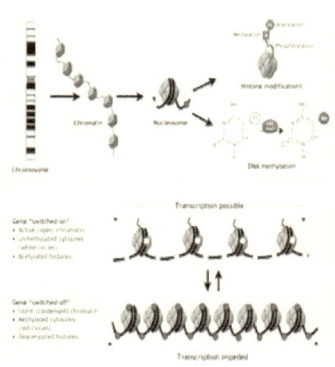

Reversible and site-specific histone modifica-tions occur at multiple sites through acetyla-tion—replacing a hydroxyl with an acetyl group—of the histone proteins. It would seem—and this is currently an active area of research—that there is a connection between the epigenetic information writ-ten on the DNA and that written on the histones working in complementary direc-tions: Methylation of DNA turns it off while acetylation of histones turns them on. It seems that methylated DNA on non-acylated histones is hard to unwrap—so its information cannot be easily accessed—while un-methylated DNA on acetylated histones is easy to unwrap and its information is more easily accessed.

This diagram is a summary of what is cur-rently known about the mechanism of epigenetic inheritance:[193]

This mechanism of storing information about the current state of the organism is now well-established; there are probably other mechanisms at work as well.

Most of the investigations into epigenetic mechanisms are currently focused on medicine and the state of disease, such as cancer, etc. There has not been much work on how this field impacts the mechanisms of evolution but it is clear that a new principle is involved.

Information about the current state of the organism is imprinted on the genetic heritage and can be accumulated over the ages as it is passed on down a lineage.

"The field of epigenetics has gained great momentum in recent years and is now a rapidly advancing field of biological and medical research. Epigenetic changes play a key role in normal development as well as in disease. The editor of this book has assembled top-quality scientists from diverse fields of epigenetics to produce a major new volume on current epigenetics research. In this book the molecular mechanisms and biological processes in which epigenetic modifications play a primordial role are described in detail.... The final chapter describes the fascinating potential transfer of epigenetic information across generations."[194]

Recombination and Sex

Recombination, or 'crossing over,' as it is otherwise called, occurs in the generation of the sex cells where two copies of a paternal chromosome and two copies of a maternal chromosome (i.e., 8 strands of DNA) entangle and crossover their genetic material.

> Chromosomal crossover (or crossing over) is an exchange of genetic material between homologous chromosomes. It is one of final phases of genetic recombination, which occurs during prophase-1 of meiosis in a process called synapsis. Synapsis begins before the synaptonemal complex develops, and is not completed until near the end of prophase-1. Crossover usually occurs when matching regions on matching chromosomes break and then reconnect to the other chromosome.[195]

It is this mixing of the genetic material that is at the heart of sexual reproduction and, since only sexual species evolve and spin off daughter species Sex can be considered to have a central role in evolution.

In Darwinism, this process of breaking and reconnecting the DNA is considered random even though it is well-established that there are 'hot-spots' (where crossing over occurs with a high frequency) and 'cold spots' (where crossing over never occurs).

Epigenetics and Recombination

Evidence is accumulating that there is a link between 'short term' epigenetic information and recombination with its long-term consequences. This is from the report of a biology convention in 2006:

> "Carmen Sapienza (Temple University Medical School, Philadelphia, USA) reported that imprinted regions in humans are historical hotspots of recombination. Together with specific DNA sequences, epigenetic factors may have an important influence on the rate of meiotic recombination and the position of cross-overs. Using in silico and in vitro analyses, Sapienza's group have shown a relationship between increased rates of meiotic recombination and genomic imprinting. Imprinted regions showed more linkage disequilibrium, and had a significantly higher number of small haplotype blocks [passed down without recombination}, than the non-imprinted regions. Their findings suggest that several factors, including both specific DNA sequences and epigenetics, are involved in controlling meiotic recombination in humans."[196]

Other groups have also established connections between epigenetics and the rearrangement of the genome in recombination including the arabidopsis plant,[197] in humans[198], in the centromeres that control the structure of the eukaryote cell[199], and in the recombination that underlies the antibody diversity in the immune system.[200]

Why sex?

One of the 'open questions' in modern biology is: "Why sex?" The overwhelming preponderance of sexual reproduction in multicellular organisms is a puzzle because asexual reproduction is so much more efficient at generating progeny.

It is a well-known fact that, while Darwin titled his epochal work, *The Origin of Species*, he did not actually propose any mechanism for the emergence of new species. The ideas he proposed, at best, dealt with the origin of races within a species, not new species themselves. To this day, there is no consensus as to how this happens other than a process of gradual divergence and gradual infertility between races.

This does accord with what is known, however, as illustrated by the human race. There has certainly been a lot of epigenetic learning and writing to genetic memory in the many tens of thousands of years since the first humans emerged in Africa. These are the human variants we call races.

> The first theory, known as the 'Out of Africa' model, is that Homo sapiens developed first in Africa and then spread around the world between 100 and 200,000 years ago, superseding all other hominid species. The implication of this argument is that all modern people are ultimately of African descent. The other theory, known as the 'Multi-regional' Model, is that Homo sapiens evolved simultaneously in different parts of the world... Although the debate is far from concluded, it is probably fair to say that the bulk of scientists support the 'Out of Africa' hypothesis and believe that all humans share a common origin.[201]

Examples of the innovations expressed in the emergence of the human races are the ability to digest milk through adulthood (a rarity in the stay-at-home Africans; common in Europeans) and the loss of UV-protecting-but-vitamin-D discouraging melanin in the races in sun-deprived northern latitudes. For all these epigenetic and genetic changes, however, the ability of Black Africans and White Europeans to interbreed is in no way diminished.[202] In fact, a quite-opposite phenomenon is firmly established in biology: that of hybrid vigor: "An increase in the performance of hybrids over that of purebreds, most noticeably in traits such as fertility and survivability."[203]

In the theory presented here, the epigenetic-directed recombination of genetic material is the key mechanism of speciation and resultant reproductive isolation. Evidence that this might be correct is to be found in the rather odd sequence of events leading up to the formation of the haploid sex cells (with one set of chromosomes) from the diploid germ cells (with two sets of chromosomes). One obvious reason for this haploid-diploid alternation is to prevent a buildup of chromosome number that would happen if the sex cells were diploid—the children would have four sets, the grandchildren eight, the great-grandchildren sixteen, etc.

The obvious way to get two haploid cells from a diploid cell would be to have a regular cell division (mitosis) that skips the chromosome duplication step. This is not the case. The formation of the sex cells (meiosis) adds a seemingly unnecessary step that just adds to the workload. First, the two sets of chromosomes—the paternal set and the maternal set—are duplicated. The cell now has four sets of chromosomes! These all commingle into what is called the tetraplex or synaptic complex[204]—the stage when recombination and reorganization of the genetic material occurs. The four sets of chromosomes are now progressively reduced to one set by two rounds of cell division to create four haploid sex cells.

Current biology has no good rational for this complicated way of doing things as recombination is considered to be random chance-and-accident. In the perspective developed here, however, this abundance of chromosomes hints at some currently-uncharacterized mechanism for the directed reorganization of genetic material while also ensuring that the new daughter species can be 'brought to term' successfully by the mother species.[205]

If epigenetic-directed recombination turns out to be at the heart of speciation, it would provide a simple answer as to why almost all species are sexual: only sexual species can evolve, only sexual species can give rise to new species and more sophisticated organisms. The adoption of the asexual mode of reproduction, while advantageous in the moment, is an evolutionary dead end.

> The evolution of traditional, female-only asexuality typically leads to a swift extinction. We know this because although such species frequently evolve, they don't stay around for long. If you look at the tree of life, female-only asexual groups are all out on the twigs: there are no great asexual lineages equivalent to fish or birds. Instead, the asexual groups are a few species of snail here, a dandelion there.[206]

Without sex, the highest form of life would be the simple unicellular forms that predominated the first billion years of life on earth and there would have been no Cambrian Explosion of multicellular forms and certainly no humans.

Unification Thought puts sexuality at the very center of human life (and the Fall). If this perspective has any validity, it would seem that Molecular Biology has sex as the dynamo of evolution.

Memory and aging

In the perspective we have developed, DNA is a long-term digital storage medium for analog memories. RNA, being more labile, is for short-to-medium storage. If this is so, then the DNA sequence of, say, the aged human brain, should be different from the sequence from aged cheek cells. If these were just mutational errors, these differences should be randomly scattered. If, however, they are a result of accumulated memories, they will be localized in distinct areas. This is akin to the paradox in the Modern Synthesis that states that 'mutations' are random events that are confined to 'hot spots,' while the conserved 'cold spots' show very little change. The changes are clearly not random, as the living system is controlling where they are occurring.

It has been said that the changes in the DNA sequence of modern humans are too great to have occurred if there were a single First Human pair ~100,000 years ago from which all humans are descended. In the Modern Synthesis, this would necessitate a mutation rate so great that it would be inimical to any continuity. For in the Modern Synthesis view, for every random change that is 'fit to survive,' there are many more that are not.

In the view developed here, these changes are not random, they are 'ancestral wisdom' accumulated down a lineage.

*****

IMPORTANT NOTE

The following section is NOT acceptable to the Unification Thought institute and must be considered to be the conclusions of this author alone. It should not be considered the current consensus as to the Unification Thought understanding of the Fall of Man. Unification Thought has no definitive views of the origin and original good role of angels in God's creation, or of the individual with whom the first woman, Eve, had an illicit sexual relationship—a situation that arose only because the first man, Adam, did not obey God's Commandment. Neither Divine Principle nor Unification Thought is explicit about the content of this parental injunction or who conveyed it to Adam. The spiritual evil caused by the Fall was passed on to the next generation, where good and evil were symbolically separated in Abel and Cain, respectively. The good was killed and evil triumphed, beginning the evil history of mankind.

I want to make it clear that the ideas presented in the following section were not warmly received when recently presented at a UTI symposium in Tokyo, 2005, or when they were published earlier. In the 1980s, this author was denounced for teaching heresy in my church newspaper-column published in 1987 when I first started to grapple with the task of harmonizing the religious and scientific views of humanity's origins. The charge was presented at a planning breakfast for an upcoming ICUS conference presided over by Reverend Moon. His response to the charge, as reported by Professor Morton Kaplan of the University of Chicago, was: "That's fine. Let a thousand thoughts collide, the truth will eventually win out." This perspective is inherent in the foundation of the scientific method, and Dr. Kaplan was ecstatic that Reverend Moon embraced it.

It is this author's opinion that the following section embraces two fundamental principles of Unification Thought that are ignored in the current consensus:

1.   The Logos is expressed sequentially in distinct steps over time. As angels are a step below humans, they should have emerged in Creation just before Man. The only assistance that God needed in fulfilling the purpose of creation was the birthing, weaning and protection of the first humans in Africa as they grew up toward maturity. Neither God nor purely spiritual beings were capable of this physical task. Physical bodies are required. From my own experiences of the fabulous beings called angels throughout history, I opine that these angelic beings are Individual Images given form by God in the spiritual realm when required. They are not unique individuals that have existed since the Big Bang. Evil spirits were created by evil lives lived on earth and, as a collective, comprise Satan, the enemy of God.

2.   The elements of the human spirit—both good and evil—are created by the actions of the physical body. The sexual act of the Fall and the evil elements generated require, in Unification Thought, that physical, not spiritual, bodies were involved in the sexual act that created evil.

So the following section must be considered very controversial and, at best, these ideas should serve as a stimulus for vigorous debate. By no means are they to be taken as scripture or, that is, as definitive. They are only a starting point for scientific exploration.

Rather than delete the following, as requested, I shall retain it, with the above caveat, to expressly honor the request to "let a thousand thoughts collide." I have, in my humble opinion, assembled a plausible science of a theistic cosmos. In these pages I have covered mathematics, Dirac Relativity, the Quantum Wavefunction, Dark Energy and Spirit Realm, the Big Bang, cosmology, the nature and origin of Life, digital genetics and analog life forms, evolutionary theory and brain science—and have dealt with all of it in a self-consistent way. The following controversial section emerged only at the conclusion of this 500-page attempt to suggest a unified view of the Cosmos. So, I am retaining this section with the warning that I, and I alone in 2014, think that what follows is vaguely similar to what will be the consensus when the magisteria of science and religion finally merge into one. I also welcome well-reasoned critiques of what now follows.

****

ORIGIN OF MAN

Step-by-step, the Logos is sequentially expressed in living systems over time, passing through what can be colloquially called the Age of Bacteria, the Age of Protists, the Age of Worms. the Age of Fishes, the Age of Reptiles, the Age of Mammals, and the Rise of the Primates. An overview of this sequence was given at a UTI symposium in Tokyo.[207]

The focus here is on he final stage, when the Logos reaches its complete expression in the birth of the first human pair, Adam and Eve. Unification Thought states that the final stage of becoming one with God is a creative act of human will, the fulfillment of human responsibility as humans mature.

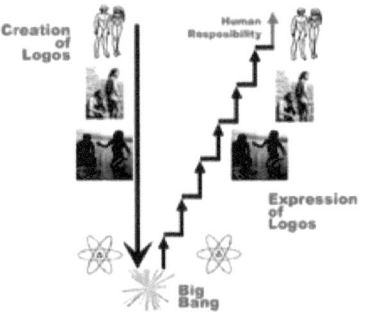

The exact sequence is still a matter of debate so, for convenience, we will equate the prehuman stage with the Neanderthals.

Materialism has that the transition from Neanderthal to human was a gradual event over an extended period. The science of Godism, as presented here, proposes that there was a specific speciation event in which epigenetic information, gleaned from the Logos via the environment, directed a reconfiguration of the Neanderthal genetic material into the human configuration. A new set of qualities was inherited from the Logos, including the knitting together of the mind into the "I Am" sense of self.

I discussed a plausible mechanism of universal speciation directed by the accumulated epigenetic information in *Volume One*.[208] The speciation event always involves a male and female, and the result was the birth of Adam and Eve, as the first humans are traditionally called. Classical science states that the transition from Neanderthal to Human was caused by random variation and survival of the fittest. A modern spokesperson for this view is Richard Dawkins. Unified Science proposes that speciation occurs by a specific mechanism directed by accumulated ancestral wisdom inherited from Logos and passed down a lineage.

Abilities that for the Neanderthals were learned with great effort became hardwired into humans. The mental abilities of a mature Neanderthal were comparable to those of a six-year-old human, and they could develop no further. In yesterday's psychology, a Neanderthal would have been classified as an *imbecile* (above an *idiot* but below a *moron*). There is evidence that various races of Neanderthals, specialized for various environments—lakes, forest, savannah, etc.—came together and 'pooled their wisdom' in the lineage leading up to humans.

Neanderthals probably communicated with a simple pidgin of nouns and verbs, had mastered fire, and were successful gatherers and hunters wielding crudely-chipped lumps of flint rock.

The time and place in which the first humans were born is traditionally called the Garden of Eden and science has a rough idea of when it occurred—the transition from the Paleolithic age (Old Stone Age) to the Neolithic age (New Stone Age).

The Paleolithic

We will equate the period in which the Neanderthals were the most sophisticated of the primates with the paleolithic age, which lasted about 2,500,000 years. The cultural changes over this vast stretch of time were very slow and incremental, with the controlled use of fire appearing just 400,000 years ago.[209] The sophistication of the stone implements, for example, hardly changed over millions of years. Stasis was the rule.

Archeological evidence exists that shows that they buried their dead and eventually had mastery of fire for cooking. They were communal (clan and tribal level) and probably had a pre-language (pidgin) of simple nouns and verbs. They fashioned simple stone and bone tools, and were successful hunters and could fend off predators such as the great cats. The fossilized Laetoli footprints left in 3,000,000-year-old volcanic ash—footprints of a male, a female and a child[210] pre-Neanderthal—suggest that pair-bonding reproduction was already established at this early stage.

The Neolithic

This million-year stasis was ended by the emergence of Man and the start of the neolithic age about 100,000 ago, and was in full swing about 50,000 ago. The stone and bone shaping of tools was much more sophisticated and decorated. They had a true language of syntax and grammar. The hunter-gatherer stage developed into that of agriculture and the domestication of animals >20,000 YBP and, most distressingly, the earliest evidence of a battle is 14,000 YBP.[211]

Habitations beyond caves were developed. The discovery of how to smelt copper from its ores and how to create its alloy, bronze, marked the end of the neolithic and the start of the Bronze Age ~15,000 YBP. Writing was developed soon after. Unlike the million-year stasis of the Old Stone Age, innovative change over thousands of years was the rule in the New Stone Age.

The location of the Origin of Man, the *Garden of Eden*, has been roughly established by three lines of evidence that are all in essential agreement. These are the study of the spread languages, the study of the female lineage using mitochondria, and the study of the male lineage using the Y-chromosome.

Female lineage

The history of the female lineage is tracked by tracing the spread of genetic markers on the mitochondrial chromosome, which is passed down the female lineage from mother to daughter. The mitochondria are not passed on by males. If a mitochondrion does make it from the sperm into the egg, it is immediately surrounded and destroyed.[212]

The pattern of human migration that emerged from these studies indicates that it started off in East Africa, then humans spread south into Africa and north to the rest of the world.[213] This original female is called "mitochondrial Eve" in the literature.

Male lineage

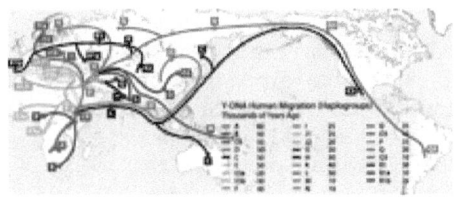

The history of the male lineage is tracked by tracing the spread of genetic markers on the Y-chromosome which is passed solely down the male lineage from father to son, and is not passed on to females. The pattern of human migration that emerged reveals that it started off in East Africa, then humans spread south into Africa and north to the rest of the world.[214] This original male is called the "Y-chromosome Adam."

The materialistic view of gradual speciation posits that the mitochondrial Eve and the Y-chromosome Adam were members of a "small breeding population." The unified view of a directed, specific mechanism for speciation affirms that this population was as small as two.

Linguistics

The study of how language has changed over time as humans migrated locates the origin of language in East Africa, from where humans spread north to the rest of the world:

> "A new linguistic analysis attempts to rewrite the story of Babel by borrowing from the methods of genetic analysis – and finds that modern language originated in sub-Saharan Africa and spread across the world with migrating human populations."[215]

All three lines of investigation suggest that the eden into which the first humans were born was in East Africa less than 100,000 years ago. Genetic analysis of the genes for skin color indicate that the first humans were black, and that the yellow and white pigmentation arose much later as human migration progressed.[216]

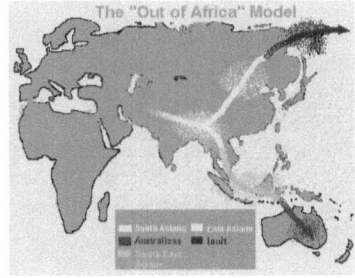

There is evidence that different races of Neanderthals commingled on the route to Humans, including races adapted to water as well as other races adapted to forest and savanna.[217] The advantage that occurs in such 'miscegenation' is known as 'hybrid vigor' in practical genetics.[218]

On purely esthetic grounds, I like to think that the Neanderthal tribes gathered, interbred, and gave rise to humans in East Africa's most dramatic and bountiful landscape between Mt. Kilimanjaro and the Great Lakes.

Complete Expression of Logos

Unificationism rejects the idea that the first humans were created fully formed, and accepts that they emerged much as children emerge today. Basic physiology was then as it is now, so we can infer much about their situation.

Putting this altogether, we conclude that the 2nd stage of Creation, the expression of the Logos, was completed with the birth of the first two humans, and their natural development to about 6 years old and to the Ne-

Now, let me talk about the history regarding Adam's birth. Did Adam have a belly button or not? You must know it. Without a belly button, where was he born from? Adam had a navel cord, and he had a mother.

Sun Myung Moon, The Blessed Family, 1999 winter issue

anderthal level of personality. Their development from then on to maturity, the ability to love as God loves, was their own creative responsibility.

Physiology

We can assume that Adam and Eve were born into the midst of a flourishing and supportive tribe of Neanderthals. They had a biological father and a biological mother and, furthermore, as these parents were at the highest level of Neanderthal development, we can assume they had a high status in the tribe, perhaps even being the leaders of the tribe.

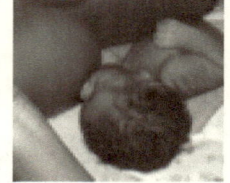

As is normal for all human beings, the first two humans emerged as helpless newborns and they were fed at their Neanderthal mother's breast.

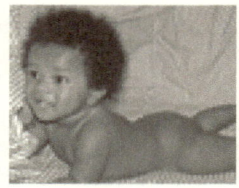

As they grew, they were protected from predators and provided with food. They played with Neanderthal children, and were taught the ways of the tribe. They were similar to their peers in their capacities until they were past the age of six. Then they left their peers far behind them.

In particular, human children have the innate capacity, inherited from the Logos, to create a true language out of a pidgin, i.e., to create a language with grammar and syntax.[219] So Adam and Eve would have created their own language, a level of sophisticated communication unavailable to the Neanderthals, the 'ur-language' as it might be called. The first humans were two golden children maturing in a world of kind and supportive imbeciles.

Original Plan

PERFECTION	Image of God
COMPLETION STAGE	Unconditional love / Divine spirit
GROWTH STAGE	Mutual love / Life spirit
FORMATION STAGE	Self-love / Form spirit

Adam and Eve were to grow to perfection, to love as God loves, with unconditional parental love for all. They were to grow through the stages of their scope and capacity to love, using their free creativity until their love was mature.

To guide them along this path without mishap, the Bible parable states that God gave a commandment to Adam and Eve, and their responsibility was simply to follow this guidance. The Commandment could have been as simple as: *Stay at home*, and it would have been quite natural for this admonition to come from the male parental Neanderthal. This is the principle behind the adage: *Two's company, three's a crowd* and why, in this age of sex abuse, priests and teachers are advised to always have another person present. It is the difference between a private and a public relationship. Indirect support for this simple Commandment is given in an early systemization of Unificationism when discussing how Lucifer's love-to-lust occurred: *If Adam had watched over Eve more closely and spent more time with her, this would not have happened.*[220]

In this balanced situation, the male parent of Adam and Eve would be dazzled by the brilliance of his children and nothing untoward could occur, and the purpose of Creation would have been completed with the uniting of Adam and Eve in True Love.

To put it simply, the Commandment was to keep them out of trouble, to keep them safe. In this situation, their characters and ability to love, their spirits, would have naturally grown from self-love, through mutual love to unconditional love, God's love. From being self-centered spirits, through what Unification Thought calls Form and Life spirit stages, to becoming Divine spirit. They would live out their days on earth, raise a true family of children and grandchildren, and then, discarding their physical bodies, pass into the spirit world where they would spend the rest of eternity with God in the realm of True Love, the kingdom of heaven.

With Adam and Eve raising their family in true love, there would be no more necessity of the Commandment since Adam and Eve were quite capable of keeping their children out of trouble.

The human race would multiply and develop true love culture as the natural leaders of the Neanderthals, who would greatly prosper under their loving and creative care. This is the dominion of true love humans were meant to exercise over Creation, and is the fulfillment of the Third Blessing.

Domestication of hominids

One rather surprising implication of this view is that the first animal to be domesticated would be the Neanderthal, not the dog. The Neanderthals would be the natural servant class, while the true love humans would be the natural aristocrats. With plenty of leisure time, humans would rapidly develop agriculture, science, and an abundance of art.

The image of a natural, hereditary aristocrat and servant class tends to be associated with the endless examples of unnatural slavery in human history. The images of White master beating Black slave, or Japanese overlord and Korean underdog are two not-so-distant examples of attempts to enslave human beings. The true love relationship is quite different to these shudder-inducing examples from fallen history.

The relationship of Human lord and domesticated Neanderthal is the ideal portrayed on American TV by the relationship between Joe and his dog Lassie, and the Lone Ranger and his horse Silver.   The image is not master-beating-slave but rather, "Dogs who talk, do dishes and can be toilet-trained."

A 'slave rebellion' would be as unthinkable as is a pedigree-dog revolt.   The true love culture would fill the earth, and then spread out to all the galaxies. We can expect that God included pathways between the stars for this expansion, and one possible way was discussed in an earlier UTI paper on physics.[221]

An Aside on Astrology

Most of this essay was written in the 1980s.

Poor President Reagan! The 40th President of the USA, and forty is usually considered such a lucky number in history. It seems our Ronald is having to weather storm after storm these days—such a difference from those early triumphant years. Hardly anybody seems to like him these days, although I have a feeling that by this time next year we will all miss him sorely. The latest goad with which the press torment him is that he and Nancy are being mocked for taking cues from astrology.

We can assume that much of this mockery derives from a politically motivated desire to discredit. As this paper is supposed to be mute about political issues, I will pass over this source of criticism.

There are, however, at least two other perspectives that cast doubt on the wisdom of heeding astrology: Science and Religion. To my mind, from a certain perspective, these rather strange bedfellows have much the same comment to make about astrology. Let's take a look at the religious perspective first.

Religion and Astrology

An almost universal concept in higher religions is that God created man to be the ruler of the physical world. It does not make sense, therefore, that man would be objective to, and ruled by, the planets. The idea of planetary influence has to be thrown out for this reason.

Another of the basic ideas in astrology that has more going for it, is that you can predict much about the characteristics of a person from the positions of the planets at the moment of their birth. This actually makes some sense, even though, as you will see, we will have to turn some concepts upside down.

A human being is created in the image of God while the rest of the universe is created as a partial reflection of God. Man is not just physical material; he also has a spirit. Just as the physical body is constructed in the womb, the Divine Principle teaches that the spirit is also coming together during this time and, at the moment of birth, with the first breath, the spirit completes itself and becomes eternal.

So here we have a situation where, at the moment of birth, a human being becomes an eternal being, a truly human being, reflecting God. God's image is fully expressed there (in potential, actually, but that's another topic) at that moment. But at that same moment, the rest of the universe is also a (partial) reflection of God's image. This means that the universe is also a partial reflection of the child at the moment when the child first becomes like God. We can speculate that people throughout the ages noticed the correlation between the planets and birth, and went on to develop the familiar theories to explain it, Voila! Astrology.

While it does not make sense to say that the planets 'influence' the newborn child, it does make sense to say that the universe is proclaiming the special qualities of the child born at that moment. There is a sound theological reason for this: The parental heart of God wants to celebrate the birth of His children—wouldn't you have designed things that way if you were a cosmic "mush-heart" parent?

That's why you can proudly say that the planets proclaimed the glory of your birth, just as they have for everyone else.

If religion can give a reason and purpose for the phenomenon, perhaps science can come up with an explanation for it, a better one than "we are ruled by the planets."

Science and Astrology

Our contemporary science seems to have little good to say about astrology. The central tenet of astrology is that the positions of the planets influence human behavior. In our science, the only interaction that both humans and planets respond to is gravitation. The gravitational forces of the planets on us, however, are extremely small. If the planets did influence us through the gravitational force, the much greater gravitational effects of local mountains and oceans could then be expected to have an even greater influence. If this were so, we would expect that people would have noticed it and developed 'geology' instead of astrology.

This is not to say, however, that science proves that astrology is bunk. Science can state that the astrologers have got their theories that explain their observations all wrong (unless there is some totally unknown interaction yet to be discovered by science—unlikely, but not impossible). A good historical illustration of this situation is that while the theories of the alchemists were way off, their descriptions of how one material could be transformed into another became the foundations of modern chemistry.

Unfortunately, I am not aware of any scientific study of the relationship between the predictions of astrology and what actually transpires. One thing is apparent, however, and that is that many people in many different cultures have placed a lot of confidence in the capabilities of astrology over thousands of years. As you are probably aware, you are not alone, Nancy.

So let's give all these people the benefit of the doubt and assume that astrology is describing a real phenomenon (even though its theories as to how it works might be wrong) and that the positions of the planets at birth do correlate with peoples' dispositions and fortunes. Assuming this, is there anything in modern science that could begin to encompass and explain the phenomenon?

Least Action

Even though we have gone through a tremendous revolution in physics this century, most scientists still think in terms of 19th century concepts of particles and forces. Astrology could never fit into that structure. There is, however, a way of thinking that is gaining hold within the mainstream of modern science that can encompass such speculative proposals as astrology. This is the "Principle of Least Action," which—even though it has only gained greater acceptance in this century because of its perfect fit with quantum mechanics—first appeared in the seventeenth century as an alternative formulation of Newton's laws of motion.

This action principle (or 'action formulation' as the scientists put it) is universally applicable: It applies to every system studied by science—classical physics, quantum mechanics and relativity included. An excellent overview of the action principle in modern science can be found in the chapter "Where the Action is Not" in the book "Fearful Symmetry: The Search for Beauty in Modern Physics" (Macmillan Pub.) by Dr. Anthony Zee. Incidentally, the formula describing the workings of the whole universe in that column was an action formulation: No other description is so succinct as to explain the universe on a table napkin—a small one at that.

The action principle is such a powerful tool and so universal in its applicability that Dr. Zee comments, "Some physicists would like to believe that the Ultimate Designer thinks in terms of action."

The principle of least action is very simple to state (and complex to calculate). For any system changing from one state to another, consider each and every way in which the change could happen. For instance, the gravitational change of falling off a wall. For each of these ways, calculate a number—there are no units involved such as miles or seconds—called the 'action' using the formula for gravitational change. The fascinating thing, and no one knows the reason why (and some scientists have waxed quite metaphysical about it), is that the change that actually happens is always the change that has the smallest number—the lowest action—associated with it (hence the name).

The great challenge of using the Principle of Least Action is, of course, being able to figure out the appropriate formula needed to calculate the action. Some have already been figured out, such as the formula for gravity and the formula for the nuclear force. Others are still unknown. Once you know the correct formulas, however, it becomes possible to deal with complicated situations, because all you have to do is add up the numbers—simple math still has a place in modern physics! For instance, consider a change that involves both gravity and electromagnetism. First, figure out all the ways the change could happen. For each way, apply the gravity formula to get one number, apply the electromagnetic formula to get another number, and add the two numbers to get the total action. The way of change with the lowest total action will be the one that actually happens. Simple.

It is an aid to humility to remember that, for all its successes, our contemporary science knows the appropriate formulas only for simple cases, but there is every reason to suspect that the same principle holds true for something as complex as childbirth.

A Common Factor

How does this relate to astrology? Is there a factor whose formula appears in both the action for planets and the action for what happens to the child at birth? If so, then we could expect a correlation between the two.

One excellent contender is gravity. Now, I know that I just wrote that gravity was out of the question, but up there I was talking about the *force* of gravity, now

I'm talking about the *principle* of gravitation, a formula describing gravity. While the force is something between two individual objects, the principle is a mathematical construct that appears in the equations for calculating action.

It is obvious that the gravitational formula appears in the action formulation of planetary motion. But does the principle of gravity appear in the action for childbirth? I think it probably does for the following reasons. There are clear rhythms, sort of built-in clocks, in all living things. Naturally enough, we humans have them as well, including daily, monthly, and yearly rhythms. Now, to be of use, these internal clocks have to be 'in sync' with the rest of the world (as any frequent flyer will testify to their clocks getting jet-lagged out of sync.). Whatever the total action is that governs these clocks, we can expect that one of the components will be the gravitational formula, as it is this formula which governs the external rhythms with which the clocks have to sync with.

In the womb we are relatively isolated from the rest of the universe, the womb being the total universe to the developing child. At birth, however, our rhythms have to sync up with the rest of the world—a situation that could be considered similar to nine months' jet lag. This adjustment process of the internal clocks has an associated action that also includes the gravitational formula.

We have the two separate phenomena of planets and one of the happenings of childbirth, both involving the gravitational formula in some way. So, it would be no surprise if there was a correlation as witnessed by the astrologers.

Please understand, however, that I'm not saying that this is the way it is, but that it is a plausible, scientific explanation of the phenomenon observed by the astrologers that has implications that can be tested—something scientists insist on.

So, there you have it. As I said at the start, from a certain perspective, both religion and science have much the same to say about astrology: The phenomenon might well be real but the astrologers have their explanations all wrong. So, Ron and Nancy, I can only hope that these speculations help soothe the sting of criticism: Perhaps you are not as far off the mark as your tormentors would have us believe.

What Went Wrong?

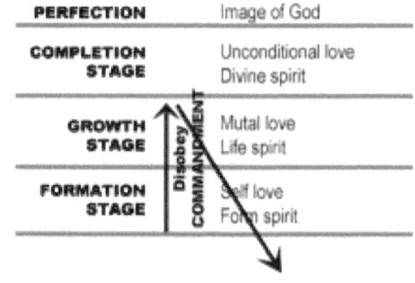

This loving, leisurely civilization never emerged. Instead a brutish culture developed where there was no true love and all humans were out for themselves. Family dysfunction was the norm—see *King Lear* or all of the classical Greek plays for examples. It was a culture into which "Thou shalt honor your father and mother" had to be injected as a divine revelation. Life was nasty, brutish and short, and the spirit of those times has been chillingly recreated in a recent movie featuring slaves and human sacrifice. [222]

Unificationism states that there were two stages in the Fall of Man:

1.   Adam did not obey the Commandment. In fact, the Bible relates that he went off naming all the animals. He did not stay home, and the father-Neanderthal was left with only Eve to fall in love with.

2. Lucifer was twisted into Satan by the misdirected power of love, and in this state had a sexual relationship with Eve.

Because Adam disobeyed, the Neanderthal male parent, called Lucifer in later ages, was left alone with an enchanting, beautiful young woman, Eve. This was an unbalanced situation and Lucifer's love for Eve became twisted into lust. His mind was filled with things that had nothing to do with the Logos which he had resonated with before.

Now, we have to ask the disquieting question: What kind of sexual relations did this parental, grown male have with the young woman? What kind of sexual relations could have so brutalized Eve that she brutalized her children and her children's children down to the present day.

For an answer to this, we turn to what is known from human psychology about the ubiquitous and deeply-disturbing prevalence and consequences of fathers raping their daughters.

Child Rape

It is only in these supposedly best-of-times that the prevalence of father-daughter rape has become well -known; we can assume that it was even more prevalent in more barbaric times.

> In North America, approximately 15% to 25% of women... were sexually abused when they were children. Most sexual abuse offenders are acquainted with their victims; approximately 30% are relatives of the child... Most child sexual abuse is committed by men... The most-often reported form of incest is father-daughter and stepfather-daughter incest...[223]

For a young girl who has been raped by a father-figure, the consequences on her psyche and spirit are extreme and devastating.

> Child rape can result in both short-term and long-term harm, including psychopathology in later life. Psychological, emotional, physical, and social effects include depression, post-traumatic stress disorder, anxiety, eating disorders, poor self-esteem, dissociative and anxiety disorders; general psychological  distress and disorders such as somatization, neurosis, chronic pain, sexualized behavior, school/learning problems; and behavior problems including substance abuse, self-destructive behavior, animal cruelty, crime in adulthood and suicide... Long term negative effects on development leading to repeated or additional victimization in adulthood are also associated with child rape. The risk of harm is greater if the abuser is a relative...[224]

It is a phenomenon of these last days that such sexual abuse by fathers, uncles, teachers, priests, etc., has come into the open, but it was probably as prevalent in the past.

> Most children are abused by someone they know and trust. A study in three states found 96% of reported rape survivors under age twelve knew the attacker. Four percent of the offenders were strangers, 20% were fathers, 16% were relatives and 50% were acquaintances or friends.[225]

Dysfunctional lineage

The abused becomes the abuser, and so it continues down the generations.

The study of the science of epigenetics was initiated in the last decade by the astonishing discovery that women who had experienced the starvation and deprivation of the Siege of Stalingrad had grandchildren with significantly shorter life expectancies.[226] So, it is not too much of a conceptual stretch to think that the far more devastating experience of child rape by a parent figure would affect all of the descendants. This epigenetic imprint is the Original Sin, the twisted wisdom of the ancestors, which has kept all people from being able to love as God loves.

The merit of the age is such that father-daughter incest is universally deplored. Even so, it occurs in even the most spiritually advanced communities, as recounted in In Jin Nim's sermon dealing with a father-daughter rape in the unification community.[227]

Epigenetics explains the seemingly-unfair warning in the Bible: "Visiting the iniquity of the fathers upon the children and the children's children to the third and the fourth generation."[228]

A consequence of the Fall, mentioned in the Bible, is the prediction that Man would now need to earn a living by great effort. *Cursed is the ground for your sake; In toil you shall eat of it.*[229] For in this broken, loveless situation there could be no natural servant class. The Neanderthals were not domesticated with God's Love; instead, they were all completely exterminated, and even eaten,[230] by brutish, fallen humans within a few thousand years.

Reversed Dominion

The human brain is organized into a hierarchy of modules that each arose in evolution as a step in the expression of the Logos. At the very bottom is the gut brain—the brain stem and the voluminous, if diffuse, network of ganglia that are spread over the internal organs. This is concerned with the survival of the self, and it arose in fish and was perfected in reptiles.

Above this is the mammalian brain with ability to love on a family level. Above this is the primate brain, with ability to love on the clan level. Above this is

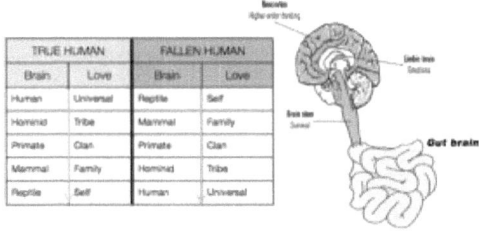

the hominid brain, with ability to love on the tribal level. At the very top of the hierarchy is the human module with the potential for universal love.

The trauma of the Fall reversed this hierarchy, and the self-centered gut brain became dominant. Incidentally, much of the gut brain is a hollow tube about the intestines, making the snake an appropriate symbol for Satan. This is why our Founder is always admonishing, "Your body is your enemy." The human level and the gut level are at war for control of the whole.

Role of angels

Many religions, including Unificationism, include angels, as well as God and humans, in the parables about human origins. Angels are usually considered to be God's helpers and assistants.

What kind of assistance would be helpful to the Creator God? Surely not in mastering calculus, or figuring out the exact balance between the electromagnetic and gravitational forces so that suns are possible. No, in all that has been said about Adam and Eve, by far the most useful things angels could do for God are the things He cannot do Himself, such as:

- Give birth to Adam and Eve
- Suckle them with mother's milk
- Protect them from lions, tigers and bears, etc.
- Wean them and then feed them nourishing food
- Teach them toilet discipline, etc.

We reach the conclusion that angels are symbols for the Neanderthals that did all of this for Adam and Eve. The concept of angels existing with God from the very beginning is in contradiction with the two-stage creation, and the stepwise expression from simple to complex. Angels, being just less than human, could not have appeared first, and Unification Thought gives angels no role to play alongside the Logos. Our mental picture of angels should be not the Sistine Chapel but rather a Neanderthal diorama.

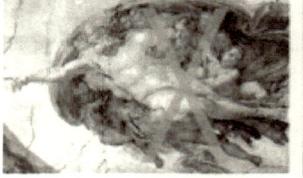

Only one other entity besides God, Adam and Eve is mentioned in the Bible—Lucifer, an archangel, a leader of the angels in a position to influence Adam and Eve. From what has gone before, we can state that Lucifer is the male Neanderthal whose mate gave birth to the first humans, and that Lucifer is a symbol for the biological father of Adam and Eve. It was he who raised the first humans, and gave them the Commandment that was to protect them.

Lucifer, the kind, imbecile father-figure, was intended in God's plan to fall totally in love with his beautiful children, to fall in love with Adam and Eve. The love that developed in the male parental Neanderthal would be balanced and healthy, just as God intended. Adam and Eve were to grow and develop in this environment to maturity, and so complete of the purpose of creation.

First, we will discuss what God intended for His first children; then we will discuss what could possibly have gone so wrong as to infect all of humanity thereafter with a broken capacity for love.

Surprising Absence

Many religious people are offended by my demotion of angels. While the *Divine Principle* demotes them to a position less that True Man, my thesis goes even further and demotes them from a class of eternal beings to a symbol in human history.

A period of reflection produced an answer to their questions by a more indirect route involving the Archangel Gabriel, a contemporary of Lucifer's, who did not Fall and has appeared throughout providential history at many important points (the logic also embraces the other unfallen Archangels such as Michael and Uriel, etc.). Given their basic similarity, we have three logical possibilities:

A. If Lucifer is a symbol, then so must be Gabriel.

B. If Lucifer is a literal, individual, eternal spirit being, then so must be Gabriel.

C. If Lucifer was created before the Big Bang and has been in the spirit world for 13.5 billion years, then so must have been Gabriel.

The converse is also true: whatever is true for Gabriel on these points must also hold for Lucifer (except for the Fall, of course). So, if we can show that Gabriel is most probably a symbol, then it is within the logic of Unification Thought to suggest that Lucifer is also a symbol.

In studying True Father's words, I realized that the solution is to be found in the curious role that Gabriel has played throughout Father's life in the contemporary providence since Jesus appeared to Father over 75 years ago. The point I wish to make is best illustrated in the adventures of Sherlock Holmes, the famous detective created by Sir Conan Doyle.

Curious absence

One of the most popular Sherlock Holmes short stories, "Silver Blaze," focuses on the disappearance of a famous race horse on the eve of an important race and the apparent murder of its trainer. The tale hinges on the "curious incident of the dog in the night":

Gregory, police detective:
"Is there any other point to which you would wish to draw my attention?"
Holmes:
"Yes, to the curious incident of the dog in the night."
Gregory:
"But the dog did nothing in the night."
Holmes:
"That was the curious incident."

Now, I have a dear friend who is a firm believer that both Lucifer and Gabriel are individual spirits who have been in existence for the last 13.5 billion years and are currently inhabitants of the spirit world, and he believes that they are distinct individuals and not symbols of anything at all like a Neanderthal.

But it was he who first researched the topic and reported the following curious fact: "In my file of 1,500 speeches by Father, I found 15 in which he mentioned Gabriel. None talk about Gabriel's providential purpose." What is so curious about this fact is that all the mentions of the Archangel Gabriel are all from the distant past—what happened 2,000 or 4,000 years ago.

Gabriel has done nothing in the contemporary providence, and Gabriel has played no role in Father's life course. What makes it so curious is that Gabriel has done nothing at all in the current era of the Providence. This is exactly what I noticed when researching Father's words:

Gabriel has played no role in the Second Coming, not a single one! Why do I find this so curious, and thus an important clue? For the following reasons:

1. We are all agreed, I think, that True Father has fulfilled the First Great Blessing of becoming one with God in heart. That he has become a Divine Spirit. That God and True Father are united as mind and body.

2. Unificationists agree, I think, that True Father has raised up True Mother to the same state, and that in their Holy Marriage they fulfilled the Second Great Blessing of becoming True Parents. They are the complete and substantial expression of God's dual nature, the substantial God on earth and in the spirit world.

3. We are all agreed, I think, that True Parents have fulfilled the Third Great Blessing, of becoming Lord of Creation—and this includes having God's authority over the angels, or at least over the unfallen ones, including Gabriel.

4. Unificationists believe that this is the time of the Second Advent and is, at the very least, as important a period to God as was the time of Jesus; and that God is exerting His maximum effort to assist True Parents rid the world of the satanic spirit and to bring in the Kingdom of Heaven world of true love.

Given these points and the assumption that the Archangel Gabriel is a literal spirit being who is close to God (point B. above), then surely we would expect, at least once, in Father's recollections to encounter a statement such as:

"I instructed Gabriel to take this crucial message to...."

or

"Gabriel brought me this warning not to ...."

or

"I asked Gabriel about what happened between Zachariah, Mary, and Elizabeth, and he told me that they....."

or, if angels have been around since the Big Bang:

"I quizzed Gabriel about what happened on Earth four-and-a-half billion years ago when the first living things emerged...."

But there is nothing of the kind to be found in Father's words. There is no contact with an individual called Gabriel. This total absence is very odd if Gabriel is an individual spirit being who is under True Parents' authority, if point B. is correct. This absence is not at all odd if Gabriel is symbol; it is to be expected. You do not do literal things with symbolic beings.

We conclude that point A. is more likely to be correct.

If Gabriel is a symbol, then Lucifer is also a symbol. Here, I suggest that Lucifer was a symbol for the Neanderthal male parent of Adam and Eve.

Because I think this is an aspect of Unification Thought that must be clear before there is any possibility of religion-science unity, I am happy to continue debating the question.

Restoration

The cycle of dysfunction, the inability to love, has been passed on from generation to generation. The history of restoration is God's effort to lead humanity out of this dreadful state, to break the cycle. The cycle of dysfunction will be broken with the advent of the True Parents and the start of the original ideal.

God's work of restoration of this ideal is constrained by the Logos, which is still running things. What was twisted must be untwisted, what was bent must be straightened. The terms 'twisted' and 'bent' do suggest some kind of rotation as well as size, which suggests that *sin*—actions not aligned with the Logos—will one day be expressed quantitatively using complex numbers. Our current mathematics is a long ways from that, as yet.

As described in the *Divine Principle*[231]—mainly in words but with a touch of mathematics—the history of restoration involves the development over time of an opportunity to unbend what was bent by a relational force, by an equal and opposite relational force. Given such an opportunity, it is the choice of a *central figure* to do the right thing, or not do the right thing. The paradigmatic example of this is the Cain/Abel situation that, in the second generation of humans, arose to correct one aspect of the fallen situation.

In the rape of Eve, the animal brain came to dominate over the human brain of the first mother-of-all. This reversal of the intended order in the Logos for the systematic hierarchy of the nervous system is called the *satanic* personality. This satanic perspective on life was created by the animal, *older* male stealing the intended position of the human, *younger* male.

This mixed human-animal nature was passed on to the children, but not in equal measure—the eldest son, Cain, was born with the animal aspect predominant; the second son, Abel, was born with the human aspect predominant. In the way of the Logos, it is natural for the eldest son to be subjective, and the younger sons to be objective to their leader. For Abel to win over his elder brother would be a force against the Logos, but opposite to that in the dynamic of the fall. If Abel had won over his brother using his human heart and abilities, the equal but opposite force generated would straighten-out that animal-over-human aspect of the fall, and the human brain would take its intended position in full control of all the lower animal levels.

Unfortunately, the opposite occurred, and the animal-Cain slaughtered the human-Abel, and the animal level dominated the human level completely for thousands of generations down the lineages of mankind as it spread out across the globe. It was only after a long and miserable history that the Cain/Abel dynamic was corrected in the Abrahamic Middle-East by Jacob winning over his elder brother Esau.

This is just a fragment of the long, complicated, miserable history of restoration outlined in the *Divine Principle*. We will leave to future generations the task of using complex numbers to make a quantitive hard science out of the, as yet, qualitative science of sin and salvation.

NOTES

[1] H. M. Georgi, "Grand Unified Theories," in The New Physics, ed. Paul Davies, Cambridge University Press, NY (1989), p. 448.

[2] Paul Davies, The Mind of God: The Scientific Basis for the Rational World, Simon & Schuster, NY (1992), p. 140.

[3] Johnjoe McFadden, *Quantum Evolution*, W. W. Norton, NY (2000), p.219

[4] Robert Rosen, *Life Itself,* Columbia University Press, NY (1991), p. 18.

[5] Edward Rubenstein, "Stages of evolution and their messengers," *Scientific American* (June 1989), p. 132.

[6] Whispered to Dustin in *The Graduate*, 1960s.

[7] R. Sheldrake *A New Science of Life* , Tarcher (1981) p 70

[8] T. E. Creighton, "The Protein Folding Problem," Science 240, 1988, p. 267, 240.

[9] Shapiro, R. (1985) "Origins: A Skeptic's Guide to the Creation of Life on Earth" Simon & Schuster, Inc. NY p. 195

[10] Rupert Sheldrake, *A New Science of Life: The Hypothesis of Formative Causation*, J. P. Tarcher Inc., Los Angeles, distributed by Houghton Mifflin Co., Boston (1981), p. 70.

[11] P. W Atkins, *Quanta* (2nd ed.), Oxford University Press, Oxford (1991), p. 348.

[12] Richard. P. Feynman, *QED: The Strange Theory of Light and Matter,* Princeton University Press (1985), p. 7.

[13] Johnjoe McFadden, *Quantum Evolution*, W. W. Norton, NY (2000), p. 219

[14] Robert Rosen, *Life Itself,* Columbia University Press, NY (1991), p. 18.

[15] Edward Rubenstein, "Stages of evolution and their messengers," *Scientific American* (June 1989), p. 132.

[16] Isaac Asimov, *The History of Science,* Walker & Co. NY (1985), pp. 378-9.

[17] Hans Christian von Baeyer, *Taming the Atom: Emergence of the Visible Microworld*, Random House, NY (1992), pp. 166–7.

[18] Philip Stehle, Order, Chaos, Order: The Transition from Classical to Quantum Physics, Oxford U. Press (1994), p. i.

[19] Ibid. p. 307.

[20] Georges Ifrah, *From One to Zero,* Viking, 1985.

[21] E. Mayor, *The Story of a Number*.

[22] Ponomarev, L. I. *The Quantum Dice*, Institute of Physicics Publishing, Bristol (1993), p. 94.

[23] A. Zee, *Fearful Symmetry,* Macmillan, NY (1986), pp. 106 - 111.

[24] A. Zee, *Fearful Symmetry: The Search for Beauty in Modern Physics,* Macmillan, NY (1986), p. 142.

[25] Hoffmann, B., *About Vectors* , Dover Publications, NY (1975), p. 101.

[26] F. David Peat , *Einstein's Moon: Bell's Theorem and the Curious Quest for Quantum Reality,* Contemporary Books, Chicago, IL (1990), p. 146.

27 J. Baggott, *The Meaning of Quantum Theory,* Oxford (1992), p. 72.

28 M. Longair, "The New Astrophysics," in *The New Physics,* ed. Paul Davies, Cambridge University Press, New York (1989), p. 199.

29 Dr. David Burton, U. of Bridgeport, CT. Personal communication, May 2005.

30 Hans Dehmelt, "Experiments on the Structure of an Individual Elementary Particle," *Science* 247 (1990), p. 544.

31 Timothy Paul Smith, *Hidden Worlds* p. 150.

32 F. David Peat, *Superstrings and the Theory of Everything,* Contemporary Books, Chicago, IL (1988), p. 97.

33 P. Knight, "Quantum Optics," in *The New Physics,* ed. Paul Davies, Cambridge University Press, New York (1989), p. 297.

34 Richard P. Feynman, *QED: The Strange Theory of Light and Matter,* Princeton University Press (1985), p. 27.

35 J-M. Lévy-Leblond & F. Balibar, *Quantics: Rudiments of Quantum Physics,* (trans.) North-Holland, Amsterdam (1990), p. 173.

36 A. G. Cairns-Smith, Evolving the Mind: on the nature of matter and the origin of consciousness, Cambridge (1996), p. 241.

37 Created with *Super MANDELZOOM 1.06* on a Mac Plus computer.

38 www.shef.ac.uk/chemistry/orbitron/AOs/4f/index

39 A. Zee, Fearful Symmetry: The Search for Beauty in Modern Physics, Macmillan, NY (1986), p. 142.

40 Edna E. Kramer, *The Nature and Growth of Modern Mathematics,* Princeton (1970), p. 259.

41 J. Baggott, *The Meaning of Quantum Theory,* Oxford, (1992), p. 185.

42 Richard P. Feynman, *QED: The Strange Theory of Light and Matter,* Princeton University Press (1985), p. 89.

43 P. Davies & J. Gribbin, *The Matter Myth,* Simon & Schuster, NY 1992.

44 S. W. Hawking, *A Brief History of Time,* Bantam, NY (1988), p. 152.

45 Edward Fitzgerald, The Rubáaiyát of Omar Khayyám (1859).

46 Edward Rubenstein, "Stages of evolution and their messengers," *Scientific American* (June 1989), p. 132.

47 E. E. Kramer, *The Nature and Growth of Modern Mathematics,* Princeton Paperbacks (1970), p. 610.

48 R. Lewis, "The Three Families of Matter," *The World & I,* April 1990, pp,. 300308.

49 Timothy Paul Smith, *Hidden Worlds,* 2003, p. 50.

50 Stephen W. Hawking, *A Brief History of Time: From the Big Bang to Black Holes,* Bantam Books, Toronto (1988), p. 81.

51 K. C. Cole, "A Theory of Everything,," *The New York Times Magazine,* Oct. 18, 1987, p.22.

52 A. Zee, Fearful Symmetry: The Search for Beauty in Modern Physics, Macmillan, NY (1986), p. 132.

53 John D. Barrow, *Pi in the Sky: Counting, Thinking and Being,* Oxford U. Press, Oxford (1992), p. 282.

[54] D. M. Considine (1983) ed. *Van Nostrand's Scientific Encyclopedia* Van Nostrand, NY (1983) p. 1067.

[55] Anthony Zee *Fearful Symmetry* Macmillan, NY (1986) p. 221

[56] While this sounds like a short period of time—and it is—we note that this is still well over a billion quantum "ticks" of the Plank Time, plenty f time for things to happen in!

[57] Christian de Deuve, *Blueprint for a Cell,* Neil Paterson Publishers, Burlington, NC (1991), p. 59.

[58] Christian de Deuve, *Vital Dust,* Basic Books, 1995, p. 168.

[59] S. J. Gould, "A Web of Tales," *Natural History* 10/88, p. 16-23

[60] Siegfried, K. G., "The Universe and Life" (1987) pp. 261-3.

[61] Tina Hesman, *Science News,* vol. 157, June 3, 2000, p. 362.

[62] For an explanation of the infinite difference between a countable and an uncountable infinity, see *To Infinity and Beyond,* Eli Maor, Birkhasuser, Boston, (1986) pp. 61-63

[63] Borisenko, A.I. & Tarapov, I>E> *Vector and Tensor Analysis,* Dover, Ny (1968), p. 1.

[64] Eldredge, N. (1985) *Time Frames,* Simon & Schuster, p.104.

[65] Dobzhansky T., Ayala F., Stebbins G. L. and Valentine J. (1977) "Evolution" (pub: W. H. Freeman, San Francisco), 1977 p.5.

[66] Ayala and Valentine, 1979, p.7.

[67] Dobzhansky, T. (1937) "Genetics and the Origin of Species" (1st Ed.); 2d Ed., 1941; 3d Ed., 1951: Columbia University Press.

[68] Stebbins, L. and Ayala, F. J. (1985) "The Evolution of Darwinism" Scientific American Vol. 253 #1, July 1985, p.13.

[69] Stebbins, L. and Ayala, F. J. (1985) "The Evolution of Darwinism" Scientific American Vol. 253 #1, July 1985, p. 72.

[70] Reviewed: Mayr, E. (1982) "Speciation and Macroevolution" *Evolution* **36(6)** p.1119-1132.

[71] Gould S. J., and Eldredge N. (1977), "Punctuated Equilibria: The Tempo And Mode Of Evolution Reconsidered" *Paleobiology* **3** pp.115-151

[72] Stanley S. (1982) "Macroevolution and the fossil record" Evolution 36(3) p.460-473

[73] Gould S. J. (1982) "Darwinism And The Expansion Of Evolutionary Theory" *Science* **216** p.380-387

[74] Goldschmidt R. (1940) "The Material Basis of Evolution," Yale University Press

[75] Robertson M. (1981) "Gene Families, Hopeful Monsters And The Selfish Genetics Of DNA" *Nature* **293** p.333-334

[76] Schopf T. (1982) "A Critical Assessment Of Punctuated Equilibria: 1. Duration Of Taxa" *Evolution* **36(6)** p.1144-1157

[77] Rose M. and Doolittle W. (1983) "Molecular Biological Mechanisms Of Speciation" *Science* **220** p.157-162

[78] Doskocil J. (1983) "A Model Of Stepwise Evolution Of Higher Eukaryotes" *Folia Biologica* (Praha) **29** 141-155

[79] Stanley S. (1981) "The New Evolutionary Timetable" Basic Books, NY, p. 135

[80] Stanley S. (1981) "The New Evolutionary Timetable" Basic Books, NY, p. 166

[81] Rhodes, F. (1983) "Gradualism, Punctuated Equilibrium and the 'Origin of Species'" *Nature* **305** p.269-272

[82] Charlesworth B., Lande R. and Slatkin M. (1982), "A Neo-darwinian Commentary On Macroevolution," *Evolution* **36(3)** pp. 474-498

[83] Eldredge, N. (1985) *Time Frames,* Simon & Schuster, pp. 69-70. Later in the book (p. 82-83), the author describes later work which developed his view that this abrupt change in the trilobites occurred over a period of thousands of years in another area with a geographic change accounting for the apparent abruptness.

[84] The use of the quantum concept is not the same way as used by George G. Simpson (Simpson 1944) where it refers to rapid, rather than sudden, evolutionary change.

[85] Darwin C. (1859) "On the Origin of Species by Means of Natural Selection or the Preservation of Favored Races in the Struggle for Life" New American Library, reprinted 1958, first ed. p. 341.

[86] Cain, 1954, as noted in Niles Eldredge's *"Time Frames,"* p. 69.

[87] Such thinking is not only relevant to the fossil record, but also in the contemporary world. If the processes of microevolution—the development of variation within a species—are the same as the processes of macroevolution—the development of species—we might reasonably expect to find a continuum of variation in the modern world. It could be argued that natural forces have worked to separate populations into the temporally real, if historically transient, 'reproductively coherent communities' we call species. But it does not seem unreasonable to expect to find areas of life where such separation of the continuum of variation is still occurring and the continuum still exists. Such is not the case, however. Clear-cut species boundaries are the rule.

[88] Darwin C. (1859) "On the Origin of Species by Means of Natural Selection or the Preservation of Favored Races in the Struggle for Life" New American Library, reprinted 1958 p. 280

[89] From the Latin *saltare* to leap

[90] Rupert Sheldrake, *A New Science of Life: The Hypothesis of Formative Causation*, J. P. Tarcher Inc. Los Angeles, distributed by Houghton Mifflin Co., Boston (1981), p. 71.

[91] Dobzhansky T., Ayala F., Stebbins G. L. and Valentine J. (1977) "Evolution" (pub: W. H. Freeman, San Francisco), 1977, p.5.

[92] Robert Rosen, *Life Itself,* Columbia University Press, NY (1991), p. 256.

[93] Christian de Duve, "Prelude to a Cell," *The Sciences,* Nov./Dec. (1990), p. 24.

[94] True in any year, I am sure.

[95] Christopher Wills, *Exons, Introns and Talking Genes.* Basic, NY (1991), p. 323

[96] Brian Goodwin, *How The Leopard Changed Its Spots: The Evolution of Complexity* Scribner's 1994, p. 176

[97] A. G. Cairns-Smith, "The First Organisms," Scientific American 252, June 1985, p. 98

[98] Cairns-Smith, A. G., (1982). *Genetic Takeover And The Mineral Origin Of Life* Cambridge University Press, p. 308

[99] Cairns-Smith, A. G., (1982). *Genetic Takeover And The Mineral Origin Of Life* Cambridge University Press. p. 310

[100] T. J. McMurry, K. N. Raymond, P. Smith, "Molecular Recognition and Metal Ion Template Synthesis," Science, 244, 1989, p. 943.

[101] P. A. Sharp & D. Eisenberg, "The Evolution of Catalytic Function," Science 238, 1987 p. 729-730, 807

[102] Joan Argetsinger Steitz, "Snurps," Scientific American, June 88, pp. 56-63.

[103] Maynard Smith, John, *Evolutionary Genetics*, Oxford University Press, 1989, pp. 217-221.

[104] Dawkins, R., (1976) *The Selfish Gene*, Oxford University Press.

[105] L. Manuelidis, 1990, "A View of Interphase Chromosomes" *Science* 250, pp. 1533-1540.

[106] Sharp, P. A. and Eisenberg, D., "The Evolution of Catalytic Function" Science238, pp. 729-730, 807.

[107] Beckmann, J. S., Brendel, V., and Trifonov, E. T. (1986) "Intervening Sequences Exhibit Distinct Vocabulary" *J. Biomolecular Structure & Dynamics* 4, pp. 391- 400.

[108] Essani, K., Goorha, R. and Granoff, A. (1987) *Virology* 161, p. 211 - 217

[109] Zeitlin, S., Parent, A., Silverstein, S. and Efstratiadis, A. (1987) "Pre-mRNA Splicing and the Nuclear Matrix" *Molecular and Cellular Biology* 7, p. 111-120

[110] Conrad, M., Brahmachari, S. K., and Sasisekharan, V., (1986) "DNA Structural Variability as A Factor in Gene Expression And Evolution" *Biosystems* **19** pp. 123-126.

[111] Wehner KA, et al, *Brain Res.* 2002 Aug 2; 945(2) pp. 160-73.

[112] Heidemann SR, Sander G, Kirschner MW. *Cell.* 1977 Mar; 10(3): pp. 337-50.

[113] Gil Ast, The Alternative Genome, *Scientific American*, April 2005, p.63.

[114] Gil Ast, The Alternative Genome, *Scientific American*, April 2005, p. 64.

[115] Kolodny GM., *Exp Cell Res.* (1971, Apr) 65(2) pp. 313-24.

[116] Ridley, M, *Genome*, Harper Collins, 1999, pp. 125-132.

[117] Ridley, M, *Genome*, Harper Collins, 1999, p. 182.

[118] Ridley, M, *Genome*, Harper Collins, 1999, p. 177.

[119] Gerald F. Joyce, "RNA evolution and the origins of life," *Nature* 338, 1989, pp. 217-224.

[120] Cairns-Smith, A. G., (1982). *Genetic Takeover And The Mineral Origin Of Life*, Cambridge University Press.

[121] Manfred Schidlowski, "A 3,800-million-year isotopic record of life from carbon in sedimentary rocks," *Nature* 333, 1988, p. 313.

[122] Ronald F. Fox "Energy and the Evolution of Life" 1988, p. 35.

[123] Shapiro, R. (1985) "Origins: A Skeptic's Guide to the Creation of Life on Earth" Simon & Schuster, Inc. NY p. 88.

[124] See *Temple of Doom* for an uncomfortably-graphic illustration of a Kali Aztec and a living victim having his heart removed.

[125] Verne Grant, *The Evolutionary Process*, Columbia University Press, NY (1985), p. 202.

[126] Examples of which are beautifully exemplified in a universally understandable tale of Scrooge and his helpers in *A Christmas Carol* by Dickens.

[127] Hyden, H., "The question of a molecular basis for the memory trace." In Pribram, K. H., & Broadbent, D. E. (eds.) *Biology of Memory*. New York: Academic Press, 1970, p. 116.

[128] Ilham A. Muslimov, Margaret Titmus, Edward Koenig, and Henri Tiedge, "Transport of Neuronal RNA in Axons," *The Journal of Neuroscience*, June 1, 2002, 22(11): 4293-4301.

[129] Sir Karl Popper, *Objective Knowledge*, Oxford University Press, Revised Edition, 1979. First published: Oxford University Press, 1972.

[130] Kerry Pobanz, my philosophical advisor, informs me that, "This is reminiscent of the treatment of habit, habituality and novelty by Rupert Sheldrake, which, in turn, is rooted in the Process Philosophy of, especially, the foremost American philosopher, Charles S. Pierce and the thought of Alfred N. Whitehead. Thanks, Kerry.

[131] See *Gateway* by F. Pohl for a SciFi prophecy on this matter.

[132] There are (at least) two excellent introductions to this subject:

Derbyshire, John (2003-04-15). *Prime Obsession*. Joseph Henry Press.

Rockmore, Dan (2007-12-18). *Stalking the Riemann Hypothesis: The Quest to Find the Hidden Law of Prime Numbers*. Vintage.

[133] Illustration from: Rockmore, Dan (2007). *Stalking the Riemann Hypothesis: The Quest to Find the Hidden Law of Prime Numbers* (p. 88). Vintage.

[134] Derbyshire, John (2003-04-15). *Prime Obsession* (Kindle Locations 5365-5366). Joseph Henry Press. Kindle Edition.

[135] P. W. Atkins, Quanta (2nd ed.), Oxford University Press, Oxford (1991), p. 348.

[136] Richard P. Feynman, *QED: The Strange Theory of Light and Matter*, Princeton University Press (1985), p. 7.

[137] A. Zee, *Fearful Symmetry, The Search for Beauty in Modern Physics*, Macmillan, NY (1986), pp. 106 - 111.

[138] A. Zee, *Fearful Symmetry*, Macmillan, NY (1986), p. 142.

[139] Johnjoe McFadden, *Quantum Evolution*, W. W. Norton, NY (2000), p. 219

[140] http://www.antapex.org/large_scale_universe.htm

[141] http://en.wikipedia.org/wiki/Superfluid

[142] http://en.wikipedia.org/wiki/Silicon

[143] Robert Rosen, *Life Itself*, Columbia University Press, NY (1991), p. 18.

[144] Philip Stehle, *Order, Chaos, Order: The Transition from Classical to Quantum Physics*, Oxford U. Press (1994), p. 307.

[145] http://www.xent.com/pipermail/fork/Week-of-Mon-20031103/026883.html

[146] Color printing also uses black ink with zero color, the K in CYMK, to replace equal amounts of CMY. So the color {50%C, 40%M, 60%Y} would actually be printed as {10%C, 0%M, 20% Y, 40%K}, a dark yellow-green.

[147] http://www.answers.com/topic/god-is-left-handed

[148] "False vacuum inflation with Einstein gravity" at: http://prola.aps.org/abstract/PRD/v49/i12/p6410_1

[149] This is a well-defined sphere centered on the earth. It has a radius that is the age of the universe in light years, i.e., about 13.3 billion light years. While the universe is probably much larger (although for theological reasons, not infinite), it is impossible to observe from our current location.

[150] Cheon Seong Gyeong – Sun Myung Moon, Book Six - Our Life And The Spiritual Realm, Chapter Two - What Kind of Place Is the Spirit World? Section 1. The Reality of the Spirit World and Its Laws, 1.1. The spirit world is an infinite world that transcends time and space

[151] C. S. Lewis, *The Great Divorce* (Glasgow: William Collins, 1946), 18-20.

[152] Ibid, 26.

[153] http://anthony3741.tripod.com/lifeintheworldunseen/id17.html

[154] From the obituary at:
www.sirjohntempletonobituary.org/templeton_report/20080709/tr20080709.pdf

[155] David Kraus, *Concepts in Modern Biology*, 1981, Globe Book Co, p. 41.

[156] Berg J.M, Tymoczko J.L., Stryer L. *Biochemistry*. 5th edition. New York: W H Freeman; 2002. "Section 8.4, The Michaelis-Menten Model Accounts for the Kinetic Properties of Many Enzymes". Available from: http://www.ncbi.nlm.nih.gov/books/NBK22430/

[157] http://www.molecularstation.com/wiki/Enzyme#_note-52.

[158] Leonid Mirny et al (2009) J. Phys. A: Math. Theor. 42.

[159] Study Bolsters Quantum Vibration Scent Theory
http://www.scientificamerican.com/article.cfm?id=study-bolsters-quantum-vibration-scent-theory

[160] http://en.wikipedia.org/wiki/Transfer_RNA

[161] Ma L, Bajic V, Zhang Z. On the classification of long non-coding RNAs. RNA Biology 2013; 10:917 - 926; PMID: 23696037; http://dx.doi.org/10.4161/rna.24604

[162] Roth, A.; Breaker, R. R. (2009). "The Structural and Functional Diversity of Metabolite-Binding Riboswitches". Annual Review of Biochemistry 78: 305–334.

[163] http://www.sciencedirect.com/science/article/pii/S0168952503003238

[164] Reed JL, Vo TD, Schilling CH, Palsson BO (2003). "An expanded genome-scale model of Escherichia coli K-12 " Genome Biol. 4 (9): R54.

[165] Claudia Fernanda Dick, André Luiz Araújo Dos-Santos, and José Roberto Meyer-Fernandes, "Inorganic Phosphate as an Important Regulator of Phosphatases," Enzyme Research, vol. 2011, Article ID 103980, 7 pages, 2011. doi:10.4061/2011/103980

[166] Philip Ball, DNA at 60: Still Much to Learn, Nature, April 28, 2013.
http://www.scientificamerican.com/article.cfm?id=dna-at-60-still-much-to-learn

[167] Morris KL (2008). "Epigenetic Regulation of Gene Expression". RNA and the Regulation of Gene Expression: A Hidden Layer of Complexity. Norfolk, England: Caister Academic Press.

[168] Josep Casadesús and David Low, Epigenetic Gene Regulation in the Bacterial World, Microbiol. Mol. Biol. Rev. September 2006 vol. 70 no. 3, 830-856

[169] http://en.wikipedia.org/wiki/RNA_world_hypothesis

[170] Gamow, George, 1947. One Two Three . . . Infinity: Facts and Speculations of Science, Dover Publications, NY.

[171] http://www.cornwalls.co.uk/photos/china-clay-mountains-nr-st-austell-2947.htm

[172] Proskurowski Giora et al. (2008). "Abiogenic Hydrocarbon Production at *Lost City* Hydrothermal Field". *Science* 319 (5863): 604–607.

[173] A. Vaccari, Clays and catalysis: a promising future, *Applied Clay Science*, vol. 14; 4, Elsevier, NY. 1999, pp. 161-198

[174] http://en.wikipedia.org/wiki/Chemiosmosis

[175] C. A. Villee, *Organization and change in Eukaryote cells*, ICUS, Houston TX, 1985., 3-11.

[176] Chichinadze K, Lazarashvili A, Tkemaladze J.,
2013, http://www.ncbi.nlm.nih.gov/pubmed

[177] Mark OJ Olson (2010), Nucleolus: Structure and
Function http://www.els.net/WileyCDA/ElsArticle/refId-a0005975.html

[178] Karsten Rippe (2012) "The Nucleolus Affects Cell Cycle and Cell Fate" in Genome Organization and Function in The Cell Nucleus, Wiley-VCH, Germany p.290

[179] Karsten Rippe (2012) "The Nucleolus as a Stress Sensor" *ibid* p.294

[180] Alberts B., Bray D., Lewis J., et al. Molecular Biology of the Cell. 3rd edition. Garland Science, NY; 1994. Ch. 21.

[181] Andrew Koob, 2009, *Roots of Thought: Unlocking Glia*, FT Press, NJ.

[182] Koob, p. 38

[183] *ibid*, pp. 59-63

[184] *ibid*, p. 68

[185] http://en.wikipedia.org/wiki/Neuroplasticity

[186] http://www.pbs.org/wgbh/nova/sciencenow/3411/02.html

[187] http://www.sciencedaily.com/releases/2009/05/090518111723.htm

[188] A. Otani, "Beyond Darwinism: Towards Unification Science" Chapter 2 (UTI, 2009)

[189] http://www.eskimo.com/~billb/weird/end.html

[190] http://www.nature.com/scitable/topicpage/Genomic-Imprinting-and-Patterns-of-Disease-Inheritance-899

[191] https://notes.utk.edu/Bio/greenberg.nsf/0/b360905554fdb7d985256ec5006a7755

[192] http://www.newgeology.us/HistoneH1.jpg

[193] http://cnx.org/content/m26565/latest/

[194] Jörg Tost. *Epigenetics* Caster Academic Press, Norfolk, UK, 2008 Review on Google Books.

[195] http://en.wikipedia.org/wiki/Chromosomal_crossover

[196] Advances in the genetics and epigenetics of gene regulation and human disease: A report on the Human Genome Organisation (HUGO) 11th Human Genome Meeting, Helsinki, Finland, 31 May-3 June 2006. http://genomebiology.com/2006/7/8/325

[197] Haibo Yin, Xia Zhang, Jun Liu, Youqun Wang, Junna He, Tao Yang, Xuhui Hong, Qing Yang, and Zhizhong Gong. Epigenetic regulation, somatic homologous recombination, and abscisic acid signaling are influenced by DNA polymerase epsilon mutation in arabidopsis. Plant, Cell, 21(2):386‚Äì402, February 2009.

[198] HapMap methylation-associated SNPs, markers of germline DNA methylation, positively correlate with regional levels of human meiotic recombination Genome Research, Vol. 19, No. 4. (26 April 2009), pp. 581-589. by Martin I. Sigurdsson, Albert V. Smith, Hans T. Bjornsson, Jon J. Jonsson

[199] http://www.epidna.com/showabstract.php?pmid=18541703

[200] http://www.reeis.usda.gov/web/crisprojectpages/207316.html

[201] http://www.bbc.co.uk/worldservice/specials/1624_story_of_africa/page92.shtml

[202] Experimentation in this area has begun. For example, see "Hybrid Vigor and Transgenerational Epigenetic Effects on Early Mouse Embryo Phenotype" at: http://www.biolreprod.org/content/79/4/638.abstract

[203] http://www.alpacas.com/AlpacaLibrary/GlossaryGL.aspx

[204] http://www.nature.com/nrm/journal/v4/n11/glossary/nrm1241_glossary.html

[205] I have speculated on a possible mechanism for such a speciation mechanism in *Volume One*, also published as *Do Proteins Teleport in an RNA World?* A PDF can be obtained by sending an email request to: RICHARDLLL@MAC.COM.

[206] http://judson.blogs.nytimes.com/2008/09/23/evolving-the-single-daddy/

[207] Richard Lewis, *Unification Science: The workings of the Logos from Inanimate to Life,* 2009, UTI-Tokyo. Available in PDF form at: https://files.me.com/richardlll/cbe22v

[208] also published under the title: *Do Proteins Teleport in an RNA World, p. 205,* UTI 2005. Available in PDF format at
https://files.me.com/richardlll/uy15st

[209]http://en.wikipedia.org/wiki/Control_of_fire_by_early_humans

[210] http://www.pbs.org/wgbh/evolution/library/07/1/l_071_03.html

[211] http://en.wikipedia.org/wiki/Cemetery_117

[212] http://www.nature.com/nature/journal/v402/n6760/full/402371a0.html

[213] http://marioarland.edublogs.org/2008/06/26/recent-single-origin-hypothesis/

[214] http://t2.gstatic.com/images?q=tbn:ANd9GcQvRNF8yIFl_XD5ugG9Ui5Smza--sXw85sY8ZFdhbm-vnXVSqprwQ

[215] http://aminotes.tumblr.com/post/4633090702/evolution-of-language-tested-with-genetic

[216] http://barclay1720.tripod.com/hist/origin/outafrica.htm

[217] http://en.wikipedia.org/wiki/Aquatic_ape_hypothesis

[218] http://www.thefreedictionary.com/hybrid+vigor

[219] Steven Pinker *The Language Instinct,* 1994.

[220] Young Whi Kim, *Divine Principle Study Guide - Part 1* , 1973, p.83.
At: http://www.tparents.org/Library/Unification/Books/DpStudy1.pdf

[221] Richard Lewis *Unification Physics,* p. 29. PDF available at: RICHARDLLL@MAC.COM

[222] Mel Gibson *Apocalypto* (2006)

[223] http://en.wikipedia.org/wiki/Child_sexual_abuse

[224] *ibid*

[225] Advocates for Youth, 1995

[226] http://chd.ucsd.edu/seminar/documents/Morgan.08.pdf

[227] Lovin' Life Ministry, Manhattan Center, July 28, 2011

[228] Exodus 34:6-8

[229] Genesis 3:17

[230] http://kasamaproject.org/2010/12/28/when-red-haired-neanderthals-were-eaten/

[231] Holy Spirit Association for the Unification of World Christianity. *Exposition of the Divine Principle* . New York: HSA-UWC, 1996.